Gravitational Theory

" The Weakest of The Four Fundamental Forces "

Edited by Paul F. Kisak

Contents

CONTENTS

iii

Chapter 1

Gravity

Gravity or **gravitation** is a natural phenomenon by which all things with mass are brought towards (or 'gravitate'

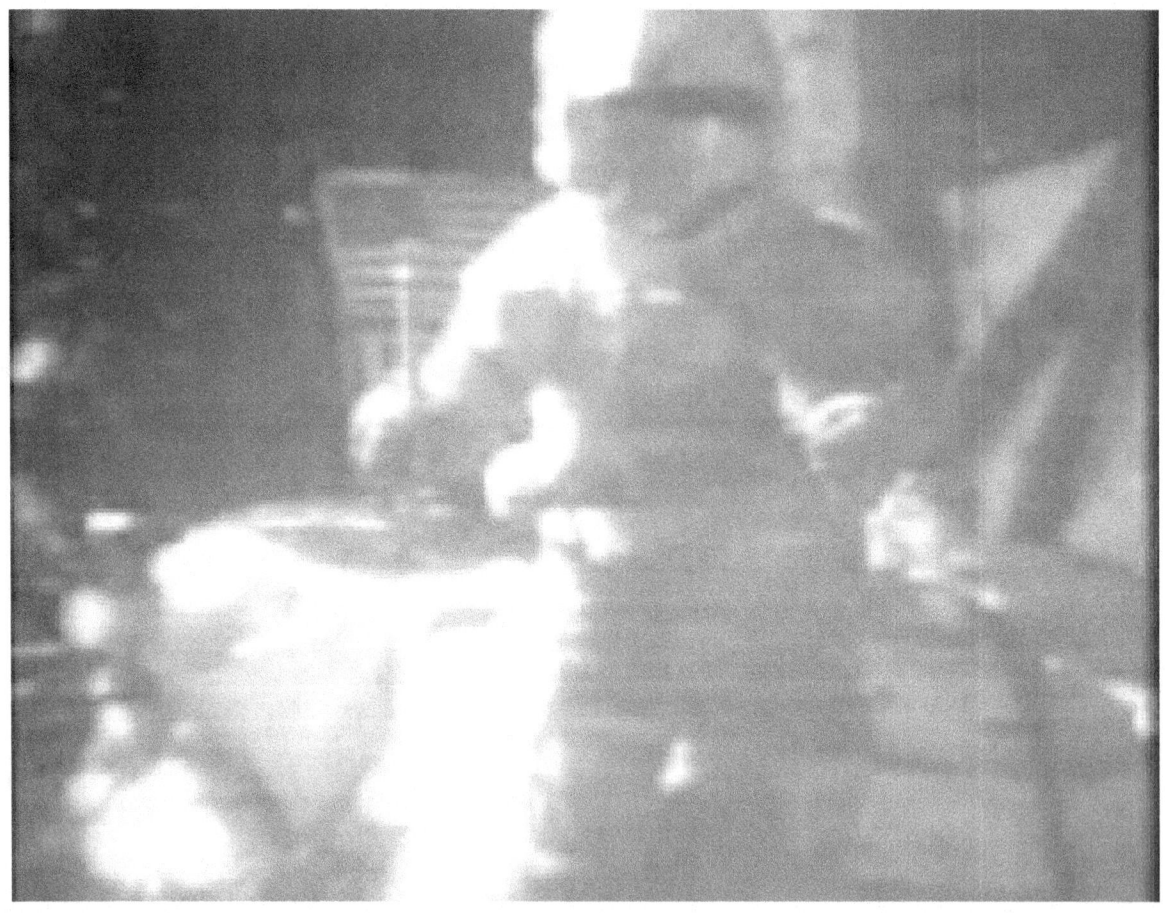

Hammer and feather drop: Apollo 15 astronaut David Scott on the Moon enacting the legend of Galileo's gravity experiment. (1.38 MB, ogg/Theora format).

towards) one another including stars, planets, galaxies and even light and sub-atomic particles. Gravity is responsible for the complexity in the universe, by creating spheres of hydrogen, igniting them under pressure to form stars and grouping them into galaxies. Without gravity, the universe would be an uncomplicated one, existing without thermal energy and composed only of equally spaced particles. On Earth, gravity gives weight to physical objects and causes the tides. Gravity has an infinite range, and it cannot be absorbed, transformed, or shielded against.

Gravity is most accurately described by the general theory of relativity (proposed by Albert Einstein in 1915) which describes gravity, not as a force, but as a consequence of the curvature of spacetime caused by the uneven distribution of mass/energy; and resulting in time dilation, where time lapses more slowly in strong gravitation. However, for most applications, gravity is well approximated by Newton's law of universal gravitation, which postulates that gravity is a force where two bodies of mass are directly drawn (or 'attracted') to each other according to a mathematical relationship, where the attractive force is proportional to the product of their masses and inversely proportional to the square of the distance between them. This is considered to occur over an infinite range, such that all bodies (with mass) in the universe are drawn to each other no matter how far they are apart.

Gravity is the weakest of the four fundamental interactions of nature. The gravitational attraction is approximately 10^{-38} times the strength of the strong force (i.e. gravity is 38 orders of magnitude weaker), 10^{-36} times the strength of the electromagnetic force, and 10^{-29} times the strength of the weak force. As a consequence, gravity has a negligible influence on the behavior of sub-atomic particles, and plays no role in determining the internal properties of everyday matter (but see quantum gravity). On the other hand, gravity is the dominant force at the macroscopic scale, that is the cause of the formation, shape, and trajectory (orbit) of astronomical bodies, including those of asteroids, comets, planets, stars, and galaxies. It is responsible for causing the Earth and the other planets to orbit the Sun; for causing the Moon to orbit the Earth; for the formation of tides; for natural convection, by which fluid flow occurs under the influence of a density gradient and gravity; for heating the interiors of forming stars and planets to very high temperatures; for solar system, galaxy, stellar formation and evolution; and for various other phenomena observed on Earth and throughout the universe.

In pursuit of a theory of everything, the merging of general relativity and quantum mechanics (or quantum field theory) into a more general theory of quantum gravity has become an area of research.

1.1   History of gravitational theory

Main article: History of gravitational theory

1.1.1   Scientific revolution

Modern work on gravitational theory began with the work of Galileo Galilei in the late 16th and early 17th centuries. In his famous (though possibly apocryphal[1]) experiment dropping balls from the Tower of Pisa, and later with careful measurements of balls rolling down inclines, Galileo showed that gravity accelerates all objects at the same rate. This was a major departure from Aristotle's belief that heavier objects accelerate faster.[2] Galileo postulated air resistance as the reason that lighter objects may fall more slowly in an atmosphere. Galileo's work set the stage for the formulation of Newton's theory of gravity.

1.1.2   Newton's theory of gravitation

Main article: Newton's law of universal gravitation

In 1687, English mathematician Sir Isaac Newton published *Principia*, which hypothesizes the inverse-square law of universal gravitation. In his own words, "I deduced that the forces which keep the planets in their orbs must [be] reciprocally as the squares of their distances from the centers about which they revolve: and thereby compared the force requisite to keep the Moon in her Orb with the force of gravity at the surface of the Earth; and found them answer pretty nearly."[3] The equation is the following:

$$F = G\frac{m_1 m_2}{r^2}$$

Sir Isaac Newton, an English physicist who lived from 1642 to 1727

Where F is the force, m_1 and m_2 are the masses of the objects interacting, r is the distance between the centers of the masses and G is the gravitational constant.

Newton's theory enjoyed its greatest success when it was used to predict the existence of Neptune based on motions of Uranus that could not be accounted for by the actions of the other planets. Calculations by both John Couch Adams and

Urbain Le Verrier predicted the general position of the planet, and Le Verrier's calculations are what led Johann Gottfried Galle to the discovery of Neptune.

A discrepancy in Mercury's orbit pointed out flaws in Newton's theory. By the end of the 19th century, it was known that its orbit showed slight perturbations that could not be accounted for entirely under Newton's theory, but all searches for another perturbing body (such as a planet orbiting the Sun even closer than Mercury) had been fruitless. The issue was resolved in 1915 by Albert Einstein's new theory of general relativity, which accounted for the small discrepancy in Mercury's orbit.

Although Newton's theory has been superseded by the Einstein's general relativity, most modern non-relativistic gravitational calculations are still made using the Newton's theory because it is simpler to work with and it gives sufficiently accurate results for most applications involving sufficiently small masses, speeds and energies.

1.1.3   Equivalence principle

The equivalence principle, explored by a succession of researchers including Galileo, Loránd Eötvös, and Einstein, expresses the idea that all objects fall in the same way. The simplest way to test the weak equivalence principle is to drop two objects of different masses or compositions in a vacuum and see whether they hit the ground at the same time. Such experiments demonstrate that all objects fall at the same rate when other forces (such as air resistance and electromagnetic effects) are negligible. More sophisticated tests use a torsion balance of a type invented by Eötvös. Satellite experiments, for example STEP, are planned for more accurate experiments in space.[4]

Formulations of the equivalence principle include:

- The weak equivalence principle: *The trajectory of a point mass in a gravitational field depends only on its initial position and velocity, and is independent of its composition.*[5]

- The Einsteinian equivalence principle: *The outcome of any local non-gravitational experiment in a freely falling laboratory is independent of the velocity of the laboratory and its location in spacetime.*[6]

- The strong equivalence principle requiring both of the above.

1.1.4   General relativity

See also: Introduction to general relativity

In general relativity, the effects of gravitation are ascribed to spacetime curvature instead of a force. The starting point for general relativity is the equivalence principle, which equates free fall with inertial motion and describes free-falling inertial objects as being accelerated relative to non-inertial observers on the ground.[7][8] In Newtonian physics, however, no such acceleration can occur unless at least one of the objects is being operated on by a force.

Einstein proposed that spacetime is curved by matter, and that free-falling objects are moving along locally straight paths in curved spacetime. These straight paths are called geodesics. Like Newton's first law of motion, Einstein's theory states that if a force is applied on an object, it would deviate from a geodesic. For instance, we are no longer following geodesics while standing because the mechanical resistance of the Earth exerts an upward force on us, and we are non-inertial on the ground as a result. This explains why moving along the geodesics in spacetime is considered inertial.

Einstein discovered the field equations of general relativity, which relate the presence of matter and the curvature of spacetime and are named after him. The Einstein field equations are a set of 10 simultaneous, non-linear, differential equations. The solutions of the field equations are the components of the metric tensor of spacetime. A metric tensor describes a geometry of spacetime. The geodesic paths for a spacetime are calculated from the metric tensor.

Notable solutions of the Einstein field equations include:

- The Schwarzschild solution, which describes spacetime surrounding a spherically symmetric non-rotating uncharged massive object. For compact enough objects, this solution generated a black hole with a central singularity. For radial distances from the center which are much greater than the Schwarzschild radius, the accelerations predicted by the Schwarzschild solution are practically identical to those predicted by Newton's theory of gravity.

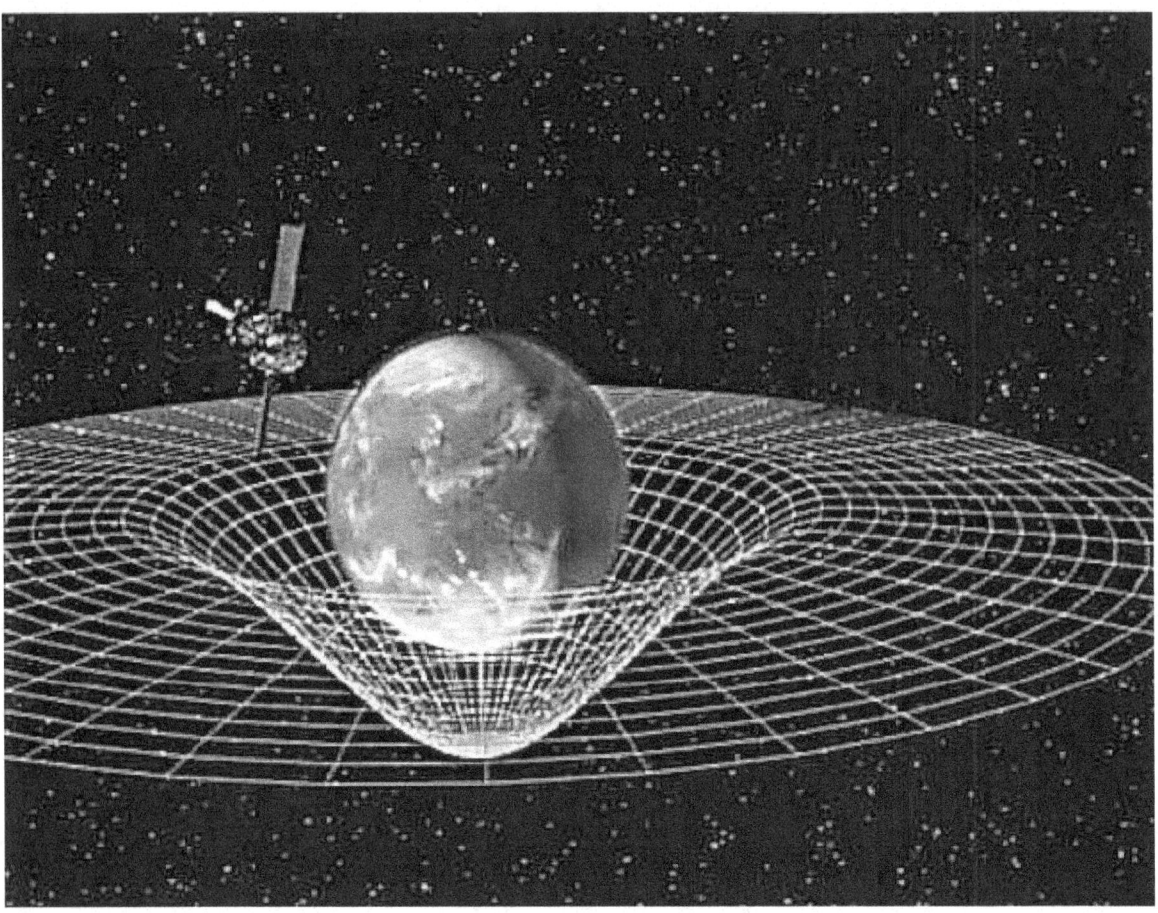

Two-dimensional analogy of spacetime distortion generated by the mass of an object. Matter changes the geometry of spacetime, this (curved) geometry being interpreted as gravity. White lines do not represent the curvature of space but instead represent the coordinate system imposed on the curved spacetime, which would be rectilinear in a flat spacetime.

- The Reissner-Nordström solution, in which the central object has an electrical charge. For charges with a geometrized length which are less than the geometrized length of the mass of the object, this solution produces black holes with two event horizons.

- The Kerr solution for rotating massive objects. This solution also produces black holes with multiple event horizons.

- The Kerr-Newman solution for charged, rotating massive objects. This solution also produces black holes with multiple event horizons.

- The cosmological Friedmann-Lemaître-Robertson-Walker solution, which predicts the expansion of the universe.

The tests of general relativity included the following:[9]

- General relativity accounts for the anomalous perihelion precession of Mercury.[10]

- The prediction that time runs slower at lower potentials has been confirmed by the Pound–Rebka experiment, the Hafele–Keating experiment, and the GPS.

- The prediction of the deflection of light was first confirmed by Arthur Stanley Eddington from his observations during the Solar eclipse of May 29, 1919.[11][12] Eddington measured starlight deflections twice those predicted by Newtonian corpuscular theory, in accordance with the predictions of general relativity. However, his interpretation of the results was later disputed.[13] More recent tests using radio interferometric measurements of quasars passing behind the Sun have more accurately and consistently confirmed the deflection of light to the degree predicted by general relativity.[14] See also gravitational lens.

- The time delay of light passing close to a massive object was first identified by Irwin I. Shapiro in 1964 in interplanetary spacecraft signals.

- Gravitational radiation has been indirectly confirmed through studies of binary pulsars.

- Alexander Friedmann in 1922 found that Einstein equations have non-stationary solutions (even in the presence of the cosmological constant). In 1927 Georges Lemaître showed that static solutions of the Einstein equations, which are possible in the presence of the cosmological constant, are unstable, and therefore the static universe envisioned by Einstein could not exist. Later, in 1931, Einstein himself agreed with the results of Friedmann and Lemaître. Thus general relativity predicted that the Universe had to be non-static—it had to either expand or contract. The expansion of the universe discovered by Edwin Hubble in 1929 confirmed this prediction.[15]

- The theory's prediction of frame dragging was consistent with the recent Gravity Probe B results.[16]

- General relativity predicts that light should lose its energy when travelling away from the massive bodies. The group of Radek Wojtak of the Niels Bohr Institute at the University of Copenhagen collected data from 8000 galaxy clusters and found that the light coming from the cluster centers tended to be red-shifted compared to the cluster edges, confirming the energy loss due to gravity.[17]

1.1.5   Gravity and quantum mechanics

Main articles: Graviton and Quantum gravity

In the decades after the discovery of general relativity, it was realized that general relativity is incompatible with quantum mechanics.[18] It is possible to describe gravity in the framework of quantum field theory like the other fundamental forces, such that the attractive force of gravity arises due to exchange of virtual gravitons, in the same way as the electromagnetic force arises from exchange of virtual photons.[19][20] This reproduces general relativity in the classical limit. However, this approach fails at short distances of the order of the Planck length,[18] where a more complete theory of quantum gravity (or a new approach to quantum mechanics) is required.

1.2   Specifics

1.2.1   Earth's gravity

Main article: Earth's gravity

Every planetary body (including the Earth) is surrounded by its own gravitational field, which exerts an attractive force on all objects. Assuming a spherically symmetrical planet, the strength of this field at any given point above the surface is proportional to the planetary body's mass and inversely proportional to the square of the distance from the center of the body.

The strength of the gravitational field is numerically equal to the acceleration of objects under its influence. The rate of acceleration of falling objects near the Earth's surface varies very slightly depending on elevation, latitude, and other factors. For purposes of weights and measures, a standard gravity value is defined by the International Bureau of Weights and Measures, under the International System of Units (SI).

That value, denoted g, is $g = 9.80665$ m/s^2 (32.1740 ft/s^2).[21][22]

The standard value of 9.80665 m/s^2 is the one originally adopted by the International Committee on Weights and Measures in 1901 for 45° latitude, even though it has been shown to be too high by about five parts in ten thousand.[23] This value has persisted in meteorology and in some standard atmospheres as the value for 45° latitude even though it applies more precisely to latitude of 45°32'33".[24]

Assuming the standardized value for g and ignoring air resistance, this means that an object falling freely near the Earth's surface increases its velocity by 9.80665 m/s (32.1740 ft/s or 22 mph) for each second of its descent. Thus, an object starting from rest will attain a velocity of 9.80665 m/s (32.1740 ft/s) after one second, approximately 19.62 m/s (64.4 ft/s) after two seconds, and so on, adding 9.80665 m/s (32.1740 ft/s) to each resulting velocity. Also, again ignoring air resistance, any and all objects, when dropped from the same height, will hit the ground at the same time. It is relevant to note that Earth's gravity doesn't have exactly the same value in all regions. There are slight variations in different parts of the globe due to latitude, surface features such as mountains and ridges, and perhaps unusually high or low sub-surface densities.[25]

According to Newton's 3rd Law, the Earth itself experiences a force equal in magnitude and opposite in direction to that which it exerts on a falling object. This means that the Earth also accelerates towards the object until they collide. Because the mass of the Earth is huge, however, the acceleration imparted to the Earth by this opposite force is negligible in comparison to the object's. If the object doesn't bounce after it has collided with the Earth, each of them then exerts a repulsive contact force on the other which effectively balances the attractive force of gravity and prevents further acceleration.

The force of gravity on Earth is the resultant (vector sum) of two forces: (a) The gravitational attraction in accordance with Newton's universal law of gravitation, and (b) the centrifugal force, which results from the choice of an earthbound, rotating frame of reference. At the equator, the force of gravity is the weakest due to the centrifugal force caused by the Earth's rotation. The force of gravity varies with latitude and increases from about 9.780 m/s^2 at the Equator to about 9.832 m/s^2 at the poles.

1.2.2 Equations for a falling body near the surface of the Earth

Main article: Equations for a falling body

Under an assumption of constant gravitational attraction, Newton's law of universal gravitation simplifies to $F = mg$, where m is the mass of the body and g is a constant vector with an average magnitude of 9.81 m/s^2 on Earth. This resulting force is the object's weight. The acceleration due to gravity is equal to this g. An initially stationary object which is allowed to fall freely under gravity drops a distance which is proportional to the square of the elapsed time. The image on the right, spanning half a second, was captured with a stroboscopic flash at 20 flashes per second. During the first $1/20$ of a second the ball drops one unit of distance (here, a unit is about 12 mm); by $2/20$ it has dropped at total of 4 units; by $3/20$, 9 units and so on.

Under the same constant gravity assumptions, the potential energy, Ep, of a body at height h is given by $Ep = mgh$ (or $Ep = Wh$, with W meaning weight). This expression is valid only over small distances h from the surface of the Earth. Similarly, the expression $h = \frac{v^2}{2g}$ for the maximum height reached by a vertically projected body with initial velocity v is useful for small heights and small initial velocities only.

1.2.3   Gravity and astronomy

The application of Newton's law of gravity has enabled the acquisition of much of the detailed information we have about the planets in our solar system, the mass of the Sun, and details of quasars; even the existence of dark matter is inferred using Newton's law of gravity. Although we have not traveled to all the planets nor to the Sun, we know their masses. These masses are obtained by applying the laws of gravity to the measured characteristics of the orbit. In space an object maintains its orbit because of the force of gravity acting upon it. Planets orbit stars, stars orbit Galactic Centers, galaxies orbit a center of mass in clusters, and clusters orbit in superclusters. The force of gravity exerted on one object by another is directly proportional to the product of those objects' masses and inversely proportional to the square of the distance between them.

1.2.4   Gravitational radiation

Main article: Gravitational wave

In general relativity, gravitational radiation is generated in situations where the curvature of spacetime is oscillating, such as is the case with co-orbiting objects. The gravitational radiation emitted by the Solar System is far too small to measure. However, gravitational radiation has been indirectly observed as an energy loss over time in binary pulsar systems such as PSR B1913+16. It is believed that neutron star mergers and black hole formation may create detectable amounts of gravitational radiation. Gravitational radiation observatories such as the Laser Interferometer Gravitational Wave Observatory (LIGO) have been created to study the problem. No confirmed detections have been made of this hypothetical radiation.

1.2.5   Speed of gravity

Main article: Speed of gravity

In December 2012, a research team in China announced that it had produced measurements of the phase lag of Earth tides during full and new moons which seem to prove that the speed of gravity is equal to the speed of light.[27] This means that if the Sun suddenly disappeared, the Earth would keep orbiting it normally for 8 minutes, which is the time light takes to travel that distance. The team's findings were released in the Chinese Science Bulletin in February 2013.[28]

1.3   Anomalies and discrepancies

There are some observations that are not adequately accounted for, which may point to the need for better theories of gravity or perhaps be explained in other ways.

- **Extra-fast stars**: Stars in galaxies follow a distribution of velocities where stars on the outskirts are moving faster than they should according to the observed distributions of normal matter. Galaxies within galaxy clusters show a similar pattern. Dark matter, which would interact gravitationally but not electromagnetically, would account for the discrepancy. Various modifications to Newtonian dynamics have also been proposed.

- **Flyby anomaly**: Various spacecraft have experienced greater acceleration than expected during gravity assist maneuvers.

- **Accelerating expansion**: The metric expansion of space seems to be speeding up. Dark energy has been proposed to explain this. A recent alternative explanation is that the geometry of space is not homogeneous (due to clusters of galaxies) and that when the data are reinterpreted to take this into account, the expansion is not speeding up after all,[29] however this conclusion is disputed.[30]

- **Anomalous increase of the astronomical unit**: Recent measurements indicate that planetary orbits are widening faster than if this were solely through the sun losing mass by radiating energy.

- **Extra energetic photons**: Photons travelling through galaxy clusters should gain energy and then lose it again on the way out. The accelerating expansion of the universe should stop the photons returning all the energy, but even taking this into account photons from the cosmic microwave background radiation gain twice as much energy as expected. This may indicate that gravity falls off *faster* than inverse-squared at certain distance scales.[31]

- **Extra massive hydrogen clouds**: The spectral lines of the Lyman-alpha forest suggest that hydrogen clouds are more clumped together at certain scales than expected and, like dark flow, may indicate that gravity falls off *slower* than inverse-squared at certain distance scales.[31]

- **Power**: Proposed extra dimensions could explain why the gravity force is so weak.[32]

1.4 Alternative theories

Main article: Alternatives to general relativity

1.4.1 Historical alternative theories

- Aristotelian theory of gravity

- Le Sage's theory of gravitation (1784) also called LeSage gravity, proposed by Georges-Louis Le Sage, based on a fluid-based explanation where a light gas fills the entire universe.

- Ritz's theory of gravitation, *Ann. Chem. Phys.* 13, 145, (1908) pp. 267–271, Weber-Gauss electrodynamics applied to gravitation. Classical advancement of perihelia.

- Nordström's theory of gravitation (1912, 1913), an early competitor of general relativity.

- Kaluza Klein theory (1921)

- Whitehead's theory of gravitation (1922), another early competitor of general relativity.

1.4.2 Recent alternative theories

- Brans–Dicke theory of gravity (1961) [33]

- Induced gravity (1967), a proposal by Andrei Sakharov according to which general relativity might arise from quantum field theories of matter

- $f(R)$ gravity (1970)

- Horndeski theory (1974) [34]

- Supergravity (1976)

- String theory

- In the modified Newtonian dynamics (MOND) (1981), Mordehai Milgrom proposes a modification of Newton's Second Law of motion for small accelerations [35]

- The self-creation cosmology theory of gravity (1982) by G.A. Barber in which the Brans-Dicke theory is modified to allow mass creation

- Loop quantum gravity (1988) by Carlo Rovelli, Lee Smolin, and Abhay Ashtekar

- Nonsymmetric gravitational theory (NGT) (1994) by John Moffat

- Tensor–vector–scalar gravity (TeVeS) (2004), a relativistic modification of MOND by Jacob Bekenstein

- Gravity as an entropic force, gravity arising as an emergent phenomenon from the thermodynamic concept of entropy.

- In the superfluid vacuum theory the gravity and curved space-time arise as a collective excitation mode of non-relativistic background superfluid.

- Chameleon theory (2004) by Justin Khoury and Amanda Weltman.

- Pressuron theory (2013) by Olivier Minazzoli and Aurélien Hees.

1.5   See also

- Angular momentum

- Anti-gravity, the idea of neutralizing or repelling gravity

- Artificial gravity

- Birkeland current

- Gravitational wave

- Gravitational wave background

- Cosmic gravitational wave background

- Einstein–Infeld–Hoffmann equations

- Escape velocity, the minimum velocity needed to escape from a gravity well

- g-force, a measure of acceleration

- Gauge gravitation theory

- Gauss's law for gravity

- Gravitational binding energy

- Gravity assist

- Gravity gradiometry

- Gravity Recovery and Climate Experiment

- Gravity Research Foundation

- Jovian–Plutonian gravitational effect

- Kepler's third law of planetary motion

- Lagrangian point

- Micro-g environment, also called microgravity

- Mixmaster dynamics

- n-body problem

- Newton's laws of motion

- Pioneer anomaly

- Scalar theories of gravitation

- Speed of gravity

- Standard gravitational parameter

- Standard gravity

- Weightlessness

1.6 Footnotes

[1] Ball, Phil (June 2005). "Tall Tales". *Nature News.* doi:10.1038/news050613-10.

[2] Galileo (1638), *Two New Sciences*, First Day Salviati speaks: "If this were what Aristotle meant you would burden him with another error which would amount to a falsehood; because, since there is no such sheer height available on earth, it is clear that Aristotle could not have made the experiment; yet he wishes to give us the impression of his having performed it when he speaks of such an effect as one which we see."

[3]   • Chandrasekhar, Subrahmanyan (2003). *Newton's Principia for the common reader.* Oxford: Oxford University Press. (pp.1–2). The quotation comes from a memorandum thought to have been written about 1714. As early as 1645 Ismaël Bullialdus had argued that any force exerted by the Sun on distant objects would have to follow an inverse-square law. However, he also dismissed the idea that any such force did exist. See, for example,

Linton, Christopher M. (2004). *From Eudoxus to Einstein—A History of Mathematical Astronomy.* Cambridge: Cambridge University Press. p. 225. ISBN 978-0-521-82750-8.

[4] M.C.W.Sandford (2008). "STEP: Satellite Test of the Equivalence Principle". Rutherford Appleton Laboratory. Retrieved 2011-10-14.

[5] Paul S Wesson (2006). *Five-dimensional Physics.* World Scientific. p. 82. ISBN 981-256-661-9.

[6] Haugen, Mark P.; C. Lämmerzahl (2001). *Principles of Equivalence: Their Role in Gravitation Physics and Experiments that Test Them.* Springer. arXiv:gr-qc/0103067. ISBN 978-3-540-41236-6.

[7] "Gravity and Warped Spacetime". black-holes.org. Retrieved 2010-10-16.

[8] Dmitri Pogosyan. "Lecture 20: Black Holes—The Einstein Equivalence Principle". University of Alberta. Retrieved 2011-10-14.

[9] Pauli, Wolfgang Ernst (1958). "Part IV. General Theory of Relativity". *Theory of Relativity.* Courier Dover Publications. ISBN 978-0-486-64152-2.

[10] Max Born (1924), *Einstein's Theory of Relativity* (The 1962 Dover edition, page 348 lists a table documenting the observed and calculated values for the precession of the perihelion of Mercury, Venus, and Earth.)

[11] Dyson, F.W.; Eddington, A.S.; Davidson, C.R. (1920). "A Determination of the Deflection of Light by the Sun's Gravitational Field, from Observations Made at the Total Eclipse of May 29, 1919". *Phil. Trans. Roy. Soc. A* **220** (571–581): 291–333. Bibcode:1920RSPTA.220..291D. doi:10.1098/rsta.1920.0009.. Quote, p. 332: "Thus the results of the expeditions to Sobral and Principe can leave little doubt that a deflection of light takes place in the neighbourhood of the sun and that it is of the amount demanded by Einstein's generalised theory of relativity, as attributable to the sun's gravitational field."

[12] Weinberg, Steven (1972). *Gravitation and cosmology.* John Wiley & Sons.. Quote, p. 192: "About a dozen stars in all were studied, and yielded values 1.98 ± 0.11" and 1.61 ± 0.31", in substantial agreement with Einstein's prediction θ☉ = 1.75"."

[13] Earman, John; Glymour. Clark (1980). "Relativity and Eclipses: The British eclipse expeditions of 1919 and their predecessors". *Historical Studies in the Physical Sciences* **11**: 49–85. doi:10.2307/27757471.

[14] Weinberg, Steven (1972). *Gravitation and cosmology.* John Wiley & Sons. p. 194.

[15] See W.Pauli, 1958. pp.219–220

[16] "NASA's Gravity Probe B Confirms Two Einstein Space-Time Theories". Nasa.gov. Retrieved 2013-07-23.

[17] Bhattacharjee, Yudhijit. "Galaxy Clusters Validate Einstein's Theory". News.sciencemag.org. Retrieved 2013-07-23.

[18] Randall, Lisa (2005). *Warped Passages: Unraveling the Universe's Hidden Dimensions*. Ecco. ISBN 0-06-053108-8.

[19] Feynman, R. P.; Morinigo, F. B.; Wagner, W. G.; Hatfield, B. (1995). *Feynman lectures on gravitation*. Addison-Wesley. ISBN 0-201-62734-5.

[20] Zee, A. (2003). *Quantum Field Theory in a Nutshell*. Princeton University Press. ISBN 0-691-01019-6.

[21] Bureau International des Poids et Mesures (2006). "The International System of Units (SI)" (PDF) (8th ed.). p. 131. Retrieved 2009-11-25. Unit names are normally printed in Roman (upright) type ... Symbols for quantities are generally single letters set in an italic font, although they may be qualified by further information in subscripts or superscripts or in brackets.

[22] "SI Unit rules and style conventions". National Institute For Standards and Technology (USA). September 2004. Retrieved 2009-11-25. Variables and quantity symbols are in italic type. Unit symbols are in Roman type.

[23] List, R. J. editor, 1968, Acceleration of Gravity, *Smithsonian Meteorological Tables*, Sixth Ed. Smithsonian Institution, Washington, D.C., p. 68.

[24] U.S. Standard Atmosphere, 1976, U.S. Government Printing Office, Washington, D.C., 1976. (Linked file is very large.)

[25] "Astronomy Picture of the Day".

[26] "Milky Way Emerges as Sun Sets over Paranal". *www.eso.org*. European Southern Obseevatory. Retrieved 29 April 2015.

[27] Chinese scientists find evidence for speed of gravity, astrowatch.com, 12/28/12.

[28] TANG, Ke Yun; HUA ChangCai; WEN Wu; CHI ShunLiang; YOU QingYu; YU Dan (February 2013). "Observational evidences for the speed of the gravity based on the Earth tide" (PDF). *Chinese Science Bulletin* **58** (4-5): 474–477. doi:10.1007/s11434-012-5603-3. Retrieved 12 June 2013.

[29] Dark energy may just be a cosmic illusion, *New Scientist*, issue 2646, 7 March 2008.

[30] Swiss-cheese model of the cosmos is full of holes, *New Scientist*, issue 2678, 18 October 2008.

[31] Chown, Marcus (16 March 2009). "Gravity may venture where matter fears to tread". *New Scientist* (2699). Retrieved 4 August 2013.

[32] CERN (20 January 2012). "Extra dimensions, gravitons, and tiny black holes".

[33] Brans, C.H. (Mar 2014). "Jordan-Brans-Dicke Theory". *Scholarpedia* **9**: 31358. Bibcode:2014Schpj...931358B. doi:10.4249/scholarpedia.31358.

[34] Horndeski, G.W. (Sep 1974). "Second-Order Scalar-Tensor Field Equations in a Four-Dimensional Space". *International Journal of Theoretical Physics* **88** (10): 363–384. Bibcode:1974IJTP...10..363H. doi:10.1007/BF01807638.

[35] Milgrom, M. (Jun 2014). "The MOND paradigm of modified dynamics". *Scholarpedia* **9**: 31410. Bibcode:2014SchpJ...931410M. doi:10.4249/scholarpedia.31410.

1.7 References

- Halliday, David; Robert Resnick; Kenneth S. Krane (2001). *Physics v. 1*. New York: John Wiley & Sons. ISBN 0-471-32057-9.

- Serway, Raymond A.; Jewett, John W. (2004). *Physics for Scientists and Engineers* (6th ed.). Brooks/Cole. ISBN 0-534-40842-7.

- Tipler, Paul (2004). *Physics for Scientists and Engineers: Mechanics, Oscillations and Waves, Thermodynamics* (5th ed.). W. H. Freeman. ISBN 0-7167-0809-4.

1.8 Further reading

- Thorne, Kip S.; Misner, Charles W.; Wheeler, John Archibald (1973). *Gravitation*. W.H. Freeman. ISBN 0-7167-0344-0.

1.9 External links

- Hazewinkel, Michiel, ed. (2001), "Gravitation", *Encyclopedia of Mathematics*, Springer, ISBN 978-1-55608-010-4

- Hazewinkel, Michiel, ed. (2001), "Gravitation, theory of", *Encyclopedia of Mathematics*, Springer, ISBN 978-1-55608-010-4

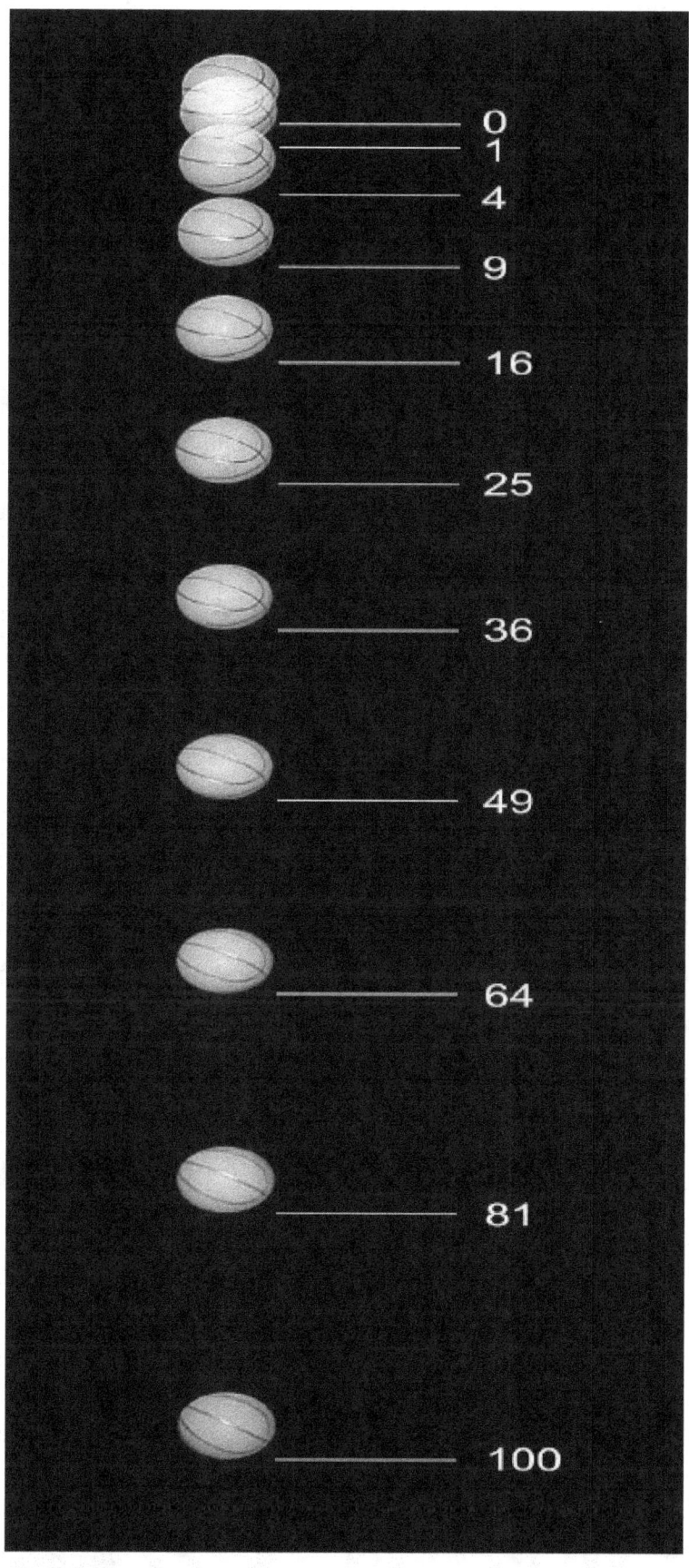

Ball falling freely under gravity. See text for description.

Gravity acts on stars that conform our Milky Way.[26]

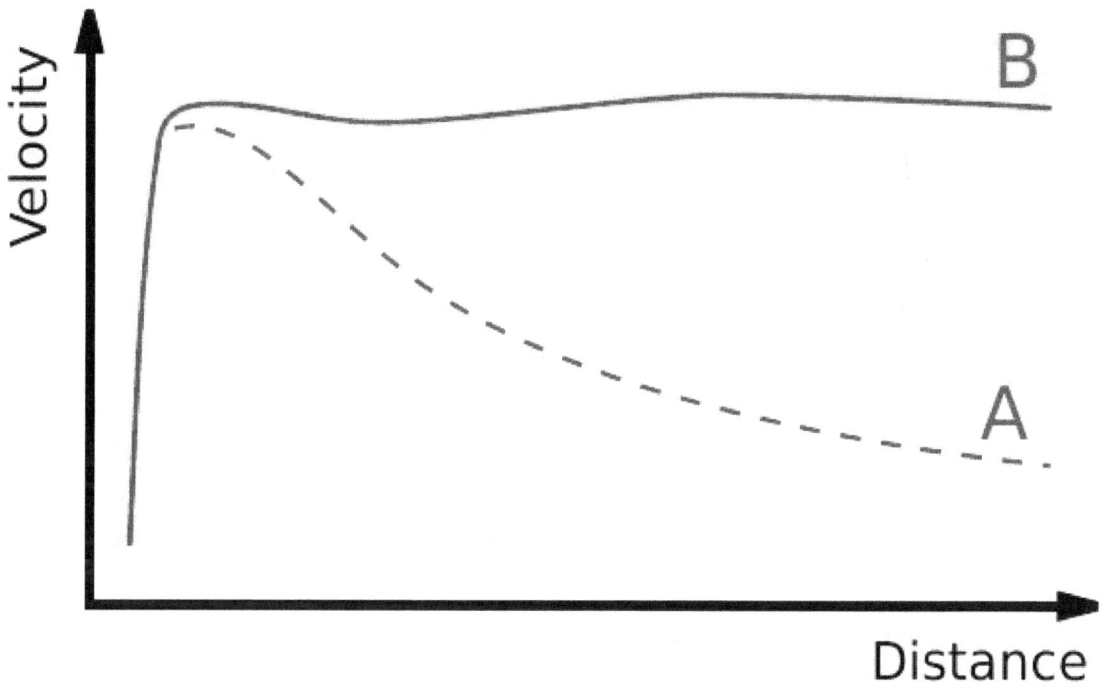

*Rotation curve of a typical spiral galaxy: predicted (**A**) and observed (**B**). The discrepancy between the curves is attributed to dark matter.*

Chapter 2

History of gravitational theory

In physics, **theories of gravitation** postulate mechanisms of interaction governing the movements of bodies with mass. There have been numerous theories of gravitation since ancient times.

2.1 Antiquity

See also: Aristotelian physics

In the 4th century BC, the Greek philosopher Aristotle believed that there is no effect or motion without a cause. The cause of the downward motion of heavy bodies, such as the element earth, was related to their nature, which caused them to move downward toward the center of the universe, which was their natural place. Conversely, light bodies such as the element fire, move by their nature upward toward the inner surface of the sphere of the Moon. Thus in Aristotle's system heavy bodies are not attracted to the earth by an external force of gravity, but tend toward the center of the universe because of an inner *gravitas* or heaviness.[1][2]

In Book VII of his *De Architectura*, the Roman engineer and architect Vitruvius contends that gravity is not dependent on a substance's "weight" but rather on its "nature" (cf. specific gravity).

> If the quicksilver is poured into a vessel, and a stone weighing one hundred pounds is laid upon it, the stone swims on the surface, and cannot depress the liquid, nor break through, nor separate it. If we remove the hundred pound weight, and put on a scruple of gold, it will not swim, but will sink to the bottom of its own accord. Hence, it is undeniable that the gravity of a substance depends not on the amount of its weight, but on its nature.[3]

Brahmagupta, the ancient Indian astronomer and mathematician, held the view that the earth was spherical and that it attracts things. Al Hamdānī and Al Biruni quote Brahmagupta saying "Disregarding this, we say that the earth on all its sides is the same; all people on the earth stand upright, and all heavy things fall down to the earth by a law of nature, for it is the nature of the earth to attract and to keep things, as it is the nature of water to flow, that of fire to burn, and that of the wind to set in motion. If a thing wants to go deeper down than the earth, let it try. The earth is the only *low* thing, and seeds always return to it, in whatever direction you may throw them away, and never rise upwards from the earth."[4][5]

2.2 Modern era (Origin of Gravitation)

During the 17th century, Galileo found that, counter to Aristotle's teachings, all objects accelerated equally when falling.

In the late 17th century, as a result of Robert Hooke's suggestion that there is a gravitational force which depends on the inverse square of the distance,[6] Isaac Newton was able to mathematically derive Kepler's three kinematic laws of planetary motion, including the elliptical orbits for the six then known planets and the Moon:

> "I deduced that the forces which keep the planets in their orbs must be reciprocally as the squares of their distances from the centres about which they revolve, and thereby compared the force requisite to keep the moon in her orb with the force of gravity at the surface of the earth and found them to answer pretty nearly."
> — Isaac Newton, 1666

So Newton's original formula was:

$$\text{Force of gravity} \propto \frac{\text{mass of object 1} \times \text{mass of object 2}}{\text{distance from centers}^2}$$

where the symbol $\propto$ means "is proportional to".

To make this into an equal-sided formula or equation, there needed to be a multiplying factor or constant that would give the correct force of gravity no matter the value of the masses or distance between them. This gravitational constant was first measured in 1797 by Henry Cavendish.

In 1907 Albert Einstein, in what was described by him as *"the happiest thought of my life"*, realized that an observer who is falling from the roof of a house experiences no gravitational field. In other words, gravitation was exactly equivalent to acceleration. Between 1911 and 1915 this idea, initially stated as the Equivalence principle, was formally developed into Einstein's theory of general relativity.

2.2.1 Newton's theory of gravitation

Main article: Law of universal gravitation

In 1687, English mathematician Sir Isaac Newton published *Principia*, which hypothesizes the inverse-square law of universal gravitation. In his own words, "I deduced that the forces which keep the planets in their orbs must be reciprocally as the squares of their distances from the centers about which they revolve; and thereby compared the force requisite to keep the Moon in her orb with the force of gravity at the surface of the Earth; and found them answer pretty nearly."

Newton's theory enjoyed its greatest success when it was used to predict the existence of Neptune based on motions of Uranus that could not be accounted by the actions of the other planets. Calculations by John Couch Adams and Urbain Le Verrier both predicted the general position of the planet, and Le Verrier's calculations are what led Johann Gottfried Galle to the discovery of Neptune.

Years later, it was another discrepancy in a planet's orbit that showed Newton's theory to be inaccurate. By the end of the 19th century, it was known that the orbit of Mercury could not be accounted for entirely under Newtonian gravity, and all searches for another perturbing body (such as a planet orbiting the Sun even closer than Mercury) have been fruitless. This issue was resolved in 1915 by Albert Einstein's new general theory of relativity, which accounted for the discrepancy in Mercury's orbit.

Paul Dirac developed the hypothesis that gravitation should have slowly and steadily decreased over the course of the history of the universe.[7]

Although Newton's theory has been superseded, most modern non-relativistic gravitational calculations still use it because it is much easier to work with and is sufficiently accurate for most applications.

2.2.2 Mechanical explanations of gravitation

Main article: Mechanical explanations of gravitation

The mechanical theories or explanations of the gravitation are attempts to explain the law of gravity by aid of basic mechanical processes, such as pushes, and without the use of any action at a distance. These theories were developed from the 16th until the 19th century in connection with the aether theories.[8]

René Descartes (1644) and Christiaan Huygens (1690) used vortices to explain gravitation. Robert Hooke (1671) and James Challis (1869) assumed, that every body emits waves which lead to an attraction of other bodies. Nicolas Fatio de Duillier (1690) and Georges-Louis Le Sage (1748) proposed a corpuscular model, using some sort of screening or shadowing mechanism. Later a similar model was created by Hendrik Lorentz, who used electromagnetic radiation instead of the corpuscles. Isaac Newton (1675) and Bernhard Riemann (1853) argued that aether streams carry all bodies to each other. Newton (1717) and Leonhard Euler (1760) proposed a model, in which the aether loses density near the masses, leading to a net force directing to the bodies. Lord Kelvin (1871) proposed that every body pulsates, which might be an explanation of gravitation and the electric charges.

However, those models were overthrown because most of them lead to an unacceptable amount of drag, which is not observed. Other models are violating the energy conservation law and are incompatible with modern thermodynamics.[9]

2.2.3 General relativity

Main article: Introduction to general relativity

In **general relativity**, the effects of gravitation are ascribed to spacetime curvature instead of to a force. The starting point for general relativity is the equivalence principle, which equates free fall with inertial motion. The issue that this creates is that free-falling objects can accelerate with respect to each other. In Newtonian physics, no such acceleration can occur unless at least one of the objects is being operated on by a force (and therefore is not moving inertially).

To deal with this difficulty, Einstein proposed that spacetime is curved by matter, and that free-falling objects are moving along locally straight paths in curved spacetime. (This type of path is called a geodesic). More specifically, Einstein and Hilbert discovered the field equations of general relativity, which relate the presence of matter and the curvature of spacetime and are named after Einstein. The Einstein field equations are a set of 10 simultaneous, non-linear, differential equations. The solutions of the field equations are the components of the metric tensor of spacetime. A metric tensor describes the geometry of spacetime. The geodesic paths for a spacetime are calculated from the metric tensor.

Notable solutions of the Einstein field equations include:

- The Schwarzschild solution, which describes spacetime surrounding a spherically symmetric non-rotating uncharged massive object. For compact enough objects, this solution generated a black hole with a central singularity. For radial distances from the center which are much greater than the Schwarzschild radius, the accelerations predicted by the Schwarzschild solution are practically identical to those predicted by Newton's theory of gravity.

- The Reissner–Nordström solution, in which the central object has an electrical charge. For charges with a geometrized length which are less than the geometrized length of the mass of the object, this solution produces black holes with an event horizon surrounding a Cauchy horizon.

- The Kerr solution for rotating massive objects. This solution also produces black holes with multiple horizons.

- The cosmological Robertson–Walker solution, which predicts the expansion of the universe.

General relativity has enjoyed much success because of the way its predictions of phenomena which are not called for by the older theory of gravity have been regularly confirmed. For example:

- General relativity accounts for the anomalous perihelion precession of the planet Mercury.

- The prediction that time runs slower at lower potentials has been confirmed by the Pound–Rebka experiment, the Hafele–Keating experiment, and the GPS.

- The prediction of the deflection of light was first confirmed by Arthur Eddington in 1919, and has more recently been strongly confirmed through the use of a quasar which passes behind the Sun as seen from the Earth. See also gravitational lensing.

- The time delay of light passing close to a massive object was first identified by Irwin Shapiro in 1964 in interplanetary spacecraft signals.

- Gravitational radiation has been indirectly confirmed through studies of binary pulsars.

- The expansion of the universe (predicted by the Robertson–Walker metric) was confirmed by Edwin Hubble in 1929.

2.2.4 Gravity and quantum mechanics

Main articles: Graviton and Quantum gravity

Several decades after the discovery of general relativity it was realized that it cannot be the complete theory of gravity because it is incompatible with quantum mechanics.[10] Later it was understood that it is possible to describe gravity in the framework of quantum field theory like the other fundamental forces. In this framework the attractive force of gravity arises due to exchange of virtual gravitons, in the same way as the electromagnetic force arises from exchange of virtual photons.[11][12] This reproduces general relativity in the classical limit. However, this approach fails at short distances of the order of the Planck length.[10]

It is notable that in general relativity, gravitational radiation, which under the rules of quantum mechanics must be composed of gravitons, is created only in situations where the curvature of spacetime is oscillating, such as is the case with co-orbiting objects. The amount of gravitational radiation emitted by the solar system is far too small to measure. However, gravitational radiation has been indirectly observed as an energy loss over time in binary pulsar systems such as PSR 1913+16. It is believed that neutron star mergers and black hole formation may create detectable amounts of gravitational radiation. Gravitational radiation observatories such as LIGO have been created to study the problem. No confirmed detections have been made of this hypothetical radiation, but as the science behind LIGO is refined and as the instruments themselves are endowed with greater sensitivity over the next decade, this may change.

2.3   References

[1] Edward Grant, *The Foundations of Modern Science in the Middle Ages*, (Cambridge: Cambridge Univ. Pr., 1996), pp. 60-1.

[2] Olaf Pedersen, *Early Physics and Astronomy*, (Cambridge: Cambridge Univ. Pr., 1993), p. 130

[3] Vitruvius, Marcus Pollio (1914). "7". In Alfred A. Howard. *De Architectura libri decem [Ten Books on Architecture]*. VII. Herbert Langford Warren, Nelson Robinson (illus), Morris Hicky Morgan (Harvard University, Cambridge: Harvard University Press). p. 215.

[4] *Alberani's India*. London : Kegan Paul, Trench, Trübner & Co., 1910.Electronic reproduction. Vol. 1 and 2. New York, N.Y. : Columbia University Libraries, 2006. p. 272. Retrieved 3 June 2014.

[5] *Kitāb al-Jawharatayn al-ʿatīqatayn al-māʾiʿatayn min al-ṣafrāʾ wa-al-bayḍāʾ : al-dhahab wa-al-fiḍḍah*. Cairo : Maṭbaʿat Dār al-Kutub wa-al-Wathāʾiq al-Qawmīyah bi-al-Qāhirah (Arabic:كتاب الجوهرتين العتيقتين المائعتين من الصفراء والبيضاء : الذهب والفضة). 2004. pp. 43–44.87. Retrieved 22 August 2014.

[6] Cohen, I. Bernard; George Edwin Smith (2002). *The Cambridge Companion to Newton*. Cambridge University Press. pp. 11–12. ISBN 978-0-521-65696-2.

[7] Haber, Heinz (1967) [1965]. "Die Expansion der Erde" [The expansion of the Earth]. *Unser blauer Planet [Our blue planet]*. Rororo Sachbuch [Rororo nonfiction] (in German) (Rororo Taschenbuch Ausgabe [Rororo pocket edition] ed.). Reinbek: Rowohlt Verlag. p. 52. Der englische Physiker und Nobelpreisträger Dirac hat [...] vor über dreißig Jahren die Vermutung begründet, dass sich das universelle Maß der Schwerkraft im Laufe der Geschichte des Universums außerordentlich langsam, aber stetig verringert. **English:** "The English physicist and Nobel laureate Dirac has [...], more than thirty years ago, substantiated the assumption that the universal strength of gravity decreases very slowly, but steadily over the course of the history of the universe.

[8] Taylor, W. B. (1876). "Kinetic Theories of Gravitation". *Smithsonian*: 205–282.

[9] Zenneck, J. (1903). "Gravitation". *Encyklopädie der mathematischen Wissenschaften mit Einschluss ihrer Anwendungen* (Leipzig) **5** (1): 25–67. doi:10.1007/978-3-663-16016-8_2.

[10] Randall, Lisa (2005). *Warped Passages: Unraveling the Universe's Hidden Dimensions.* Ecco. ISBN.

[11] Feynman, R. P.; Morinigo, F. B.; Wagner, W. G.; Hatfield, B. (1995). *Feynman lectures on gravitation.* Addison-Wesley. ISBN 0-201-62734-5.

[12] Zee, A. (2003). *Quantum Field Theory in a Nutshell.* Princeton University Press. ISBN.

Chapter 3

Newton's law of universal gravitation

Professor Walter Lewin explains Newton's law of gravitation during the 1999 MIT Physics course 8.01 [1]

Newton's law of universal gravitation states that any two bodies in the universe attract each other with a force that is directly proportional to the product of their masses and inversely proportional to the square of the distance between them.[note 1] This is a general physical law derived from empirical observations by what Isaac Newton called induction.[2] It is a part of classical mechanics and was formulated in Newton's work *Philosophiæ Naturalis Principia Mathematica* ("the *Principia*"), first published on 5 July 1687. (When Newton's book was presented in 1686 to the Royal Society, Robert Hooke made a claim that Newton had obtained the inverse square law from him; see the History section below.)

In modern language, the law states: Every point mass attracts every single other point mass by a force pointing along the line intersecting both points. The force is proportional to the product of the two masses and inversely proportional to the square of the distance between them.[3] The first test of Newton's theory of gravitation between masses in the laboratory was the Cavendish experiment conducted by the British scientist Henry Cavendish in 1798.[4] It took place 111 years after the publication of Newton's *Principia* and 71 years after his death.

Newton's law of gravitation resembles Coulomb's law of electrical forces, which is used to calculate the magnitude of electrical force arising between two charged bodies. Both are inverse-square laws, where force is inversely proportional to the square of the distance between the bodies. Coulomb's law has the product of two charges in place of the product of the masses, and the electrostatic constant in place of the gravitational constant.

Newton's law has since been superseded by Einstein's theory of general relativity, but it continues to be used as an excellent approximation of the effects of gravity in most applications. Relativity is required only when there is a need for extreme precision, or when dealing with very strong gravitational fields, such as those found near extremely massive and dense objects, or at very close distances (such as Mercury's orbit around the sun).

3.1 History

3.1.1 Early history

A recent assessment (by Ofer Gal) about the early history of the inverse square law is "by the late 1660s", the assumption of an "inverse proportion between gravity and the square of distance was rather common and had been advanced by a number of different people for different reasons". The same author does credit Hooke with a significant and even seminal contribution, but he treats Hooke's claim of priority on the inverse square point as uninteresting since several individuals besides Newton and Hooke had at least suggested it, and he points instead to the idea of "compounding the celestial motions" and the conversion of Newton's thinking away from "centrifugal" and towards "centripetal" force as Hooke's significant contributions.

3.1.2 Plagiarism dispute

In 1686, when the first book of Newton's *Principia* was presented to the Royal Society, Robert Hooke accused Newton of plagiarism by claiming that he had taken from him the "notion" of "the rule of the decrease of Gravity, being reciprocally as the squares of the distances from the Center". At the same time (according to Edmond Halley's contemporary report) Hooke agreed that "the Demonstration of the Curves generated thereby" was wholly Newton's.[5]

In this way the question arose as to what, if anything, Newton owed to Hooke. This is a subject extensively discussed since that time and on which some points continue to excite some controversy.

3.1.3 Hooke's work and claims

Robert Hooke published his ideas about the "System of the World" in the 1660s, when he read to the Royal Society on 21 March 1666 a paper "On gravity", "concerning the inflection of a direct motion into a curve by a supervening attractive principle", and he published them again in somewhat developed form in 1674, as an addition to "An Attempt to Prove the Motion of the Earth from Observations".[6] Hooke announced in 1674 that he planned to "explain a System of the World differing in many particulars from any yet known", based on three "Suppositions": that "all Celestial Bodies whatsoever, have an attraction or gravitating power towards their own Centers" [and] "they do also attract all the other Celestial Bodies that are within the sphere of their activity";[7] that "all bodies whatsoever that are put into a direct and simple motion, will so continue to move forward in a straight line, till they are by some other effectual powers deflected and bent..."; and that "these attractive powers are so much the more powerful in operating, by how much the nearer the body wrought upon is to their own Centers". Thus Hooke clearly postulated mutual attractions between the Sun and planets, in a way that increased with nearness to the attracting body, together with a principle of linear inertia.

Hooke's statements up to 1674 made no mention, however, that an inverse square law applies or might apply to these attractions. Hooke's gravitation was also not yet universal, though it approached universality more closely than previous hypotheses.[8] He also did not provide accompanying evidence or mathematical demonstration. On the latter two aspects, Hooke himself stated in 1674: "Now what these several degrees [of attraction] are I have not yet experimentally verified"; and as to his whole proposal: "This I only hint at present", "having my self many other things in hand which I would first compleat, and therefore cannot so well attend it" (i.e. "prosecuting this Inquiry").[6] It was later on, in writing on 6 January 1679|80[9] to Newton, that Hooke communicated his "supposition ... that the Attraction always is in a duplicate proportion to the Distance from the Center Reciprocall, and Consequently that the Velocity will be in a subduplicate proportion to the Attraction and Consequently as Kepler Supposes Reciprocall to the Distance."[10] (The inference about the velocity was incorrect.[11])

Hooke's correspondence of 1679-1680 with Newton mentioned not only this inverse square supposition for the decline of attraction with increasing distance, but also, in Hooke's opening letter to Newton, of 24 November 1679, an approach of "compounding the celestial motions of the planets of a direct motion by the tangent & an attractive motion towards the central body".[12]

3.1.4   Newton's work and claims

Newton, faced in May 1686 with Hooke's claim on the inverse square law, denied that Hooke was to be credited as author of the idea. Among the reasons, Newton recalled that the idea had been discussed with Sir Christopher Wren previous to Hooke's 1679 letter.[13] Newton also pointed out and acknowledged prior work of others,[14] including Bullialdus,[15] (who suggested, but without demonstration, that there was an attractive force from the Sun in the inverse square proportion to the distance), and Borelli[16] (who suggested, also without demonstration, that there was a centrifugal tendency in counterbalance with a gravitational attraction towards the Sun so as to make the planets move in ellipses). D T Whiteside has described the contribution to Newton's thinking that came from Borelli's book, a copy of which was in Newton's library at his death.[17]

Newton further defended his work by saying that had he first heard of the inverse square proportion from Hooke, he would still have some rights to it in view of his demonstrations of its accuracy. Hooke, without evidence in favor of the supposition, could only guess that the inverse square law was approximately valid at great distances from the center. According to Newton, while the 'Principia' was still at pre-publication stage, there were so many a-priori reasons to doubt the accuracy of the inverse-square law (especially close to an attracting sphere) that "without my (Newton's) Demonstrations, to which Mr Hooke is yet a stranger, it cannot believed by a judicious Philosopher to be any where accurate."[18]

This remark refers among other things to Newton's finding, supported by mathematical demonstration, that if the inverse square law applies to tiny particles, then even a large spherically symmetrical mass also attracts masses external to its surface, even close up, exactly as if all its own mass were concentrated at its center. Thus Newton gave a justification, otherwise lacking, for applying the inverse square law to large spherical planetary masses as if they were tiny particles.[19] In addition, Newton had formulated in Propositions 43-45 of Book 1,[20] and associated sections of Book 3, a sensitive test of the accuracy of the inverse square law, in which he showed that only where the law of force is accurately as the inverse square of the distance will the directions of orientation of the planets' orbital ellipses stay constant as they are observed to do apart from small effects attributable to inter-planetary perturbations.

In regard to evidence that still survives of the earlier history, manuscripts written by Newton in the 1660s show that Newton himself had arrived by 1669 at proofs that in a circular case of planetary motion, "endeavour to recede" (what was later called centrifugal force) had an inverse-square relation with distance from the center.[21] After his 1679-1680 correspondence with Hooke, Newton adopted the language of inward or centripetal force. According to Newton scholar J. Bruce Brackenridge, although much has been made of the change in language and difference of point of view, as between centrifugal or centripetal forces, the actual computations and proofs remained the same either way. They also involved the combination of tangential and radial displacements, which Newton was making in the 1660s. The lesson offered by Hooke to Newton here, although significant, was one of perspective and did not change the analysis.[22] This background shows there was basis for Newton to deny deriving the inverse square law from Hooke.

3.1.5 Newton's acknowledgment

On the other hand, Newton did accept and acknowledge, in all editions of the 'Principia', that Hooke (but not exclusively Hooke) had separately appreciated the inverse square law in the solar system. Newton acknowledged Wren, Hooke and Halley in this connection in the Scholium to Proposition 4 in Book 1.[23] Newton also acknowledged to Halley that his correspondence with Hooke in 1679-80 had reawakened his dormant interest in astronomical matters, but that did not mean, according to Newton, that Hooke had told Newton anything new or original: "yet am I not beholden to him for any light into that business but only for the diversion he gave me from my other studies to think on these things & for his dogmaticalness in writing as if he had found the motion in the Ellipsis, which inclined me to try it ..."[14]

3.1.6 Modern controversy

Since the time of Newton and Hooke, scholarly discussion has also touched on the question of whether Hooke's 1679 mention of 'compounding the motions' provided Newton with something new and valuable, even though that was not a claim actually voiced by Hooke at the time. As described above, Newton's manuscripts of the 1660s do show him actually combining tangential motion with the effects of radially directed force or endeavour, for example in his derivation of the inverse square relation for the circular case. They also show Newton clearly expressing the concept of linear inertia—for which he was indebted to Descartes' work, published in 1644 (as Hooke probably was).[24] These matters do not appear to have been learned by Newton from Hooke.

Nevertheless, a number of authors have had more to say about what Newton gained from Hooke and some aspects remain controversial.[25] The fact that most of Hooke's private papers had been destroyed or have disappeared does not help to establish the truth.

Newton's role in relation to the inverse square law was not as it has sometimes been represented. He did not claim to think it up as a bare idea. What Newton did was to show how the inverse-square law of attraction had many necessary mathematical connections with observable features of the motions of bodies in the solar system; and that they were related in such a way that the observational evidence and the mathematical demonstrations, taken together, gave reason to believe that the inverse square law was not just approximately true but exactly true (to the accuracy achievable in Newton's time and for about two centuries afterwards – and with some loose ends of points that could not yet be certainly examined, where the implications of the theory had not yet been adequately identified or calculated).[26][27]

About thirty years after Newton's death in 1727, Alexis Clairaut, a mathematical astronomer eminent in his own right in the field of gravitational studies, wrote after reviewing what Hooke published, that "One must not think that this idea ... of Hooke diminishes Newton's glory"; and that "the example of Hooke" serves "to show what a distance there is between a truth that is glimpsed and a truth that is demonstrated".[28][29]

3.2   Modern form

In modern language, the law states the following:

Assuming SI units, F is measured in newtons (N), m_1 and m_2 in kilograms (kg), r in meters (m), and the constant G is approximately equal to 6.674×10^{-11} N m^2 kg^{-2}.[30] The value of the constant G was first accurately determined from the results of the Cavendish experiment conducted by the British scientist Henry Cavendish in 1798, although Cavendish did not himself calculate a numerical value for G.[4] This experiment was also the first test of Newton's theory of gravitation between masses in the laboratory. It took place 111 years after the publication of Newton's *Principia* and 71 years after Newton's death, so none of Newton's calculations could use the value of G; instead he could only calculate a force relative to another force.

3.3   Bodies with spatial extent

If the bodies in question have spatial extent (rather than being theoretical point masses), then the gravitational force between them is calculated by summing the contributions of the notional point masses which constitute the bodies. In

Gravitational Field Strenth:
Inside the Earth

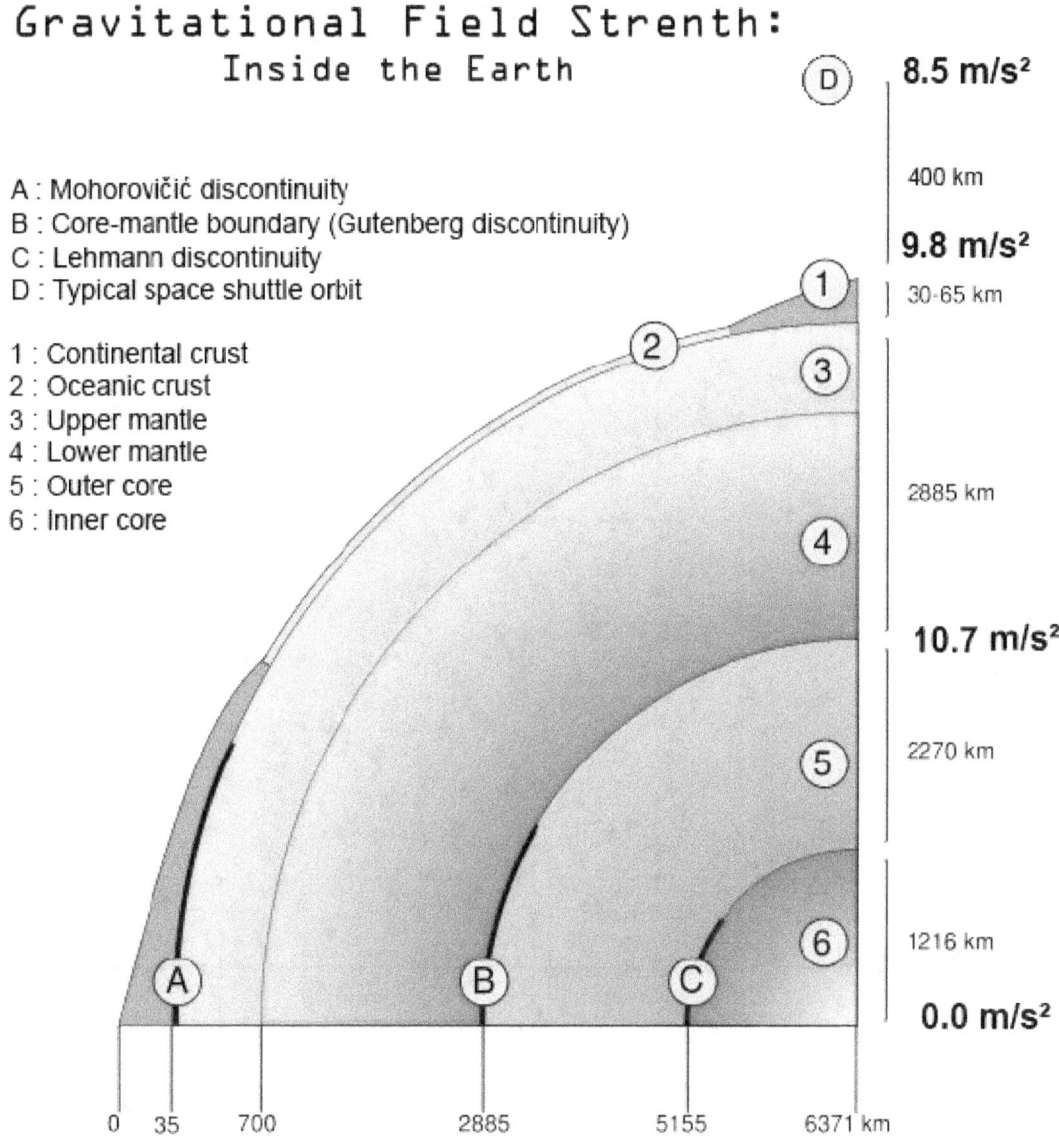

A : Mohorovičić discontinuity
B : Core-mantle boundary (Gutenberg discontinuity)
C : Lehmann discontinuity
D : Typical space shuttle orbit

1 : Continental crust
2 : Oceanic crust
3 : Upper mantle
4 : Lower mantle
5 : Outer core
6 : Inner core

(D) 8.5 m/s²

400 km

(1) 9.8 m/s²

30-65 km

(2) (3)

2885 km

(4)

10.7 m/s²

2270 km

(5)

1216 km

(6)

0.0 m/s²

0   35   700   2885   5155   6371 km

Gravitational field strength within the Earth.

the limit, as the component point masses become "infinitely small", this entails integrating the force (in vector form, see below) over the extents of the two bodies.

In this way it can be shown that an object with a spherically-symmetric distribution of mass exerts the same gravitational attraction on external bodies as if all the object's mass were concentrated at a point at its centre.[3] (This is not generally true for non-spherically-symmetrical bodies.)

For points *inside* a spherically-symmetric distribution of matter, Newton's Shell theorem can be used to find the gravitational force. The theorem tells us how different parts of the mass distribution affect the gravitational force measured at a point located a distance r_0 from the center of the mass distribution:[31]

- The portion of the mass that is located at radii $r < r_0$ causes the same force at r_0 as if all of the mass enclosed within a sphere of radius r_0 was concentrated at the center of the mass distribution (as noted above).

- The portion of the mass that is located at radii $r > r_0$ exerts *no net* gravitational force at the distance r_0 from the

near Earth surface

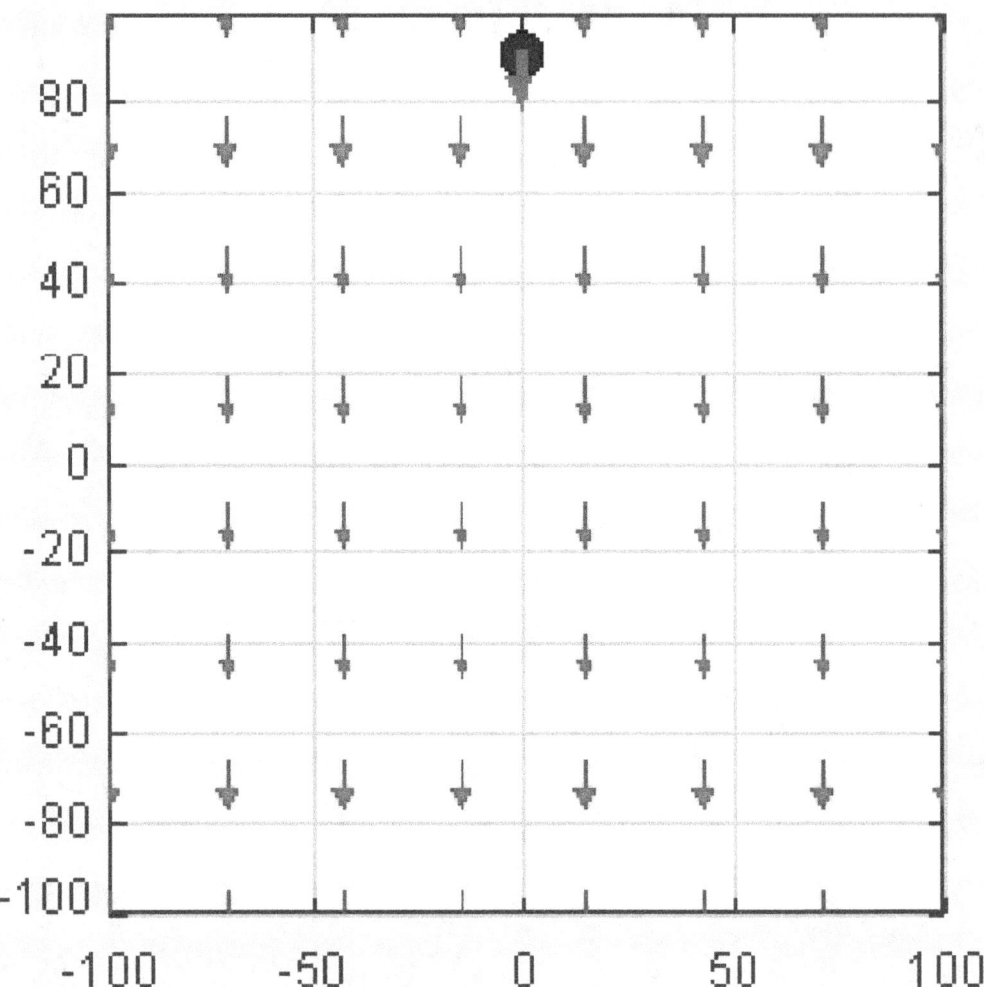

Gravity field near earth at 1,2 and A.

center. That is, the individual gravitational forces exerted by the elements of the sphere out there, on the point at $\mathbf{r}_0$, cancel each other out.

As a consequence, for example, within a shell of uniform thickness and density there is *no net* gravitational acceleration anywhere within the hollow sphere.

Furthermore, inside a uniform sphere the gravity increases linearly with the distance from the center; the increase due to the additional mass is 1.5 times the decrease due to the larger distance from the center. Thus, if a spherically symmetric body has a uniform core and a uniform mantle with a density that is less than 2/3 of that of the core, then the gravity initially decreases outwardly beyond the boundary, and if the sphere is large enough, further outward the gravity increases again, and eventually it exceeds the gravity at the core/mantle boundary. The gravity of the Earth may be highest at the core/mantle boundary.

3.4   Vector form

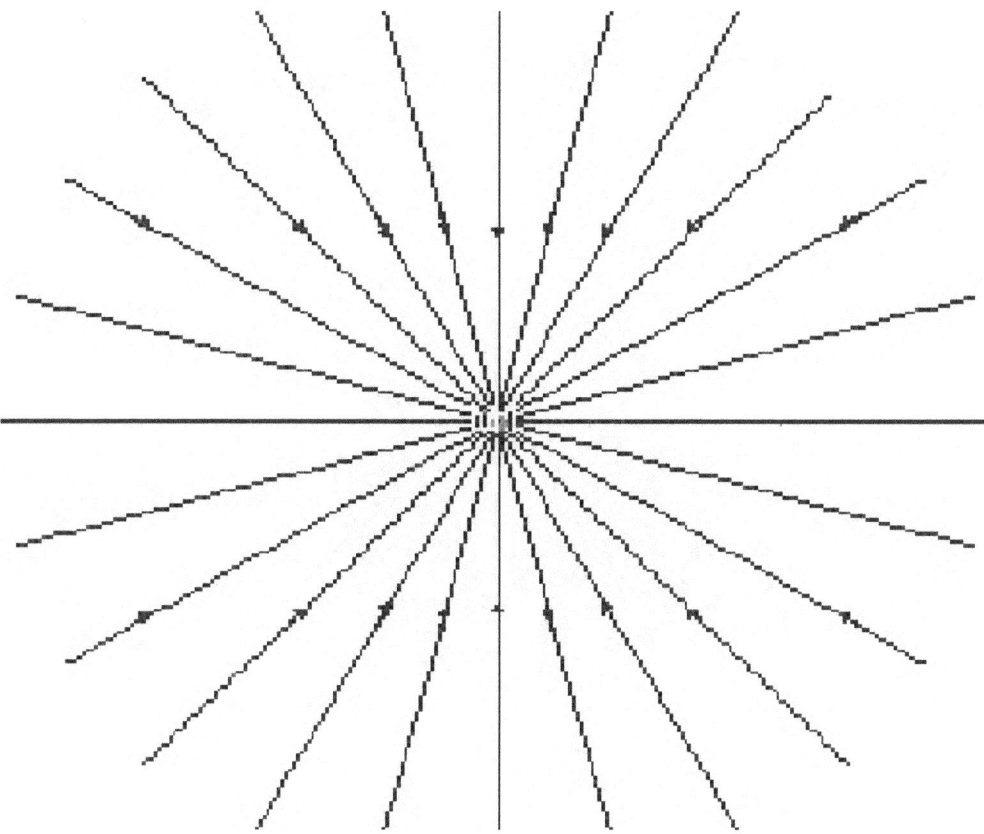

Field lines drawn for a point mass using 24 field lines

Newton's law of universal gravitation can be written as a vector equation to account for the direction of the gravitational force as well as its magnitude. In this formula, quantities in bold represent vectors.

$$\mathbf{F}_{12} = -G\frac{m_1 m_2}{|\mathbf{r}_{12}|^2}\hat{\mathbf{r}}_{12}$$

where

> $\mathbf{F}_{12}$ is the force applied on object 2 due to object 1,
>
> G is the gravitational constant,
>
> m_1 and m_2 are respectively the masses of objects 1 and 2,
>
> $|\mathbf{r}_{12}| = |\mathbf{r}_2 - \mathbf{r}_1|$ is the distance between objects 1 and 2, and
>
> $\hat{\mathbf{r}}_{12} \stackrel{\text{def}}{=} \frac{\mathbf{r}_2 - \mathbf{r}_1}{|\mathbf{r}_2 - \mathbf{r}_1|}$ is the unit vector from object 1 to 2.

It can be seen that the vector form of the equation is the same as the scalar form given earlier, except that $\mathbf{F}$ is now a vector quantity, and the right hand side is multiplied by the appropriate unit vector. Also, it can be seen that $\mathbf{F}_{12} = -\mathbf{F}_{21}$.

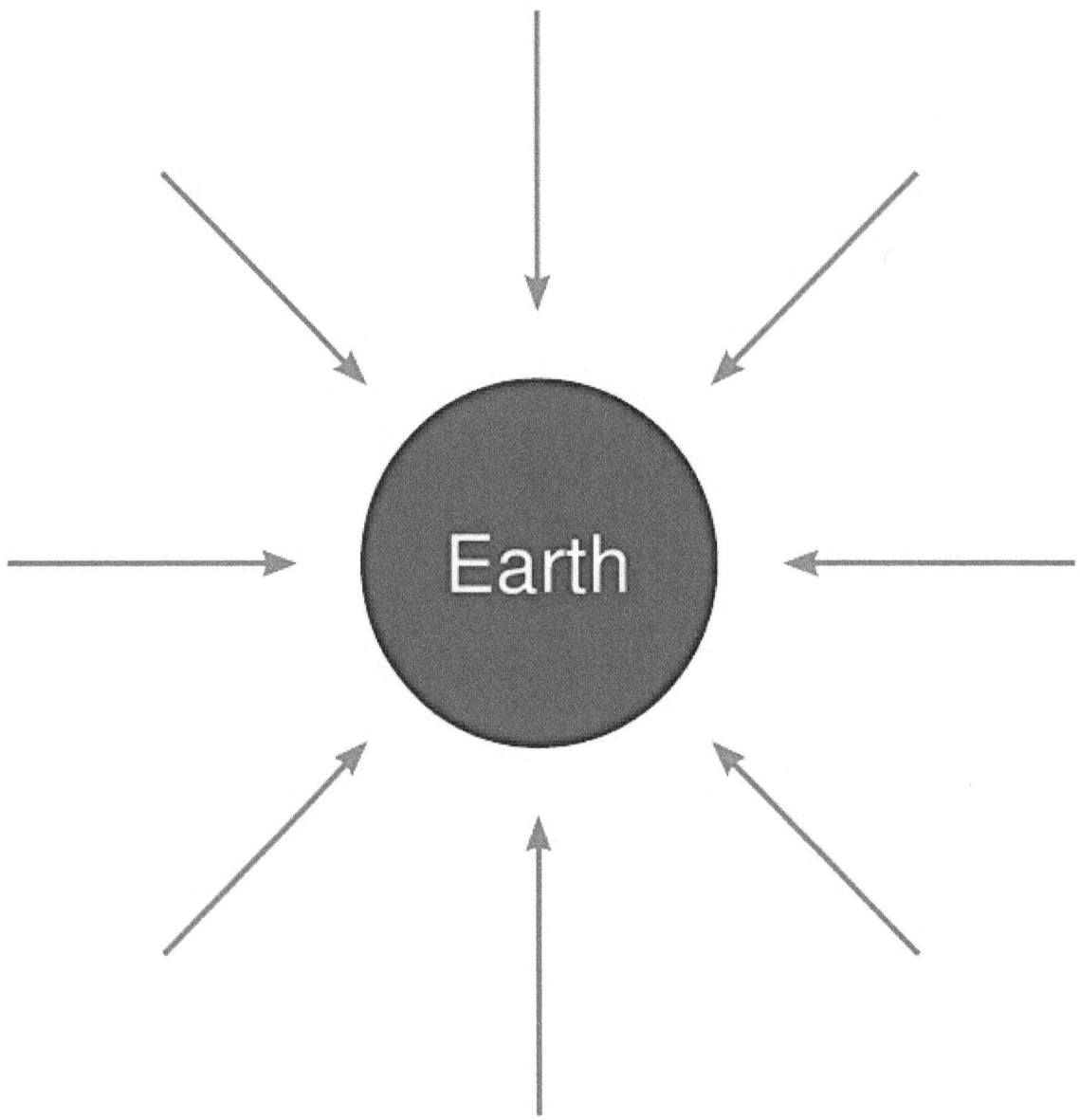

Gravity field surrounding Earth from a macroscopic perspective.

3.5   Gravitational field

Main article: Gravitational field

The **gravitational field** is a vector field that describes the gravitational force which would be applied on an object in any given point in space, per unit mass. It is actually equal to the gravitational acceleration at that point.

It is a generalization of the vector form, which becomes particularly useful if more than 2 objects are involved (such as a rocket between the Earth and the Moon). For 2 objects (e.g. object 2 is a rocket, object 1 the Earth), we simply write $\mathbf{r}$ instead of $\mathbf{r}_{12}$ and m instead of m_2 and define the gravitational field $\mathbf{g}(\mathbf{r})$ as:

$$\mathbf{g}(\mathbf{r}) = -G\frac{m_1}{|\mathbf{r}|^2}\hat{\mathbf{r}}$$

near Earth surface

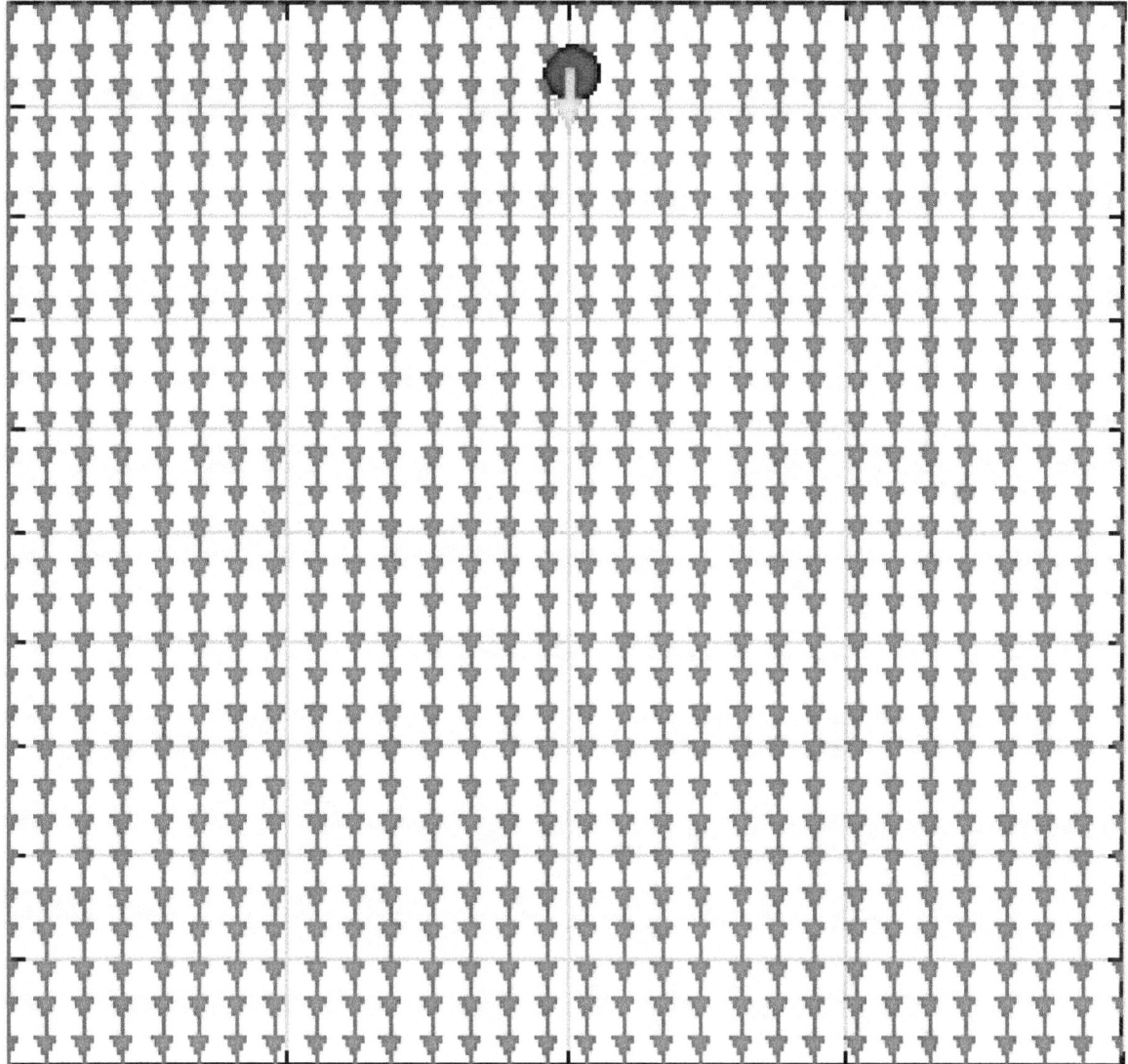

Gravity field lines representation is arbitrary as illustrated here represented in 30x30 grid to 0x0 grid and almost being parallel and pointing straight down to the center of the Earth

so that we can write:

$$\mathbf{F}(\mathbf{r}) = m\mathbf{g}(\mathbf{r}).$$

This formulation is dependent on the objects causing the field. The field has units of acceleration; in SI, this is m/s^2.

Gravitational fields are also conservative; that is, the work done by gravity from one position to another is path-independent. This has the consequence that there exists a gravitational potential field $V(\mathbf{r})$ such that

$$\mathbf{g}(\mathbf{r}) = -\nabla V(\mathbf{r}).$$

If m_1 is a point mass or the mass of a sphere with homogeneous mass distribution, the force field $\mathbf{g}(\mathbf{r})$ outside the sphere

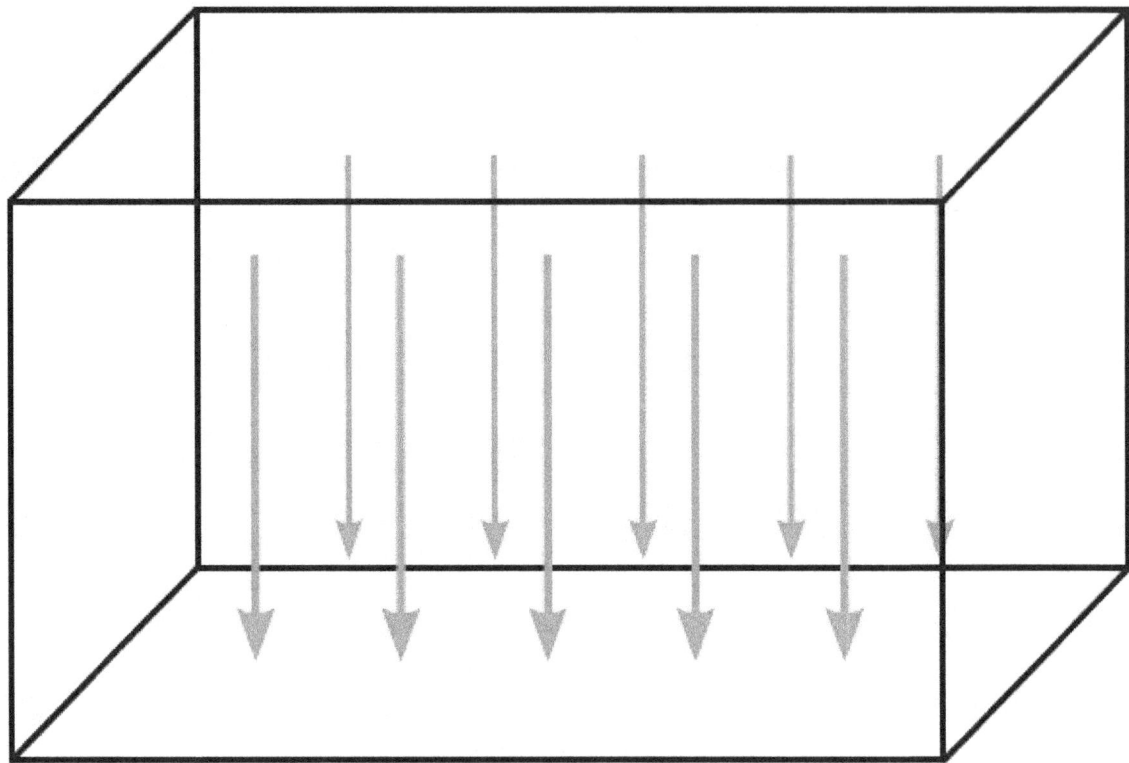

Gravity in a room: the curvature of the Earth is negligible at this scale, and the force lines can be approximated as being parallel and pointing straight down to the center of the Earth

is isotropic, i.e., depends only on the distance r from the center of the sphere. In that case

$$V(r) = -G\frac{m_1}{r}.$$

the gravitational field is on, inside and outside of symmetric masses.

As per Gauss Law, field in a symmetric body can be found by the mathematical equation:

$$\oiint_{\partial V} \mathbf{g}(\mathbf{r}) \cdot d\mathbf{A} = -4\pi G M_{enc}$$

where ∂V is a closed surface and M_{enc} is the mass enclosed by the surface.

Hence, for a hollow sphere of radius R and total mass M,

$$|\mathbf{g}(\mathbf{r})| = \begin{cases} 0, & \text{if } r < R \\[2mm] \dfrac{GM}{r^2}, & \text{if } r \geq R \end{cases}$$

For a uniform solid sphere of radius R and total mass M,

$$|\mathbf{g}(\mathbf{r})| = \begin{cases} \dfrac{GMr}{R^3}, & \text{if } r < R \\[2em] \dfrac{GM}{r^2}, & \text{if } r \geq R \end{cases}$$

3.6   Problematic aspects

Newton's description of gravity is sufficiently accurate for many practical purposes and is therefore widely used. Deviations from it are small when the dimensionless quantities φ/c^2 and $(v/c)^2$ are both much less than one, where φ is the gravitational potential, v is the velocity of the objects being studied, and c is the speed of light.[32] For example, Newtonian gravity provides an accurate description of the Earth/Sun system, since

$$\frac{\Phi}{c^2} = \frac{GM_{sun}}{r_{orbit} c^2} \sim 10^{-8}, \quad \left(\frac{v_{Earth}}{c}\right)^2 = \left(\frac{2\pi r_{orbit}}{(1 \text{ yr})c}\right)^2 \sim 10^{-8}$$

where r_{orbit} is the radius of the Earth's orbit around the Sun.

In situations where either dimensionless parameter is large, then general relativity must be used to describe the system. General relativity reduces to Newtonian gravity in the limit of small potential and low velocities, so Newton's law of gravitation is often said to be the low-gravity limit of general relativity.

3.6.1   Theoretical concerns with Newton's expression

- There is no immediate prospect of identifying the mediator of gravity. Attempts by physicists to identify the relationship between the gravitational force and other known fundamental forces are not yet resolved, although considerable headway has been made over the last 50 years (See: Theory of everything and Standard Model). Newton himself felt that the concept of an inexplicable *action at a distance* was unsatisfactory (see "Newton's reservations" below), but that there was nothing more that he could do at the time.

- Newton's theory of gravitation requires that the gravitational force be transmitted instantaneously. Given the classical assumptions of the nature of space and time before the development of General Relativity, a significant propagation delay in gravity leads to unstable planetary and stellar orbits.

3.6.2   Observations conflicting with Newton's formula

- Newton's Theory does not fully explain the precession of the perihelion of the orbits of the planets, especially of planet Mercury, which was detected long after the life of Newton.[33] There is a 43 arcsecond per century discrepancy between the Newtonian calculation, which arises only from the gravitational attractions from the other planets, and the observed precession, made with advanced telescopes during the 19th Century.

- The predicted angular deflection of light rays by gravity that is calculated by using Newton's Theory is only one-half of the deflection that is actually observed by astronomers. Calculations using General Relativity are in much closer agreement with the astronomical observations.

- In spiral galaxies the orbiting of stars around their centers seems to strongly disobey to Newton's law of universal gravitation. Astrophysicists, however, explain this spectacular phenomenon in the framework of the Newton's laws, with the presence of large amounts of Dark matter.

The observed fact that the *gravitational mass* and the *inertial mass* is the same for all objects is unexplained within Newton's Theories. General Relativity takes this as a basic principle. See the Equivalence Principle. In point of fact, the experiments of Galileo Galilei, decades before Newton, established that objects that have the same air or fluid resistance are accelerated by the force of the Earth's gravity equally, regardless of their different *inertial* masses. Yet, the forces and energies that are required to accelerate various masses is completely dependent upon their different *inertial* masses, as can be seen from Newton's Second Law of Motion, F = ma.

3.6.3 Newton's reservations

While Newton was able to formulate his law of gravity in his monumental work, he was deeply uncomfortable with the notion of "action at a distance" which his equations implied. In 1692, in his third letter to Bentley, he wrote: *"That one body may act upon another at a distance through a vacuum without the mediation of anything else, by and through which their action and force may be conveyed from one another, is to me so great an absurdity that, I believe, no man who has in philosophic matters a competent faculty of thinking could ever fall into it."*

He never, in his words, "assigned the cause of this power". In all other cases, he used the phenomenon of motion to explain the origin of various forces acting on bodies, but in the case of gravity, he was unable to experimentally identify the motion that produces the force of gravity (although he invented two mechanical hypotheses in 1675 and 1717). Moreover, he refused to even offer a hypothesis as to the cause of this force on grounds that to do so was contrary to sound science. He lamented that "philosophers have hitherto attempted the search of nature in vain" for the source of the gravitational force, as he was convinced "by many reasons" that there were "causes hitherto unknown" that were fundamental to all the "phenomena of nature". These fundamental phenomena are still under investigation and, though hypotheses abound, the definitive answer has yet to be found. And in Newton's 1713 *General Scholium* in the second edition of *Principia*: *"I have not yet been able to discover the cause of these properties of gravity from phenomena and I feign no hypotheses... It is enough that gravity does really exist and acts according to the laws I have explained, and that it abundantly serves to account for all the motions of celestial bodies."*[34]

3.6.4 Einstein's solution

These objections were explained by Einstein's theory of general relativity, in which gravitation is an attribute of curved spacetime instead of being due to a force propagated between bodies. In Einstein's theory, energy and momentum distort spacetime in their vicinity, and other particles move in trajectories determined by the geometry of spacetime. This allowed a description of the motions of light and mass that was consistent with all available observations. In general relativity, the gravitational force is a fictitious force due to the curvature of spacetime, because the gravitational acceleration of a body in free fall is due to its world line being a geodesic of spacetime.

3.7 Extensions

Newton was the first to consider in his Principia an extended expression of his law of gravity including an inverse-cube term of the form

$$F = G\frac{m_1 m_2}{r^2} + B\frac{m_1 m_2}{r^3}$$

attempting to explain the Moon's apsidal motion. Other extensions were proposed by Laplace (around 1790) and Decombes (1913):[35]

$$F(r) = k\frac{m_1 m_2}{r^2}\exp(-\alpha r)$$

$$F(r) = k\frac{m_1 m_2}{r^2}\left(1 + \frac{\alpha}{r^3}\right)$$

In recent years quests for non-inverse square terms in the law of gravity have been carried out by neutron interferometry.[36]

3.8 Solutions of Newton's law of universal gravitation

Main article: n-body problem

The n-body problem is an ancient, classical problem[37] of predicting the individual motions of a group of celestial objects interacting with each other gravitationally. Solving this problem — from the time of the Greeks and on — has been motivated by the desire to understand the motions of the Sun, planets and the visible stars. In the 20th century, understanding the dynamics of globular cluster star systems became an important n-body problem too.[38] The n-body problem in general relativity is considerably more difficult to solve.

The classical physical problem can be informally stated as: *given the quasi-steady orbital properties (instantaneous position, velocity and time)*[39] *of a group of celestial bodies, predict their interactive forces; and consequently, predict their true orbital motions for all future times.*[40]

The two-body problem has been completely solved, as has the *Restricted 3-Body Problem.*[41]

3.9 See also

- Bentley's paradox
- Gauss's law for gravity
- Jordan and Einstein frames
- Kepler orbit
- Newton's cannonball
- Newton's laws of motion
- Static forces and virtual-particle exchange

3.10 Notes

[1] It was shown separately that large, spherically symmetrical masses attract and are attracted as if all their mass were concentrated at their centers.

3.11 References

[1] Walter Lewin (October 4, 1999). *Work, Energy, and Universal GravitatioT Coarse 8.01: Classical Mechanics. Lecture 11* (ogg) (videotape). Cambridge, MA USA: MIT OCW. Event occurs at 1:21-10:10. Retrieved December 23, 2010.

[2] Isaac Newton: "In [experimental] philosophy particular propositions are inferred from the phenomena and afterwards rendered general by induction": "Principia", Book 3, General Scholium, at p.392 in Volume 2 of Andrew Motte's English translation published 1729.

[3] - Proposition 75, Theorem 35: p.956 - I.Bernard Cohen and Anne Whitman, translators: Isaac Newton, *The Principia*: Mathematical Principles of Natural Philosophy. Preceded by *A Guide to Newton's Principia*, by I.Bernard Cohen. University of California Press 1999 ISBN 0-520-08816-6 ISBN 0-520-08817-4

[4] The Michell-Cavendish Experiment, Laurent Hodges

[5] H W Turnbull (ed.), Correspondence of Isaac Newton, Vol 2 (1676-1687), (Cambridge University Press, 1960), giving the Halley-Newton correspondence of May to July 1686 about Hooke's claims at pp.431-448, see particularly page 431.

[6] Hooke's 1674 statement in "An Attempt to Prove the Motion of the Earth from Observations" is available in online facsimile here.

[7] Purrington, Robert D. (2009). *The First Professional Scientist: Robert Hooke and the Royal Society of London*. Springer. p. 168. ISBN 3-0346-0036-4., Extract of page 168

[8] See page 239 in Curtis Wilson (1989). "The Newtonian achievement in astronomy", ch.13 (pages 233-274) in "Planetary astronomy from the Renaissance to the rise of astrophysics: 2A: Tycho Brahe to Newton", CUP 1989.

[9] Calendar (New Style) Act 1750

[10] Page 309 in H W Turnbull (ed.), Correspondence of Isaac Newton, Vol 2 (1676-1687), (Cambridge University Press, 1960), document #239.

[11] See Curtis Wilson (1989) at page 244.

[12] Page 297 in H W Turnbull (ed.), Correspondence of Isaac Newton, Vol 2 (1676-1687), (Cambridge University Press, 1960), document #235, 24 November 1679.

[13] Page 433 in H W Turnbull (ed.), Correspondence of Isaac Newton, Vol 2 (1676-1687), (Cambridge University Press, 1960), document #286, 27 May 1686.

[14] Pages 435-440 in H W Turnbull (ed.), Correspondence of Isaac Newton, Vol 2 (1676-1687), (Cambridge University Press, 1960), document #288, 20 June 1686.

[15] Bullialdus (Ismael Bouillau) (1645). "Astronomia philolaica", Paris, 1645.

[16] Borelli, G. A., "Theoricae Mediceorum Planetarum ex causis physicis deductae", Florence, 1666.

[17] D T Whiteside, "Before the Principia: the maturing of Newton's thoughts on dynamical astronomy, 1664-1684", Journal for the History of Astronomy, i (1970), pages 5-19; especially at page 13.

[18] Page 436, Correspondence, Vol.2, already cited.

[19] Propositions 70 to 75 in Book 1, for example in the 1729 English translation of the *Principia*, start at page 263.

[20] Propositions 43 to 45 in Book 1, in the 1729 English translation of the *Principia*, start at page 177.

[21] D T Whiteside, "The pre-history of the 'Principia' from 1664 to 1686", Notes and Records of the Royal Society of London, 45 (1991), pages 11-61; especially at 13-20.

[22] See J. Bruce Brackenridge, "The key to Newton's dynamics: the Kepler problem and the Principia", (University of California Press, 1995), especially at pages 20-21.

[23] See for example the 1729 English translation of the *Principia*, at page 66.

[24] See page 10 in D T Whiteside, "Before the Principia: the maturing of Newton's thoughts on dynamical astronomy, 1664-1684", Journal for the History of Astronomy, i (1970), pages 5-19.

[25] Discussion points can be seen for example in the following papers: N Guicciardini, "Reconsidering the Hooke-Newton debate on Gravitation: Recent Results", in Early Science and Medicine, 10 (2005), 511-517; Ofer Gal, "The Invention of Celestial Mechanics", in Early Science and Medicine, 10 (2005), 529-534; M Nauenberg, "Hooke's and Newton's Contributions to the Early Development of Orbital mechanics and Universal Gravitation", in Early Science and Medicine, 10 (2005), 518-528.

[26] See for example the results of Propositions 43-45 and 70-75 in Book 1, cited above.

[27] See also G E Smith, in Stanford Encyclopedia of Philosophy, "Newton's Philosophiae Naturalis Principia Mathematica".

[28] The second extract is quoted and translated in W.W. Rouse Ball, "An Essay on Newton's 'Principia'" (London and New York: Macmillan, 1893), at page 69.

[29] The original statements by Clairaut (in French) are found (with orthography here as in the original) in "Explication abregée du systême du monde, et explication des principaux phénomenes astronomiques tirée des Principes de M. Newton" (1759), at Introduction (section IX), page 6: "Il ne faut pas croire que cette idée ... de Hook diminue la gloire de M. Newton", [and] "L'exemple de Hook" [serve] "à faire voir quelle distance il y a entre une vérité entrevue & une vérité démontrée".

[30] Mohr, Peter J.; Taylor, Barry N.; Newell, David B. (2008). "CODATA Recommended Values of the Fundamental Physical Constants: 2006". *Rev. Mod. Phys.* **80** (2): 633–730. arXiv:0801.0028. Bibcode:2008RvMP...80..633M. doi:10.1103/RevMod Phys.80.633.Direct link to value..

[31] Equilibrium State

[32] Misner, Charles W.; Thorne, Kip S.; Wheeler, John Archibald (1973). *Gravitation*. New York: W. H.Freeman and Company. ISBN 0-7167-0344-0 Page 1049.

[33] - Max Born (1924), *Einstein's Theory of Relativity* (The 1962 Dover edition, page 348 lists a table documenting the observed and calculated values for the precession of the perihelion of Mercury, Venus, and the Earth.)

[34] - *The Construction of Modern Science: Mechanisms and Mechanics*, by Richard S. Westfall. Cambridge University Press. 1978

[35] http://physicsessays.org/doi/abs/10.4006/1.3038751?journalCode=phes

[36] http://journals.aps.org/prc/abstract/10.1103/PhysRevC.75.015501

[37] Leimanis and Minorsky: Our interest is with Leimanis, who first discusses some history about the *n*-body problem, especially Ms. Kovalevskaya's ~1868-1888, twenty-year complex-variables approach, failure: **Section 1: The Dynamics of Rigid Bodies and Mathematical Exterior Ballistics** (Chapter 1, *the motion of a rigid body about a fixed point* (**Euler** and **Poisson** *equations*); Chapter 2, *Mathematical Exterior Ballistics*), good precursor background to the *n*-body problem: **Section 2: Celestial Mechanics** (Chapter 1, *The Uniformization of the Three-body Problem* (Restricted Three-body Problem); Chapter 2, *Capture in the Three-Body Problem*; Chapter 3, *Generalized n-body Problem*).

[38] See References sited for Heggie and Hut. This Wikipedia page has made their approach obsolete.

[39] *Quasi-steady* loads refers to the instantaneous inertial loads generated by instantaneous angular velocities and accelerations, as well as translational accelerations (9 variables). It is as though one took a photograph, which also recorded the instantaneous position and properties of motion. In contrast, a *steady-state* condition refers to a system's state being invariant to time; otherwise, the first derivatives and all higher derivatives are zero.

[40] R. M. Rosenberg states the *n*-body problem similarly (see References): *Each particle in a system of a finite number of particles is subjected to a Newtonian gravitational attraction from all the other particles, and to no other forces. If the initial state of the system is given, how will the particles move?* Rosenberg failed to realize, like everyone else, that it is necessary to determine the forces *first* before the motions can be determined.

[41] A general, classical solution in terms of first integrals is known to be impossible. An exact theoretical solution for arbitrary *n* can be approximated via Taylor series, but in practice such an infinite series must be truncated, giving at best only an approximate solution; and an approach now obsolete. In addition, the *n*-body problem may be solved using numerical integration, but these, too, are approximate solutions; and again obsolete. See Sverre J. Aarseth's book **Gravitational N-body Simulations** listed in the References.

3.12   External links

- Feather & Hammer Drop on Moon on YouTube

- Newton's Law of Universal Gravitation Javascript calculator

Chapter 4

Semiclassical gravity

Semiclassical gravity is the approximation to the theory of quantum gravity in which one treats matter fields as being quantum and the gravitational field as being classical.

In semiclassical gravity, matter is represented by quantum matter fields that propagate according to the theory of quantum fields in curved spacetime. The spacetime in which the fields propagate is classical but dynamical. The curvature of the spacetime is given by the *semiclassical Einstein equations*, which relate the curvature of the spacetime, given by the Einstein tensor $G_{\mu\nu}$, to the expectation value of the energy–momentum tensor operator, $T_{\mu\nu}$, of the matter fields:

$$G_{\mu\nu} = \frac{8\pi G}{c^4} \left\langle \hat{T}_{\mu\nu} \right\rangle_\psi$$

where G is Newton's constant and ψ indicates the quantum state of the matter fields.

4.1 Stress–energy tensor

There is some ambiguity in regulating the stress–energy tensor, and this depends upon the curvature. This ambiguity can be absorbed into the cosmological constant, Newton's constant, and the quadratic couplings[1]

$$\int d^d x \sqrt{-g} R^2 \text{ and } \int d^d x \sqrt{-g} R^{\mu\nu} R_{\mu\nu} .$$

There's also the other quadratic term

$$\int d^d x \sqrt{-g} R^{\mu\nu\rho\sigma} R_{\mu\nu\rho\sigma}$$

but (in 4-dimensions) this term is a linear combination of the other two terms and a surface term. See Gauss–Bonnet gravity for more details.

Since the theory of quantum gravity is not yet known, it is difficult to say what is the regime of validity of semiclassical gravity. However, one can formally show that semiclassical gravity could be deduced from quantum gravity by considering N copies of the quantum matter fields, and taking the limit of N going to infinity while keeping the product GN constant. At diagrammatic level, semiclassical gravity corresponds to summing all Feynman diagrams which do not have loops of gravitons (but have an arbitrary number of matter loops). Semiclassical gravity can also be deduced from an axiomatic approach.

4.2   Experimental status

There are cases where semiclassical gravity breaks down. For instance,[2] if M is a huge mass, then the superposition

$$\frac{1}{\sqrt{2}}\left(|M \text{ at } A\rangle + |M \text{ at } B\rangle\right)$$

where A and B are widely separated, then the expectation value of the stress–energy tensor is $M/2$ at A and $M/2$ at B, but we would never observe the metric sourced by such a distribution. Instead, we decohere into a state with the metric sourced at A and another sourced at B with a 50% chance each.

4.3   Applications

The most important applications of semiclassical gravity are to understand the Hawking radiation of black holes and the generation of random gaussian-distributed perturbations in the theory of cosmic inflation, which is thought to occur at the very beginnings of the big bang.

4.4   Notes

[1] See Wald (1994) Chapter 4, section 6 "The Stress-Energy Tensor".

[2] See Page and Geilker; Eppley and Hannah; Albers, Kiefer, and Reginatto.

4.5   References

- Birrell, N. D. and Davies, P. C. W., *Quantum fields in curved space*, (Cambridge University Press, Cambridge, UK, 1982).

- Don N. Page, and C. D. Geilker, "Indirect Evidence for Quantum Gravity." *Phys. Rev. Lett.* **47** (1981) 979–982. DOI:10.1103/PhysRevLett.47.979

- K. Eppley and E. Hannah, "The necessity of quantizing the gravitational field." *Found. Phys.* **7** (1977) 51–68. doi:10.1007/BF00715241

- Mark Albers, Claus Kiefer, Marcel Reginatto, "Measurement Analysis and Quantum Gravity." *Phys.Rev.D* **78** 6 (2008) 064051, DOI:10.1103/PhysRevD.78.064051. Eprint arXiv:0802.1978 [gr-qc].

- Robert M. Wald, *Quantum Field Theory in Curved Spacetime and Black Hole Thermodynamics*. University of Chicago Press, 1994.

- Semiclassical gravity on arxiv.org

Chapter 5

Introduction to general relativity

This article is a non-technical introduction to the subject. For the main encyclopedia article, see General relativity.

General relativity is a theory of gravitation that was developed by Albert Einstein between 1907 and 1915. According to general relativity, the observed gravitational effect between masses results from their warping of spacetime.

By the beginning of the 20th century, Newton's law of universal gravitation had been accepted for more than two hundred years as a valid description of the gravitational force between masses. In Newton's model, gravity is the result of an attractive force between massive objects. Although even Newton was troubled by the unknown nature of that force,[1] the basic framework was extremely successful at describing motion.

Experiments and observations show that Einstein's description of gravitation accounts for several effects that are unexplained by Newton's law, such as minute anomalies in the orbits of Mercury and other planets. General relativity also predicts novel effects of gravity, such as gravitational waves, gravitational lensing and an effect of gravity on time known as gravitational time dilation. Many of these predictions have been confirmed by experiment, while others are the subject of ongoing research. For example, although there is indirect evidence for gravitational waves, direct evidence of their existence is still being sought by several teams of scientists in experiments such as the LIGO and GEO600 projects.

General relativity has developed into an essential tool in modern astrophysics. It provides the foundation for the current understanding of black holes, regions of space where the gravitational effect is so strong that even light cannot escape. Their strong gravity is thought to be responsible for the intense radiation emitted by certain types of astronomical objects (such as active galactic nuclei or microquasars). General relativity is also part of the framework of the standard Big Bang model of cosmology.

Although general relativity is not the only relativistic theory of gravity, it is the simplest such theory that is consistent with the experimental data. Nevertheless, a number of open questions remain, the most fundamental of which is how general relativity can be reconciled with the laws of quantum physics to produce a complete and self-consistent theory of quantum gravity.

5.1 From special to general relativity

In September 1905, Albert Einstein published his theory of special relativity, which reconciles Newton's laws of motion with electrodynamics (the interaction between objects with electric charge). Special relativity introduced a new framework for all of physics by proposing new concepts of space and time. Some then-accepted physical theories were inconsistent with that framework; a key example was Newton's theory of gravity, which describes the mutual attraction experienced by bodies due to their mass.

Several physicists, including Einstein, searched for a theory that would reconcile Newton's law of gravity and special relativity. Only Einstein's theory proved to be consistent with experiments and observations. To understand the theory's basic ideas, it is instructive to follow Einstein's thinking between 1907 and 1915, from his simple thought experiment involving an observer in free fall to his fully geometric theory of gravity.[2]

5.1.1 Equivalence principle

Main article: Equivalence principle

A person in a free-falling elevator experiences weightlessness and objects either float motionless or drift at constant speed. Since everything in the elevator is falling together, no gravitational effect can be observed. In this way, the experiences of an observer in free fall are indistinguishable from those of an observer in deep space, far from any significant source of gravity. Such observers are the privileged ("inertial") observers Einstein described in his theory of special relativity: observers for whom light travels along straight lines at constant speed.[3]

Einstein hypothesized that the similar experiences of weightless observers and inertial observers in special relativity represented a fundamental property of gravity, and he made this the cornerstone of his theory of general relativity, formalized in his equivalence principle. Roughly speaking, the principle states that a person in a free-falling elevator cannot tell that they are in free fall. Every experiment in such a free-falling environment has the same results as it would for an observer at rest or moving uniformly in deep space, far from all sources of gravity.[4]

5.1.2 Gravity and acceleration

Most effects of gravity vanish in free fall, but effects that seem the same as those of gravity can be *produced* by an accelerated frame of reference. An observer in a closed room cannot tell which of the following is true:

- Objects are falling to the floor because the room is resting on the surface of the Earth and the objects are being pulled down by gravity.

- Objects are falling to the floor because the room is aboard a rocket in space, which is accelerating at 9.81 m/s² and is far from any source of gravity. The objects are being pulled towards the floor by the same "inertial force" that presses the driver of an accelerating car into the back of his seat.

Conversely, any effect observed in an accelerated reference frame should also be observed in a gravitational field of corresponding strength. This principle allowed Einstein to predict several novel effects of gravity in 1907, as explained in the next section.

An observer in an accelerated reference frame must introduce what physicists call fictitious forces to account for the acceleration experienced by himself and objects around him. One example, the force pressing the driver of an accelerating car into his or her seat, has already been mentioned; another is the force you can feel pulling your arms up and out if you attempt to spin around like a top. Einstein's master insight was that the constant, familiar pull of the Earth's gravitational field is fundamentally the same as these fictitious forces.[5] The apparent magnitude of the fictitious forces always appears to be proportional to the mass of any object on which they act - for instance, the driver's seat exerts just enough force to accelerate the driver at the same rate as the car. By analogy, Einstein proposed that an object in a gravitational field should feel a gravitational force proportional to its mass, as embodied in Newton's law of gravitation.[6]

5.1.3 Physical consequences

In 1907, Einstein was still eight years away from completing the general theory of relativity. Nonetheless, he was able to make a number of novel, testable predictions that were based on his starting point for developing his new theory: the equivalence principle.[7]

The first new effect is the gravitational frequency shift of light. Consider two observers aboard an accelerating rocket-ship. Aboard such a ship, there is a natural concept of "up" and "down": the direction in which the ship accelerates is "up", and unattached objects accelerate in the opposite direction, falling "downward". Assume that one of the observers is "higher up" than the other. When the lower observer sends a light signal to the higher observer, the acceleration causes the light to be red-shifted, as may be calculated from special relativity; the second observer will measure a lower frequency for the light than the first. Conversely, light sent from the higher observer to the lower is blue-shifted, that is, shifted towards higher frequencies.[8] Einstein argued that such frequency shifts must also be observed in a gravitational field. This is illustrated in the figure at left, which shows a light wave that is gradually red-shifted as it works its way upwards against the gravitational acceleration. This effect has been confirmed experimentally, as described below.

This gravitational frequency shift corresponds to a gravitational time dilation: Since the "higher" observer measures the same light wave to have a lower frequency than the "lower" observer, time must be passing faster for the higher observer. Thus, time runs more slowly for observers who are lower in a gravitational field.

It is important to stress that, for each observer, there are no observable changes of the flow of time for events or processes that are at rest in his or her reference frame. Five-minute-eggs as timed by each observer's clock have the same consistency; as one year passes on each clock, each observer ages by that amount; each clock, in short, is in perfect agreement with all processes happening in its immediate vicinity. It is only when the clocks are compared between separate observers that one can notice that time runs more slowly for the lower observer than for the higher.[9] This effect is minute, but it too has been confirmed experimentally in multiple experiments, as described below.

In a similar way, Einstein predicted the gravitational deflection of light: in a gravitational field, light is deflected downward. Quantitatively, his results were off by a factor of two; the correct derivation requires a more complete formulation of the theory of general relativity, not just the equivalence principle.[10]

5.1.4   Tidal effects

The equivalence between gravitational and inertial effects does not constitute a complete theory of gravity. When it comes to explaining gravity near our own location on the Earth's surface, noting that our reference frame is not in free fall, so that fictitious forces are to be expected, provides a suitable explanation. But a freely falling reference frame on one side of the Earth cannot explain why the people on the opposite side of the Earth experience a gravitational pull in the opposite direction.

A more basic manifestation of the same effect involves two bodies that are falling side by side towards the Earth. In a reference frame that is in free fall alongside these bodies, they appear to hover weightlessly – but not exactly so. These bodies are not falling in precisely the same direction, but towards a single point in space: namely, the Earth's center of gravity. Consequently, there is a component of each body's motion towards the other (see the figure). In a small environment such as a freely falling lift, this relative acceleration is minuscule, while for skydivers on opposite sides of the Earth, the effect is large. Such differences in force are also responsible for the tides in the Earth's oceans, so the term "tidal effect" is used for this phenomenon.

The equivalence between inertia and gravity cannot explain tidal effects – it cannot explain variations in the gravitational field.[11] For that, a theory is needed which describes the way that matter (such as the large mass of the Earth) affects the inertial environment around it.

5.1.5   From acceleration to geometry

In exploring the equivalence of gravity and acceleration as well as the role of tidal forces, Einstein discovered several analogies with the geometry of surfaces. An example is the transition from an inertial reference frame (in which free particles coast along straight paths at constant speeds) to a rotating reference frame (in which extra terms corresponding to fictitious forces have to be introduced in order to explain particle motion): this is analogous to the transition from a Cartesian coordinate system (in which the coordinate lines are straight lines) to a curved coordinate system (where coordinate lines need not be straight).

A deeper analogy relates tidal forces with a property of surfaces called *curvature*. For gravitational fields, the absence or presence of tidal forces determines whether or not the influence of gravity can be eliminated by choosing a freely falling reference frame. Similarly, the absence or presence of curvature determines whether or not a surface is equivalent to a plane. In the summer of 1912, inspired by these analogies, Einstein searched for a geometric formulation of gravity.[12]

The elementary objects of geometry – points, lines, triangles – are traditionally defined in three-dimensional space or on two-dimensional surfaces. In 1907, Hermann Minkowski, Einstein's former mathematics professor at the Swiss Federal Polytechnic, introduced a geometric formulation of Einstein's special theory of relativity where the geometry included not only space but also time. The basic entity of this new geometry is four-dimensional spacetime. The orbits of moving bodies are curves in spacetime; the orbits of bodies moving at constant speed without changing direction correspond to straight lines.[13]

For surfaces, the generalization from the geometry of a plane – a flat surface – to that of a general curved surface had been described in the early 19th century by Carl Friedrich Gauss. This description had in turn been generalized to higher-dimensional spaces in a mathematical formalism introduced by Bernhard Riemann in the 1850s. With the help of Riemannian geometry, Einstein formulated a geometric description of gravity in which Minkowski's spacetime is replaced by distorted, curved spacetime, just as curved surfaces are a generalization of ordinary plane surfaces. **Embedding Diagrams** are used to illustrate curved spacetime in educational contexts.[14][15]

After he had realized the validity of this geometric analogy, it took Einstein a further three years to find the missing cornerstone of his theory: the equations describing how matter influences spacetime's curvature. Having formulated what are now known as Einstein's equations (or, more precisely, his field equations of gravity), he presented his new theory of gravity at several sessions of the Prussian Academy of Sciences in late 1915, culminating in his final presentation on November 25, 1915.[16]

5.2   Geometry and gravitation

Paraphrasing John Wheeler, Einstein's geometric theory of gravity can be summarized thus: **spacetime tells matter how to move; matter tells spacetime how to curve**.[17] What this means is addressed in the following three sections, which explore the motion of so-called test particles, examine which properties of matter serve as a source for gravity, and, finally, introduce Einstein's equations, which relate these matter properties to the curvature of spacetime.

5.2.1   Probing the gravitational field

In order to map a body's gravitational influence, it is useful to think about what physicists call probe or test particles: particles that are influenced by gravity, but are so small and light that we can neglect their own gravitational effect. In the absence of gravity and other external forces, a test particle moves along a straight line at a constant speed. In the language of spacetime, this is equivalent to saying that such test particles move along straight world lines in spacetime. In the presence of gravity, spacetime is non-Euclidean, or curved, and in curved spacetime straight world lines may not exist. Instead, test particles move along lines called geodesics, which are "as straight as possible", that is, they follow the shortest path between starting and ending points, taking the curvature into consideration.

A simple analogy is the following: In geodesy, the science of measuring Earth's size and shape, a geodesic (from Greek "geo", Earth, and "daiein", to divide) is the shortest route between two points on the Earth's surface. Approximately, such a route is a segment of a great circle, such as a line of longitude or the equator. These paths are certainly not straight, simply because they must follow the curvature of the Earth's surface. But they are as straight as is possible subject to this constraint.

The properties of geodesics differ from those of straight lines. For example, on a plane, parallel lines never meet, but this is not so for geodesics on the surface of the Earth: for example, lines of longitude are parallel at the equator, but intersect at the poles. Analogously, the world lines of test particles in free fall are spacetime geodesics, the straightest possible lines in spacetime. But still there are crucial differences between them and the truly straight lines that can be traced out in the gravity-free spacetime of special relativity. In special relativity, parallel geodesics remain parallel. In a gravitational field with tidal effects, this will not, in general, be the case. If, for example, two bodies are initially at rest relative to each other, but are then dropped in the Earth's gravitational field, they will move towards each other as they fall towards the Earth's center.[18]

Compared with planets and other astronomical bodies, the objects of everyday life (people, cars, houses, even mountains) have little mass. Where such objects are concerned, the laws governing the behavior of test particles are sufficient to describe what happens. Notably, in order to deflect a test particle from its geodesic path, an external force must be applied. A person sitting on a chair is trying to follow a geodesic, that is, to fall freely towards the center of the Earth. But the chair applies an external upwards force preventing the person from falling. In this way, general relativity explains the daily experience of gravity on the surface of the Earth *not* as the downwards pull of a gravitational force, but as the upwards push of external forces. These forces deflect all bodies resting on the Earth's surface from the geodesics they would otherwise follow.[19] For matter objects whose own gravitational influence cannot be neglected, the laws of motion are somewhat more complicated than for test particles, although it remains true that spacetime tells matter how to move.[20]

5.2.2 Sources of gravity

In Newton's description of gravity, the gravitational force is caused by matter. More precisely, it is caused by a specific property of material objects: their mass. In Einstein's theory and related theories of gravitation, curvature at every point in spacetime is also caused by whatever matter is present. Here, too, mass is a key property in determining the gravitational influence of matter. But in a relativistic theory of gravity, mass cannot be the only source of gravity. Relativity links mass with energy, and energy with momentum.

The equivalence between mass and energy, as expressed by the formula $E = mc^2$, is the most famous consequence of special relativity. In relativity, mass and energy are two different ways of describing one physical quantity. If a physical system has energy, it also has the corresponding mass, and vice versa. In particular, all properties of a body that are associated with energy, such as its temperature or the binding energy of systems such as nuclei or molecules, contribute to that body's mass, and hence act as sources of gravity.[21]

In special relativity, energy is closely connected to momentum. Just as space and time are, in that theory, different aspects of a more comprehensive entity called spacetime, energy and momentum are merely different aspects of a unified, four-dimensional quantity that physicists call four-momentum. In consequence, if energy is a source of gravity, momentum must be a source as well. The same is true for quantities that are directly related to energy and momentum, namely internal pressure and tension. Taken together, in general relativity it is mass, energy, momentum, pressure and tension that serve as sources of gravity: they are how matter tells spacetime how to curve. In the theory's mathematical formulation, all these quantities are but aspects of a more general physical quantity called the energy–momentum tensor.[22]

5.2.3 Einstein's equations

Einstein's equations are the centerpiece of general relativity. They provide a precise formulation of the relationship between spacetime geometry and the properties of matter, using the language of mathematics. More concretely, they are formulated using the concepts of Riemannian geometry, in which the geometric properties of a space (or a spacetime) are described by a quantity called a metric. The metric encodes the information needed to compute the fundamental geometric notions of distance and angle in a curved space (or spacetime).

A spherical surface like that of the Earth provides a simple example. The location of any point on the surface can be described by two coordinates: the geographic latitude and longitude. Unlike the Cartesian coordinates of the plane, coordinate differences are not the same as distances on the surface, as shown in the diagram on the right: for someone at the equator, moving 30 degrees of longitude westward (magenta line) corresponds to a distance of roughly 3,300 kilometers (2,100 mi). On the other hand, someone at a latitude of 55 degrees, moving 30 degrees of longitude westward (blue line) covers a distance of merely 1,900 kilometers (1,200 mi). Coordinates therefore do not provide enough information to describe the geometry of a spherical surface, or indeed the geometry of any more complicated space or spacetime. That information is precisely what is encoded in the metric, which is a function defined at each point of the surface (or space, or spacetime) and relates coordinate differences to differences in distance. All other quantities that are of interest in geometry, such as the length of any given curve, or the angle at which two curves meet, can be computed from this metric function.[23]

The metric function and its rate of change from point to point can be used to define a geometrical quantity called the Riemann curvature tensor, which describes exactly how the space or spacetime is curved at each point. In general relativity, the metric and the Riemann curvature tensor are quantities defined at each point in spacetime. As has already been mentioned, the matter content of the spacetime defines another quantity, the energy–momentum tensor **T**, and the principle that "spacetime tells matter how to move, and matter tells spacetime how to curve" means that these quantities must be related to each other. Einstein formulated this relation by using the Riemann curvature tensor and the metric to define another geometrical quantity **G**, now called the Einstein tensor, which describes some aspects of the way spacetime is curved. *Einstein's equation* then states that

$$\mathbf{G} = \frac{8\pi G}{c^4}\mathbf{T},$$

i.e. up to a constant multiple, the quantity **G** (which measures curvature) is equated with the quantity **T** (which measures matter content). Here, G is the gravitational constant of Newtonian gravity, and c is the speed of light from special relativity.

This equation is often referred to in the plural as *Einstein's equations*, since the quantities **G** and **T** are each determined by several functions of the coordinates of spacetime, and the equations equate each of these component functions.[24] A solution of these equations describes a particular geometry of spacetime; for example, the Schwarzschild solution describes the geometry around a spherical, non-rotating mass such as a star or a black hole, whereas the Kerr solution describes a rotating black hole. Still other solutions can describe a gravitational wave or, in the case of the Friedmann–Lemaître– Robertson–Walker solution, an expanding universe. The simplest solution is the uncurved Minkowski spacetime, the spacetime described by special relativity.[25]

5.3   Experiments

No scientific theory is apodictically true; each is a model that must be checked by experiment. Newton's law of gravity was accepted because it accounted for the motion of planets and moons in the solar system with considerable accuracy. As the precision of experimental measurements gradually improved, some discrepancies with Newton's predictions were observed, and these were accounted for in the general theory of relativity. Similarly, the predictions of general relativity must also be checked with experiment, and Einstein himself devised three tests now known as the classical tests of the theory:

- Newtonian gravity predicts that the orbit which a single planet traces around a perfectly spherical star should be an ellipse. Einstein's theory predicts a more complicated curve: the planet behaves as if it were travelling around an ellipse, but at the same time, the ellipse as a whole is rotating slowly around the star. In the diagram on the right, the ellipse predicted by Newtonian gravity is shown in red, and part of the orbit predicted by Einstein in blue. For a planet orbiting the Sun, this deviation from Newton's orbits is known as the anomalous perihelion shift. The first measurement of this effect, for the planet Mercury, dates back to 1859. The most accurate results for Mercury and for other planets to date are based on measurements which were undertaken between 1966 and 1990, using radio telescopes.[26] General relativity predicts the correct anomalous perihelion shift for all planets where this can be measured accurately (Mercury, Venus and the Earth).

- According to general relativity, light does not travel along straight lines when it propagates in a gravitational field. Instead, it is deflected in the presence of massive bodies. In particular, starlight is deflected as it passes near the Sun, leading to apparent shifts of up 1.75 arc seconds in the stars' positions in the sky (an arc second is equal to 1/3600 of a degree). In the framework of Newtonian gravity, a heuristic argument can be made that leads to light deflection by half that amount. The different predictions can be tested by observing stars that are close to the Sun during a solar eclipse. In this way, a British expedition to West Africa in 1919, directed by Arthur Eddington, confirmed that Einstein's prediction was correct, and the Newtonian predictions wrong, via observation of the May 1919 eclipse. Eddington's results were not very accurate; subsequent observations of the deflection of the light of distant quasars by the Sun, which utilize highly accurate techniques of radio astronomy, have confirmed Eddington's results with significantly better precision (the first such measurements date from 1967, the most recent comprehensive analysis from 2004).[27]

- Gravitational redshift was first measured in a laboratory setting in 1959 by Pound and Rebka. It is also seen in astrophysical measurements, notably for light escaping the white dwarf Sirius B. The related gravitational time dilation effect has been measured by transporting atomic clocks to altitudes of between tens and tens of thousands of kilometers (first by Hafele and Keating in 1971; most accurately to date by Gravity Probe A launched in 1976).[28]

Of these tests, only the perihelion advance of Mercury was known prior to Einstein's final publication of general relativity in 1916. The subsequent experimental confirmation of his other predictions, especially the first measurements of the deflection of light by the sun in 1919, catapulted Einstein to international stardom.[29] These three experiments justified adopting general relativity over Newton's theory and, incidentally, over a number of alternatives to general relativity that had been proposed.

Further tests of general relativity include precision measurements of the Shapiro effect or gravitational time delay for light, most recently in 2002 by the Cassini space probe. One set of tests focuses on effects predicted by general relativity for the behavior of gyroscopes travelling through space. One of these effects, geodetic precession, has been tested with the Lunar Laser Ranging Experiment (high-precision measurements of the orbit of the Moon). Another, which is related to rotating masses, is called frame-dragging. The geodetic and frame-dragging effects were both tested by the Gravity Probe B satellite experiment launched in 2004, with results confirming relativity to within 0.5% and 15%, respectively, as of December 2008.[30]

By cosmic standards, gravity throughout the solar system is weak. Since the differences between the predictions of Einstein's and Newton's theories are most pronounced when gravity is strong, physicists have long been interested in testing various relativistic effects in a setting with comparatively strong gravitational fields. This has become possible thanks to precision observations of binary pulsars. In such a star system, two highly compact neutron stars orbit each other. At least one of them is a pulsar – an astronomical object that emits a tight beam of radiowaves. These beams strike the Earth at very regular intervals, similarly to the way that the rotating beam of a lighthouse means that an observer sees the lighthouse blink, and can be observed as a highly regular series of pulses. General relativity predicts specific deviations from the regularity of these radio pulses. For instance, at times when the radio waves pass close to the other neutron star, they should be deflected by the star's gravitational field. The observed pulse patterns are impressively close to those predicted by general relativity.[31]

One particular set of observations is related to eminently useful practical applications, namely to satellite navigation systems such as the Global Positioning System that are used both for precise positioning and timekeeping. Such systems rely on two sets of atomic clocks: clocks aboard satellites orbiting the Earth, and reference clocks stationed on the Earth's surface. General relativity predicts that these two sets of clocks should tick at slightly different rates, due to their different motions (an effect already predicted by special relativity) and their different positions within the Earth's gravitational field. In order to ensure the system's accuracy, the satellite clocks are either slowed down by a relativistic factor, or that same factor is made part of the evaluation algorithm. In turn, tests of the system's accuracy (especially the very thorough measurements that are part of the definition of universal coordinated time) are testament to the validity of the relativistic predictions.[32]

A number of other tests have probed the validity of various versions of the equivalence principle; strictly speaking, all measurements of gravitational time dilation are tests of the weak version of that principle, not of general relativity itself. So far, general relativity has passed all observational tests.[33]

5.4 Astrophysical applications

Models based on general relativity play an important role in astrophysics; the success of these models is further testament to the theory's validity.

5.4.1 Gravitational lensing

Since light is deflected in a gravitational field, it is possible for the light of a distant object to reach an observer along two or more paths. For instance, light of a very distant object such as a quasar can pass along one side of a massive galaxy and be deflected slightly so as to reach an observer on Earth, while light passing along the opposite side of that same galaxy is deflected as well, reaching the same observer from a slightly different direction. As a result, that particular observer will see one astronomical object in two different places in the night sky. This kind of focussing is well-known when it comes to optical lenses, and hence the corresponding gravitational effect is called gravitational lensing.[34]

Observational astronomy uses lensing effects as an important tool to infer properties of the lensing object. Even in cases where that object is not directly visible, the shape of a lensed image provides information about the mass distribution responsible for the light deflection. In particular, gravitational lensing provides one way to measure the distribution of dark matter, which does not give off light and can be observed only by its gravitational effects. One particularly interesting application are large-scale observations, where the lensing masses are spread out over a significant fraction of the observable universe, and can be used to obtain information about the large-scale properties and evolution of our cosmos.[35]

5.4.2   Gravitational waves

Gravitational waves, a direct consequence of Einstein's theory, are distortions of geometry that propagate at the speed of light, and can be thought of as ripples in spacetime. They should not be confused with the gravity waves of fluid dynamics, which are a different concept.

Indirectly, the effect of gravitational waves has been detected in observations of specific binary stars. Such pairs of stars orbit each other and, as they do so, gradually lose energy by emitting gravitational waves. For ordinary stars like the Sun, this energy loss would be too small to be detectable, but this energy loss was observed in 1974 in a binary pulsar called PSR1913+16. In such a system, one of the orbiting stars is a pulsar. This has two consequences: a pulsar is an extremely dense object known as a neutron star, for which gravitational wave emission is much stronger than for ordinary stars. Also, a pulsar emits a narrow beam of electromagnetic radiation from its magnetic poles. As the pulsar rotates, its beam sweeps over the Earth, where it is seen as a regular series of radio pulses, just as a ship at sea observes regular flashes of light from the rotating light in a lighthouse. This regular pattern of radio pulses functions as a highly accurate "clock". It can be used to time the double star's orbital period, and it reacts sensitively to distortions of spacetime in its immediate neighborhood.

The discoverers of PSR1913+16, Russell Hulse and Joseph Taylor, were awarded the Nobel Prize in Physics in 1993. Since then, several other binary pulsars have been found. The most useful are those in which both stars are pulsars, since they provide the most accurate tests of general relativity.[36]

Currently, one major goal of research in relativity is the direct detection of gravitational waves. To this end, a number of land-based gravitational wave detectors are in operation, and a mission to launch a space-based detector, LISA, is currently under development, with a precursor mission (LISA Pathfinder) due for launch in 2015. If gravitational waves are detected, they could be used to obtain information about compact objects such as neutron stars and black holes, and also to probe the state of the early universe fractions of a second after the Big Bang.[37]

5.4.3   Black holes

When mass is concentrated into a sufficiently compact region of space, general relativity predicts the formation of a black hole – a region of space with a gravitational effect so strong that not even light can escape. Certain types of black holes are thought to be the final state in the evolution of massive stars. On the other hand, supermassive black holes with the mass of millions or billions of Suns are assumed to reside in the cores of most galaxies, and they play a key role in current models of how galaxies have formed over the past billions of years.[38]

Matter falling onto a compact object is one of the most efficient mechanisms for releasing energy in the form of radiation, and matter falling onto black holes is thought to be responsible for some of the brightest astronomical phenomena imaginable. Notable examples of great interest to astronomers are quasars and other types of active galactic nuclei. Under the right conditions, falling matter accumulating around a black hole can lead to the formation of jets, in which focused beams of matter are flung away into space at speeds near that of light.[39]

There are several properties that make black holes most promising sources of gravitational waves. One reason is that black holes are the most compact objects that can orbit each other as part of a binary system; as a result, the gravitational waves emitted by such a system are especially strong. Another reason follows from what are called black-hole uniqueness theorems: over time, black holes retain only a minimal set of distinguishing features (these theorems have become known as "no-hair" theorems, since different hairstyles are a crucial part of what gives different people their different appearances). For instance, in the long term, the collapse of a hypothetical matter cube will not result in a cube-shaped black hole. Instead, the resulting black hole will be indistinguishable from a black hole formed by the collapse of a spherical mass, but with one important difference: in its transition to a spherical shape, the black hole formed by the collapse of a cube will emit gravitational waves.[40]

5.4.4 Cosmology

One of the most important aspects of general relativity is that it can be applied to the universe as a whole. A key point is that, on large scales, our universe appears to be constructed along very simple lines: all current observations suggest that, on average, the structure of the cosmos should be approximately the same, regardless of an observer's location or direction of observation: the universe is approximately homogeneous and isotropic. Such comparatively simple universes can be described by simple solutions of Einstein's equations. The current cosmological models of the universe are obtained by combining these simple solutions to general relativity with theories describing the properties of the universe's matter content, namely thermodynamics, nuclear- and particle physics. According to these models, our present universe emerged from an extremely dense high-temperature state – the Big Bang – roughly 14 billion years ago and has been expanding ever since.[41]

Einstein's equations can be generalized by adding a term called the cosmological constant. When this term is present, empty space itself acts as a source of attractive (or, less commonly, repulsive) gravity. Einstein originally introduced this term in his pioneering 1917 paper on cosmology, with a very specific motivation: contemporary cosmological thought held the universe to be static, and the additional term was required for constructing static model universes within the framework of general relativity. When it became apparent that the universe is not static, but expanding, Einstein was quick to discard this additional term. Since the end of the 1990s, however, astronomical evidence indicating an accelerating expansion consistent with a cosmological constant – or, equivalently, with a particular and ubiquitous kind of dark energy – has steadily been accumulating.[42]

5.5 Modern research

General relativity is very successful in providing a framework for accurate models which describe an impressive array of physical phenomena. On the other hand, there are many interesting open questions, and in particular, the theory as a whole is almost certainly incomplete.[43]

In contrast to all other modern theories of fundamental interactions, general relativity is a classical theory: it does not include the effects of quantum physics. The quest for a quantum version of general relativity addresses one of the most fundamental open questions in physics. While there are promising candidates for such a theory of quantum gravity, notably string theory and loop quantum gravity, there is at present no consistent and complete theory. It has long been hoped that a theory of quantum gravity would also eliminate another problematic feature of general relativity: the presence of spacetime singularities. These singularities are boundaries ("sharp edges") of spacetime at which geometry becomes ill-defined, with the consequence that general relativity itself loses its predictive power. Furthermore, there are so-called singularity theorems which predict that such singularities *must* exist within the universe if the laws of general relativity were to hold without any quantum modifications. The best-known examples are the singularities associated with the model universes that describe black holes and the beginning of the universe.[44]

Other attempts to modify general relativity have been made in the context of cosmology. In the modern cosmological models, most energy in the universe is in forms that have never been detected directly, namely dark energy and dark matter. There have been several controversial proposals to obviate the need for these enigmatic forms of matter and energy, by modifying the laws governing gravity and the dynamics of cosmic expansion, for example modified Newtonian dynamics.[45]

Beyond the challenges of quantum effects and cosmology, research on general relativity is rich with possibilities for further exploration: mathematical relativists explore the nature of singularities and the fundamental properties of Einstein's equations,[46] ever more comprehensive computer simulations of specific spacetimes (such as those describing merging black holes) are run.[47] and the race for the first direct detection of gravitational waves continues apace.[48] More than ninety years after the theory was first published, research is more active than ever.[49]

5.6 See also

- General relativity

- Introduction to mathematics of general relativity

- Introduction to special relativity

- History of general relativity

- Tests of general relativity

- Numerical relativity

- Derivations of the Lorentz transformations

5.7 Notes

[1] - *The Construction of Modern Science: Mechanisms and Mechanics*, by Richard S. Westfall. Cambridge University Press, 1978

[2] This development is traced e.g. in Renn 2005, p. 110ff., in chapters 9 through 15 of Pais 1982, and in Janssen 2005. A precis of Newtonian gravity can be found in Schutz 2003, chapters 2–4. It is impossible to say whether the problem of Newtonian gravity crossed Einstein's mind before 1907, but, by his own admission, his first serious attempts to reconcile that theory with special relativity date to that year, cf. Pais 1982, p. 178.

[3] This is described in detail in chapter 2 of Wheeler 1990.

[4] While the equivalence principle is still part of modern expositions of general relativity, there are some differences between the modern version and Einstein's original concept, cf. Norton 1985.

[5] E. g. Janssen 2005, p. 64f. Einstein himself also explains this in section XX of his non-technical book Einstein 1961. Following earlier ideas by Ernst Mach, Einstein also explored centrifugal forces and their gravitational analogue, cf. Stachel 1989.

[6] Einstein explained this in section XX of Einstein 1961. He considered an object "suspended" by a rope from the ceiling of a room aboard an accelerating rocket: from inside the room it looks as if gravitation is pulling the object down with a force proportional to its mass, but from outside the rocket it looks as if the rope is simply transferring the acceleration of the rocket to the object, and must therefore exert just the "force" to do so.

[7] More specifically, Einstein's calculations, which are described in chapter 11b of Pais 1982, use the equivalence principle, the equivalence of gravity and inertial forces, and the results of special relativity for the propagation of light and for accelerated observers (the latter by considering, at each moment, the instantaneous inertial frame of reference associated with such an accelerated observer).

[8] This effect can be derived directly within special relativity, either by looking at the equivalent situation of two observers in an accelerated rocket-ship or by looking at a falling elevator; in both situations, the frequency shift has an equivalent description as a Doppler shift between certain inertial frames. For simple derivations of this, see Harrison 2002.

[9] See chapter 12 of Mermin 2005.

[10] Cf. Ehlers & Rindler 1997; for a non-technical presentation, see Pössel 2007.

[11] These and other tidal effects are described in Wheeler 1990, pp. 83–91.

[12] Tides and their geometric interpretation are explained in chapter 5 of Wheeler 1990. This part of the historical development is traced in Pais 1982, section 12b.

[13] For elementary presentations of the concept of spacetime, see the first section in chapter 2 of Thorne 1994, and Greene 2004, p. 47–61. More complete treatments on a fairly elementary level can be found e.g. in Mermin 2005 and in Wheeler 1990, chapters 8 and 9.

[14] Donald Marolf: *Spacetime Embedding Diagrams for Black Holes*. General Relativity and Gravitation 31, 1999, 919–944, arXiv:gr-qc/9806123.

[15] See Wheeler 1990, chapters 8 and 9 for vivid illustrations of curved spacetime.

[16] Einstein's struggle to find the correct field equations is traced in chapters 13–15 of Pais 1982.

[17] E.g. p. xi in Wheeler 1990.

[18] A thorough, yet accessible account of basic differential geometry and its application in general relativity can be found in Geroch 1978.

[19] See chapter 10 of Wheeler 1990.

[20] In fact, when starting from the complete theory, Einstein's equation can be used to derive these more complicated laws of motion for matter as a consequence of geometry, but deriving from this the motion of idealized test particles is a highly non-trivial task, cf. Poisson 2004.

[21] A simple explanation of mass–energy equivalence can be found in sections 3.8 and 3.9 of Giulini 2005.

[22] See chapter 6 of Wheeler 1990.

[23] For a more detailed definition of the metric, but one that is more informal than a textbook presentation, see chapter 14.4 of Penrose 2004.

[24] The geometrical meaning of Einstein's equations is explored in chapters 7 and 8 of Wheeler 1990; cf. box 2.6 in Thorne 1994. An introduction using only very simple mathematics is given in chapter 19 of Schutz 2003.

[25] The most important solutions are listed in every textbook on general relativity; for a (technical) summary of our current understanding, see Friedrich 2005.

[26] More precisely, these are VLBI measurements of planetary positions; see chapter 5 of Will 1993 and section 3.5 of Will 2006.

[27] For the historical measurements, see Hartl 2005, Kennefick 2005, and Kennefick 2007; Soldner's original derivation in the framework of Newton's theory is von Soldner 1804. For the most precise measurements to date, see Bertotti 2005.

[28] See Kennefick 2005 and chapter 3 of Will 1993. For the Sirius B measurements, see Trimble & Barstow 2007.

[29] Pais 1982, Mercury on pp. 253–254, Einstein's rise to fame in sections 16b and 16c.

[30] Everitt, C.W.F.; Parkinson, B.W. (2009), *Gravity Probe B Science Results—NASA Final Report* (PDF), retrieved 2009-05-02

[31] Kramer 2004.

[32] An accessible account of relativistic effects in the global positioning system can be found in Ashby 2002; details are given in Ashby 2003.

[33] An accessible introduction to tests of general relativity is Will 1993; a more technical, up-to-date account is Will 2006.

[34] The geometry of such situations is explored in chapter 23 of Schutz 2003.

[35] Introductions to gravitational lensing and its applications can be found on the webpages Newbury 1997 and Lochner 2007.

[36] Schutz 2003, pp. 317–321; Bartusiak 2000, pp. 70–86.

[37] The ongoing search for gravitational waves is described vividly in Bartusiak 2000 and in Blair & McNamara 1997.

[38] For an overview of the history of black hole physics from its beginnings in the early 20th century to modern times, see the very readable account by Thorne 1994. For an up-to-date account of the role of black holes in structure formation, see Springel et al. 2005; a brief summary can be found in the related article Gnedin 2005.

[39] See chapter 8 of Sparke & Gallagher 2007 and Disney 1998. A treatment that is more thorough, yet involves only comparatively little mathematics can be found in Robson 1996.

[40] An elementary introduction to the black hole uniqueness theorems can be found in Chrusciel 2006 and in Thorne 1994, pp. 272–286.

[41] Detailed information can be found in Ned Wright's Cosmology Tutorial and FAQ, Wright 2007; a very readable introduction is Hogan 1999. Using undergraduate mathematics but avoiding the advanced mathematical tools of general relativity, Berry 1989 provides a more thorough presentation.

[42] Einstein's original paper is Einstein 1917; good descriptions of more modern developments can be found in Cowen 2001 and Caldwell & Crittenden 2004.

[43] Cf. Maddox 1998, pp. 52–59 and 98–122; Penrose 2004, section 34.1 and chapter 30.

[44] With a focus on string theory, the search for quantum gravity is described in Greene 1999; for an account from the point of view of loop quantum gravity, see Smolin 2001.

[45] For dark matter, see Milgrom 2002; for dark energy, Caldwell & Crittenden 2004

[46] See Friedrich 2005.

[47] A review of the various problems and the techniques being developed to overcome them, see Lehner 2002.

[48] See Bartusiak 2000 for an account up to that year; up-to-date news can be found on the websites of major detector collaborations such as GEO 600 and LIGO.

[49] A good starting point for a snapshot of present-day research in relativity is the electronic review journal Living Reviews in Relativity.

5.8 References

• Ashby, Neil (2002),"Relativity and the Global Positioning System"(PDF),*Physics Today***55**(5): 41–47,Bibcode:2 doi:10.1063/1.1485583

• Ashby, Neil (2003),"Relativity in the Global Positioning System", *Living Reviews in Relativity***6**: 1,Bibcode:2003 doi:10.12942/lrr-2003-1, retrieved 2007-07-06

• Bartusiak, Marcia (2000), *Einstein's Unfinished Symphony: Listening to the Sounds of Space-Time*, Berkley, ISBN 978-0-425-18620-6

• Berry, Michael V. (1989), *Principles of Cosmology and Gravitation* (2nd ed.), Institute of Physics Publishing, ISBN 0-85274-037-9

• Bertotti, Bruno (2005), "The Cassini Experiment: Investigating the Nature of Gravity", in Renn, Jürgen, *One hundred authors for Einstein*, Wiley-VCH, pp. 402–405, ISBN 3-527-40574-7

• Blair, David; McNamara, Geoff (1997), *Ripples on a Cosmic Sea. The Search for Gravitational Waves*, Perseus, ISBN 0-7382-0137-5

• Caldwell, Robert R.; Crittenden, R (2004), "Dark Energy", *Physics World*, 17(5) (6969): 37–42, arXiv:astro-ph/0305001, Bibcode:2004Natur.427...45B, doi:10.1038/nature02139, PMID 14702078

• Chrusciel, Piotr (2006), "How many different kinds of black hole are there?", *[http://www.einstein-online.info Einstein Online]*, retrieved 2007-07-15

• Cowen, Ron (2001), "A Dark Force in the Universe", *Science News* (Society for Science &#38) **159** (14): 218, doi:10.2307/3981642, JSTOR 3981642

• Disney, Michael (1998), "A New Look at Quasars",*Scientific American***6**(6): 52–57,doi:10.1038/scientificameric 52

• Ehlers, Jürgen; Rindler, Wolfgang (1997), "Local and Global Light Bending in Einstein's and other Gravitational Theories",*General Relativity and Gravitation***29**(4): 519–529,Bibcode:1997GReGr..29..519E,doi:10.1023/A:1

• Einstein, Albert (1917), "Kosmologische Betrachtungen zur allgemeinen Relativitätstheorie", *Sitzungsberichte der Preußischen Akademie der Wissenschaften*: 142

• Einstein, Albert (1961), *Relativity. The special and general theory*, Crown Publishers

• Friedrich, Helmut (2005), "Is general relativity 'essentially understood'?", *Annalen Phys.* **15** (1–2): 84–108, arXiv:gr-qc/0508016, Bibcode:2006AnP...518...84F, doi:10.1002/andp.200510173

- Geroch, Robert (1978), *General relativity from A to B*, University of Chicago Press, ISBN 0-226-28864-1

- Giulini, Domenico (2005), *Special relativity. A first encounter*, Oxford University Press, ISBN 0-19-856746-4

- Gnedin, Nickolay Y. (2005), "Digitizing the Universe",*Nature***435**(7042): 572–573,Bibcode:2005Natur.435..5 doi:10.1038/435572a, PMID 15931201

- Greene, Brian (1999), *The Elegant Universe: Superstrings, Hidden Dimensions, and the Quest for the Ultimate Theory*, Vintage, ISBN 0-375-70811-1

- Greene, Brian(2004),*The Fabric of the Cosmos: Space, Time, and the Texture of Reality*, A. A. Knopf,Bibcode: ISBN 0-375-41288-3

- Harrison, David M. (2002), *A Non-mathematical Proof of Gravitational Time Dilation* (PDF), retrieved 2007-05-06

- Hartl, Gerhard (2005), "The Confirmation of the General Theory of Relativity by the British Eclipse Expedition of 1919", in Renn, Jürgen, *One hundred authors for Einstein*, Wiley-VCH, pp. 182–187, ISBN 3-527-40574-7

- Hogan, Craig J. (1999), *The Little Book of the Big Bang. A Cosmic Primer*, Springer, ISBN 0-387-98385-6

- Janssen, Michel (2005), "Of pots and holes: Einstein's bumpy road to general relativity" (PDF), *Ann. Phys. (Leipzig)* **14** (S1): 58–85, Bibcode:2005AnP...517S..58J, doi:10.1002/andp.200410130

- Kennefick, Daniel (2005), "Astronomers Test General Relativity: Light-bending and the Solar Redshift", in Renn, Jürgen, *One hundred authors for Einstein*, Wiley-VCH, pp. 178–181, ISBN 3-527-40574-7

- Kennefick, Daniel (2007), "Not Only Because of Theory: Dyson, Eddington and the Competing Myths of the 1919 Eclipse Expedition", *Proceedings of the 7th Conference on the History of General Relativity, Tenerife, 2005* **0709**, p. 685, arXiv:0709.0685, Bibcode:2007arXiv0709.0685K

- Kramer, Michael (2004), "Millisecond Pulsars as Tools of Fundamental Physics", in Karshenboim, S. G.; Peik, E., *Astrophysics, Clocks and Fundamental Constants (Lecture Notes in Physics Vol. 648)*, Springer, pp. 33–54 (E-Print at astro-ph/0405178)

- Lehner, Luis (2002), "Numerical Relativity: Status and Prospects", *Proceedings of the 16th International Conference on General Relativity and Gravitation. held 15–21 July 2001 in Durban*, p. 210, arXiv:gr-qc/0202055, Bibcode:2002grg..conf..210L, doi:10.1142/9789812776556_0010, ISBN 978-981-238-171-2

- Lochner, Jim, ed. (2007), "Gravitational Lensing", *Imagine the Universe website* (NASA GSFC), retrieved 2007-06-12

- Maddox, John (1998), *What Remains To Be Discovered*, Macmillan, ISBN 0-684-82292-X

- Mermin, N. David (2005), *It's About Time. Understanding Einstein's Relativity*, Princeton University Press, ISBN 0-691-12201-6

- Milgrom, Mordehai (2002),"Does dark matter really exist?",*Scientific American***287**(2): 30–37,doi:10.1038/scient 42

- Norton, John D. (1985), "What was Einstein's principle of equivalence?" (PDF), *Studies in History and Philosophy of Science* **16** (3): 203–246, doi:10.1016/0039-3681(85)90002-0, retrieved 2007-06-11

- Newbury, Pete (1997), *Gravitational lensing webpages*, retrieved 2007-06-12

- Nieto, Michael Martin (2006), "The quest to understand the Pioneer anomaly" (PDF), *EurophysicsNews* **37** (6): 30–34, Bibcode:2006ENews..37...30N, doi:10.1051/epn:2006604

- Pais, Abraham (1982), *'Subtle is the Lord ...' The Science and life of Albert Einstein*, Oxford University Press, ISBN 0-19-853907-X

- Penrose, Roger (2004), *The Road to Reality*, A. A. Knopf, ISBN 0-679-45443-8

- Pössel, M. (2007), "The equivalence principle and the deflection of light", *[http://www.einstein-online.info Einstein Online]*, archived from the original on 2007-05-03, retrieved 2007-05-06

- Poisson, Eric (2004), "The Motion of Point Particles in Curved Spacetime", *Living Rev. Relativity* **7**, doi:10.12942/lrr-2004-6, retrieved 2007-06-13

- Renn, Jürgen, ed. (2005), *Albert Einstein – Chief Engineer of the Universe: Einstein's Life and Work in Context*, Berlin: Wiley-VCH, ISBN 3-527-40571-2

- Robson, Ian (1996), *Active galactic nuclei*, John Wiley, ISBN 0-471-95853-0

- Schutz, Bernard F. (2003), *Gravity from the ground up*, Cambridge University Press, ISBN 0-521-45506-5

- Smolin, Lee (2001), *Three Roads to Quantum Gravity*, Basic, ISBN 0-465-07835-4

- von Soldner, Johann Georg (1804), "Ueber die Ablenkung eines Lichtstrals von seiner geradlinigen Bewegung, durch die Attraktion eines Weltkörpers, an welchem er nahe vorbei geht", *Berliner Astronomisches Jahrbuch*: 161–172.

- Sparke, Linda S.; Gallagher, John S. (2007), *Galaxies in the universe – An introduction*, Cambridge University Press, ISBN 0-521-85593-4

- Springel, Volker; White, Simon D. M.; Jenkins, Adrian; Frenk, Carlos S.; Yoshida, N; Gao, L; Navarro, J; Thacker, R; Croton, D et al. (2005), "Simulations of the formation, evolution and clustering of galaxies and quasars", *Nature* **435** (7042): 629–636, arXiv:astro-ph/0504097, Bibcode:2005Natur.435..629S, doi:10.1038/nature03597, PMID 15931216

- Stachel, John (1989), "The Rigidly Rotating Disk as the 'Missing Link in the History of General Relativity'", in Howard, D.; Stachel, J., *Einstein and the History of General Relativity (Einstein Studies, Vol. 1)*, Birkhäuser, pp. 48–62, ISBN 0-8176-3392-8

- Thorne, Kip (1994), *Black Holes and Time Warps: Einstein's Outrageous Legacy*, W W Norton & Company, ISBN 0-393-31276-3

- Trimble, Virginia; Barstow, Martin (2007), "Gravitational redshift and White Dwarf stars", *[http://www.einstein-on info Einstein Online]*, retrieved 2007-06-13

- Wheeler, John A. (1990), *A Journey Into Gravity and Spacetime*, Scientific American Library, San Francisco: W. H. Freeman, ISBN 0-7167-6034-7

- Will, Clifford M. (1993), *Was Einstein Right?*, Oxford University Press, ISBN 0-19-286170-0

- Will, Clifford M. (2006), "The Confrontation between General Relativity and Experiment", *Living Rev. Relativity* **9**: 3, arXiv:gr-qc/0510072, Bibcode:2006LRR.....9....3W, doi:10.12942/lrr-2006-3, retrieved 2007-06-12

- Wright, Ned (2007), *Cosmology tutorial and FAQ*, University of California at Los Angeles, retrieved 2007-06-12

5.9 External links

Additional resources, including more advanced material, can be found in General relativity resources.

- Einstein Online. Website featuring articles on a variety of aspects of relativistic physics for a general audience, hosted by the Max Planck Institute for Gravitational Physics

- NCSA Spacetime Wrinkles. Website produced by the numerical relativity group at the National Center for Supercomputing Applications, featuring an elementary introduction to general relativity, black holes and gravitational waves

High-precision test of general relativity by the Cassini space probe (artist's impression): radio signals sent between the Earth and the probe (green wave) are delayed by the warping of spacetime (blue lines) due to the Sun's mass.

Albert Einstein, pictured here in 1921, developed the theories of special and general relativity.

Ball falling to the floor in an accelerating rocket (left) and on Earth (right).

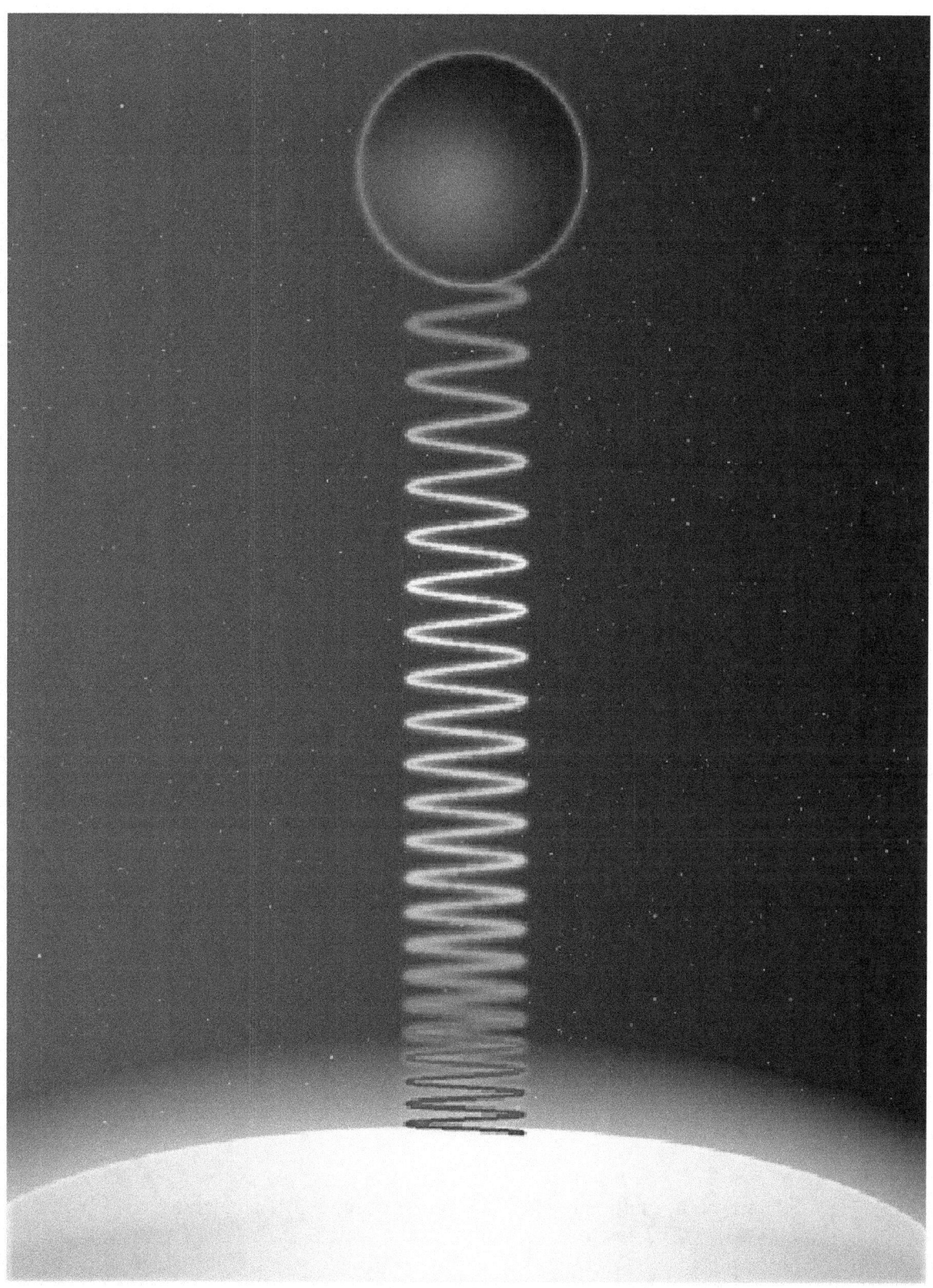

The gravitational redshift of a light wave as it moves upwards against a gravitational field (caused by the yellow star below).

Two bodies falling towards the center of the Earth accelerate
towards each other as they fall.

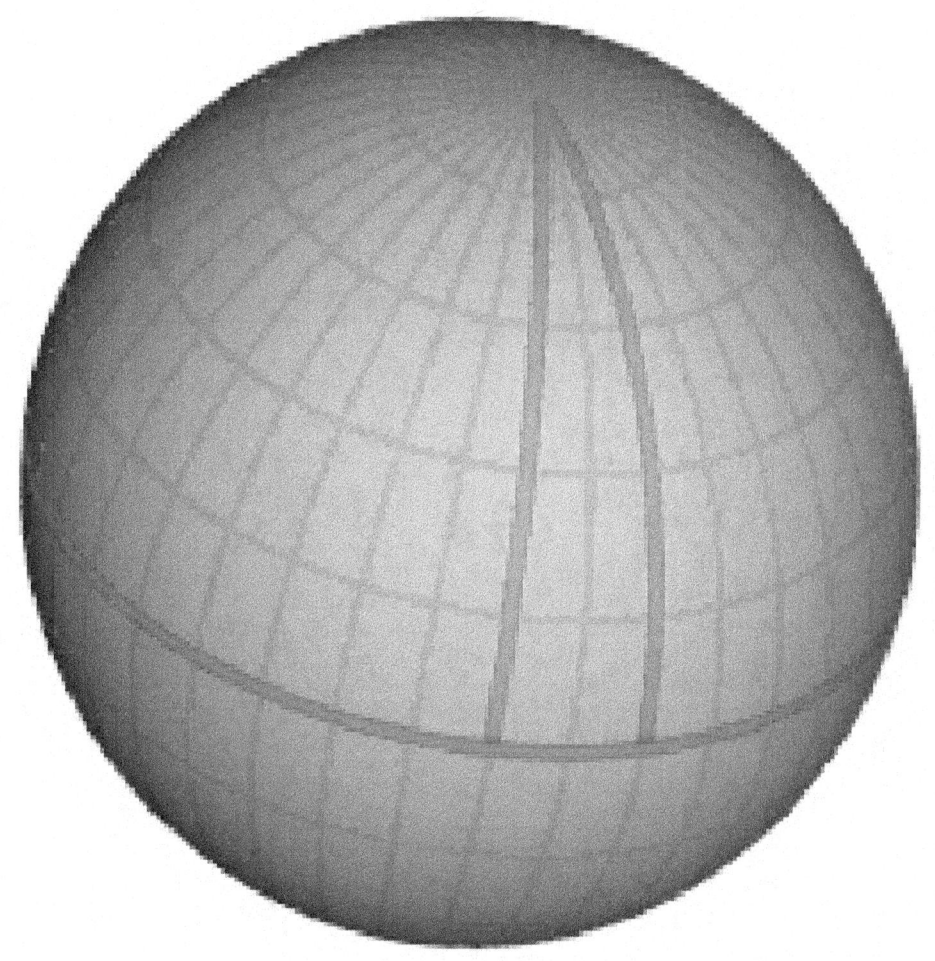

Converging geodesics: two lines of longitude (green) that start out in parallel at the equator (red) but converge to meet at the pole.

Distances, at different latitudes, corresponding to 30 degrees difference in longitude.

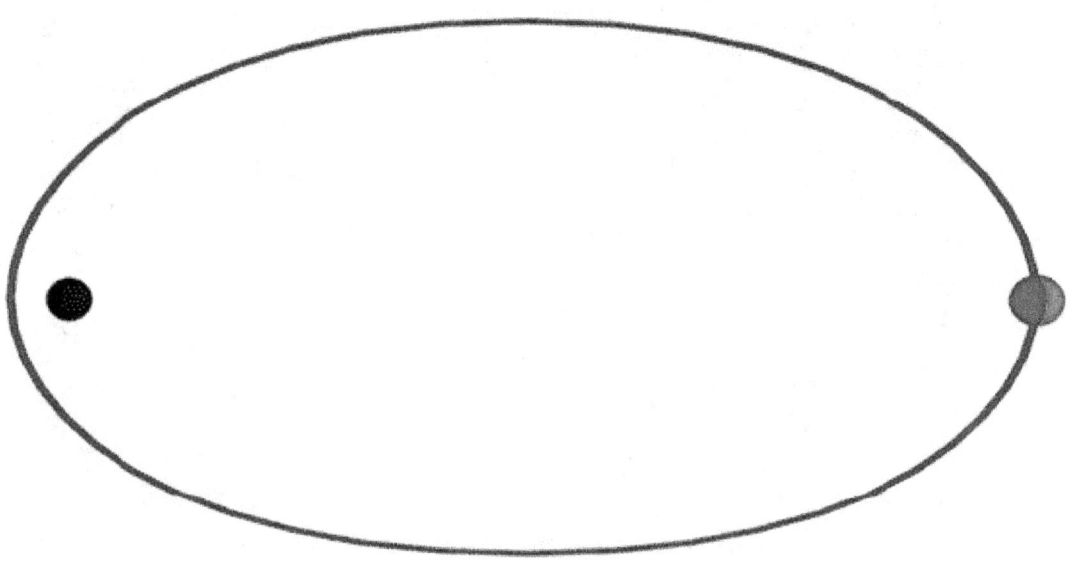

Newtonian (red) vs. Einsteinian orbit (blue) of a single planet orbiting a spherical star.

Gravity Probe B with solar panels folded.

Einstein cross: four images of the same astronomical object, produced by a gravitational lens.

Black hole-powered jet emanating from the central region of the galaxy M87.

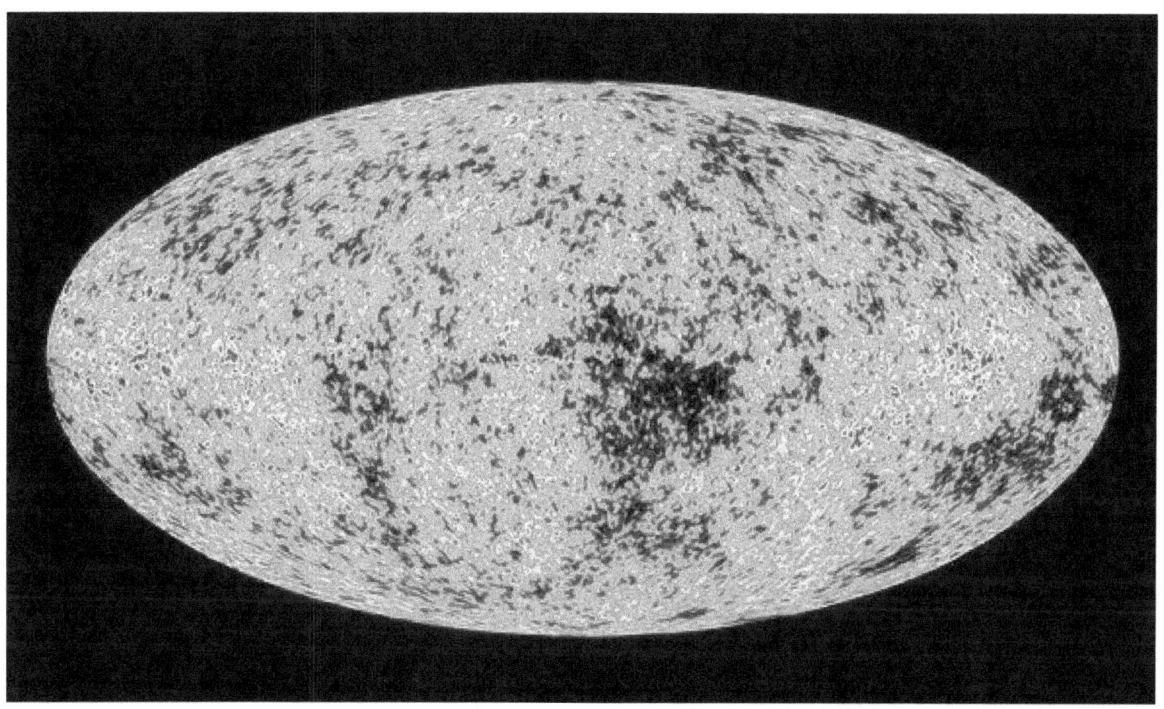

An image, created using data from the WMAP satellite telescope, of the radiation emitted no more than a few hundred thousand years after the Big Bang.

Chapter 6

History of general relativity

See also: history of special relativity

General relativity (GR) is a theory of gravitation that was developed by Albert Einstein between 1907 and 1915, with contributions by many others after 1915. According to general relativity, the observed gravitational attraction between masses results from the warping of space and time by those masses.

Before the advent of general relativity, Newton's law of universal gravitation had been accepted for more than two hundred years as a valid description of the gravitational force between masses, even though Newton himself did not regard the theory as the final word on the nature of gravity. Within a century of Newton's formulation, careful astronomical observation revealed unexplainable variations between the theory and the observations. Under Newton's model, gravity was the result of an attractive force between massive objects. Although even Newton was bothered by the unknown nature of that force, the basic framework was extremely successful at describing motion.

However, experiments and observations show that Einstein's description accounts for several effects that are unexplained by Newton's law, such as minute anomalies in the orbits of Mercury and other planets. General relativity also predicts novel effects of gravity, such as gravitational waves, gravitational lensing and an effect of gravity on time known as gravitational time dilation. Many of these predictions have been confirmed by experiment, while others are the subject of ongoing research. For example, although there is indirect evidence for gravitational waves, direct evidence of their existence is still being sought by several teams of scientists in experiments such as the LIGO and GEO 600 projects.

General relativity has developed into an essential tool in modern astrophysics. It provides the foundation for the current understanding of black holes, regions of space where gravitational attraction is so strong that not even light can escape. Their strong gravity is thought to be responsible for the intense radiation emitted by certain types of astronomical objects (such as active galactic nuclei or microquasars). General relativity is also part of the framework of the standard Big Bang model of cosmology.

6.1 Creation of general relativity

6.1.1 Early investigations

As Einstein later said, the reason for the development of general relativity was the preference of inertial motion within special relativity, while a theory which from the outset prefers no state of motion (even accelerated ones) appeared more satisfactory to him.[1] So, while still working at the patent office in 1907, Einstein had what he would call his "happiest thought". He realized that the principle of relativity could be extended to gravitational fields.

Consequently, in 1907 (published 1908) he wrote an article on acceleration under special relativity.[2] In that article, he argued that free fall is really inertial motion, and that for a freefalling observer the rules of special relativity must apply. This argument is called the Equivalence principle. In the same article, Einstein also predicted the phenomenon of

Albert Einstein developed the theories of special and general relativity. Picture from 1921.

gravitational time dilation.

In 1911, Einstein published another article expanding on the 1907 article.[3] There, he thought about the case of a uniformly accelerated box not in a gravitational field, and noted that it would be indistinguishable from a box sitting still in an unchanging gravitational field. He used special relativity to see that the rate of clocks at the top of a box accelerating upward would be faster than the rate of clocks at the bottom. He concludes that the rates of clocks depend on their position in a gravitational field, and that the difference in rate is proportional to the gravitational potential to first approximation.

Also the deflection of light by massive bodies was predicted. Although the approximation was crude, it allowed him to calculate that the deflection is nonzero. German astronomer Erwin Finlay-Freundlich publicized Einstein's challenge to scientists around the world.[4] This urged astronomers to detect the deflection of light during a solar eclipse, and gave Einstein confidence that the scalar theory of gravity proposed by Gunnar Nordström was incorrect. But the actual value for the deflection that he calculated was too small by a factor of two, because the approximation he used doesn't work well for things moving at near the speed of light. When Einstein finished the full theory of general relativity, he would rectify this error and predict the correct amount of light deflection by the sun.

Another of Einstein's notable thought experiments about the nature of the gravitational field is that of the rotating disk (a variant of the Ehrenfest paradox). He imagined an observer performing experiments on a rotating turntable. He noted that such an observer would find a different value for the mathematical constant π than the one predicted by Euclidean geometry. The reason is that the radius of a circle would be measured with an uncontracted ruler, but, according to special relativity, the circumference would seem to be longer because the ruler would be contracted. Since Einstein believed that the laws of physics were local, described by local fields, he concluded from this that spacetime could be locally curved. This led him to study Riemannian geometry, and to formulate general relativity in this language.

6.1.2   Developing general relativity

In 1912, Einstein returned to Switzerland to accept a professorship at his *alma mater*, the ETH. Once back in Zurich, he immediately visited his old ETH classmate Marcel Grossmann, now a professor of mathematics, who introduced him to Riemannian geometry and, more generally, to differential geometry. On the recommendation of Italian mathematician Tullio Levi-Civita, Einstein began exploring the usefulness of general covariance (essentially the use of tensors) for his gravitational theory. For a while Einstein thought that there were problems with the approach, but he later returned to it and, by late 1915, had published his general theory of relativity in the form in which it is used today.[5] This theory explains gravitation as distortion of the structure of spacetime by matter, affecting the inertial motion of other matter.

During World War I, the work of Central Powers scientists was available only to Central Powers academics, for national security reasons. Some of Einstein's work did reach the United Kingdom and the United States through the efforts of the Austrian Paul Ehrenfest and physicists in the Netherlands, especially 1902 Nobel Prize-winner Hendrik Lorentz and Willem de Sitter of Leiden University. After the war ended, Einstein maintained his relationship with Leiden University, accepting a contract as an *Extraordinary Professor*; for ten years, from 1920 to 1930, he travelled to Holland regularly to lecture.[6]

In 1917, several astronomers accepted Einstein's 1911 challenge from Prague. The Mount Wilson Observatory in California, U.S., published a solar spectroscopic analysis that showed no gravitational redshift.[7] In 1918, the Lick Observatory, also in California, announced that it too had disproved Einstein's prediction, although its findings were not published.[8]

However, in May 1919, a team led by the British astronomer Arthur Stanley Eddington claimed to have confirmed Einstein's prediction of gravitational deflection of starlight by the Sun while photographing a solar eclipse with dual expeditions in Sobral, northern Brazil, and Príncipe, a west African island.[4] Nobel laureate Max Born praised general relativity as the "greatest feat of human thinking about nature";[9] fellow laureate Paul Dirac was quoted saying it was "probably the greatest scientific discovery ever made".[10] The international media guaranteed Einstein's global renown.

There have been claims that scrutiny of the specific photographs taken on the Eddington expedition showed the experimental uncertainty to be comparable to the same magnitude as the effect Eddington claimed to have demonstrated, and that a 1962 British expedition concluded that the method was inherently unreliable.[11] The deflection of light during a solar eclipse was confirmed by later, more accurate observations.[12] Some resented the newcomer's fame, notably among some German physicists, who later started the *Deutsche Physik* (German Physics) movement.[13][14]

Eddington's photograph of a solar eclipse, which confirmed Einstein's theory that light "bends".

6.1.3 General covariance and the hole argument

By 1912, Einstein was actively seeking a theory in which gravitation was explained as a geometric phenomenon. At the urging of Tullio Levi-Civita, Einstein began by exploring the use of general covariance (which is essentially the use of curvature tensors) to create a gravitational theory. However, in 1913 Einstein abandoned that approach, arguing that it is inconsistent based on the "hole argument". In 1914 and much of 1915, Einstein was trying to create field equations based on another approach. When that approach was proven to be inconsistent, Einstein revisited the concept of general covariance and discovered that the hole argument was flawed.

6.1.4 The development of the Einstein field equations

Main article: Einstein field equations

When Einstein realized that general covariance was actually tenable, he quickly completed the development of the field equations that are named after him. However, he made a now-famous mistake. The field equations he published in October 1915 were

$$R_{\mu\nu} = T_{\mu\nu}$$

where $R_{\mu\nu}$ is the Ricci tensor, and $T_{\mu\nu}$ the energy–momentum tensor. This predicted the non-Newtonian perihelion precession of Mercury, and so had Einstein very excited. However, it was soon realized that they were inconsistent with the local conservation of energy–momentum unless the universe had a constant density of mass–energy–momentum. In other words, air, rock and even a vacuum should all have the same density. This inconsistency with observation sent Einstein back to the drawing board. However, the solution was all but obvious, and on November 25, 1915 Einstein presented the actual Einstein field equations to the Prussian Academy of Sciences:[15]

$$R_{\mu\nu} - \frac{1}{2} R g_{\mu\nu} = T_{\mu\nu}$$

where R is the Ricci scalar and $g_{\mu\nu}$ the metric tensor. With the publication of the field equations, the issue became one of solving them for various cases and interpreting the solutions. This and experimental verification have dominated general relativity research ever since.

6.1.5 Einstein and Hilbert

See also: Relativity priority dispute

Although Einstein is credited with finding the field equations, the German mathematician David Hilbert published them in an article before Einstein's article. This has resulted in accusations of plagiarism against Einstein, although not from Hilbert, and assertions that the field equations should be called the "Einstein–Hilbert field equations". However, Hilbert did not press his claim for priority and some have asserted that Einstein submitted the correct equations before Hilbert amended his own work to include them. This suggests that Einstein developed the correct field equations first, though Hilbert may have reached them later independently (or even learned of them afterwards through his correspondence with Einstein).[16] However, others have criticized those assertions.[17]

6.1.6 Sir Arthur Eddington

In the early years after Einstein's theory was published, Sir Arthur Eddington lent his considerable prestige in the British scientific establishment in an effort to champion the work of this German scientist. Because the theory was so complex and abstruse (even today it is popularly considered the pinnacle of scientific thinking; in the early years it was even more so), it was rumored that only three people in the world understood it. There was an illuminating, though probably apoc-ryphal, anecdote about this. As related by Ludwik Silberstein,[18] during one of Eddington's lectures he asked "Professor Eddington, you must be one of three persons in the world who understands general relativity." Eddington paused, unable to answer. Silberstein continued "Don't be modest, Eddington!" Finally, Eddington replied "On the contrary, I'm trying to think who the third person is."

6.2 Solutions

6.2.1 The Schwarzschild solution

Since the field equations are non-linear, Einstein assumed that they were unsolvable. However, in 1915 Karl Schwarzschild discovered an exact solution for the case of a spherically symmetric spacetime surrounding a massive object in spherical coordinates. This is now known as the Schwarzschild solution. Since then, many other exact solutions have been found.

6.2.2 The expanding universe and the cosmological constant

Main article: Cosmological constant

In 1922, Alexander Friedmann found a solution in which the universe may expand or contract, and later Georges Lemaître derived a solution for an expanding universe. However, Einstein believed that the universe was apparently static, and since a static cosmology was not supported by the general relativistic field equations, he added a cosmological constant Λ to the field equations, which became

$$R_{\mu\nu} - \frac{1}{2}Rg_{\mu\nu} + \Lambda g_{\mu\nu} = T_{\mu\nu}$$

This permitted the creation of steady-state solutions, but they were unstable: the slightest perturbation of a static state would result in the universe expanding or contracting. In 1929, Edwin Hubble found evidence for the idea that the universe is expanding. This resulted in Einstein dropping the cosmological constant, referring to it as "the biggest blunder in my career". At the time, it was an ad hoc hypothesis to add in the cosmological constant, as it was only intended to justify one result (a static universe).

6.2.3 More exact solutions

Progress in solving the field equations and understanding the solutions has been ongoing. The solution for a spherically symmetric charged object was discovered by Reissner and later rediscovered by Nordström, and is called the Reissner–Nordström solution. The black hole aspect of the Schwarzschild solution was very controversial, and Einstein did not believe that singularities could be real. However, in 1957 (two years after Einstein's death in 1955), Martin Kruskal published a proof that black holes are called for by the Schwarzschild Solution. Additionally, the solution for a rotating massive object was obtained by Kerr in the 1960s and is called the Kerr solution. The Kerr–Newman solution for a rotating, charged massive object was published a few years later.

6.3 Testing the theory

Main article: Tests of general relativity

The perihelion precession of Mercury was the first evidence that general relativity is correct. Sir Arthur Stanley Eddington's 1919 expedition in which he confirmed Einstein's prediction for the deflection of light by the Sun during the total solar eclipse of 29 May 1919 helped to cement the status of general relativity as a likely true theory. Since then many observations have confirmed the correctness of general relativity. These include studies of binary pulsars, observations of radio signals passing the limb of the Sun, and even the GPS system.

6.4 Alternative theories

Main article: Alternatives to general relativity

There have been various attempts to find modifications to general relativity. The most famous of these are the Brans–Dicke theory (also known as scalar-tensor theory), and Rosen's bimetric theory. Both of these theories proposed changes to the field equations of general relativity, and both suffer from these changes permitting the presence of bipolar gravitational radiation. As a result, Rosen's original theory has been refuted by observations of binary pulsars. As for Brans–Dicke (which has a tunable parameter ω such that $\omega = \infty$ is the same as general relativity), the amount by which it can differ from general relativity has been severely constrained by these observations.

In addition, general relativity is inconsistent with quantum mechanics, the physical theory that describes the wave–particle duality of matter, and quantum mechanics does not currently describe gravitational attraction at relevant (microscopic) scales. There is a great deal of speculation in the physics community as to the modifications that might be needed to both general relativity and quantum mechanics in order to unite them consistently. The speculative theory that unites general relativity and quantum mechanics is usually called quantum gravity, prominent examples of which include String Theory and Loop Quantum Gravity.

6.5 More about GR history

Kip Thorne identifies the "golden age of general relativity" as the period roughly from 1960 to 1975 during which the study of general relativity,[19] which had previously been regarded as something of a curiosity, entered the mainstream of theoretical physics. During this period, many of the concepts and terms which continue to inspire the imagination of gravitation researchers and the general public were introduced, including black holes and 'gravitational singularity'. At the same time, in a closely related development, the study of physical cosmology entered the mainstream and the Big Bang became well established.

6.6 See also

- Contributors to general relativity

- Golden age of physics

- Golden age of cosmology

6.7 Notes

[1] Albert Einstein, Nobel lecture in 1921

[2] Einstein, A., "Relativitätsprinzip und die aus demselben gezogenen Folgerungen (On the Relativity Principle and the Conclusions Drawn from It)", *Jahrbuch der Radioaktivität (Yearbook of Radioactivity)* 4: 411–462 page 454 (Wir betrachen zwei Bewegung systeme ...)

[3] Einstein, Albert (1911), "Einfluss der Schwerkraft auf die Ausbreitung des Lichtes (On the Influence of Gravity on the Propagation of Light)", *Annalen der Physik* 35: 898–908, Bibcode:1911AnP...340..898E, doi:10.1002/andp.19113401005 (also in *Collected Papers* Vol. 3, document 23)

[4] Crelinsten, Jeffrey. "Einstein's Jury: The Race to Test Relativity". *Princeton University Press*. 2006. Retrieved on 13 March 2007. ISBN 978-0-691-12310-3

[5] O'Connor, J.J. and E.F. Robertson (1996). "General relativity". *Mathematical Physics index*, School of Mathematics and Statistics, University of St. Andrews, Scotland, May, 1996. Retrieved 2015-02-04.

[6] *Two friends in Leiden*, retrieved 11 June 2007

[7] Crelinsten, Jeffrey (2006), *Einstein's Jury: The Race to Test Relativity*, Princeton University Press, pp. 103–108, ISBN 978-0-691-12310-3, retrieved 13 March 2007

[8] Crelinsten, Jeffrey (2006), *Einstein's Jury: The Race to Test Relativity*, Princeton University Press, pp. 114–119, ISBN 978-0-691-12310-3, retrieved 13 March 2007

[9] Smith, PD (17 September 2005), *The genius of space and time*, London: The Guardian, retrieved 31 March 2007

[10] Jürgen Schmidhuber. "Albert Einstein (1879–1955) and the 'Greatest Scientific Discovery Ever'". 2006. Retrieved on 4 October 2006.

[11] Andrzej, Stasiak (2003), "Myths in science", *EMBO Reports* 4 (3): 236, doi:10.1038/sj.embor.embor779, retrieved 31 March 2007

[12] See the table in MathPages Bending Light

[13] Hentschel, Klaus and Ann M. (1996), *Physics and National Socialism: An Anthology of Primary Sources*, Birkhaeuser Verlag, xxi, ISBN 3-7643-5312-0

[14] For a discussion of astronomers' attitudes and debates about relativity, see Crelinsten, Jeffrey (2006), *Einstein's Jury: The Race to Test Relativity*, Princeton University Press, ISBN 0-691-12310-1, especially chapters 6, 9, 10 and 11.

[15] Pais, Abraham (1982). "14. The Field Equations of Gravitation". *Subtle is the Lord : The Science and the Life of Albert Einstein: The Science and the Life of Albert Einstein*. Oxford University Press. p. 239. ISBN 9780191524028.

[16] Leo Corry, Jürgen Renn, John Stachel: "Belated Decision in the Hilbert-Einstein Priority Dispute", SCIENCE, Vol. 278, 14 November 1997 - article text

[17] Friedwart Winterberg's response to the Cory-Renn-Stachel paper as printed in "Zeitschrift für Naturforschung" 59a, 715-719.

[18] John Waller (2002), *Einstein's Luck*, Oxford University Press, ISBN 0-19-860719-9

[19] Thorne, Kip (2003). "Warping spacetime". *The future of theoretical physics and cosmology: celebrating Stephen Hawking's 60th birthday*. Cambridge University Press. p. 74. ISBN 0-521-82081-2., Extract of page 74

6.8 References

- Pais, Abraham (1982). *Subtle is the lord: the science and life of Albert Einstein*. Oxford: Oxford University Press. ISBN 0-19-853907-X.

- Einstein, A.; Grossmann, M. (1913). "Entwurf einer verallgemeinerten Relativitätstheorie und einer Theorie der Gravitation" [Outline of a Generalized Theory of Relativity and of a Theory of Gravitation]. *Zeitschrift für Mathematik und Physik* **62**: 225–261.

- *Einstein and the Changing Worldviews of Physics* (editors—Lehner C., Renn J., Schemmel M.) 2012 (Birkhäuser).

- Genesis of general relativity series

Chapter 7

Mathematics of general relativity

For a more accessible and less technical introduction to this topic, see Introduction to mathematics of general relativity.

The **mathematics of general relativity** refers to various mathematical structures and techniques that are used in studying and formulating Albert Einstein's theory of general relativity. The main tools used in this geometrical theory of gravitation are tensor fields defined on a Lorentzian manifold representing spacetime. This article is a general description of the mathematics of general relativity.

Note: General relativity articles using tensors will use the abstract index notation.

7.1 Why tensors?

The principle of general covariance states that the laws of physics should take the same mathematical form in all reference frames and was one of the central principles in the development of general relativity. The term 'general covariance' was used in the early formulation of general relativity, but is now referred to by many as diffeomorphism covariance. Although *diffeomorphism covariance is not the defining feature of general relativity*[1], and controversies remain regarding its present status in GR, the invariance property of physical laws implied in the principle coupled with the fact that the theory is essentially geometrical in character (making use of geometries which are not Euclidean) suggested that general relativity be formulated using the language of tensors. This will be discussed further below.

7.2 Spacetime as a manifold

Main articles: Spacetime and Spacetime topology

Most modern approaches to mathematical general relativity begin with the concept of a manifold. More precisely, the basic physical construct representing gravitation - a curved spacetime - is modelled by a four-dimensional, smooth, connected, Lorentzian manifold. Other physical descriptors are represented by various tensors, discussed below.

The rationale for choosing a manifold as the fundamental mathematical structure is to reflect desirable physical properties. For example, in the theory of manifolds, each point is contained in a (by no means unique) coordinate chart, and this chart can be thought of as representing the 'local spacetime' around the observer (represented by the point). The principle of local Lorentz covariance, which states that the laws of special relativity hold locally about each point of spacetime, lends further support to the choice of a manifold structure for representing spacetime, as locally around a point on a general manifold, the region 'looks like', or approximates very closely Minkowski space (flat spacetime).

The idea of coordinate charts as 'local observers who can perform measurements in their vicinity' also makes good physical sense, as this is how one actually collects physical data - locally. For cosmological problems, a coordinate chart may be quite large.

7.2.1   Local versus global structure

An important distinction in physics is the difference between local and global structures. Measurements in physics are performed in a relatively small region of spacetime and this is one reason for studying the local structure of spacetime in general relativity, whereas determining the global spacetime structure is important, especially in cosmological problems.

An important problem in general relativity is to tell when two spacetimes are 'the same', at least locally. This problem has its roots in manifold theory where determining if two Riemannian manifolds of the same dimension are locally isometric ('locally the same'). This latter problem has been solved and its adaptation for general relativity is called the Cartan–Karlhede algorithm.

7.3   Tensors in general relativity

Further information: Tensor

One of the profound consequences of relativity theory was the abolition of privileged reference frames. The description of physical phenomena should not depend upon who does the measuring - one reference frame should be as good as any other. Special relativity demonstrated that no inertial reference frame was preferential to any other inertial reference frame, but preferred inertial reference frames over noninertial reference frames. General relativity eliminated preference for inertial reference frames by showing that there is no preferred reference frame (inertial or not) for describing nature.

Any observer can make measurements and the precise numerical quantities obtained only depend on the coordinate system used. This suggested a way of formulating relativity using 'invariant structures', those that are independent of the coordinate system (represented by the observer) used, yet still have an independent existence. The most suitable mathematical structure seemed to be a tensor. For example, when measuring the electric and magnetic fields produced by an accelerating charge, the values of the fields will depend on the coordinate system used, but the fields are regarded as having an independent existence, this independence represented by the electromagnetic tensor .

Mathematically, tensors are generalised linear operators - multilinear maps. As such, the ideas of linear algebra are employed to study tensors.

At each point p of a manifold, the tangent and cotangent spaces to the manifold at that point may be constructed. Vectors (sometimes referred to as contravariant vectors) are defined as elements of the tangent space and covectors (sometimes termed covariant vectors, but more commonly dual vectors or one-forms) are elements of the cotangent space.

At p , these two vector spaces may be used to construct type (r,s) tensors, which are real-valued multilinear maps acting on the direct sum of r copies of the cotangent space with s copies of the tangent space. The set of all such multilinear maps forms a vector space, called the tensor product space of type (r,s) at p and denoted by $(T_p)^r{}_s M$. If the tangent space is n-dimensional, it can be shown that $\dim(T_p)^r{}_s M = n^{r+s}$.

In the general relativity literature, it is conventional to use the component syntax for tensors.

A type (r,s) tensor may be written as

$$T = T^{a_1 \ldots a_r}{}_{b_1 \ldots b_s} \frac{\partial}{\partial x^{a_1}} \otimes \ldots \otimes \frac{\partial}{\partial x^{a_r}} \otimes dx^{b_1} \otimes \ldots \otimes dx^{b_s}$$

where $\frac{\partial}{\partial x^{a_i}}$ is a basis for the i-th tangent space and dx^{b_j} a basis for the j-th cotangent space.

As spacetime is assumed to be four-dimensional, each index on a tensor can be one of four values. Hence, the total number of elements a tensor possesses equals 4^R, where R is the sum of the numbers of covariant and contravariant indices on the tensor (a number called the rank of the tensor).

7.3.1 Symmetric and antisymmetric tensors

Main articles: Antisymmetric tensor and Symmetric tensor

Some physical quantities are represented by tensors not all of whose components are independent. Important examples of such tensors include symmetric and antisymmetric tensors. Antisymmetric tensors are commonly used to represent rotations (for example, the vorticity tensor).

Although a generic rank R tensor in 4 dimensions has 4^R components, constraints on the tensor such as symmetry or antisymmetry serve to reduce the number of distinct components. For example, a symmetric rank two tensor T satisfies $T_{ab} = T_{ba}$ and possesses 10 independent components, whereas an antisymmetric (skew-symmetric) rank two tensor P satisfies $P_{ab} = -P_{ba}$ and has 6 independent components. For ranks greater than two, the symmetric or antisymmetric index pairs must be explicitly identified.

Antisymmetric tensors of rank 2 play important roles in relativity theory. The set of all such tensors - often called bivectors - forms a vector space of dimension 6, sometimes called bivector space.

7.3.2 The metric tensor

Main article: Metric tensor (general relativity)

The metric tensor is a central object in general relativity that describes the local geometry of spacetime (as a result of solving the Einstein field equations). Using the weak-field approximation, the metric can also be thought of as representing the 'gravitational potential'. The metric tensor is often just called 'the metric'.

The metric is a symmetric tensor and is an important mathematical tool. As well as being used to raise and lower tensor indices, it also generates the connections which are used to construct the geodesic equations of motion and the Riemann curvature tensor.

A convenient means of expressing the metric tensor in combination with the incremental intervals of coordinate distance that it relates to is through the line element:

$$ds^2 = g_{ab}\, dx^a\, dx^b$$

This way of expressing the metric was used by the pioneers of differential geometry. While some relativists consider the notation to be somewhat old-fashioned, many readily switch between this and the alternative notation:

$$g = g_{ab}\, dx^a \otimes dx^b$$

The metric tensor is commonly written as a 4 by 4 matrix. This matrix is symmetric and thus has 10 independent components.

7.3.3 Invariants

One of the central features of GR is the idea of invariance of physical laws. This invariance can be described in many ways, for example, in terms of local Lorentz covariance, the general principle of relativity, or diffeomorphism covariance.

A more explicit description can be given using tensors. The crucial feature of tensors used in this approach is the fact that (once a metric is given) the operation of contracting a tensor of rank R over all R indices gives a number - an *invariant* - that is independent of the coordinate chart one uses to perform the contraction. Physically, this means that if the invariant is calculated by any two observers, they will get the same number, thus suggesting that the invariant has some independent significance. Some important invariants in relativity include:

- The Ricci scalar: $R = R^{ab} g_{ab}$.

- The Kretschmann scalar: $K = R^{abcd} R_{abcd}$.

Other examples of invariants in relativity include the electromagnetic invariants, and various other curvature invariants, some of the latter finding application in the study of gravitational entropy and the Weyl curvature hypothesis.

7.3.4 Tensor classifications

The classification of tensors is a purely mathematical problem. In GR, however, certain tensors that have a physical interpretation can be classified with the different forms of the tensor usually corresponding to some physics. Examples of tensor classifications useful in general relativity include the Segre classification of the energy–momentum tensor and the Petrov classification of the Weyl tensor. There are various methods of classifying these tensors, some of which use tensor invariants.

7.4 Tensor fields in general relativity

Main article: Tensor field

Tensor fields on a manifold are maps which attach a tensor to each point of the manifold. This notion can be made more precise by introducing the idea of a fibre bundle, which in the present context means to collect together all the tensors at all points of the manifold, thus 'bundling' them all into one grand object called the tensor bundle. A tensor field is then defined as a map from the manifold to the tensor bundle, each point p being associated with a tensor at p .

The notion of a tensor field is of major importance in GR. For example, the geometry around a star is described by a metric tensor at each point, so at each point of the spacetime the value of the metric should be given to solve for the paths of material particles. Another example is the values of the electric and magnetic fields (given by the electromagnetic field tensor) and the metric at each point around a charged black hole to determine the motion of a charged particle in such a field.

Vector fields are contravariant rank one tensor fields. Important vector fields in relativity include the four-velocity, $U^a = \dot{x}^a$, which is the coordinate distance travelled per unit of proper time, the four-acceleration $A^a = \ddot{x}^a$ and the four-current J^a describing the charge and current densities. Other physically important tensor fields in relativity include the following:

- The stress–energy tensor T^{ab} , a symmetric rank-two tensor.

- The electromagnetic field tensor F^{ab} , a rank-two antisymmetric tensor.

Although the word 'tensor' refers to an object at a point, it is common practice to refer to tensor fields on a spacetime (or a region of it) as just 'tensors'.

At each point of a spacetime on which a metric is defined, the metric can be reduced to the Minkowski form using Sylvester's law of inertia.

7.5 Tensorial derivatives

Before the advent of general relativity, changes in physical processes were generally described by partial derivatives, for example, in describing changes in electromagnetic fields (see Maxwell's equations). Even in special relativity, the partial derivative is still sufficient to describe such changes. However, in general relativity, it is found that derivatives which are also tensors must be used. The derivatives have some common features including that they are derivatives along integral curves of vector fields.

The problem in defining derivatives on manifolds that are not flat is that there is no natural way to compare vectors at different points. An extra structure on a general manifold is required to define derivatives. Below are described two important derivatives that can be defined by imposing an additional structure on the manifold in each case.

7.5.1 Affine connections

Main article: Affine connection

The curvature of a spacetime can be characterised by taking a vector at some point and parallel transporting it along a curve on the spacetime. An affine connection is a rule which describes how to legitimately move a vector along a curve on the manifold without changing its direction.

By definition, an affine connection is a bilinear map $\Gamma(TM) \times \Gamma(TM) \to \Gamma(TM)$, where $\Gamma(TM)$ is a space of all vector fields on the spacetime. This bilinear map can be described in terms of a set of *connection coefficients* (also known as Christoffel symbols) specifying what happens to components of basis vectors under infinitesimal parallel transport:

$$\nabla_{e_i} e_j = \Gamma_{ji}^k e_k$$

Despite their appearance, the **connection coefficients are not the components of a tensor**.

Generally speaking, there are D^3 independent connection coefficients at each point of spacetime. The connection is called *symmetric* or *torsion-free*, if $\Gamma_{ji}^k = \Gamma_{ij}^k$. A symmetric connection has at most $\frac{1}{2}D^2(D+1)$ unique coefficients.

For any curve γ and two points $A = \gamma(0)$ and $B = \gamma(t)$ on this curve, an affine connection gives rise to a map of vectors in the tangent space at A into vectors in the tangent space at B:

$$X(t) = \Pi_{0,t,\gamma} X(0)$$

and $X(t)$ can be computed component-wise by solving the differential equation

$$\frac{d}{dt} X^i(t) = \nabla_{C(t)} X^i(t) = \Gamma_{jk}^i X^j(t) C^k(t)$$

$C^i(t)$ being the vector tangent to the curve at the point $\gamma(t)$.

An important affine connection in general relativity is the Levi-Civita connection, which is a symmetric connection obtained from parallel transporting a tangent vector along a curve whilst keeping the inner product of that vector constant along the curve. The resulting connection coefficients (Christoffel symbols) can be calculated directly from the metric. For this reason, this type of connection is often called a *metric connection*.

7.5.2 The covariant derivative

Main article: Covariant derivative

Let x be a point, $\vec{A}$ a vector located at x, and $\vec{B}$ a vector field. The idea of differentiating $\vec{B}$ at x along the direction of $\vec{A}$ in a physically meaningful way can be made sense of by choosing an affine connection and a parameterized smooth curve $\gamma(t)$ such that $x = \gamma(0)$ and $\vec{A} = \frac{d}{dt}\gamma(0)$. The formula

$$\nabla_{\vec{A}} \vec{B}(X) = \lim_{\epsilon \to 0} \frac{1}{\epsilon} \left[\Pi_{(\epsilon,0,\gamma)} \vec{B}(\gamma[\epsilon]) - \vec{B}(X) \right]$$

for a *covariant derivative of $\vec{B}$ along $\vec{A}$ associated with connection* 11 *turns out to give curve-independent results and can be used as a "physical definition" of a covariant derivative.*

It can be expressed using connection coefficients:

$$\nabla_{\vec{Y}}\vec{X} = X^a{}_{;b}Y^b\frac{\partial}{\partial x^a} = (X^a{}_{,b} + \Gamma^a_{bc}X^c)Y^b\frac{\partial}{\partial x^a}$$

The expression in brackets, called a *covariant derivative of* X *(with respect to the connection)* and denoted by $\nabla\vec{X}$, is more often used in calculations:

$$\nabla\vec{X} = X^a{}_{;b}\frac{\partial}{\partial x^a}\otimes dx^b = (X^a{}_{,b} + \Gamma^a_{bc}X^c)\frac{\partial}{\partial x^a}\otimes dx^b$$

A covariant derivative of X can thus be viewed as a differential operator acting on a vector field sending it to a type (1, 1) tensor ('increasing the covariant index by 1') and can be generalised to act on type (r, s) tensor fields sending them to type (r, s + 1) tensor fields. Notions of parallel transport can then be defined similarly as for the case of vector fields. By definition, a covariant derivative of a scalar field is equal to the regular derivative of the field.

In the literature, there are three common methods of denoting covariant differentiation:

$$D_a T^{b...c}_{d...e} = \nabla_a T^{b...c}_{d...e} = T^{b...c}_{d...e;a}$$

Many standard properties of regular partial derivatives also apply to covariant derivatives:

$$\nabla_a(X^b + Y^b) = \nabla_a X^b + \nabla_a Y^b$$
$$\nabla_a(X^b Y^c) = Y^c(\nabla_a X^b) + X^b(\nabla_a Y^c)$$
$$\nabla_a(f(x)X^b) = f\nabla_a X^b + X^b\nabla_a f = f\nabla_a X^b + X^b\frac{\partial f}{\partial x^a}$$
$$\nabla_a(cX^b) = c\nabla_a X^b, \quad c\text{constant is}$$

In general relativity, one usually refers to "the" covariant derivative, which is the one associated with Levi-Civita affine connection. By definition, Levi-Civita connection preserves the metric under parallel transport, therefore, the covariant derivative gives zero when acting on a metric tensor (as well as its inverse). It means that we can take the (inverse) metric tensor in and out of the derivative and use it to raise and lower indices:

$$\nabla_a T^b = \nabla_a(T_c g^{bc}) = g^{bc}\nabla_a T_c$$

7.5.3   The Lie derivative

Main articles: Lie derivative and Spacetime symmetries

Another important tensorial derivative is the Lie derivative. Unlike the covariant derivative, the Lie derivative is independent of the metric, although in general relativity one usually uses an expression that seemingly depends on the metric through the affine connection. Whereas the covariant derivative required an affine connection to allow comparison between vectors at different points, the Lie derivative uses a congruence from a vector field to achieve the same purpose. The idea of Lie dragging a function along a congruence leads to a definition of the Lie derivative, where the dragged function is compared with the value of the original function at a given point. The Lie derivative can be defined for type (r, s) tensor fields and in this respect can be viewed as a map that sends a type (r, s) to a type (r, s) tensor.

The Lie derivative is usually denoted by $\mathcal{L}_X$, where X is the vector field along whose congruence the Lie derivative is taken.

The Lie derivative of any tensor along a vector field can be expressed through the covariant derivatives of that tensor and vector field. The Lie derivative of a scalar is just the directional derivative:

$$\mathcal{L}_X \phi = X^a \nabla_a \phi = X^a \frac{\partial \phi}{\partial x^a}$$

Higher rank objects pick up additional terms when the Lie derivative is taken. For example, the Lie derivative of a type (0, 2) tensor is

$$\mathcal{L}_X T_{ab} = X^c \nabla_c T_{ab} + (\nabla_a X^c) T_{cb} + (\nabla_b X^c) T_{ac} = X^c T_{ab,c} + X^c_{,a} T_{cb} + X^c_{,b} T_{ac}$$

More generally,

$$\mathcal{L}_X T^{a_1 \ldots a_r}{}_{b_1 \ldots b_s} = X^c (\nabla_c T^{a_1 \ldots a_r}{}_{b_1 \ldots b_s}) -$$
$$(\nabla_c X^{a_1}) T^{c \ldots a_r}{}_{b_1 \ldots b_s} - \ldots - (\nabla_c X^{a_r}) T^{a_1 \ldots a_{r-1} c}{}_{b_1 \ldots b_s} +$$
$$(\nabla_{b_1} X^c) T^{a_1 \ldots a_r}{}_{c \ldots b_s} + \ldots + (\nabla_{b_s} X^c) T^{a_1 \ldots a_r}{}_{b_1 \ldots b_{s-1} c}$$

In fact in the above expression, one can replace the covariant derivative ∇_a with *any* torsion free connection $\hat{\nabla}_a$ or locally, with the coordinate dependent derivative ∂_a , showing that the Lie derivative is independent of the metric. The covariant derivative is convenient however because it commutes with raising and lowering indices.

One of the main uses of the Lie derivative in general relativity is in the study of spacetime symmetries where tensors or other geometrical objects are preserved. In particular, Killing symmetry (symmetry of the metric tensor under Lie dragging) occurs very often in the study of spacetimes. Using the formula above, we can write down the condition that must be satisfied for a vector field to generate a Killing symmetry:

$$\mathcal{L}_X g_{ab} = 0$$
$$\Leftrightarrow \nabla_a X_b + \nabla_b X_a = 0$$
$$\Leftrightarrow X^c g_{ab,c} + X^c_{,a} g_{bc} + X^c_{,b} g_{ac} = 0$$

7.6 The Riemann curvature tensor

Main article: Riemann tensor (general relativity)

A crucial feature of general relativity is the concept of a curved manifold. A useful way of measuring the curvature of a manifold is with an object called the Riemann (curvature) tensor.

This tensor measures curvature by use of an affine connection by considering the effect of parallel transporting a vector between two points along two curves. The discrepancy between the results of these two parallel transport routes is essentially quantified by the Riemann tensor.

This property of the Riemann tensor can be used to describe how initially parallel geodesics diverge. This is expressed by the equation of geodesic deviation and means that the tidal forces experienced in a gravitational field are a result of the curvature of spacetime.

Using the above procedure, the Riemann tensor is defined as a type (1, 3) tensor and when fully written out explicitly contains the Christoffel symbols and their first partial derivatives. The Riemann tensor has 20 independent components. The vanishing of all these components over a region indicates that the spacetime is flat in that region. From the viewpoint of geodesic deviation, this means that initially parallel geodesics in that region of spacetime will stay parallel.

The Riemann tensor has a number of properties sometimes referred to as the symmetries of the Riemann tensor. Of particular relevance to general relativity are the algebraic and differential Bianchi identities.

The connection and curvature of any Riemannian manifold are closely related, the theory of holonomy groups, which are formed by taking linear maps defined by parallel transport around curves on the manifold, providing a description of this relationship.

What the Riemann Tensor allows us to do is tell, mathematically, whether a space is flat or, if curved, how much curvature takes place in any given region. In order to derive the Riemann curvature tensor we must first recall the definition of the covariant derivative of a Tensor with one and 2 indices;

1. : $\nabla_\mu V_\nu = \partial_\mu V_\nu - \Gamma^\rho_{\mu\nu} V_\rho$

2. : $\nabla_m [V_{\mu\nu}] = \partial_m V_{\mu\nu} - \Gamma^\rho_{m\nu} V_\rho - \Gamma^\rho_{m\mu} V_\rho$

For the formation of the Riemann tensor, the covariant derivative is taken twice with the respects to a tensor of rank one. The equation is set up as follows;

$$\text{Additive Property of } \nabla_\sigma$$
$$\text{Second Rank Tensor Rule}$$

$$
\begin{aligned}
\nabla_{\sigma,\mu} V_\nu &= \nabla_\sigma [\nabla_\mu V_\nu] \\
&= \nabla_\sigma [\partial_\mu V_\nu - \Gamma^\rho_{\mu\nu} V_\rho] \\
&= \nabla_\sigma [\partial_\mu V_\nu] - \nabla_\sigma [\Gamma^\rho_{\mu\nu} V_\rho] \\
&= (\partial_\sigma [\partial_\mu V_\nu] - \Gamma^\rho_{\mu\nu} \partial_\sigma V_\rho - \Gamma^\rho_{\sigma\nu} \partial_\mu V_\rho - \Gamma^\rho_{\sigma\mu} \partial_\rho V_\nu) - (\partial_\sigma [\Gamma^\rho_{\mu\nu} V_\rho] - \Gamma^\alpha_{\sigma\nu} \Gamma^\rho_{\alpha\mu} V_\rho - \Gamma^\alpha_{\sigma\mu} \Gamma_{\rho\alpha d} V\rho \\
&= \partial_\sigma [\partial_\mu V_\nu] - \Gamma^\rho_{\mu\nu} \partial_\sigma V_\rho - \Gamma^\rho_{\sigma\nu} \partial_\mu V_\rho - \Gamma^\rho_{\sigma\mu} \partial_\rho V_\nu - \partial_\sigma [\Gamma^\rho_{\mu\nu} V_\rho] + \Gamma^\alpha_{\sigma\nu} \Gamma^\rho_{\alpha\mu} V_\rho + \Gamma^\alpha_{\sigma\mu} \Gamma^\rho_{\alpha\nu} V_\rho
\end{aligned}
$$

Similarly we have:

$$\nabla_{\mu,\sigma} V_\nu = \partial_\mu [\partial_\sigma V_\nu] - \Gamma^\rho_{\sigma\nu} \partial_\mu V_\rho - \Gamma^\rho_{\mu\nu} \partial_\mu V_\rho - \Gamma^\rho_{\mu\sigma} \partial_\rho V_\nu - \partial_\mu [\Gamma^\rho_{\sigma\nu} V_\rho] + \Gamma^\alpha_{\mu\nu} \Gamma^\rho_{\alpha\sigma} V_\rho + \Gamma^\alpha_{\mu\sigma} \Gamma^\rho_{\alpha\nu} V_\rho$$

Now subtracting the two equations and using the symmetry of Christoffel symbols we arrive at:

$$\nabla_{\sigma,\mu} V_\nu - \nabla_{\mu,\sigma} V_\nu = (\partial_\mu \Gamma^\rho_{\sigma\nu} - \partial_\sigma \Gamma^\rho_{\mu\nu} + \Gamma^\alpha_{\sigma\nu} \Gamma^\rho_{\alpha\mu} - \Gamma^\alpha_{\mu\nu} \Gamma^\rho_{\alpha\sigma}) V_\rho$$

You'll notice that the left side of the equation has 3 indices and the right side has 4, so we will have to sum over a pair of indices.

$$R^\rho_{\sigma\mu\nu} V_\rho = (\partial_\mu \Gamma^\rho_{\sigma\nu} - \partial_\sigma \Gamma^\rho_{\mu\nu} + \Gamma^\alpha_{\sigma\nu} \Gamma^\rho_{\alpha\mu} - \Gamma^\alpha_{\mu\nu} \Gamma^\rho_{\alpha\sigma}) V_\rho$$

Finally the Riemann curvature tensor is written as;

$$R^\rho_{\sigma\mu\nu} = \partial_\mu \Gamma^\rho_{\sigma\nu} - \partial_\sigma \Gamma^\rho_{\mu\nu} + \Gamma^\alpha_{\sigma\nu} \Gamma^\rho_{\alpha\mu} - \Gamma^\alpha_{\mu\nu} \Gamma^\rho_{\alpha\sigma}$$

You can contract indices to make the tensor covariant simply by multiplying by the metric, which will be useful when working with Einstein's field equations.

$$g_{\rho\lambda} R^\lambda_{\sigma\mu\nu} = R_{\rho\sigma\mu\nu}$$

and by further decomposition,

$$g^{\rho\mu} R_{\rho\sigma\mu\nu} = R_{\sigma\nu}$$

This tensor is called the Ricci tensor which can also be derived by setting ρ and μ in the Riemann tensor to the same indice and summing over them. Then the curvature scalar can be found by going one step further.

$$g^{\sigma\nu} R_{\sigma\nu} = R$$

So now we have 3 different objects.

1. the Riemann curvature tensor: $R^\rho_{\sigma\mu\nu}$ or $R_{\rho\sigma\mu\nu}$
2. the Ricci tensor: $R_{\sigma\nu}$
3. the scalar curvature: R

all of which are useful in calculating solutions to Einstein's field equations.

7.7 The energy–momentum tensor

Main article: Energy-momentum tensor (general relativity)

The sources of any gravitational field (matter and energy) are represented in relativity by a type (0, 2) symmetric tensor called the energy–momentum tensor. It is closely related to the Ricci tensor. Being a second rank tensor in four dimensions, the energy–momentum tensor may be viewed as a 4 by 4 matrix. The various admissible matrix types, called Jordan forms cannot all occur, as the energy conditions that the energy–momentum tensor is forced to satisfy rule out certain forms.

7.7.1 Energy conservation

In GR, there is a *local* law for the conservation of energy–momentum. It can be succinctly expressed by the tensor equation:

$$T^{ab}{}_{;b} = 0$$

The corresponding statement of local energy conservation in special relativity is:

$$T^{ab}{}_{,b} = 0$$

This illustrates the rule of thumb that 'partial derivatives go to covariant derivatives'.

7.8 The Einstein field equations

Main article: Einstein field equations
See also: Solutions of the Einstein field equations

The Einstein field equations (EFE) are the core of general relativity theory. The EFE describe how mass and energy (as represented in the stress–energy tensor) are related to the curvature of space-time (as represented in the Einstein tensor). In abstract index notation, the EFE reads as follows:

$$G_{ab} + \Lambda g_{ab} = \frac{8\pi G}{c^4} T_{ab}$$

where G_{ab} is the Einstein tensor, Λ is the cosmological constant, c is the speed of light in a vacuum and G is the gravitational constant, which comes from Newton's law of universal gravitation.

The solutions of the EFE are metric tensors. The EFE, being non-linear differential equations for the metric, are often difficult to solve. There are a number of strategies used to solve them. For example, one strategy is to start with an ansatz (or an educated guess) of the final metric, and refine it until it is specific enough to support a coordinate system but still general enough to yield a set of simultaneous differential equations with unknowns that can be solved for. Metric tensors resulting from cases where the resultant differential equations can be solved exactly for a physically reasonable distribution of energy–momentum are called exact solutions. Examples of important exact solutions include the Schwarzschild solution and the Friedman-Lemaître-Robertson–Walker solution.

The EIH approximation plus other references (e.g. Geroch and Jang, 1975 - 'Motion of a body in general relativity', JMP, Vol. 16 Issue 1).

7.9   The geodesic equations

Main article: Geodesic (general relativity)

Once the EFE are solved to obtain a metric, it remains to determine the motion of inertial objects in the spacetime. In general relativity, it is assumed that inertial motion occurs along timelike and null geodesics of spacetime as parameterized by proper time. Geodesics are curves that parallel transport their own tangent vector U; i.e., $\nabla_U U = 0$. This condition, the geodesic equation, can be written in terms of a coordinate system x^a with the tangent vector $U^a = \frac{dx^a}{d\tau}$:

$$\ddot{x}^a + \Gamma^a_{bc} \dot{x}^b \dot{x}^c = 0$$

where denotes the derivative by proper time, $d/d\tau$, with τ parametrising proper time along the curve and making manifest the presence of the Christoffel symbols.

A principal feature of general relativity is to determine the paths of particles and radiation in gravitational fields. This is accomplished by solving the geodesic equations.

The EFE relate the total matter (energy) distribution to the curvature of spacetime. Their nonlinearity leads to a problem in determining the precise motion of matter in the resultant spacetime. For example, in a system composed of one planet orbiting a star, the motion of the planet is determined by solving the field equations with the energy–momentum tensor the sum of that for the planet and the star. The gravitational field of the planet affects the total spacetime geometry and hence the motion of objects. It is therefore reasonable to suppose that the field equations can be used to derive the geodesic equations.

When the energy–momentum tensor for a system is that of dust, it may be shown by using the local conservation law for the energy–momentum tensor that the geodesic equations are satisfied exactly.

7.10   Lagrangian formulation

Main article: Variational methods in general relativity

The issue of deriving the equations of motion or the field equations in any physical theory is considered by many researchers to be appealing. A fairly universal way of performing these derivations is by using the techniques of variational calculus, the main objects used in this being Lagrangians.

Many consider this approach to be an elegant way of constructing a theory, others as merely a formal way of expressing a theory (usually, the Lagrangian construction is performed *after* the theory has been developed).

7.11 Mathematical techniques for analysing spacetimes

Having outlined the basic mathematical structures used in formulating the theory, some important mathematical techniques that are employed in investigating spacetimes will now be discussed.

7.11.1 Frame fields

Main article: Frame fields in general relativity

A frame field is an orthonormal set of 4 vector fields (1 timelike, 3 spacelike) defined on a spacetime. Each frame field can be thought of as representing an observer in the spacetime moving along the integral curves of the timelike vector field. Every tensor quantity can be expressed in terms of a frame field, in particular, the metric tensor takes on a particularly convenient form. When allied with coframe fields, frame fields provide a powerful tool for analysing spacetimes and physically interpreting the mathematical results.

7.11.2 Symmetry vector fields

Main article: Spacetime symmetries

Some modern techniques in analysing spacetimes rely heavily on using spacetime symmetries, which are infinitesimally generated by vector fields (usually defined locally) on a spacetime that preserve some feature of the spacetime. The most common type of such *symmetry vector fields* include Killing vector fields (which preserve the metric structure) and their generalisations called *generalised Killing vector fields*. Symmetry vector fields find extensive application in the study of exact solutions in general relativity and the set of all such vector fields usually forms a finite-dimensional Lie algebra.

7.11.3 The Cauchy problem

Main article: Initial value formulation (general relativity)

The Cauchy problem (sometimes called the initial value problem) is the attempt at finding a solution to a differential equation given initial conditions. In the context of general relativity, it means the problem of finding solutions to Einstein's field equations - a system of hyperbolic partial differential equations - given some initial data on a hypersurface. Studying the Cauchy problem allows one to formulate the concept of causality in general relativity, as well as 'parametrising' solutions of the field equations. Ideally, one desires *global solutions*, but usually *local solutions* are the best that can be hoped for. Typically, solving this initial value problem requires selection of particular coordinate conditions.

7.11.4 Spinor formalism

Spinors find several important applications in relativity. Their use as a method of analysing spacetimes using tetrads, in particular, in the Newman–Penrose formalism is important.

Another appealing feature of spinors in general relativity is the condensed way in which some tensor equations may be written using the spinor formalism. For example, in classifying the Weyl tensor, determining the various Petrov types becomes much easier when compared with the tensorial counterpart.

7.11.5 Regge calculus

Main article: Regge calculus

Regge calculus is a formalism which chops up a Lorentzian manifold into discrete 'chunks' (four-dimensional simplicial blocks) and the block edge lengths are taken as the basic variables. A discrete version of the Einstein–Hilbert action is obtained by considering so-called *deficit angles* of these blocks, a zero deficit angle corresponding to no curvature. This novel idea finds application in approximation methods in numerical relativity and quantum gravity, the latter using a generalisation of Regge calculus.

7.11.6 Singularity theorems

Main article: Penrose–Hawking singularity theorems

In general relativity, it was noted that, under fairly generic conditions, gravitational collapse will inevitably result in a so-called singularity. A singularity is a point where the solutions to the equations become infinite, indicating that the theory has been probed at inappropriate ranges.

7.11.7 Numerical relativity

Main article: Numerical relativity

Numerical relativity is the sub-field of general relativity which seeks to solve Einstein's equations through the use of numerical methods. Finite difference, finite element and pseudo-spectral methods are used to approximate the solution to the partial differential equations which arise. Novel techniques developed by numerical relativity include the excision method and the puncture method for dealing with the singularities arising in black hole spacetimes. Common research topics include black holes and neutron stars.

7.11.8 Perturbation methods

Main article: Perturbation methods in general relativity

The nonlinearity of the Einstein field equations often leads one to consider approximation methods in solving them. For example, an important approach is to linearise the field equations. Techniques from perturbation theory find ample application in such areas.

7.12 See also

- Ricci calculus

7.13 Notes

[1] The defining feature (central physical idea) of general relativity is that matter and energy cause the surrounding space-time geometry to be curved.

7.14 References

- Einstein, A. (1961). *Relativity: The Special and General Theory*. New York: Crown. ISBN 0-517-02961-8.

- Misner, Charles; Thorne, Kip S. & Wheeler, John Archibald (1973). *Gravitation*. San Francisco: W. H. Freeman. ISBN 0-7167-0344-0.

- Landau, L. D. and Lifshitz, E. M. (1975). *Classical Theory of Fields (Fourth Revised English Edition)*. Oxford: Pergamon. ISBN 0-08-018176-7.

Chapter 8

General relativity

For the book by Robert Wald, see General Relativity (book).

For a more accessible and less technical introduction to this topic, see Introduction to general relativity.

General relativity, also known as the **general theory of relativity**, is the geometric theory of gravitation published

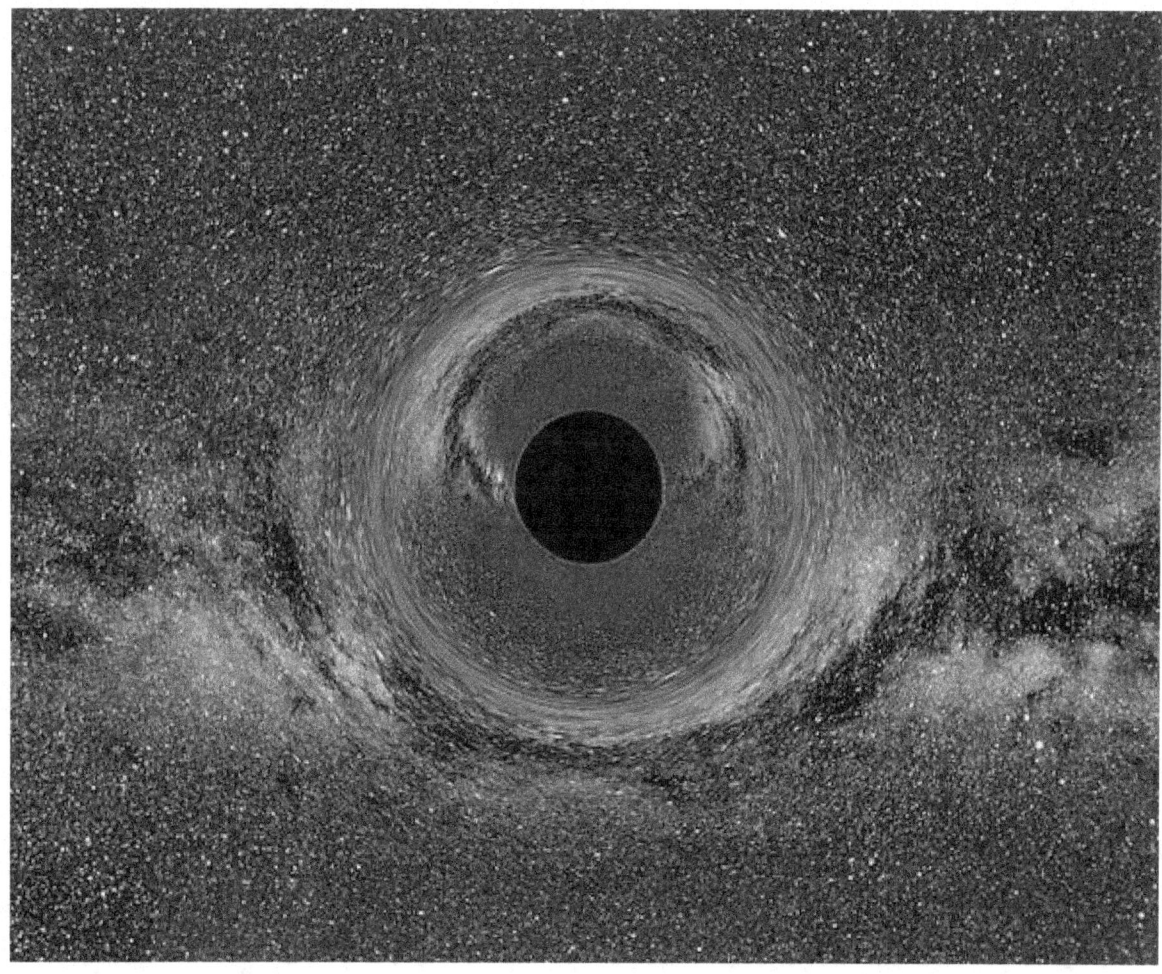

A simulated black hole of 10 solar masses within the Milky Way, seen from a distance of 600 kilometers.

by Albert Einstein in 1915[1] and the current description of gravitation in modern physics. General relativity generalizes special relativity and Newton's law of universal gravitation, providing a unified description of gravity as a geometric property of space and time, or spacetime. In particular, the curvature of spacetime is directly related to the energy and momentum of whatever matter and radiation are present. The relation is specified by the Einstein field equations, a system of partial differential equations.

Some predictions of general relativity differ significantly from those of classical physics, especially concerning the passage of time, the geometry of space, the motion of bodies in free fall, and the propagation of light. Examples of such differences include gravitational time dilation, gravitational lensing, the gravitational redshift of light, and the gravitational time delay. The predictions of general relativity have been confirmed in all observations and experiments to date. Although general relativity is not the only relativistic theory of gravity, it is the simplest theory that is consistent with experimental data. However, unanswered questions remain, the most fundamental being how general relativity can be reconciled with the laws of quantum physics to produce a complete and self-consistent theory of quantum gravity.

Einstein's theory has important astrophysical implications. For example, it implies the existence of black holes—regions of space in which space and time are distorted in such a way that nothing, not even light, can escape—as an end-state for massive stars. There is ample evidence that the intense radiation emitted by certain kinds of astronomical objects is due to black holes; for example, microquasars and active galactic nuclei result from the presence of stellar black holes and black holes of a much more massive type, respectively. The bending of light by gravity can lead to the phenomenon of gravitational lensing, in which multiple images of the same distant astronomical object are visible in the sky. General relativity also predicts the existence of gravitational waves, which have since been observed indirectly; a direct measurement is the aim of projects such as LIGO and NASA/ESA Laser Interferometer Space Antenna and various pulsar timing arrays. In addition, general relativity is the basis of current cosmological models of a consistently expanding universe.

8.1 History

Main articles: History of general relativity and Classical theories of gravitation

Soon after publishing the special theory of relativity in 1905, Einstein started thinking about how to incorporate gravity into his new relativistic framework. In 1907, beginning with a simple thought experiment involving an observer in free fall, he embarked on what would be an eight-year search for a relativistic theory of gravity. After numerous detours and false starts, his work culminated in the presentation to the Prussian Academy of Science in November 1915 of what are now known as the Einstein field equations. These equations specify how the geometry of space and time is influenced by whatever matter and radiation are present, and form the core of Einstein's general theory of relativity.[2]

The Einstein field equations are nonlinear and very difficult to solve. Einstein used approximation methods in working out initial predictions of the theory. But as early as 1916, the astrophysicist Karl Schwarzschild found the first non-trivial exact solution to the Einstein field equations, the so-called Schwarzschild metric. This solution laid the groundwork for the description of the final stages of gravitational collapse, and the objects known today as black holes. In the same year, the first steps towards generalizing Schwarzschild's solution to electrically charged objects were taken, which eventually resulted in the Reissner–Nordström solution, now associated with electrically charged black holes.[3] In 1917, Einstein applied his theory to the universe as a whole, initiating the field of relativistic cosmology. In line with contemporary thinking, he assumed a static universe, adding a new parameter to his original field equations—the cosmological constant— to match that observational presumption.[4] By 1929, however, the work of Hubble and others had shown that our universe is expanding. This is readily described by the expanding cosmological solutions found by Friedmann in 1922, which do not require a cosmological constant. Lemaître used these solutions to formulate the earliest version of the Big Bang models, in which our universe has evolved from an extremely hot and dense earlier state.[5] Einstein later declared the cosmological constant the biggest blunder of his life.[6]

During that period, general relativity remained something of a curiosity among physical theories. It was clearly superior to Newtonian gravity, being consistent with special relativity and accounting for several effects unexplained by the Newtonian theory. Einstein himself had shown in 1915 how his theory explained the anomalous perihelion advance of the planet Mercury without any arbitrary parameters ("fudge factors").[7] Similarly, a 1919 expedition led by Eddington confirmed general relativity's prediction for the deflection of starlight by the Sun during the total solar eclipse of May 29, 1919,[8] making Einstein instantly famous.[9] Yet the theory entered the mainstream of theoretical physics and astrophysics only with the developments between approximately 1960 and 1975, now known as the golden age of general relativity.[10] Physicists began to understand the concept of a black hole, and to identify quasars as one of these objects' astrophysical manifestations.[11] Ever more precise solar system tests confirmed the theory's predictive power,[12] and relativistic

Albert Einstein developed the theories of special and general relativity. Picture from 1921.

cosmology, too, became amenable to direct observational tests.[13]

8.2 From classical mechanics to general relativity

General relativity can be understood by examining its similarities with and departures from classical physics. The first step is the realization that classical mechanics and Newton's law of gravity admit a geometric description. The combination of this description with the laws of special relativity results in a heuristic derivation of general relativity.[14]

8.2.1 Geometry of Newtonian gravity

According to general relativity, objects in a gravitational field behave similarly to objects within an accelerating enclosure. For example, an observer will see a ball fall the same way in a rocket (left) as it does on Earth (right), provided that the acceleration of the rocket is equal to 9.8 m/s² (the acceleration due to gravity at the surface of the Earth).

At the base of classical mechanics is the notion that a body's motion can be described as a combination of free (or inertial) motion, and deviations from this free motion. Such deviations are caused by external forces acting on a body in accordance with Newton's second law of motion, which states that the net force acting on a body is equal to that body's (inertial) mass multiplied by its acceleration.[15] The preferred inertial motions are related to the geometry of space and time: in the standard reference frames of classical mechanics, objects in free motion move along straight lines at constant speed. In modern parlance, their paths are geodesics, straight world lines in curved spacetime.[16]

Conversely, one might expect that inertial motions, once identified by observing the actual motions of bodies and making allowances for the external forces (such as electromagnetism or friction), can be used to define the geometry of space, as well as a time coordinate. However, there is an ambiguity once gravity comes into play. According to Newton's law of gravity, and independently verified by experiments such as that of Eötvös and its successors (see Eötvös experiment), there is a universality of free fall (also known as the weak equivalence principle, or the universal equality of inertial and passive-gravitational mass): the trajectory of a test body in free fall depends only on its position and initial speed, but not on any of its material properties.[17] A simplified version of this is embodied in Einstein's elevator experiment, illustrated in the figure on the right: for an observer in a small enclosed room, it is impossible to decide, by mapping the trajectory of bodies such as a dropped ball, whether the room is at rest in a gravitational field, or in free space aboard a rocket that is accelerating at a rate equal to that of the gravitational field.[18]

Given the universality of free fall, there is no observable distinction between inertial motion and motion under the influence of the gravitational force. This suggests the definition of a new class of inertial motion, namely that of objects in free fall under the influence of gravity. This new class of preferred motions, too, defines a geometry of space and time—in mathematical terms, it is the geodesic motion associated with a specific connection which depends on the gradient of the gravitational potential. Space, in this construction, still has the ordinary Euclidean geometry. However, space*time* as a whole is more complicated. As can be shown using simple thought experiments following the free-fall trajectories of different test particles, the result of transporting spacetime vectors that can denote a particle's velocity (time-like vectors) will vary with the particle's trajectory; mathematically speaking, the Newtonian connection is not integrable. From this, one can deduce that spacetime is curved. The result is a geometric formulation of Newtonian gravity using only covariant concepts, i.e. a description which is valid in any desired coordinate system.[19] In this geometric description, tidal effects—the relative acceleration of bodies in free fall—are related to the derivative of the connection, showing how the modified geometry is caused by the presence of mass.[20]

8.2.2 Relativistic generalization

As intriguing as geometric Newtonian gravity may be, its basis, classical mechanics, is merely a limiting case of (special) relativistic mechanics.[21] In the language of symmetry: where gravity can be neglected, physics is Lorentz invariant as in special relativity rather than Galilei invariant as in classical mechanics. (The defining symmetry of special relativity is the Poincaré group, which includes translations and rotations.) The differences between the two become significant when dealing with speeds approaching the speed of light, and with high-energy phenomena.[22]

With Lorentz symmetry, additional structures come into play. They are defined by the set of light cones (see image). The light-cones define a causal structure: for each event A, there is a set of events that can, in principle, either influence or be influenced by A via signals or interactions that do not need to travel faster than light (such as event B in the image), and a set of events for which such an influence is impossible (such as event C in the image). These sets are observer-independent.[23] In conjunction with the world-lines of freely falling particles, the light-cones can be used to reconstruct the space–time's semi-Riemannian metric, at least up to a positive scalar factor. In mathematical terms, this defines a conformal structure.[24]

Special relativity is defined in the absence of gravity, so for practical applications, it is a suitable model whenever gravity can be neglected. Bringing gravity into play, and assuming the universality of free fall, an analogous reasoning as in the previous section applies: there are no global inertial frames. Instead there are approximate inertial frames moving alongside freely falling particles. Translated into the language of spacetime: the straight time-like lines that define a gravity-free inertial frame are deformed to lines that are curved relative to each other, suggesting that the inclusion of gravity necessitates a change in spacetime geometry.[25]

A priori, it is not clear whether the new local frames in free fall coincide with the reference frames in which the laws of special relativity hold—that theory is based on the propagation of light, and thus on electromagnetism, which could have a different set of preferred frames. But using different assumptions about the special-relativistic frames (such as their being earth-fixed, or in free fall), one can derive different predictions for the gravitational redshift, that is, the way in which the frequency of light shifts as the light propagates through a gravitational field (cf. below). The actual measurements show that free-falling frames are the ones in which light propagates as it does in special relativity.[26] The generalization of this statement, namely that the laws of special relativity hold to good approximation in freely falling (and non-rotating) reference frames, is known as the Einstein equivalence principle, a crucial guiding principle for generalizing special-relativistic physics to include gravity.[27]

The same experimental data shows that time as measured by clocks in a gravitational field—proper time, to give the technical term—does not follow the rules of special relativity. In the language of spacetime geometry, it is not measured by the Minkowski metric. As in the Newtonian case, this is suggestive of a more general geometry. At small scales, all reference frames that are in free fall are equivalent, and approximately Minkowskian. Consequently, we are now dealing with a curved generalization of Minkowski space. The metric tensor that defines the geometry—in particular, how lengths and angles are measured—is not the Minkowski metric of special relativity, it is a generalization known as a semi- or pseudo-Riemannian metric. Furthermore, each Riemannian metric is naturally associated with one particular kind of connection, the Levi-Civita connection, and this is, in fact, the connection that satisfies the equivalence principle and makes space locally Minkowskian (that is, in suitable locally inertial coordinates, the metric is Minkowskian, and its first partial derivatives and the connection coefficients vanish).[28]

8.2.3 Einstein's equations

Main articles: Einstein field equations and Mathematics of general relativity

Having formulated the relativistic, geometric version of the effects of gravity, the question of gravity's source remains. In Newtonian gravity, the source is mass. In special relativity, mass turns out to be part of a more general quantity called the energy–momentum tensor, which includes both energy and momentum densities as well as stress (that is, pressure and shear).[29] Using the equivalence principle, this tensor is readily generalized to curved space-time. Drawing further upon the analogy with geometric Newtonian gravity, it is natural to assume that the field equation for gravity relates this tensor and the Ricci tensor, which describes a particular class of tidal effects: the change in volume for a small cloud of test particles that are initially at rest, and then fall freely. In special relativity, conservation of energy–momentum corresponds to the statement that the energy–momentum tensor is divergence-free. This formula, too, is readily generalized to curved spacetime by replacing partial derivatives with their curved-manifold counterparts, covariant derivatives studied in differential geometry. With this additional condition—the covariant divergence of the energy–momentum tensor, and hence of whatever is on the other side of the equation, is zero— the simplest set of equations are what are called Einstein's (field) equations:

On the left-hand side is the Einstein tensor, a specific divergence-free combination of the Ricci tensor $R_{\mu\nu}$ and the metric. Where $G_{\mu\nu}$ is symmetric. In particular,

$$R = g^{\mu\nu} R_{\mu\nu}$$

is the curvature scalar. The Ricci tensor itself is related to the more general Riemann curvature tensor as

$$R_{\mu\nu} = R^{\alpha}{}_{\mu\alpha\nu}.$$

On the right-hand side, $T_{\mu\nu}$ is the energy–momentum tensor. All tensors are written in abstract index notation.[30] Matching the theory's prediction to observational results for planetary orbits (or, equivalently, assuring that the weak-gravity, low-speed limit is Newtonian mechanics), the proportionality constant can be fixed as $\kappa = 8\pi G/c^4$, with G the gravitational constant and c the speed of light.[31] When there is no matter present, so that the energy–momentum tensor vanishes, the results are the vacuum Einstein equations.

$$R_{\mu\nu} = 0.$$

There are alternatives to general relativity built upon the same premises, which include additional rules and/or constraints, leading to different field equations. Examples are Brans–Dicke theory, teleparallelism, and Einstein–Cartan theory.[32]

8.3 Definition and basic applications

See also: Mathematics of general relativity and Physical theories modified by general relativity

The derivation outlined in the previous section contains all the information needed to define general relativity, describe its key properties, and address a question of crucial importance in physics, namely how the theory can be used for model-building.

8.3.1   Definition and basic properties

General relativity is a metric theory of gravitation. At its core are Einstein's equations, which describe the relation between the geometry of a four-dimensional, pseudo-Riemannian manifold representing spacetime, and the energy–momentum contained in that spacetime.[33] Phenomena that in classical mechanics are ascribed to the action of the force of gravity (such as free-fall, orbital motion, and spacecraft trajectories), correspond to inertial motion within a curved geometry of spacetime in general relativity; there is no gravitational force deflecting objects from their natural, straight paths. Instead, gravity corresponds to changes in the properties of space and time, which in turn changes the straightest-possible paths that objects will naturally follow.[34] The curvature is, in turn, caused by the energy–momentum of matter. Paraphrasing the relativist John Archibald Wheeler, spacetime tells matter how to move; matter tells spacetime how to curve.[35]

While general relativity replaces the scalar gravitational potential of classical physics by a symmetric rank-two tensor, the latter reduces to the former in certain limiting cases. For weak gravitational fields and slow speed relative to the speed of light, the theory's predictions converge on those of Newton's law of universal gravitation.[36]

As it is constructed using tensors, general relativity exhibits general covariance: its laws—and further laws formulated within the general relativistic framework—take on the same form in all coordinate systems.[37] Furthermore, the theory does not contain any invariant geometric background structures, i.e. it is background independent. It thus satisfies a more stringent general principle of relativity, namely that the laws of physics are the same for all observers.[38] Locally, as expressed in the equivalence principle, spacetime is Minkowskian, and the laws of physics exhibit local Lorentz invariance.[39]

8.3.2   Model-building

The core concept of general-relativistic model-building is that of a solution of Einstein's equations. Given both Einstein's equations and suitable equations for the properties of matter, such a solution consists of a specific semi-Riemannian manifold (usually defined by giving the metric in specific coordinates), and specific matter fields defined on that manifold. Matter and geometry must satisfy Einstein's equations, so in particular, the matter's energy–momentum tensor must be divergence-free. The matter must, of course, also satisfy whatever additional equations were imposed on its properties. In short, such a solution is a model universe that satisfies the laws of general relativity, and possibly additional laws governing whatever matter might be present.[40]

Einstein's equations are nonlinear partial differential equations and, as such, difficult to solve exactly.[41] Nevertheless, a number of exact solutions are known, although only a few have direct physical applications.[42] The best-known exact solutions, and also those most interesting from a physics point of view, are the Schwarzschild solution, the Reissner–Nordström solution and the Kerr metric, each corresponding to a certain type of black hole in an otherwise empty universe,[43] and the Friedmann–Lemaître–Robertson–Walker and de Sitter universes, each describing an expanding cosmos.[44] Exact solutions of great theoretical interest include the Gödel universe (which opens up the intriguing possibility of time travel in curved spacetimes), the Taub-NUT solution (a model universe that is homogeneous, but anisotropic), and anti-de Sitter space (which has recently come to prominence in the context of what is called the Maldacena conjecture).[45]

Given the difficulty of finding exact solutions, Einstein's field equations are also solved frequently by numerical integration on a computer, or by considering small perturbations of exact solutions. In the field of numerical relativity, powerful computers are employed to simulate the geometry of spacetime and to solve Einstein's equations for interesting situations such as two colliding black holes.[46] In principle, such methods may be applied to any system, given sufficient computer resources, and may address fundamental questions such as naked singularities. Approximate solutions may also be found by perturbation theories such as linearized gravity[47] and its generalization, the post-Newtonian expansion, both of which were developed by Einstein. The latter provides a systematic approach to solving for the geometry of a spacetime that contains a distribution of matter that moves slowly compared with the speed of light. The expansion involves a series of terms; the first terms represent Newtonian gravity, whereas the later terms represent ever smaller corrections to Newton's theory due to general relativity.[48] An extension of this expansion is the parametrized post-Newtonian (PPN) formalism, which allows quantitative comparisons between the predictions of general relativity and alternative theories.[49]

8.4 Consequences of Einstein's theory

General relativity has a number of physical consequences. Some follow directly from the theory's axioms, whereas others have become clear only in the course of many years of research that followed Einstein's initial publication.

8.4.1 Gravitational time dilation and frequency shift

Main article: Gravitational time dilation

Assuming that the equivalence principle holds,[50] gravity influences the passage of time. Light sent down into a gravity well is blueshifted, whereas light sent in the opposite direction (i.e., climbing out of the gravity well) is redshifted; collectively, these two effects are known as the gravitational frequency shift. More generally, processes close to a massive body run more slowly when compared with processes taking place farther away; this effect is known as gravitational time dilation.[51]

Gravitational redshift has been measured in the laboratory[52] and using astronomical observations.[53] Gravitational time dilation in the Earth's gravitational field has been measured numerous times using atomic clocks,[54] while ongoing validation is provided as a side effect of the operation of the Global Positioning System (GPS).[55] Tests in stronger gravitational fields are provided by the observation of binary pulsars.[56] All results are in agreement with general relativity.[57] However, at the current level of accuracy, these observations cannot distinguish between general relativity and other theories in which the equivalence principle is valid.[58]

8.4.2 Light deflection and gravitational time delay

Main articles: Kepler problem in general relativity, Gravitational lens and Shapiro delay

General relativity predicts that the path of light is bent in a gravitational field; light passing a massive body is deflected towards that body. This effect has been confirmed by observing the light of stars or distant quasars being deflected as it passes the Sun.[59]

This and related predictions follow from the fact that light follows what is called a light-like or null geodesic—a generalization of the straight lines along which light travels in classical physics. Such geodesics are the generalization of the invariance of lightspeed in special relativity.[60] As one examines suitable model spacetimes (either the exterior Schwarzschild solution or, for more than a single mass, the post-Newtonian expansion),[61] several effects of gravity on light propagation emerge. Although the bending of light can also be derived by extending the universality of free fall to light,[62] the angle of deflection resulting from such calculations is only half the value given by general relativity.[63]

Closely related to light deflection is the gravitational time delay (or Shapiro delay), the phenomenon that light signals take longer to move through a gravitational field than they would in the absence of that field. There have been numerous successful tests of this prediction.[64] In the parameterized post-Newtonian formalism (PPN), measurements of both the deflection of light and the gravitational time delay determine a parameter called γ, which encodes the influence of gravity on the geometry of space.[65]

8.4.3 Gravitational waves

Main article: Gravitational wave

One of several analogies between weak-field gravity and electromagnetism is that, analogous to electromagnetic waves, there are gravitational waves: ripples in the metric of spacetime that propagate at the speed of light.[66] The simplest type of such a wave can be visualized by its action on a ring of freely floating particles. A sine wave propagating through such a ring towards the reader distorts the ring in a characteristic, rhythmic fashion (animated image to the right).[67] Since Einstein's equations are non-linear, arbitrarily strong gravitational waves do not obey linear superposition, making their description difficult. However, for weak fields, a linear approximation can be made. Such linearized gravitational waves are sufficiently accurate to describe the exceedingly weak waves that are expected to arrive here on Earth from far-off cosmic events, which typically result in relative distances increasing and decreasing by 10^{-21} or less. Data analysis methods routinely make use of the fact that these linearized waves can be Fourier decomposed.[68]

Some exact solutions describe gravitational waves without any approximation, e.g., a wave train traveling through empty space[69] or so-called Gowdy universes, varieties of an expanding cosmos filled with gravitational waves.[70] But for gravitational waves produced in astrophysically relevant situations, such as the merger of two black holes, numerical methods are presently the only way to construct appropriate models.[71]

8.4.4   Orbital effects and the relativity of direction

Main article: Kepler problem in general relativity

General relativity differs from classical mechanics in a number of predictions concerning orbiting bodies. It predicts an overall rotation (precession) of planetary orbits, as well as orbital decay caused by the emission of gravitational waves and effects related to the relativity of direction.

Precession of apsides

In general relativity, the apsides of any orbit (the point of the orbiting body's closest approach to the system's center of mass) will precess—the orbit is not an ellipse, but akin to an ellipse that rotates on its focus, resulting in a rose curve-like shape (see image). Einstein first derived this result by using an approximate metric representing the Newtonian limit and treating the orbiting body as a test particle. For him, the fact that his theory gave a straightforward explanation of the anomalous perihelion shift of the planet Mercury, discovered earlier by Urbain Le Verrier in 1859, was important evidence that he had at last identified the correct form of the gravitational field equations.[72]

The effect can also be derived by using either the exact Schwarzschild metric (describing spacetime around a spherical mass)[73] or the much more general post-Newtonian formalism.[74] It is due to the influence of gravity on the geometry of space and to the contribution of self-energy to a body's gravity (encoded in the nonlinearity of Einstein's equations).[75] Relativistic precession has been observed for all planets that allow for accurate precession measurements (Mercury, Venus, and Earth),[76] as well as in binary pulsar systems, where it is larger by five orders of magnitude.[77]

Orbital decay

According to general relativity, a binary system will emit gravitational waves, thereby losing energy. Due to this loss, the distance between the two orbiting bodies decreases, and so does their orbital period. Within the Solar System or for ordinary double stars, the effect is too small to be observable. This is not the case for a close binary pulsar, a system of two orbiting neutron stars, one of which is a pulsar: from the pulsar, observers on Earth receive a regular series of radio pulses that can serve as a highly accurate clock, which allows precise measurements of the orbital period. Because neutron stars are very compact, significant amounts of energy are emitted in the form of gravitational radiation.[79]

The first observation of a decrease in orbital period due to the emission of gravitational waves was made by Hulse and Taylor, using the binary pulsar PSR1913+16 they had discovered in 1974. This was the first detection of gravitational waves, albeit indirect, for which they were awarded the 1993 Nobel Prize in physics.[80] Since then, several other binary pulsars have been found, in particular the double pulsar PSR J0737-3039, in which both stars are pulsars.[81]

Geodetic precession and frame-dragging

Main articles: Geodetic precession and Frame dragging

Several relativistic effects are directly related to the relativity of direction.[82] One is geodetic precession: the axis direction of a gyroscope in free fall in curved spacetime will change when compared, for instance, with the direction of light received from distant stars—even though such a gyroscope represents the way of keeping a direction as stable as possible ("parallel transport").[83] For the Moon–Earth system, this effect has been measured with the help of lunar laser ranging.[84] More recently, it has been measured for test masses aboard the satellite Gravity Probe B to a precision of better than 0.3%.[85][86]

Near a rotating mass, there are so-called gravitomagnetic or frame-dragging effects. A distant observer will determine that objects close to the mass get "dragged around". This is most extreme for rotating black holes where, for any object entering a zone known as the ergosphere, rotation is inevitable.[87] Such effects can again be tested through their influence on the orientation of gyroscopes in free fall.[88] Somewhat controversial tests have been performed using the LAGEOS satellites, confirming the relativistic prediction.[89] Also the Mars Global Surveyor probe around Mars has been used.[90][91]

8.5 Astrophysical applications

8.5.1 Gravitational lensing

Main article: Gravitational lensing
The deflection of light by gravity is responsible for a new class of astronomical phenomena. If a massive object is situated between the astronomer and a distant target object with appropriate mass and relative distances, the astronomer will see multiple distorted images of the target. Such effects are known as gravitational lensing.[92] Depending on the configuration, scale, and mass distribution, there can be two or more images, a bright ring known as an Einstein ring, or partial rings called arcs.[93] The earliest example was discovered in 1979;[94] since then, more than a hundred gravitational lenses have been observed.[95] Even if the multiple images are too close to each other to be resolved, the effect can still be measured, e.g., as an overall brightening of the target object; a number of such "microlensing events" have been observed.[96]

Gravitational lensing has developed into a tool of observational astronomy. It is used to detect the presence and distribution of dark matter, provide a "natural telescope" for observing distant galaxies, and to obtain an independent estimate of the Hubble constant. Statistical evaluations of lensing data provide valuable insight into the structural evolution of galaxies.[97]

8.5.2 Gravitational wave astronomy

Main articles: Gravitational wave and Gravitational wave astronomy
Observations of binary pulsars provide strong indirect evidence for the existence of gravitational waves (see Orbital decay, above). However, gravitational waves reaching us from the depths of the cosmos have not been detected directly. Such detection is a major goal of current relativity-related research.[98] Several land-based gravitational wave detectors are currently in operation, most notably the interferometric detectors GEO 600, LIGO (two detectors), TAMA 300 and VIRGO.[99] Various pulsar timing arrays are using millisecond pulsars to detect gravitational waves in the 10^{-9} to 10^{-6} Hertz frequency range, which originate from binary supermassive blackholes.[100] European space-based detector, eLISA / NGO, is currently under development,[101] with a precursor mission (LISA Pathfinder) due for launch in 2015.[102]

Observations of gravitational waves promise to complement observations in the electromagnetic spectrum.[103] They are expected to yield information about black holes and other dense objects such as neutron stars and white dwarfs, about certain kinds of supernova implosions, and about processes in the very early universe, including the signature of certain types of hypothetical cosmic string.[104]

8.5.3 Black holes and other compact objects

Main article: Black hole

Whenever the ratio of an object's mass to its radius becomes sufficiently large, general relativity predicts the formation of a black hole, a region of space from which nothing, not even light, can escape. In the currently accepted models of stellar evolution, neutron stars of around 1.4 solar masses, and stellar black holes with a few to a few dozen solar masses, are thought to be the final state for the evolution of massive stars.[105] Usually a galaxy has one supermassive black hole with a few million to a few billion solar masses in its center,[106] and its presence is thought to have played an important role in the formation of the galaxy and larger cosmic structures.[107]

Astronomically, the most important property of compact objects is that they provide a supremely efficient mechanism for converting gravitational energy into electromagnetic radiation.[108] Accretion, the falling of dust or gaseous matter onto stellar or supermassive black holes, is thought to be responsible for some spectacularly luminous astronomical objects, notably diverse kinds of active galactic nuclei on galactic scales and stellar-size objects such as microquasars.[109] In particular, accretion can lead to relativistic jets, focused beams of highly energetic particles that are being flung into space at almost light speed.[110] General relativity plays a central role in modelling all these phenomena,[111] and observations provide strong evidence for the existence of black holes with the properties predicted by the theory.[112]

Black holes are also sought-after targets in the search for gravitational waves (cf. Gravitational waves, above). Merging black hole binaries should lead to some of the strongest gravitational wave signals reaching detectors here on Earth, and the phase directly before the merger ("chirp") could be used as a "standard candle" to deduce the distance to the merger events–and hence serve as a probe of cosmic expansion at large distances.[113] The gravitational waves produced as a stellar black hole plunges into a supermassive one should provide direct information about the supermassive black hole's geometry.[114]

8.5.4   Cosmology

Main article: Physical cosmology

The current models of cosmology are based on Einstein's field equations, which include the cosmological constant Λ since it has important influence on the large-scale dynamics of the cosmos,

$$ R_{\mu\nu} - \frac{1}{2} R \, g_{\mu\nu} + \Lambda \, g_{\mu\nu} = \frac{8\pi G}{c^4} \, T_{\mu\nu} $$

where $g_{\mu\nu}$ is the spacetime metric.[115] Isotropic and homogeneous solutions of these enhanced equations, the Friedmann–Lemaître–Robertson–Walker solutions,[116] allow physicists to model a universe that has evolved over the past 14 billion years from a hot, early Big Bang phase.[117] Once a small number of parameters (for example the universe's mean matter density) have been fixed by astronomical observation,[118] further observational data can be used to put the models to the test.[119] Predictions, all successful, include the initial abundance of chemical elements formed in a period of primordial nucleosynthesis,[120] the large-scale structure of the universe,[121] and the existence and properties of a "thermal echo" from the early cosmos, the cosmic background radiation.[122]

Astronomical observations of the cosmological expansion rate allow the total amount of matter in the universe to be estimated, although the nature of that matter remains mysterious in part. About 90% of all matter appears to be so-called dark matter, which has mass (or, equivalently, gravitational influence), but does not interact electromagnetically and, hence, cannot be observed directly.[123] There is no generally accepted description of this new kind of matter, within the framework of known particle physics[124] or otherwise.[125] Observational evidence from redshift surveys of distant supernovae and measurements of the cosmic background radiation also show that the evolution of our universe is significantly influenced by a cosmological constant resulting in an acceleration of cosmic expansion or, equivalently, by a form of energy with an unusual equation of state, known as dark energy, the nature of which remains unclear.[126]

A so-called inflationary phase,[127] an additional phase of strongly accelerated expansion at cosmic times of around 10^{-33} seconds, was hypothesized in 1980 to account for several puzzling observations that were unexplained by classical cosmological models, such as the nearly perfect homogeneity of the cosmic background radiation.[128] Recent measurements of the cosmic background radiation have resulted in the first evidence for this scenario.[129] However, there is a bewildering variety of possible inflationary scenarios, which cannot be restricted by current observations.[130] An even larger question is the physics of the earliest universe, prior to the inflationary phase and close to where the classical models predict the big bang singularity. An authoritative answer would require a complete theory of quantum gravity, which has not yet been developed[131] (cf. the section on quantum gravity, below).

8.5.5 Time travel

Kurt Gödel showed that solutions to Einstein's equations exist that contain closed timelike curves (CTCs), which allow for loops in time. The solutions require extreme physical conditions unlikely ever to occur in practice, and it remains an open question whether further laws of physics will eliminate them completely. Since then other—similarly impractical—GR solutions containing CTCs have been found, such as the Tipler cylinder and traversable wormholes.

8.6 Advanced concepts

8.6.1 Causal structure and global geometry

Main article: Causal structure

In general relativity, no material body can catch up with or overtake a light pulse. No influence from an event A can reach any other location X before light sent out at A to X. In consequence, an exploration of all light worldlines (null geodesics) yields key information about the spacetime's causal structure. This structure can be displayed using Penrose–Carter diagrams in which infinitely large regions of space and infinite time intervals are shrunk ("compactified") so as to fit onto a finite map, while light still travels along diagonals as in standard spacetime diagrams.[132]

Aware of the importance of causal structure, Roger Penrose and others developed what is known as global geometry. In global geometry, the object of study is not one particular solution (or family of solutions) to Einstein's equations. Rather, relations that hold true for all geodesics, such as the Raychaudhuri equation, and additional non-specific assumptions about the nature of matter (usually in the form of so-called energy conditions) are used to derive general results.[133]

8.6.2 Horizons

Main articles: Horizon (general relativity), No hair theorem and Black hole mechanics

Using global geometry, some spacetimes can be shown to contain boundaries called horizons, which demarcate one region from the rest of spacetime. The best-known examples are black holes: if mass is compressed into a sufficiently compact region of space (as specified in the hoop conjecture, the relevant length scale is the Schwarzschild radius[134]), no light from inside can escape to the outside. Since no object can overtake a light pulse, all interior matter is imprisoned as well. Passage from the exterior to the interior is still possible, showing that the boundary, the black hole's *horizon*, is not a physical barrier.[135]

Early studies of black holes relied on solutions of Einstein's equations, notably the spherically symmetric Schwarzschild solution(used to describe a static black hole) and the axisymmetric Kerr solution(used to describe a rotating,stationary black hole,and introducing interesting features such as the ergosphere). Using global geometry, later studies have revealed more general properties of black holes.In the long run,they are rather simple objects characterized by eleven parameters specifying energy, linear momentum, angular momentum, location at a specified time and electric charge. This is stated by the black hole uniqueness theorems: "black holes have no hair",that is,no distinguishing marks like the hairstyles of humans. Irrespective of the complexity of a gravitating object collapsing to form a black hole, the object that results (having emitted gravitational waves) is very simple.[136]

Even more remarkably, there is a general set of laws known as black hole mechanics, which is analogous to the laws of thermodynamics. For instance, by the second law of black hole mechanics, the area of the event horizon of a general black hole will never decrease with time, analogous to the entropy of a thermodynamic system. This limits the energy that can be extracted by classical means from a rotating black hole (e.g. by the Penrose process).[137] There is strong evidence that the laws of black hole mechanics are, in fact, a subset of the laws of thermodynamics, and that the black hole area is proportional to its entropy.[138] This leads to a modification of the original laws of black hole mechanics: for instance, as the second law of black hole mechanics becomes part of the second law of thermodynamics, it is possible for black hole area to decrease—as long as other processes ensure that, overall, entropy increases. As thermodynamical objects with non-zero temperature, black holes should emit thermal radiation. Semi-classical calculations indicate that indeed they do, with the surface gravity playing the role of temperature in Planck's law. This radiation is known as Hawking radiation (cf. the quantum theory section, below).[139]

There are other types of horizons. In an expanding universe, an observer may find that some regions of the past cannot be observed ("particle horizon"), and some regions of the future cannot be influenced (event horizon).[140] Even in flat Minkowski space, when described by an accelerated observer (Rindler space), there will be horizons associated with a semi-classical radiation known as Unruh radiation.[141]

8.6.3   Singularities

Main article: Spacetime singularity

Another general feature of general relativity is the appearance of spacetime boundaries known as singularities. Spacetime can be explored by following up on timelike and lightlike geodesics—all possible ways that light and particles in free fall can travel. But some solutions of Einstein's equations have "ragged edges"—regions known as spacetime singularities, where the paths of light and falling particles come to an abrupt end, and geometry becomes ill-defined. In the more interesting cases, these are "curvature singularities", where geometrical quantities characterizing spacetime curvature, such as the Ricci scalar, take on infinite values.[142] Well-known examples of spacetimes with future singularities—where worldlines end—are the Schwarzschild solution, which describes a singularity inside an eternal static black hole,[143] or the Kerr solution with its ring-shaped singularity inside an eternal rotating black hole.[144] The Friedmann–Lemaître–Robertson–Walker solutions and other spacetimes describing universes have past singularities on which worldlines begin, namely Big Bang singularities, and some have future singularities (Big Crunch) as well.[145]

Given that these examples are all highly symmetric—and thus simplified—it is tempting to conclude that the occurrence of singularities is an artifact of idealization.[146] The famous singularity theorems, proved using the methods of global geometry, say otherwise: singularities are a generic feature of general relativity, and unavoidable once the collapse of an object with realistic matter properties has proceeded beyond a certain stage[147] and also at the beginning of a wide class of expanding universes.[148] However, the theorems say little about the properties of singularities, and much of current research is devoted to characterizing these entities' generic structure (hypothesized e.g. by the so-called BKL conjecture).[149] The cosmic censorship hypothesis states that all realistic future singularities (no perfect symmetries, matter with realistic properties) are safely hidden away behind a horizon, and thus invisible to all distant observers. While no formal proof yet exists, numerical simulations offer supporting evidence of its validity.[150]

8.6.4   Evolution equations

Main article: Initial value formulation (general relativity)

Each solution of Einstein's equation encompasses the whole history of a universe — it is not just some snapshot of how things are, but a whole, possibly matter-filled, spacetime. It describes the state of matter and geometry everywhere and at every moment in that particular universe. Due to its general covariance, Einstein's theory is not sufficient by itself to determine the time evolution of the metric tensor. It must be combined with a coordinate condition, which is analogous to gauge fixing in other field theories.[151]

To understand Einstein's equations as partial differential equations, it is helpful to formulate them in a way that describes the evolution of the universe over time. This is done in so-called "3+1" formulations, where spacetime is split into three space dimensions and one time dimension. The best-known example is the ADM formalism.[152] These decompositions show that the spacetime evolution equations of general relativity are well-behaved: solutions always exist, and are uniquely defined, once suitable initial conditions have been specified.[153] Such formulations of Einstein's field equations are the basis of numerical relativity.[154]

8.6.5   Global and quasi-local quantities

Main article: Mass in general relativity

The notion of evolution equations is intimately tied in with another aspect of general relativistic physics. In Einstein's theory, it turns out to be impossible to find a general definition for a seemingly simple property such as a system's total mass (or energy). The main reason is that the gravitational field—like any physical field—must be ascribed a certain energy, but that it proves to be fundamentally impossible to localize that energy.[155]

Nevertheless, there are possibilities to define a system's total mass, either using a hypothetical "infinitely distant observer" (ADM mass)[156] or suitable symmetries (Komar mass).[157] If one excludes from the system's total mass the energy being carried away to infinity by gravitational waves, the result is the so-called Bondi mass at null infinity.[158] Just as in classical physics, it can be shown that these masses are positive.[159] Corresponding global definitions exist for momentum and angular momentum.[160] There have also been a number of attempts to define *quasi-local* quantities, such as the mass of an isolated system formulated using only quantities defined within a finite region of space containing that system. The hope is to obtain a quantity useful for general statements about isolated systems, such as a more precise formulation of the hoop conjecture.[161]

8.7 Relationship with quantum theory

If general relativity were considered to be one of the two pillars of modern physics, then quantum theory, the basis of understanding matter from elementary particles to solid state physics, would be the other.[162] However, how to reconcile quantum theory with general relativity is still an open question.

8.7.1 Quantum field theory in curved spacetime

Main article: Quantum field theory in curved spacetime

Ordinary quantum field theories, which form the basis of modern elementary particle physics, are defined in flat Minkowski space, which is an excellent approximation when it comes to describing the behavior of microscopic particles in weak gravitational fields like those found on Earth.[163] In order to describe situations in which gravity is strong enough to influence (quantum) matter, yet not strong enough to require quantization itself, physicists have formulated quantum field theories in curved spacetime. These theories rely on general relativity to describe a curved background spacetime, and define a generalized quantum field theory to describe the behavior of quantum matter within that spacetime.[164] Using this formalism, it can be shown that black holes emit a blackbody spectrum of particles known as Hawking radiation, leading to the possibility that they evaporate over time.[165] As briefly mentioned above, this radiation plays an important role for the thermodynamics of black holes.[166]

8.7.2 Quantum gravity

Main article: Quantum gravity
See also: String theory, Canonical general relativity, Loop quantum gravity, Causal Dynamical Triangulations and Causal sets

The demand for consistency between a quantum description of matter and a geometric description of spacetime,[167] as well as the appearance of singularities (where curvature length scales become microscopic), indicate the need for a full theory of quantum gravity: for an adequate description of the interior of black holes, and of the very early universe, a theory is required in which gravity and the associated geometry of spacetime are described in the language of quantum physics.[168] Despite major efforts, no complete and consistent theory of quantum gravity is currently known, even though a number of promising candidates exist.[169]

Attempts to generalize ordinary quantum field theories, used in elementary particle physics to describe fundamental interactions, so as to include gravity have led to serious problems. At low energies, this approach proves successful, in that it results in an acceptable effective (quantum) field theory of gravity.[170] At very high energies, however, the result are models devoid of all predictive power ("non-renormalizability").[171]

One attempt to overcome these limitations is string theory, a quantum theory not of point particles, but of minute one-dimensional extended objects.[172] The theory promises to be a unified description of all particles and interactions, including gravity;[173] the price to pay is unusual features such as six extra dimensions of space in addition to the usual three.[174] In what is called the second superstring revolution, it was conjectured that both string theory and a unification of general relativity and supersymmetry known as supergravity[175] form part of a hypothesized eleven-dimensional model known as M-theory, which would constitute a uniquely defined and consistent theory of quantum gravity.[176]

Another approach starts with the canonical quantization procedures of quantum theory. Using the initial-value-formulation of general relativity (cf. evolution equations above), the result is the Wheeler–deWitt equation (an analogue of the Schrödinger equation) which, regrettably, turns out to be ill-defined.[177] However, with the introduction of what are now known as Ashtekar variables,[178] this leads to a promising model known as loop quantum gravity. Space is represented by a web-like structure called a spin network, evolving over time in discrete steps.[179]

Depending on which features of general relativity and quantum theory are accepted unchanged, and on what level changes are introduced,[180] there are numerous other attempts to arrive at a viable theory of quantum gravity, some examples being dynamical triangulations,[181] causal sets,[182] twistor models[183] or the path-integral based models of quantum cosmology.[184]

All candidate theories still have major formal and conceptual problems to overcome. They also face the common problem that, as yet, there is no way to put quantum gravity predictions to experimental tests (and thus to decide between the candidates where their predictions vary), although there is hope for this to change as future data from cosmological observations and particle physics experiments becomes available.[185]

8.8   Current status

General relativity has emerged as a highly successful model of gravitation and cosmology, which has so far passed many unambiguous observational and experimental tests. However, there are strong indications the theory is incomplete.[186] The problem of quantum gravity and the question of the reality of spacetime singularities remain open.[187] Observational data that is taken as evidence for dark energy and dark matter could indicate the need for new physics.[188] Even taken as is, general relativity is rich with possibilities for further exploration. Mathematical relativists seek to understand the nature of singularities and the fundamental properties of Einstein's equations.[189] and increasingly powerful computer simulations (such as those describing merging black holes) are run.[190] The race for the first direct detection of gravitational waves continues,[191] in the hope of creating opportunities to test the theory's validity for much stronger gravitational fields than has been possible to date.[192] Almost a hundred years after its publication, general relativity remains a highly active area of research.[193]

8.9   See also

- Center of mass (relativistic)

- Contributors to general relativity

- Derivations of the Lorentz transformations

- Ehrenfest paradox

- Einstein–Hilbert action

- Introduction to mathematics of general relativity

- Relativity priority dispute

- Ricci calculus

- Tests of general relativity

- Timeline of gravitational physics and relativity

- Two-body problem in general relativity

8.10 Notes

[1] O'Connor, J.J. and E.F. Robertson (1996), "General relativity". *Mathematical Physics index*, School of Mathematics and Statistics, University of St. Andrews, Scotland, May, 1996. Retrieved 2015-02-04.

[2] Pais 1982, ch. 9 to 15, Janssen 2005; an up-to-date collection of current research, including reprints of many of the original articles, is Renn 2007; an accessible overview can be found in Renn 2005, pp. 110ff. An early key article is Einstein 1907, cf. Pais 1982, ch. 9. The publication featuring the field equations is Einstein 1915, cf. Pais 1982, ch. 11–15

[3] Schwarzschild 1916a, Schwarzschild 1916b and Reissner 1916 (later complemented in Nordström 1918)

[4] Einstein 1917, cf. Pais 1982, ch. 15e

[5] Hubble's original article is Hubble 1929; an accessible overview is given in Singh 2004, ch. 2–4

[6] As reported in Gamow 1970. Einstein's condemnation would prove to be premature, cf. the section Cosmology, below

[7] Pais 1982, pp. 253–254

[8] Kennefick 2005, Kennefick 2007

[9] Pais 1982, ch. 16

[10] Thorne, Kip (2003). "Warping spacetime". *The future of theoretical physics and cosmology: celebrating Stephen Hawking's 60th birthday*. Cambridge University Press. p. 74. ISBN 0-521-82081-2., Extract of page 74

[11] Israel 1987, ch. 7.8–7.10, Thorne 1994, ch. 3–9

[12] Sections Orbital effects and the relativity of direction, Gravitational time dilation and frequency shift and Light deflection and gravitational time delay, and references therein

[13] Section Cosmology and references therein; the historical development is in Overbye 1999

[14] The following exposition re-traces that of Ehlers 1973, sec. 1

[15] Arnold 1989, ch. 1

[16] Ehlers 1973, pp. 5f

[17] Will 1993, sec. 2.4, Will 2006, sec. 2

[18] Wheeler 1990, ch. 2

[19] Ehlers 1973, sec. 1.2, Havas 1964, Künzle 1972. The simple thought experiment in question was first described in Heckmann & Schücking 1959

[20] Ehlers 1973, pp. 10f

[21] Good introductions are, in order of increasing presupposed knowledge of mathematics, Giulini 2005, Mermin 2005, and Rindler 1991; for accounts of precision experiments, cf. part IV of Ehlers & Lämmerzahl 2006

[22] An in-depth comparison between the two symmetry groups can be found in Giulini 2006a

[23] Rindler 1991, sec. 22, Synge 1972, ch. 1 and 2

[24] Ehlers 1973, sec. 2.3

[25] Ehlers 1973, sec. 1.4, Schutz 1985, sec. 5.1

[26] Ehlers 1973, pp. 17ff; a derivation can be found in Mermin 2005, ch. 12. For the experimental evidence, cf. the section Gravitational time dilation and frequency shift, below

[27] Rindler 2001, sec. 1.13; for an elementary account, see Wheeler 1990, ch. 2; there are, however, some differences between the modern version and Einstein's original concept used in the historical derivation of general relativity, cf. Norton 1985

[28] Ehlers 1973, sec. 1.4 for the experimental evidence, see once more section Gravitational time dilation and frequency shift. Choosing a different connection with non-zero torsion leads to a modified theory known as Einstein–Cartan theory

[29] Ehlers 1973, p. 16, Kenyon 1990, sec. 7.2, Weinberg 1972, sec. 2.8

[30] Ehlers 1973, pp. 19–22; for similar derivations, see sections 1 and 2 of ch. 7 in Weinberg 1972. The Einstein tensor is the only divergence-free tensor that is a function of the metric coefficients, their first and second derivatives at most, and allows the spacetime of special relativity as a solution in the absence of sources of gravity, cf. Lovelock 1972. The tensors on both side are of second rank, that is, they can each be thought of as 4×4 matrices, each of which contains ten independent terms; hence, the above represents ten coupled equations. The fact that, as a consequence of geometric relations known as Bianchi identities, the Einstein tensor satisfies a further four identities reduces these to six independent equations, e.g. Schutz 1985, sec. 8.3

[31] Kenyon 1990, sec. 7.4

[32] Brans & Dicke 1961, Weinberg 1972, sec. 3 in ch. 7, Goenner 2004, sec. 7.2, and Trautman 2006, respectively

[33] Wald 1984, ch. 4, Weinberg 1972, ch. 7 or, in fact, any other textbook on general relativity

[34] At least approximately, cf. Poisson 2004

[35] Wheeler 1990, p. xi

[36] Wald 1984, sec. 4.4

[37] Wald 1984, sec. 4.1

[38] For the (conceptual and historical) difficulties in defining a general principle of relativity and separating it from the notion of general covariance, see Giulini 2006b

[39] section 5 in ch. 12 of Weinberg 1972

[40] Introductory chapters of Stephani et al. 2003

[41] A review showing Einstein's equation in the broader context of other PDEs with physical significance is Geroch 1996

[42] For background information and a list of solutions, cf. Stephani et al. 2003; a more recent review can be found in MacCallum 2006

[43] Chandrasekhar 1983, ch. 3,5,6

[44] Narlikar 1993, ch. 4, sec. 3.3

[45] Brief descriptions of these and further interesting solutions can be found in Hawking & Ellis 1973, ch. 5

[46] Lehner 2002

[47] For instance Wald 1984, sec. 4.4

[48] Will 1993, sec. 4.1 and 4.2

[49] Will 2006, sec. 3.2, Will 1993, ch. 4

[50] Rindler 2001, pp. 24–26 vs. pp. 236–237 and Ohanian & Ruffini 1994, pp. 164–172. Einstein derived these effects using the equivalence principle as early as 1907, cf. Einstein 1907 and the description in Pais 1982, pp. 196–198

[51] Rindler 2001, pp. 24–26; Misner, Thorne & Wheeler 1973, § 38.5

[52] Pound–Rebka experiment, see Pound & Rebka 1959, Pound & Rebka 1960; Pound & Snider 1964; a list of further experiments is given in Ohanian & Ruffini 1994, table 4.1 on p. 186

[53] Greenstein, Oke & Shipman 1971; the most recent and most accurate Sirius B measurements are published in Barstow, Bond et al. 2005.

[54] Starting with the Hafele–Keating experiment, Hafele & Keating 1972a and Hafele & Keating 1972b, and culminating in the Gravity Probe A experiment; an overview of experiments can be found in Ohanian & Ruffini 1994, table 4.1 on p. 186

[55] GPS is continually tested by comparing atomic clocks on the ground and aboard orbiting satellites; for an account of relativistic effects, see Ashby 2002 and Ashby 2003

[56] Stairs 2003 and Kramer 2004

[57] General overviews can be found in section 2.1. of Will 2006; Will 2003, pp. 32–36; Ohanian & Ruffini 1994, sec. 4.2

[58] Ohanian & Ruffini 1994, pp. 164–172

[59] Cf. Kennefick 2005 for the classic early measurements by the Eddington expeditions; for an overview of more recent measurements, see Ohanian & Ruffini 1994, ch. 4.3. For the most precise direct modern observations using quasars, cf. Shapiro et al. 2004

[60] This is not an independent axiom; it can be derived from Einstein's equations and the Maxwell Lagrangian using a WKB approximation, cf. Ehlers 1973, sec. 5

[61] Blanchet 2006, sec. 1.3

[62] Rindler 2001, sec. 1.16; for the historical examples, Israel 1987, pp. 202–204; in fact, Einstein published one such derivation as Einstein 1907. Such calculations tacitly assume that the geometry of space is Euclidean, cf. Ehlers & Rindler 1997

[63] From the standpoint of Einstein's theory, these derivations take into account the effect of gravity on time, but not its consequences for the warping of space, cf. Rindler 2001, sec. 11.11

[64] For the Sun's gravitational field using radar signals reflected from planets such as Venus and Mercury, cf. Shapiro 1964, Weinberg 1972, ch. 8, sec. 7; for signals actively sent back by space probes (transponder measurements), cf. Bertotti, Iess & Tortora 2003; for an overview, see Ohanian & Ruffini 1994, table 4.4 on p. 200; for more recent measurements using signals received from a pulsar that is part of a binary system, the gravitational field causing the time delay being that of the other pulsar, cf. Stairs 2003, sec. 4.4

[65] Will 1993, sec. 7.1 and 7.2

[66] These have been indirectly observed through the loss of energy in binary pulsar systems such as the Hulse–Taylor binary, the subject of the 1993 Nobel Prize in physics. A number of projects are underway to attempt to observe directly the effects of gravitational waves. For an overview, see Misner, Thorne & Wheeler 1973, part VIII. Unlike electromagnetic waves, the dominant contribution for gravitational waves is not the dipole, but the quadrupole; see Schutz 2001

[67] Most advanced textbooks on general relativity contain a description of these properties, e.g. Schutz 1985, ch. 9

[68] For example Jaranowski & Królak 2005

[69] Rindler 2001, ch. 13

[70] Gowdy 1971, Gowdy 1974

[71] See Lehner 2002 for a brief introduction to the methods of numerical relativity, and Seidel 1998 for the connection with gravitational wave astronomy

[72] Schutz 2003, pp. 48–49, Pais 1982, pp. 253–254

[73] Rindler 2001, sec. 11.9

[74] Will 1993, pp. 177–181

[75] In consequence, in the parameterized post-Newtonian formalism (PPN), measurements of this effect determine a linear combination of the terms β and γ, cf. Will 2006, sec. 3.5 and Will 1993, sec. 7.3

[76] The most precise measurements are VLBI measurements of planetary positions; see Will 1993, ch. 5, Will 2006, sec. 3.5, Anderson et al. 1992; for an overview, Ohanian & Ruffini 1994, pp. 406–407

[77] Kramer et al. 2006

[78] A figure that includes error bars is fig. 7 in Will 2006, sec. 5.1

[79] Stairs 2003, Schutz 2003, pp. 317–321, Bartusiak 2000, pp. 70–86

[80] Weisberg & Taylor 2003; for the pulsar discovery, see Hulse & Taylor 1975; for the initial evidence for gravitational radiation, see Taylor 1994

[81] Kramer 2004

[82] Penrose 2004, §14.5, Misner, Thorne & Wheeler 1973, §11.4

[83] Weinberg 1972, sec. 9.6, Ohanian & Ruffini 1994, sec. 7.8

[84] Bertotti, Ciufolini & Bender 1987, Nordtvedt 2003

[85] Kahn 2007

[86] A mission description can be found in Everitt et al. 2001; a first post-flight evaluation is given in Everitt, Parkinson & Kahn 2007; further updates will be available on the mission website Kahn 1996–2012.

[87] Townsend 1997, sec. 4.2.1, Ohanian & Ruffini 1994, pp. 469–471

[88] Ohanian & Ruffini 1994, sec. 4.7, Weinberg 1972, sec. 9.7; for a more recent review, see Schäfer 2004

[89] Ciufolini & Pavlis 2004, Ciufolini, Pavlis & Peron 2006, Iorio 2009

[90] Iorio L. (August 2006), "COMMENTS, REPLIES AND NOTES: A note on the evidence of the gravitomagnetic field of Mars", *Classical Quantum Gravity* **23** (17): 5451–5454, arXiv:gr-qc/0606092, Bibcode:2006CQGra..23.5451I, doi:10.1088/0264-9381/23/17/N01

[91] Iorio L. (June 2010), "On the Lense–Thirring test with the Mars Global Surveyor in the gravitational field of Mars", *Central European Journal of Physics* **8** (3): 509–513, arXiv:gr-qc/0701146, Bibcode:2010CEJPh...8..509I, doi:10.2478/s11534-009-0117-6

[92] For overviews of gravitational lensing and its applications, see Ehlers, Falco & Schneider 1992 and Wambsganss 1998

[93] For a simple derivation, see Schutz 2003, ch. 23; cf. Narayan & Bartelmann 1997, sec. 3

[94] Walsh, Carswell & Weymann 1979

[95] Images of all the known lenses can be found on the pages of the CASTLES project, Kochanek et al. 2007

[96] Roulet & Mollerach 1997

[97] Narayan & Bartelmann 1997, sec. 3.7

[98] Barish 2005, Bartusiak 2000, Blair & McNamara 1997

[99] Hough & Rowan 2000

[100] Hobbs, George; Archibald, A.; Arzoumanian, Z.; Backer, D.; Bailes, M.; Bhat, N. D. R.; Burgay, M.; Burke-Spolaor, S.; et al. (2010), "The international pulsar timing array project: using pulsars as a gravitational wave detector", *Classical and Quantum Gravity* **27** (8): 084013, arXiv:0911.5206, Bibcode:2010CQGra..27h4013H, doi:10.1088/0264-9381/27/8/084013

[101] Danzmann & Rüdiger 2003

[102] "LISA pathfinder overview", ESA. Retrieved 2012-04-23.

[103] Thorne 1995

[104] Cutler & Thorne 2002

[105] Miller 2002, lectures 19 and 21

[106] Celotti, Miller & Sciama 1999, sec. 3

[107] Springel et al. 2005 and the accompanying summary Gnedin 2005

[108] Blandford 1987, sec. 8.2.4

[109] For the basic mechanism, see Carroll & Ostlie 1996, sec. 17.2; for more about the different types of astronomical objects associated with this, cf. Robson 1996

[110] For a review, see Begelman, Blandford & Rees 1984. To a distant observer, some of these jets even appear to move faster than light; this, however, can be explained as an optical illusion that does not violate the tenets of relativity, see Rees 1966

[111] For stellar end states, cf. Oppenheimer & Snyder 1939 or, for more recent numerical work, Font 2003, sec. 4.1; for supernovae, there are still major problems to be solved, cf. Buras et al. 2003; for simulating accretion and the formation of jets, cf. Font 2003, sec. 4.2. Also, relativistic lensing effects are thought to play a role for the signals received from X-ray pulsars, cf. Kraus 1998

[112] The evidence includes limits on compactness from the observation of accretion-driven phenomena ("Eddington luminosity"), see Celotti, Miller & Sciama 1999, observations of stellar dynamics in the center of our own Milky Way galaxy, cf. Schödel et al. 2003, and indications that at least some of the compact objects in question appear to have no solid surface, which can be deduced from the examination of X-ray bursts for which the central compact object is either a neutron star or a black hole; cf. Remillard et al. 2006 for an overview, Narayan 2006, sec. 5. Observations of the "shadow" of the Milky Way galaxy's central black hole horizon are eagerly sought for, cf. Falcke, Melia & Agol 2000

[113] Dalal et al. 2006

[114] Barack & Cutler 2004

[115] Originally Einstein 1917; cf. Pais 1982, pp. 285–288

[116] Carroll 2001, ch. 2

[117] Bergström & Goobar 2003, ch. 9–11; use of these models is justified by the fact that, at large scales of around hundred million light-years and more, our own universe indeed appears to be isotropic and homogeneous, cf. Peebles et al. 1991

[118] E.g. with WMAP data, see Spergel et al. 2003

[119] These tests involve the separate observations detailed further on, see, e.g., fig. 2 in Bridle et al. 2003

[120] Peebles 1966; for a recent account of predictions, see Coc, Vangioni-Flam et al. 2004; an accessible account can be found in Weiss 2006; compare with the observations in Olive & Skillman 2004, Bania, Rood & Balser 2002, O'Meara et al. 2001, and Charbonnel & Primas 2005

[121] Lahav & Suto 2004, Bertschinger 1998, Springel et al. 2005

[122] Alpher & Herman 1948, for a pedagogical introduction, see Bergström & Goobar 2003, ch. 11; for the initial detection, see Penzias & Wilson 1965 and, for precision measurements by satellite observatories, Mather et al. 1994 (COBE) and Bennett et al. 2003 (WMAP). Future measurements could also reveal evidence about gravitational waves in the early universe; this additional information is contained in the background radiation's polarization, cf. Kamionkowski, Kosowsky & Stebbins 1997 and Seljak & Zaldarriaga 1997

[123] Evidence for this comes from the determination of cosmological parameters and additional observations involving the dynamics of galaxies and galaxy clusters cf. Peebles 1993, ch. 18, evidence from gravitational lensing, cf. Peacock 1999, sec. 4.6, and simulations of large-scale structure formation, see Springel et al. 2005

[124] Peacock 1999, ch. 12, Peskin 2007; in particular, observations indicate that all but a negligible portion of that matter is not in the form of the usual elementary particles ("non-baryonic matter"), cf. Peacock 1999, ch. 12

[125] Namely, some physicists have questioned whether or not the evidence for dark matter is, in fact, evidence for deviations from the Einsteinian (and the Newtonian) description of gravity cf. the overview in Mannheim 2006, sec. 9

[126] Carroll 2001; an accessible overview is given in Caldwell 2004. Here, too, scientists have argued that the evidence indicates not a new form of energy, but the need for modifications in our cosmological models, cf. Mannheim 2006, sec. 10; aforementioned modifications need not be modifications of general relativity, they could, for example, be modifications in the way we treat the inhomogeneities in the universe, cf. Buchert 2007

[127] A good introduction is Linde 1990; for a more recent review, see Linde 2005

[128] More precisely, these are the flatness problem, the horizon problem, and the monopole problem; a pedagogical introduction can be found in Narlikar 1993, sec. 6.4, see also Börner 1993, sec. 9.1

[129] Spergel et al. 2007, sec. 5.6

[130] More concretely, the potential function that is crucial to determining the dynamics of the inflaton is simply postulated, but not derived from an underlying physical theory

[131] Brandenberger 2007, sec. 2

[132] Frauendiener 2004, Wald 1984, sec. 11.1, Hawking & Ellis 1973, sec. 6.8, 6.9

[133] Wald 1984, sec. 9.2–9.4 and Hawking & Ellis 1973, ch. 6

[134] Thorne 1972; for more recent numerical studies, see Berger 2002, sec. 2.1

[135] Israel 1987. A more exact mathematical description distinguishes several kinds of horizon, notably event horizons and apparent horizons cf. Hawking & Ellis 1973, pp. 312–320 or Wald 1984, sec. 12.2; there are also more intuitive definitions for isolated systems that do not require knowledge of spacetime properties at infinity, cf. Ashtekar & Krishnan 2004

[136] For first steps, cf. Israel 1971; see Hawking & Ellis 1973, sec. 9.3 or Heusler 1996, ch. 9 and 10 for a derivation, and Heusler 1998 as well as Beig & Chruściel 2006 as overviews of more recent results

[137] The laws of black hole mechanics were first described in Bardeen, Carter & Hawking 1973; a more pedagogical presentation can be found in Carter 1979; for a more recent review, see Wald 2001, ch. 2. A thorough, book-length introduction including an introduction to the necessary mathematics Poisson 2004. For the Penrose process, see Penrose 1969

[138] Bekenstein 1973, Bekenstein 1974

[139] The fact that black holes radiate, quantum mechanically, was first derived in Hawking 1975; a more thorough derivation can be found in Wald 1975. A review is given in Wald 2001, ch. 3

[140] Narlikar 1993, sec. 4.4.4, 4.4.5

[141] Horizons: cf. Rindler 2001, sec. 12.4. Unruh effect: Unruh 1976, cf. Wald 2001, ch. 3

[142] Hawking & Ellis 1973, sec. 8.1, Wald 1984, sec. 9.1

[143] Townsend 1997, ch. 2; a more extensive treatment of this solution can be found in Chandrasekhar 1983, ch. 3

[144] Townsend 1997, ch. 4; for a more extensive treatment, cf. Chandrasekhar 1983, ch. 6

[145] Ellis & Van Elst 1999; a closer look at the singularity itself is taken in Börner 1993, sec. 1.2

[146] Here one should remind to the well-known fact that the important "quasi-optical" singularities of the so-called eikonal approximations of many wave-equations, namely the "caustics", are resolved into finite peaks beyond that approximation.

[147] Namely when there are trapped null surfaces, cf. Penrose 1965

[148] Hawking 1966

[149] The conjecture was made in Belinskii, Khalatnikov & Lifschitz 1971; for a more recent review, see Berger 2002. An accessible exposition is given by Garfinkle 2007

[150] The restriction to future singularities naturally excludes initial singularities such as the big bang singularity, which in principle be visible to observers at later cosmic time. The cosmic censorship conjecture was first presented in Penrose 1969; a textbook-level account is given in Wald 1984, pp. 302–305. For numerical results, see the review Berger 2002, sec. 2.1

[151] Hawking & Ellis 1973, sec. 7.1

[152] Arnowitt, Deser & Misner 1962; for a pedagogical introduction, see Misner, Thorne & Wheeler 1973, §21.4–§21.7

[153] Fourès-Bruhat 1952 and Bruhat 1962; for a pedagogical introduction, see Wald 1984, ch. 10; an online review can be found in Reula 1998

[154] Gourgoulhon 2007; for a review of the basics of numerical relativity, including the problems arising from the peculiarities of Einstein's equations, see Lehner 2001

[155] Misner, Thorne & Wheeler 1973, §20.4

[156] Arnowitt, Deser & Misner 1962

[157] Komar 1959; for a pedagogical introduction, see Wald 1984, sec. 11.2; although defined in a totally different way, it can be shown to be equivalent to the ADM mass for stationary spacetimes, cf. Ashtekar & Magnon-Ashtekar 1979

[158] For a pedagogical introduction, see Wald 1984, sec. 11.2

[159] Wald 1984, p. 295 and refs therein; this is important for questions of stability—if there were negative mass states, then flat, empty Minkowski space, which has mass zero, could evolve into these states

[160] Townsend 1997, ch. 5

[161] Such quasi-local mass–energy definitions are the Hawking energy, Geroch energy, or Penrose's quasi-local energy–momentum based on twistor methods; cf. the review article Szabados 2004

[162] An overview of quantum theory can be found in standard textbooks such as Messiah 1999; a more elementary account is given in Hey & Walters 2003

[163] Ramond 1990, Weinberg 1995, Peskin & Schroeder 1995; a more accessible overview is Auyang 1995

[164] Wald 1994, Birrell & Davies 1984

[165] For Hawking radiation Hawking 1975, Wald 1975; an accessible introduction to black hole evaporation can be found in Traschen 2000

[166] Wald 2001, ch. 3

[167] Put simply, matter is the source of spacetime curvature, and once matter has quantum properties, we can expect spacetime to have them as well. Cf. Carlip 2001, sec. 2

[168] Schutz 2003, p. 407

[169] A timeline and overview can be found in Rovelli 2000

[170] Donoghue 1995

[171] In particular, a technique known as renormalization, an integral part of deriving predictions which take into account higher-energy contributions, cf. Weinberg 1996, ch. 17, 18, fails in this case; cf. Goroff & Sagnotti 1985

[172] An accessible introduction at the undergraduate level can be found in Zwiebach 2004; more complete overviews can be found in Polchinski 1998a and Polchinski 1998b

[173] At the energies reached in current experiments, these strings are indistinguishable from point-like particles, but, crucially, different modes of oscillation of one and the same type of fundamental string appear as particles with different (electric and other) charges, e.g. Ibanez 2000. The theory is successful in that one mode will always correspond to a graviton, the messenger particle of gravity, e.g. Green, Schwarz & Witten 1987, sec. 2.3, 5.3

[174] Green, Schwarz & Witten 1987, sec. 4.2

[175] Weinberg 2000, ch. 31

[176] Townsend 1996, Duff 1996

[177] Kuchař 1973, sec. 3

[178] These variables represent geometric gravity using mathematical analogues of electric and magnetic fields; cf. Ashtekar 1986, Ashtekar 1987

[179] For a review, see Thiemann 2006; more extensive accounts can be found in Rovelli 1998, Ashtekar & Lewandowski 2004 as well as in the lecture notes Thiemann 2003

[180] Isham 1994, Sorkin 1997

[181] Loll 1998

[182] Sorkin 2005

[183] Penrose 2004, ch. 33 and refs therein

[184] Hawking 1987

[185] Ashtekar 2007, Schwarz 2007

[186] Maddox 1998, pp. 52–59, 98–122; Penrose 2004, sec. 34.1, ch. 30

[187] section Quantum gravity, above

[188] section Cosmology, above

[189] Friedrich 2005

[190] A review of the various problems and the techniques being developed to overcome them, see Lehner 2002

[191] See Bartusiak 2000 for an account up to that year; up-to-date news can be found on the websites of major detector collaborations such as GEO 600 and LIGO

[192] For the most recent papers on gravitational wave polarizations of inspiralling compact binaries, see Blanchet et al. 2008, and Arun et al. 2007; for a review of work on compact binaries, see Blanchet 2006 and Futamase & Itoh 2006; for a general review of experimental tests of general relativity, see Will 2006

[193] See, e.g., the electronic review journal Living Reviews in Relativity

8.11   References

• Alpher, R. A.; Herman, R. C. (1948), "Evolution of the universe", *Nature* **162** (4124): 774–775, Bibcode:1948Nat doi:10.1038/162774b0

• Anderson, J. D.; Campbell, J. K.; Jurgens, R. F.; Lau, E. L. (1992), "Recent developments in solar-system tests of general relativity", in Sato, H.; Nakamura, T., *Proceedings of the Sixth Marcel Großmann Meeting on General Relativity*, World Scientific, pp. 353–355, ISBN 981-02-0950-9

• Arnold, V. I. (1989), *Mathematical Methods of Classical Mechanics*, Springer, ISBN 3-540-96890-3

• Arnowitt, Richard; Deser, Stanley; Misner, Charles W. (1962), "The dynamics of general relativity", in Witten, Louis, *Gravitation: An Introduction to Current Research*, Wiley, pp. 227–265

• Arun, K.G.; Blanchet, L.; Iyer, B. R.; Qusailah, M. S. S. (2007), "Inspiralling compact binaries in quasi-elliptical orbits: The complete 3PN energy flux", *Physical Review D* **77** (6), arXiv:0711.0302, Bibcode:2008PhRvD..77f4035A, doi:10.1103/PhysRevD.77.064035

• Ashby, Neil (2002), "Relativity and the Global Positioning System" (PDF), *Physics Today* **55** (5): 41–47, Bibcod doi:10.1063/1.1485583

• Ashby, Neil (2003), "Relativity in the Global Positioning System", *Living Reviews in Relativity* **6**, retrieved 2007-07-06 External link in |work= (help)

• Ashtekar, Abhay (1986), "New variables for classical and quantum gravity", *Phys. Rev. Lett.* **57** (18): 2244–2247, Bibcode:1986PhRvL..57.2244A, doi:10.1103/PhysRevLett.57.2244, PMID 10033673

• Ashtekar, Abhay (1987), "New Hamiltonian formulation of general relativity", *Phys. Rev.* **D36** (6): 1587–1602, Bibcode:1987PhRvD..36.1587A, doi:10.1103/PhysRevD.36.1587

• Ashtekar, Abhay (2007), "LOOP QUANTUM GRAVITY: FOUR RECENT ADVANCES AND A DOZEN FREQUENTLY ASKED QUESTIONS", *The Eleventh Marcel Grossmann Meeting - on Recent Developments in Theoretical and Experimental General Relativity, Gravitation and Relativistic Field Theories - Proceedings of the MG11 Meeting on General Relativity*, p. 126, arXiv:0705.2222, Bibcode:2008mgm..conf..126A, doi:10.1142/978981283430 0_0008, ISBN9789812834263

- Ashtekar, Abhay; Krishnan, Badri (2004), "Isolated and Dynamical Horizons and Their Applications", *Living Rev. Relativity* **7**, arXiv:gr-qc/0407042, Bibcode:2004LRR.....7...10A, doi:10.12942/lrr-2004-10, retrieved 2007-08-28

- Ashtekar, Abhay; Lewandowski, Jerzy (2004), "Background Independent Quantum Gravity: A Status Report", *Class. Quant. Grav.* **21** (15): R53–R152, arXiv:gr-qc/0404018, Bibcode:2004CQGra..21R..53A, doi:10.1088/0264-9381/21/15/R01

- Ashtekar, Abhay; Magnon-Ashtekar, Anne (1979), "On conserved quantities in general relativity", *Journal of Mathematical Physics* **20** (5): 793–800, Bibcode:1979JMP....20..793A, doi:10.1063/1.524151

- Auyang, Sunny Y. (1995), *How is Quantum Field Theory Possible?*, Oxford University Press, ISBN 0-19-509345-3

- Bania, T. M.; Rood, R. T.; Balser, D. S. (2002), "The cosmological density of baryons from observations of 3He+ in the Milky Way", *Nature* **415** (6867): 54–57, Bibcode:2002Natur.415...54B, doi:10.1038/415054a, PMID 11780112

- Barack, Leor; Cutler, Curt (2004), "LISA Capture Sources: Approximate Waveforms, Signal-to-Noise Ratios, and Parameter Estimation Accuracy", *Phys. Rev.* **D69** (8): 082005, arXiv:gr-qc/0310125, Bibcode:2004PhRvD..69h2005B, doi:10.1103/PhysRevD.69.082005

- Bardeen, J. M.; Carter, B.; Hawking, S. W. (1973), "The Four Laws of Black Hole Mechanics", *Comm. Math. Phys.* **31** (2): 161–170, Bibcode:1973CMaPh..31..161B, doi:10.1007/BF01645742

- Barish, Barry (2005), "Towards detection of gravitational waves", in Florides, P.; Nolan, B.; Ottewil, A., *General Relativity and Gravitation. Proceedings of the 17th International Conference*, World Scientific, pp. 24–34, ISBN 981-256-424-1

- Barstow, M; Bond, Howard E.; Holberg, J. B.; Burleigh, M. R.; Hubeny, I.; Koester, D. (2005), "Hubble Space Telescope Spectroscopy of the Balmer lines in Sirius B", *Mon. Not. Roy. Astron. Soc.* **362** (4): 1134–1142, arXiv:astro-ph/0506600, Bibcode:2005MNRAS.362.1134B, doi:10.1111/j.1365-2966.2005.09359.x

- Bartusiak, Marcia (2000), *Einstein's Unfinished Symphony: Listening to the Sounds of Space-Time*, Berkley, ISBN 978-0-425-18620-6

- Begelman, Mitchell C.; Blandford, Roger D.; Rees, Martin J. (1984), "Theory of extragalactic radio sources", *Rev. Mod. Phys.* **56** (2): 255–351, Bibcode:1984RvMP...56..255B, doi:10.1103/RevModPhys.56.255

- Beig, Robert; Chruściel, Piotr T. (2006), "Stationary black holes", in Françoise, J.-P.; Naber, G.; Tsou, T.S., *Encyclopedia of Mathematical Physics, Volume 2*, Elsevier, p. 2041, arXiv:gr-qc/0502041, Bibcode:2005gr.qc.....2041B, ISBN 0-12-512660-3

- Bekenstein, Jacob D. (1973), "Black Holes and Entropy", *Phys. Rev.* **D7** (8): 2333–2346, Bibcode:1973PhRvD...7 doi:10.1103/PhysRevD.7.2333 .2333B,

- Bekenstein, Jacob D. (1974), "Generalized Second Law of Thermodynamics in Black-Hole Physics", *Phys. Rev.* **D9** (12): 3292–3300, Bibcode:1974PhRvD...9.3292B, doi:10.1103/PhysRevD.9.3292

- Belinskii, V. A.; Khalatnikov, I. M.; Lifschitz, E. M. (1971), "Oscillatory approach to the singular point in relativistic cosmology", *Advances in Physics* **19** (80): 525–573, Bibcode:1970AdPhy..19..525B, doi:10.1080/00018737000101171;original paper in Russian:Belinsky,V.A.;Lifshits,I.M.;Khalatnikov,E.M. (1970), "КолебательныйРеж имПриближенияКОсобойТочкеВРелятивистскойКосмологии",*Uspekhi Fizicheskikh Nauk(УспехиФизичес кихНаук)*,102(3) (11):463–500,Bibcode:1970UsFiN.102..463B

- Bennett, C. L.; Halpern, M.; Hinshaw, G.; Jarosik, N.; Kogut, A.; Limon, M.; Meyer, S. S.; Page, L.; et al. (2003), "First Year Wilkinson Microwave Anisotropy Probe (WMAP) Observations: Preliminary Maps and Basic Results", *Astrophys. J. Suppl.* **148** (1): 1–27, arXiv:astro-ph/0302207, Bibcode:2003ApJS..148....1B, doi:10.1086/377253

- Berger, Beverly K. (2002), "Numerical Approaches to Spacetime Singularities", *Living Rev. Relativity* **5**, arXiv:gr-qc/0201056, Bibcode:2002LRR.....5....1B, doi:10.12942/lrr-2002-1, retrieved 2007-08-04

- Bergström, Lars; Goobar, Ariel (2003), *Cosmology and Particle Astrophysics* (2nd ed.), Wiley & Sons, ISBN 3-540-43128-4

- Bertotti, Bruno; Ciufolini, Ignazio; Bender, Peter L. (1987), "New test of general relativity: Measurement of de Sitter geodetic precession rate for lunar perigee", *Physical Review Letters* **58** (11): 1062–1065, Bibcode:1987PhRvL..58.1062B,doi:10.1103/PhysRevLett.58.1062,PMID10034329

- Bertotti, Bruno; Iess, L.; Tortora, P. (2003), "A test of general relativity using radio links with the Cassini spacecraft", *Nature* **425** (6956): 374–376, Bibcode:2003Natur.425..374B, doi:10.1038/nature01997, PMID 14508481

- Bertschinger, Edmund (1998), "Simulations of structure formation in the universe", *Annu. Rev. Astron. Astrophys.* **36** (1): 599–654, Bibcode:1998ARA&A..36..599B, doi:10.1146/annurev.astro.36.1.599

- Birrell, N. D.; Davies, P. C. (1984), *Quantum Fields in Curved Space*, Cambridge University Press, ISBN 0-521-27858-9

- Blair, David; McNamara, Geoff (1997), *Ripples on a Cosmic Sea. The Search for Gravitational Waves*, Perseus, ISBN 0-7382-0137-5

- Blanchet, L.; Faye, G.; Iyer, B. R.; Sinha, S. (2008), "The third post-Newtonian gravitational wave polarisations and associated spherical harmonic modes for inspiralling compact binaries in quasi-circular orbits", *Classical and Quantum Gravity* **25** (16): 165003, arXiv:0802.1249, Bibcode:2008CQGra..25p5003B, doi:10.1088/0264-9381/25/16/165003

- Blanchet, Luc (2006), "Gravitational Radiation from Post-Newtonian Sources and Inspiralling Compact Binaries", *Living Rev. Relativity* **9**, Bibcode:2006LRR.....9....4B, doi:10.12942/lrr-2006-4, retrieved 2007-08-07

- Blandford, R. D. (1987), "Astrophysical Black Holes", in Hawking, Stephen W.; Israel, Werner, *300 Years of Gravitation*, Cambridge University Press, pp. 277–329, ISBN 0-521-37976-8

- Börner, Gerhard (1993), *The Early Universe. Facts and Fiction*, Springer, ISBN 0-387-56729-1

- Brandenberger, Robert H. (2007), "Conceptual Problems of Inflationary Cosmology and a New Approach to Cosmological Structure Formation", *Inflationary Cosmology*, Lecture Notes in Physics **738**, p. 393, arXiv:hep-th/0701111, Bibcode:2008LNP...738..393B, doi:10.1007/978-3-540-74353-8_11, ISBN 978-3-540-74352-1

- Brans, C. H.; Dicke, R. H. (1961), "Mach's Principle and a Relativistic Theory of Gravitation", *Physical Review* **124** (3): 925–935, Bibcode:1961PhRv..124..925B, doi:10.1103/PhysRev.124.925

- Bridle, Sarah L.; Lahav, Ofer; Ostriker, Jeremiah P.; Steinhardt, Paul J. (2003), "Precision Cosmology? Not Just Yet", *Science* **299** (5612): 1532–1533, arXiv:astro-ph/0303180, Bibcode:2003Sci...299.1532B, doi:10.1126/science.1082158,PMID12624255

- Bruhat, Yvonne (1962), "The Cauchy Problem", in Witten, Louis, *Gravitation: An Introduction to Current Research*, Wiley, p. 130, ISBN 978-1-114-29166-9

- Buchert, Thomas (2007), "Dark Energy from Structure—A Status Report", *General Relativity and Gravitation* **40** (2–3): 467–527, arXiv:0707.2153, Bibcode:2008GReGr..40..467B, doi:10.1007/s10714-007-0554-8

- Buras, R.; Rampp, M.; Janka, H.-Th.; Kifonidis, K. (2003), "Improved Models of Stellar Core Collapse and Still no Explosions: What is Missing?", *Phys. Rev. Lett.* **90** (24): 241101, arXiv:astro-ph/0303171, Bibcode:2003PhRvL..90x1101B,doi:10.1103/PhysRevLett.90.241101,PMID12857181

- Caldwell, Robert R. (2004), "Dark Energy", *Physics World* **17** (5): 37–42

- Carlip, Steven (2001), "Quantum Gravity: a Progress Report", *Rept. Prog. Phys.* **64** (8): 885–942, arXiv:gr-qc/0108040, Bibcode:2001RPPh...64..885C, doi:10.1088/0034-4885/64/8/301

- Carroll, Bradley W.; Ostlie, Dale A. (1996), *An Introduction to Modern Astrophysics*, Addison-Wesley, ISBN 0-201-54730-9

• Carroll, Sean M.(2001), "The Cosmological Constant",*Living Rev. Relativity***4**,arXiv:astro-ph/0004075,Bibcode: doi:10.12942/lrr-2001-1, retrieved 2007-07-21

• Carter, Brandon (1979), "The general theory of the mechanical, electromagnetic and thermodynamic properties of black holes", in Hawking, S. W.; Israel, W., *General Relativity, an Einstein Centenary Survey*, Cambridge University Press, pp. 294–369 and 860–863, ISBN 0-521-29928-4

• Celotti, Annalisa; Miller, John C.; Sciama, Dennis W. (1999), "Astrophysical evidence for the existence of black holes", *Class. Quant. Grav.* **16** (12A): A3–A21, arXiv:astro-ph/9912186, doi:10.1088/0264-9381/16/12A/301

• Chandrasekhar, Subrahmanyan (1983), *The Mathematical Theory of Black Holes*, Oxford University Press, ISBN 0-19-850370-9

• Charbonnel, C.; Primas, F. (2005), "The Lithium Content of the Galactic Halo Stars", *Astronomy & Astrophysics* **442** (3): 961–992, arXiv:astro-ph/0505247, Bibcode:2005A&A...442..961C, doi:10.1051/0004-6361:20042491

• Ciufolini, Ignazio; Pavlis, Erricos C. (2004), "A confirmation of the general relativistic prediction of the Lense-Thirring effect", *Nature* **431** (7011): 958–960, Bibcode:2004Natur.431..958C, doi:10.1038/nature03007, PMID 15496915

• Ciufolini, Ignazio; Pavlis, Erricos C.; Peron, R. (2006), "Determination of frame-dragging using Earth gravity models from CHAMP and GRACE",*New Astron.***11**(8): 527–550,Bibcode:2006NewA...11..527C,doi:10.1016/j.ne

• Coc, A.; Vangioni-Flam, Elisabeth; Descouvemont, Pierre; Adahchour, Abderrahim; Angulo, Carmen (2004), "Updated Big Bang Nucleosynthesis confronted to WMAP observations and to the Abundance of Light Elements", *Astrophysical Journal***600**(2): 544–552,arXiv:astro-ph/0309480,Bibcode:2004ApJ...600..544C,doi:10.1086/3

• Cutler, Curt; Thorne, Kip S. (2002), "An overview of gravitational wave sources", in Bishop, Nigel; Maharaj, Sunil D., *Proceedings of 16th International Conference on General Relativity and Gravitation (GR16)*, World Scientific, p. 4090, arXiv:gr-qc/0204090, Bibcode:2002gr.qc.....4090C, ISBN 981-238-171-6

• Dalal, Neal; Holz, Daniel E.; Hughes, Scott A.; Jain, Bhuvnesh (2006), "Short GRB and binary black hole standard sirens as a probe of dark energy", *Phys.Rev.* **D74** (6): 063006, arXiv:astro-ph/0601275, Bibcode:2006PhRvD..74f3 006D,doi:10.1103/PhysRevD.74.063006

• Danzmann, Karsten; Rüdiger, Albrecht (2003), "LISA Technology—Concepts, Status, Prospects" (PDF), *Class. Quant. Grav.* **20** (10): S1–S9, Bibcode:2003CQGra..20S...1D, doi:10.1088/0264-9381/20/10/301

• Dirac, Paul (1996), *General Theory of Relativity*, Princeton University Press, ISBN 0-691-01146-X

• Donoghue, John F. (1995), "Introduction to the Effective Field Theory Description of Gravity", in Cornet, Fernando, *Effective Theories: Proceedings of the Advanced School, Almunecar, Spain, 26 June–1 July 1995*, Singapore: World Scientific, p. 12024, arXiv:gr-qc/9512024, Bibcode:1995gr.qc....12024D, ISBN 981-02-2908-9

• Duff, Michael (1996), "M-Theory (the Theory Formerly Known as Strings)", *Int. J. Mod. Phys.* **A11** (32): 5623–5641, arXiv:hep-th/9608117, Bibcode:1996IJMPA..11.5623D, doi:10.1142/S0217751X96002583

• Ehlers, Jürgen (1973), "Survey of general relativity theory", in Israel, Werner, *Relativity, Astrophysics and Cosmology*, D. Reidel, pp. 1–125, ISBN 90-277-0369-8

• Ehlers, Jürgen; Falco, Emilio E.; Schneider, Peter (1992), *Gravitational lenses*, Springer, ISBN 3-540-66506-4

• Ehlers, Jürgen; Lämmerzahl, Claus, eds. (2006), *Special Relativity—Will it Survive the Next 101 Years?*, Springer, ISBN 3-540-34522-1

• Ehlers, Jürgen; Rindler, Wolfgang (1997), "Local and Global Light Bending in Einstein's and other Gravitational Theories",*General Relativity and Gravitation***29**(4): 519–529,Bibcode:1997GReGr..29..519E,doi:10.1023/A:101

• Einstein, Albert (1907), "Über das Relativitätsprinzip und die aus demselben gezogene Folgerungen" (PDF), *Jahrbuch der Radioaktivität und Elektronik* **4**: 411, retrieved 2008-05-05

- Einstein, Albert (1915), "Die Feldgleichungen der Gravitation", *Sitzungsberichte der Preussischen Akademie der Wissenschaften zu Berlin*: 844–847, retrieved 2006-09-12

- Einstein, Albert (1916), "Die Grundlage der allgemeinen Relativitätstheorie", *Annalen der Physik* **49**: 769–822, Bibcode:1916AnP...354..769E, doi:10.1002/andp.19163540702, archived from the original (PDF) on 2006-08-29, retrieved 2006-09-03

- Einstein, Albert (1917), "Kosmologische Betrachtungen zur allgemeinen Relativitätstheorie", *Sitzungsberichte der Preußischen Akademie der Wissenschaften*: 142

- Ellis, George F R; Van Elst, Henk (1999), Lachièze-Rey, Marc, ed., "Theoretical and Observational Cosmology: Cosmological models (Cargèse lectures 1998)", *Theoretical and observational cosmology : proceedings of the NATO Advanced Study Institute on Theoretical and Observational Cosmology* (Kluwer): 1–116, arXiv:gr-qc/9812046, Bibcode:1999toc..conf....1E, doi:10.1007/978-94-011-4455-1_1, ISBN 978-0-7923-5946-3

- Everitt, C. W. F.; Buchman, S.; DeBra, D. B.; Keiser, G. M. (2001), "Gravity Probe B: Countdown to launch", in Lämmerzahl, C.; Everitt, C. W. F.; Hehl, F. W., *Gyros, Clocks, and Interferometers: Testing Relativistic Gravity in Space (Lecture Notes in Physics 562)*, Springer, pp. 52–82, ISBN 3-540-41236-0

- Everitt, C. W. F.; Parkinson, Bradford; Kahn, Bob (2007), *The Gravity Probe B experiment. Post Flight Analysis—Final Report (Preface and Executive Summary)* (PDF), Project Report: NASA, Stanford University and Lockheed Martin, retrieved 2007-08-05

- Falcke, Heino; Melia, Fulvio; Agol, Eric (2000), "Viewing the Shadow of the Black Hole at the Galactic Center", *Astrophysical Journal* **528** (1): L13–L16, arXiv:astro-ph/9912263, Bibcode:2000ApJ...528L..13F, doi:10.1086/312423, PMID10587484

- Flanagan, Éanna É.; Hughes, Scott A. (2005), "The basics of gravitational wave theory", *New J.Phys.* **7**: 204, arXiv:gr-qc/0501041, Bibcode:2005NJPh....7..204F, doi:10.1088/1367-2630/7/1/204

- Font, José A. (2003), "Numerical Hydrodynamics in General Relativity", *Living Rev. Relativity* **6**, doi:10.12942/lrr-2003-4, retrieved 2007-08-19

- Fourès-Bruhat, Yvonne (1952), "Théoréme d'existence pour certains systémes d'équations aux derivées partielles non linéaires", *Acta Mathematica* **88** (1): 141–225, Bibcode:1952AcM....88..141F, doi:10.1007/BF02392131

- Frauendiener, Jörg (2004), "Conformal Infinity", *Living Rev. Relativity* **7**, Bibcode:2004LRR.....7....1F, doi:10.12942/lrr-2004-1, retrieved 2007-07-21

- Friedrich, Helmut (2005), "Is general relativity 'essentially understood'?", *Annalen Phys.* **15** (1–2): 84–108, arXiv:gr-qc/0508016, Bibcode:2006AnP...518...84F, doi:10.1002/andp.200510173

- Futamase, T.; Itoh, Y. (2006), "The Post-Newtonian Approximation for Relativistic Compact Binaries", *Living Rev. Relativity* **10**, retrieved 2008-02-29

- Gamow, George (1970), *My World Line*, Viking Press, ISBN 0-670-50376-2

- Garfinkle, David (2007), "Of singularities and breadmaking", *Einstein Online*, retrieved 2007-08-03 External link in |work= (help)

- Geroch, Robert (1996), "Partial Differential Equations of Physics", arXiv:gr-qc/9602055 [gr-qc].

- Giulini, Domenico (2005), *Special Relativity: A First Encounter*, Oxford University Press, ISBN 0-19-856746-4

- Giulini, Domenico (2006a), "Algebraic and Geometric Structures in Special Relativity", in Ehlers, Jürgen; Lämmerzahl, Claus, *Special Relativity—Will it Survive the Next 101 Years?*, Springer, pp. 45–111, arXiv:math-ph/0602018, Bibcode:2006math.ph...2018G, ISBN 3-540-34522-1

- Giulini, Domenico (2006b), Stamatescu, I. O., ed., "An assessment of current paradigms in the physics of fundamental interactions: Some remarks on the notions of general covariance and background independence", *Approaches to Fundamental Physics*, Lecture Notes in Physics (Springer) **721**: 105, arXiv:gr-qc/0603087, Bibcode:2007LNP...721..105G,doi:10.1007/978-3-540-71117-9_6,ISBN978-3-540-71115-5

- Gnedin, Nickolay Y. (2005), "Digitizing the Universe", *Nature* **435** (7042): 572–573, Bibcode:2005Natur.435..572G, doi:10.1038/435572a, PMID 15931201

- Goenner, Hubert F. M. (2004), "On the History of Unified Field Theories", *Living Rev. Relativity* **7**, Bibcode:2004LR doi:10.12942/lrr-2004-2, retrieved 2008-02-28 R.....7....2G,

- Goroff, Marc H.; Sagnotti, Augusto (1985), "Quantum gravity at two loops", *Phys. Lett.* **160B** (1–3): 81–86, Bibcode:1985PhLB..160...81G, doi:10.1016/0370-2693(85)91470-4

- Gourgoulhon, Eric (2007). "3+1 Formalism and Bases of Numerical Relativity". arXiv:gr-qc/0703035 [gr-qc].

- Gowdy, Robert H. (1971), "Gravitational Waves in Closed Universes", *Phys. Rev. Lett.* **27** (12): 826–829, Bibcode:1971PhRvL...27..826G, doi:10.1103/PhysRevLett.27.826

- Gowdy, Robert H. (1974), "Vacuum spacetimes with two-parameter spacelike isometry groups and compact invariant hypersurfaces: Topologies and boundary conditions", *Ann. Phys. (N.Y.)* **83** (1): 203–241, Bibcode:1974AnPhy..83..203G,doi:10.1016/0003-4916(74)90384-4

- Green, M. B.; Schwarz, J. H.; Witten, E. (1987), *Superstring theory. Volume 1: Introduction*, Cambridge University Press, ISBN 0-521-35752-7

- Greenstein, J. L.; Oke, J. B.; Shipman, H. L. (1971), "Effective Temperature, Radius, and Gravitational Redshift of Sirius B", *Astrophysical Journal* **169**: 563, Bibcode:1971ApJ...169..563G, doi:10.1086/151174

- Hafele, J. C.; Keating, R. E. (July 14, 1972). "Around-the-World Atomic Clocks: Predicted Relativistic Time Gains". *Science* **177** (4044): 166–168. Bibcode:1972Sci...177..166H. doi:10.1126/science.177.4044.166. PMID 17779917.

- Hafele, J. C.; Keating, R. E. (July 14, 1972). "Around-the-World Atomic Clocks: Observed Relativistic Time Gains". *Science* **177** (4044): 168–170. Bibcode:1972Sci...177..168H. doi:10.1126/science.177.4044.168. PMID 17779918.

- Havas, P. (1964), "Four-Dimensional Formulation of Newtonian Mechanics and Their Relation to the Special and the General Theory of Relativity", *Rev. Mod. Phys.* **36** (4): 938–965, Bibcode:1964RvMP...36..938H, doi:10.1103/RevModPhys.36.938

- Hawking, Stephen W. (1966), "The occurrence of singularities in cosmology", *Proceedings of the Royal Society* **A294** (1439): 511–521, Bibcode:1966RSPSA.294..511H, doi:10.1098/rspa.1966.0221

- Hawking, S. W. (1975), "Particle Creation by Black Holes", *Communications in Mathematical Physics* **43** (3): 199–220, Bibcode:1975CMaPh..43..199H, doi:10.1007/BF02345020

- Hawking, Stephen W. (1987), "Quantum cosmology", in Hawking, Stephen W.; Israel, Werner, *300 Years of Gravitation*, Cambridge University Press, pp. 631–651, ISBN 0-521-37976-8

- Hawking, Stephen W.; Ellis, George F. R. (1973), *The large scale structure of space-time*, Cambridge University Press, ISBN 0-521-09906-4

- Heckmann, O. H. L.; Schücking, E. (1959), "Newtonsche und Einsteinsche Kosmologie", in Flügge, S., *Encyclopedia of Physics* **53**, p. 489

- Heusler,Markus(1998),"Stationary Black Holes:Uniqueness and Beyond",*Living Rev.Relativity***1**,doi:10.12942/1998-6, retrieved 2007-08-04

- Heusler, Markus (1996), *Black Hole Uniqueness Theorems*, Cambridge University Press, ISBN 0-521-56735-1

- Hey, Tony; Walters, Patrick (2003), *The new quantum universe*, Cambridge University Press, ISBN 0-521-56457-3

- Hough, Jim; Rowan, Sheila (2000), "Gravitational Wave Detection by Interferometry (Ground and Space)", *Living Rev. Relativity* **3**, retrieved 2007-07-21

- Hubble, Edwin (1929), "A Relation between Distance and Radial Velocity among Extra-Galactic Nebulae" (PDF), *Proc. Nat. Acad. Sci.* **15** (3): 168–173, Bibcode:1929PNAS...15..168H, doi:10.1073/pnas.15.3.168, PMC 522427, PMID 16577160

- Hulse, Russell A.; Taylor, Joseph H. (1975), "Discovery of a pulsar in a binary system", *Astrophys. J.* **195**: L51–L55, Bibcode:1975ApJ...195L..51H, doi:10.1086/181708

- Ibanez, L. E. (2000), "The second string (phenomenology) revolution", *Class. Quant. Grav.* **17** (5): 1117–1128, arXiv:hep-ph/9911499, Bibcode:2000CQGra..17.1117I, doi:10.1088/0264-9381/17/5/321

- Iorio, L. (2009), "An Assessment of the Systematic Uncertainty in Present and Future Tests of the Lense-Thirring Effect with Satellite Laser Ranging", *Space Sci. Rev.* **148** (1–4): 363, arXiv:0809.1373, Bibcode:2009SSRv..148..363I,doi:10.1007/s11214-008-9478-1

- Isham, Christopher J. (1994), "Prima facie questions in quantum gravity", in Ehlers, Jürgen; Friedrich, Helmut, *Canonical Gravity: From Classical to Quantum*, Springer, ISBN 3-540-58339-4

- Israel, Werner (1971), "Event Horizons and Gravitational Collapse", *General Relativity and Gravitation* **2** (1): 53–59, Bibcode:1971GReGr...2...53I, doi:10.1007/BF02450518

- Israel, Werner (1987), "Dark stars: the evolution of an idea", in Hawking, Stephen W.; Israel, Werner, *300 Years of Gravitation*, Cambridge University Press, pp. 199–276, ISBN 0-521-37976-8

- Janssen, Michel (2005), "Of pots and holes: Einstein's bumpy road to general relativity" (PDF), *Ann. Phys. (Leipzig)* **14** (S1): 58–85, Bibcode:2005AnP...517S..58J, doi:10.1002/andp.200410130

- Jaranowski, Piotr; Królak, Andrzej (2005), "Gravitational-Wave Data Analysis. Formalism and Sample Applications: The Gaussian Case", *Living Rev. Relativity* **8**, doi:10.12942/lrr-2005-3, retrieved 2007-07-30

- Kahn, Bob (1996–2012), *Gravity Probe B Website*, Stanford University, retrieved 2012-04-20

- Kahn, Bob (April 14, 2007), *Was Einstein right? Scientists provide first public peek at Gravity Probe B results (Stanford University Press Release)* (PDF), Stanford University News Service

- Kamionkowski, Marc; Kosowsky, Arthur; Stebbins, Albert (1997), "Statistics of Cosmic Microwave Background Polarization",*Phys. Rev.***D55**(12): 7368–7388,arXiv:astro-ph/9611125,Bibcode:1997PhRvD..55.7368K,doi:1

- Kennefick, Daniel (2005), "Astronomers Test General Relativity: Light-bending and the Solar Redshift", in Renn, Jürgen, *One hundred authors for Einstein*, Wiley-VCH, pp. 178–181, ISBN 3-527-40574-7

- Kennefick, Daniel (2007), "Not Only Because of Theory: Dyson, Eddington and the Competing Myths of the 1919 Eclipse Expedition", *Proceedings of the 7th Conference on the History of General Relativity, Tenerife, 2005* **0709**, p. 685, arXiv:0709.0685, Bibcode:2007arXiv0709.0685K

- Kenyon, I. R. (1990), *General Relativity*, Oxford University Press, ISBN 0-19-851996-6

- Kochanek, C.S.; Falco, E.E.; Impey, C.; Lehar, J. (2007), *CASTLES Survey Website*, Harvard-Smithsonian Center for Astrophysics, retrieved 2007-08-21

- Komar, Arthur (1959), "Covariant Conservation Laws in General Relativity", *Phys. Rev.* **113** (3): 934–936, Bibcode:1959PhRv..113..934K, doi:10.1103/PhysRev.113.934

- Kramer, Michael (2004), Karshenboim, S. G.; Peik, E., eds., "Astrophysics, Clocks and Fundamental Constants: Millisecond Pulsars as Tools of Fundamental Physics", *Lecture Notes in Physics* (Springer) **648**: 33–54, arXiv:astro-ph/0405178, Bibcode:2004LNP...648...33K, doi:10.1007/978-3-540-40991-5_3, ISBN 978-3-540-21967-5

- Kramer, M.; Stairs, I. H.; Manchester, R. N.; McLaughlin, M. A.; Lyne, A. G.; Ferdman, R. D.; Burgay, M.; Lorimer, D. R.; et al. (2006), "Tests of general relativity from timing the double pulsar", *Science* **314** (5796): 97–102, arXiv:astro-ph/0609417, Bibcode:2006Sci...314...97K, doi:10.1126/science.1132305, PMID 16973838

- Kraus, Ute (1998), "Light Deflection Near Neutron Stars", *Relativistic Astrophysics*, Vieweg, pp. 66–81, ISBN 3-528-06909-0

- Kuchař, Karel (1973), "Canonical Quantization of Gravity", in Israel, Werner, *Relativity, Astrophysics and Cosmology*, D. Reidel, pp. 237–288, ISBN 90-277-0369-8

- Künzle, H. P. (1972), "Galilei and Lorentz Structures on spacetime: comparison of the corresponding geometry and physics", *Ann. Inst. Henri Poincaré a* **17**: 337–362

- Lahav, Ofer; Suto, Yasushi (2004), "Measuring our Universe from Galaxy Redshift Surveys", *Living Rev. Relativity* **7**, arXiv:astro-ph/0310642, Bibcode:2004LRR.....7....8L, doi:10.12942/lrr-2004-8, retrieved 2007-08-19

- Landgraf, M.; Hechler, M.; Kemble, S. (2005), "Mission design for LISA Pathfinder", *Class. Quant. Grav.* **22** (10): S487–S492, arXiv:gr-qc/0411071, Bibcode:2005CQGra..22S.487L, doi:10.1088/0264-9381/22/10/048

- Lehner, Luis (2001), "Numerical Relativity: A review", *Class. Quant. Grav.* **18** (17): R25–R86, arXiv:gr-qc/0106072, Bibcode:2001CQGra..18R..25L, doi:10.1088/0264-9381/18/17/202

- Lehner, Luis (2002), "NUMERICAL RELATIVITY: STATUS AND PROSPECTS", *General Relativity and Gravitation - Proceedings of the 16th International Conference*, p. 210, arXiv:gr-qc/0202055, Bibcode:2002grg..conf..210L, doi:10.1142/9789812776556_0010, ISBN 9789812381712

- Linde, Andrei (1990), *Particle Physics and Inflationary Cosmology*, Harwood, p. 3203, arXiv:hep-th/0503203, Bibcode:2005hep.th....3203L, ISBN 3-7186-0489-2

- Linde, Andrei (2005), "Towards inflation in string theory", *J. Phys. Conf. Ser.* **24**: 151–160, arXiv:hep-th/0503195, Bibcode:2005JPhCS..24..151L, doi:10.1088/1742-6596/24/1/018

- Loll, Renate (1998), "Discrete Approaches to Quantum Gravity in Four Dimensions", *Living Rev. Relativity* **1**, arXiv:gr-qc/9805049, Bibcode:1998LRR.....1...13L, doi:10.12942/lrr-1998-13, retrieved 2008-03-09

- Lovelock, David (1972), "The Four-Dimensionality of Space and the Einstein Tensor", *J. Math. Phys.* **13** (6): 874–876, Bibcode:1972JMP....13..874L, doi:10.1063/1.1666069

- Ludyk, Günter (2013). *Einstein in Matrix Form* (1st ed.). Berlin: Springer. ISBN 9783642357978.

- MacCallum, M. (2006), "Finding and using exact solutions of the Einstein equations", in Mornas, L.; Alonso, J. D., *A Century of Relativity Physics (ERE05, the XXVIII Spanish Relativity Meeting)* **841**, American Institute of Physics, p. 129, arXiv:gr-qc/0601102, Bibcode:2006AIPC..841..129M, doi:10.1063/1.2218172

- Maddox, John (1998), *What Remains To Be Discovered*, Macmillan, ISBN 0-684-82292-X

- Mannheim, Philip D. (2006), "Alternatives to Dark Matter and Dark Energy", *Prog. Part. Nucl. Phys.* **56** (2): 340–445, arXiv:astro-ph/0505266, Bibcode:2006PrPNP..56..340M, doi:10.1016/j.ppnp.2005.08.001

- Mather, J. C.; Cheng, E. S.; Cottingham, D. A.; Eplee, R. E.; Fixsen, D. J.; Hewagama, T.; Isaacman, R. B.; Jensen, K. A.; et al. (1994), "Measurement of the cosmic microwave spectrum by the COBE FIRAS instrument", *Astrophysical Journal* **420**: 439–444, Bibcode:1994ApJ...420..439M, doi:10.1086/173574

- Mermin, N. David (2005), *It's About Time. Understanding Einstein's Relativity*, Princeton University Press, ISBN 0-691-12201-6

- Messiah, Albert (1999), *Quantum Mechanics*, Dover Publications, ISBN 0-486-40924-4

- Miller, Cole (2002), *Stellar Structure and Evolution (Lecture notes for Astronomy 606)*, University of Maryland, retrieved 2007-07-25

- Misner, Charles W.; Thorne, Kip. S.; Wheeler, John A. (1973), *Gravitation*, W. H. Freeman, ISBN 0-7167-0344-0

- Møller, Christian (1952), *The Theory of Relativity* (3rd ed.), Oxford University Press

- Narayan, Ramesh (2006), "Black holes in astrophysics", *New Journal of Physics* **7**: 199, arXiv:gr-qc/0506078, Bibcode:2005NJPh....7..199N, doi:10.1088/1367-2630/7/1/199

- Narayan, Ramesh; Bartelmann, Matthias (1997). "Lectures on Gravitational Lensing". arXiv:astro-ph/9606001 [astro-ph].

- Narlikar, Jayant V. (1993), *Introduction to Cosmology*, Cambridge University Press, ISBN 0-521-41250-1

- Nieto, Michael Martin (2006), "The quest to understand the Pioneer anomaly" (PDF), *EurophysicsNews* **37** (6): 30–34, Bibcode:2006ENews..37...30N, doi:10.1051/epn:2006604

- Nordström, Gunnar (1918), "On the Energy of the Gravitational Field in Einstein's Theory", *Verhandl. Koninkl. Ned. Akad. Wetenschap.*, **26**: 1238–1245

- Nordtvedt, Kenneth (2003). "Lunar Laser Ranging—a comprehensive probe of post-Newtonian gravity". arXiv:gr-qc/0301024 [gr-qc].

- Norton, John D. (1985), "What was Einstein's principle of equivalence?" (PDF), *Studies in History and Philosophy of Science* **16** (3): 203–246, doi:10.1016/0039-3681(85)90002-0, retrieved 2007-06-11

- Ohanian, Hans C.; Ruffini, Remo (1994), *Gravitation and Spacetime*, W. W. Norton & Company, ISBN 0-393-96501-5

- Olive, K. A.; Skillman, E. A. (2004), "A Realistic Determination of the Error on the Primordial Helium Abundance", *Astrophysical Journal* **617** (1): 29–49, arXiv:astro-ph/0405588, Bibcode:2004ApJ...617...29O, doi:10.10

- O'Meara, John M.; Tytler, David; Kirkman, David; Suzuki, Nao; Prochaska, Jason X.; Lubin, Dan; Wolfe, Arthur M. (2001), "The Deuterium to Hydrogen Abundance Ratio Towards a Fourth QSO: HS0105+1619", *Astrophysical Journal* **552** (2): 718–730, arXiv:astro-ph/0011179, Bibcode:2001ApJ...552..718O, doi:10.1086/320579

- Oppenheimer, J. Robert; Snyder, H. (1939), "On continued gravitational contraction", *Physical Review* **56** (5): 455–459, Bibcode:1939PhRv...56..455O, doi:10.1103/PhysRev.56.455

- Overbye, Dennis (1999), *Lonely Hearts of the Cosmos: the story of the scientific quest for the secret of the Universe*, Back Bay, ISBN 0-316-64896-5

- Pais, Abraham (1982), *'Subtle is the Lord...' The Science and life of Albert Einstein*, Oxford University Press, ISBN 0-19-853907-X

- Peacock, John A. (1999), *Cosmological Physics*, Cambridge University Press, ISBN 0-521-41072-X

- Peebles, P. J. E. (1966), "Primordial Helium abundance and primordial fireball II", *Astrophysical Journal* **146**: 542–552, Bibcode:1966ApJ...146..542P, doi:10.1086/148918

- Peebles, P. J. E. (1993), *Principles of physical cosmology*, Princeton University Press, ISBN 0-691-01933-9

- Peebles, P.J.E.; Schramm, D.N.; Turner, E.L.; Kron, R.G. (1991), "The case for the relativistic hot Big Bang cosmology", *Nature* **352** (6338): 769–776, Bibcode:1991Natur.352..769P, doi:10.1038/352769a0

- Penrose, Roger (1965), "Gravitational collapse and spacetime singularities", *Physical Review Letters* **14** (3): 57–59, Bibcode:1965PhRvL..14...57P, doi:10.1103/PhysRevLett.14.57

- Penrose, Roger (1969), "Gravitational collapse: the role of general relativity", *Rivista del Nuovo Cimento* **1**: 252–276, Bibcode:1969NCimR...1..252P

- Penrose, Roger (2004), *The Road to Reality*, A. A. Knopf, ISBN 0-679-45443-8

- Penzias, A. A.; Wilson, R. W. (1965), "A measurement of excess antenna temperature at 4080 Mc/s", *Astrophysical Journal* **142**: 419–421, Bibcode:1965ApJ...142..419P, doi:10.1086/148307

- Peskin, Michael E.; Schroeder, Daniel V. (1995), *An Introduction to Quantum Field Theory*, Addison-Wesley, ISBN 0-201-50397-2

- Peskin, Michael E. (2007), "Dark Matter and Particle Physics", *Journal of the Physical Society of Japan* **76** (11): 111017, arXiv:0707.1536, Bibcode:2007JPSJ...76k1017P, doi:10.1143/JPSJ.76.111017

- Poisson, Eric (2004), "The Motion of Point Particles in Curved Spacetime", *Living Rev. Relativity* **7**, doi:10.12942/2004-6, retrieved 2007-06-13

- Poisson, Eric (2004), *A Relativist's Toolkit. The Mathematics of Black-Hole Mechanics*, Cambridge University Press, ISBN 0-521-83091-5

- Polchinski, Joseph (1998a), *String Theory Vol. I: An Introduction to the Bosonic String*, Cambridge University Press, ISBN 0-521-63303-6

- Polchinski, Joseph (1998b), *String Theory Vol. II: Superstring Theory and Beyond*, Cambridge University Press, ISBN 0-521-63304-4

- Pound, R. V.; Rebka, G. A. (1959), "Gravitational Red-Shift in Nuclear Resonance", *Physical Review Letters* **3** (9): 439–441, Bibcode:1959PhRvL...3..439P, doi:10.1103/PhysRevLett.3.439

- Pound, R. V.; Rebka, G. A. (1960), "Apparent weight of photons", *Phys. Rev. Lett.* **4** (7): 337–341, Bibcode:1960 doi:10.1103/PhysRevLett.4.337

- Pound, R. V.; Snider, J. L. (1964), "Effect of Gravity on Nuclear Resonance", *Phys. Rev. Lett.* **13** (18): 539–540, Bibcode:1964PhRvL...13..539P, doi:10.1103/PhysRevLett.13.539

- Ramond, Pierre (1990), *Field Theory: A Modern Primer*, Addison-Wesley, ISBN 0-201-54611-6

- Rees, Martin (1966), "Appearance of Relativistically Expanding Radio Sources", *Nature* **211** (5048): 468–470, Bibcode:1966Natur.211..468R, doi:10.1038/211468a0

- Reissner, H. (1916), "Über die Eigengravitation des elektrischen Feldes nach der Einsteinschen Theorie", *Annalen der Physik* **355** (9): 106–120, Bibcode:1916AnP...355..106R, doi:10.1002/andp.19163550905

- Remillard, Ronald A.; Lin, Dacheng; Cooper, Randall L.; Narayan, Ramesh (2006), "The Rates of Type I X-Ray Bursts from Transients Observed with RXTE: Evidence for Black Hole Event Horizons", *Astrophysical Journal* **646** (1): 407–419, arXiv:astro-ph/0509758, Bibcode:2006ApJ...646..407R, doi:10.1086/504862

- Renn, Jürgen, ed. (2007), *The Genesis of General Relativity (4 Volumes)*, Dordrecht: Springer, ISBN 1-4020-3999-9

- Renn, Jürgen, ed. (2005), *Albert Einstein—Chief Engineer of the Universe: Einstein's Life and Work in Context*, Berlin: Wiley-VCH, ISBN 3-527-40571-2

- Reula, Oscar A. (1998), "Hyperbolic Methods for Einstein's Equations", *Living Rev. Relativity* **1**, Bibcode:1998LR doi:10.12942/lrr-1998-3, retrieved 2007-08-29

- Rindler, Wolfgang (2001), *Relativity. Special, General and Cosmological*, Oxford University Press, ISBN 0-19-850836-0

- Rindler, Wolfgang (1991), *Introduction to Special Relativity*, Clarendon Press, Oxford, ISBN 0-19-853952-5

- Robson, Ian (1996), *Active galactic nuclei*, John Wiley, ISBN 0-471-95853-0

- Roulet, E.; Mollerach, S. (1997), "Microlensing", *Physics Reports* **279** (2): 67–118, arXiv:astro-ph/9603119, Bibcode:1997PhR...279...67R, doi:10.1016/S0370-1573(96)00020-8

- Rovelli, Carlo (2000). "Notes for a brief history of quantum gravity". arXiv:gr-qc/0006061 [gr-qc].

- Rovelli, Carlo (1998), "Loop Quantum Gravity", *Living Rev. Relativity* **1**, doi:10.12942/lrr-1998-1, retrieved 2008-03-13

- Schäfer, Gerhard (2004), "Gravitomagnetic Effects", *General Relativity and Gravitation* **36** (10): 2223–2235, arXiv:gr-qc/0407116, Bibcode:2004GReGr..36.2223S, doi:10.1023/B:GERG.0000046180.97877.32

- Schödel, R.; Ott, T.; Genzel, R.; Eckart, A.; Mouawad, N.; Alexander, T. (2003), "Stellar Dynamics in the Central Arcsecond of Our Galaxy", *Astrophysical Journal* **596** (2): 1015–1034, arXiv:astro-ph/0306214, Bibcode:2003ApJ...596.1015S, doi:10.1086/378122

- Schutz, Bernard F. (1985), *A first course in general relativity*, Cambridge University Press, ISBN 0-521-27703-5

- Schutz, Bernard F. (2001), "Gravitational radiation", in Murdin, Paul, *Encyclopedia of Astronomy and Astrophysics*, Grove's Dictionaries, ISBN 1-56159-268-4

- Schutz, Bernard F. (2003), *Gravity from the ground up*, Cambridge University Press, ISBN 0-521-45506-5

- Schwarz, John H. (2007), "String Theory: Progress and Problems", *Progress of Theoretical Physics Supplement* **170**: 214, arXiv:hep-th/0702219, Bibcode:2007PThPS.170..214S, doi:10.1143/PTPS.170.214

- Schwarzschild, Karl (1916a), "Über das Gravitationsfeld eines Massenpunktes nach der Einsteinschen Theorie", *Sitzungsber. Preuss. Akad. D. Wiss.*: 189–196

- Schwarzschild, Karl (1916b), "Über das Gravitationsfeld eines Kugel aus inkompressibler Flüssigkeit nach der Einsteinschen Theorie", *Sitzungsber. Preuss. Akad. D. Wiss.*: 424–434

- Seidel, Edward (1998), "Numerical Relativity: Towards Simulations of 3D Black Hole Coalescence", in Narlikar, J. V.; Dadhich, N., *Gravitation and Relativity: At the turn of the millennium (Proceedings of the GR-15 Conference, held at IUCAA, Pune, India, December 16–21, 1997)*, IUCAA, p. 6088, arXiv:gr-qc/9806088, Bibcode:1998gr.qc.....6088S, ISBN 81-900378-3-8

- Seljak, Uroš; Zaldarriaga, Matias (1997), "Signature of Gravity Waves in the Polarization of the Microwave Background", *Phys. Rev. Lett.* **78** (11): 2054–2057, arXiv:astro-ph/9609169, Bibcode:1997PhRvL..78.2054S, doi:10.1103/PhysRevLett.78.2054

- Shapiro, S. S.; Davis, J. L.; Lebach, D. E.; Gregory, J. S. (2004), "Measurement of the solar gravitational deflection of radio waves using geodetic very-long-baseline interferometry data, 1979–1999", *Phys. Rev. Lett.* **92** (12): 121101, Bibcode:2004PhRvL..92l1101S, doi:10.1103/PhysRevLett.92.121101, PMID 15089661

- Shapiro, Irwin I. (1964), "Fourth test of general relativity", *Phys. Rev. Lett.* **13**(26): 789–791, Bibcode:1964PhRv doi:10.1103/PhysRevLett.13.789

- Shapiro, I. I.; Pettengill, Gordon; Ash, Michael; Stone, Melvin; Smith, William; Ingalls, Richard; Brockelman, Richard (1968), "Fourth test of general relativity: preliminary results", *Phys. Rev. Lett.* **20** (22): 1265–1269, Bibcode:1968PhRvL..20.1265S, doi:10.1103/PhysRevLett.20.1265

- Singh, Simon (2004), *Big Bang: The Origin of the Universe*, Fourth Estate, ISBN 0-00-715251-5

- Sorkin, Rafael D. (2005), "Causal Sets: Discrete Gravity", in Gomberoff, Andres; Marolf, Donald, *Lectures on Quantum Gravity*, Springer, p. 9009, arXiv:gr-qc/0309009, Bibcode:2003gr.qc.....9009S, ISBN 0-387-23995-2

- Sorkin, Rafael D. (1997), "Forks in the Road, on the Way to Quantum Gravity", *Int. J. Theor. Phys.* **36** (12): 2759–2781, arXiv:gr-qc/9706002, Bibcode:1997IJTP...36.2759S, doi:10.1007/BF02435709

- Spergel, D. N.; Verde, L.; Peiris, H. V.; Komatsu, E.; Nolta, M. R.; Bennett, C. L.; Halpern, M.; Hinshaw, G.; et al. (2003), "First Year Wilkinson Microwave Anisotropy Probe (WMAP) Observations: Determination of Cosmological Parameters", *Astrophys. J. Suppl.* **148** (1): 175–194, arXiv:astro-ph/0302209, Bibcode:2003ApJS..148..175S, doi:10.1086/377226

- Spergel, D. N.; Bean, R.; Doré, O.; Nolta, M. R.; Bennett, C. L.; Dunkley, J.; Hinshaw, G.; Jarosik, N.; et al. (2007), "Wilkinson Microwave Anisotropy Probe (WMAP) Three Year Results: Implications for Cosmology", *Astrophysical Journal Supplement* **170** (2): 377–408, arXiv:astro-ph/0603449, Bibcode:2007ApJS..170..377S, doi:10.1086/513700

- Springel, Volker; White, Simon D. M.; Jenkins, Adrian; Frenk, Carlos S.; Yoshida, Naoki; Gao, Liang; Navarro, Julio; Thacker, Robert; et al. (2005), "Simulations of the formation, evolution and clustering of galaxies and quasars", *Nature* **435** (7042): 629–636, arXiv:astro-ph/0504097, Bibcode:2005Natur.435..629S, doi:10.1038/nature03597,PMID15931216

- Stairs, Ingrid H. (2003), "Testing General Relativity with Pulsar Timing", *Living Rev. Relativity* **6**, arXiv:astro-ph/0307536, Bibcode:2003LRR.....6....5S, doi:10.12942/lrr-2003-5, retrieved 2007-07-21

- Stephani, H.; Kramer, D.; MacCallum, M.; Hoenselaers, C.; Herlt, E. (2003), *Exact Solutions of Einstein's Field Equations* (2 ed.), Cambridge University Press, ISBN 0-521-46136-7

- Synge, J. L. (1972), *Relativity: The Special Theory*, North-Holland Publishing Company, ISBN 0-7204-0064-3

- Szabados, László B. (2004), "Quasi-Local Energy-Momentum and Angular Momentum in GR", *Living Rev. Relativity* **7**, doi:10.12942/lrr-2004-4, retrieved 2007-08-23

- Taylor, Joseph H.(1994), "Binary pulsars and relativistic gravity",*Rev. Mod. Phys.***66**(3): 711–719,Bibcode:1994 doi:10.1103/RevModPhys.66.711

- Thiemann, Thomas (2006), "Approaches to Fundamental Physics: Loop Quantum Gravity: An Inside View", *Lecture Notes in Physics* **721**: 185–263, arXiv:hep-th/0608210, Bibcode:2007LNP...721..185T, doi:10.1007/978-3-540-71117-9_10, ISBN 978-3-540-71115-5

- Thiemann, Thomas (2003), "Lectures on Loop Quantum Gravity", *Lecture Notes in Physics* **631**: 41–135, arXiv:gr-qc/0210094, doi:10.1007/978-3-540-45230-0_3, ISBN 978-3-540-40810-9

- Thorne, Kip S. (1972), "Nonspherical Gravitational Collapse—A Short Review", in Klauder, J., *Magic without Magic*, W. H. Freeman, pp. 231–258

- Thorne, Kip S. (1994), *Black Holes and Time Warps: Einstein's Outrageous Legacy*, W W Norton & Company, ISBN 0-393-31276-3

- Thorne, Kip S. (1995), "Gravitational radiation", *Particle and Nuclear Astrophysics and Cosmology in the Next Millenium*: 160, arXiv:gr-qc/9506086, Bibcode:1995pnac.conf..160T, ISBN 0-521-36853-7

- Townsend, Paul K. (1997). "Black Holes (Lecture notes)". arXiv:gr-qc/9707012 [gr-qc].

- Townsend, Paul K. (1996). "Four Lectures on M-Theory". arXiv:hep-th/9612121 [hep-th].

- Traschen, Jenny (2000), Bytsenko, A.; Williams, F., eds., "An Introduction to Black Hole Evaporation", *Mathematical Methods of Physics (Proceedings of the 1999 Londrina Winter School)* (World Scientific): 180, arXiv:gr-qc/0010055, Bibcode:2000mmp..conf..180T

- Trautman, Andrzej (2006), "Einstein–Cartan theory", in Françoise, J.-P.; Naber, G. L.; Tsou, S. T., *Encyclopedia of Mathematical Physics, Vol. 2*, Elsevier, pp. 189–195, arXiv:gr-qc/0606062, Bibcode:2006gr.qc.....6062T

- Unruh, W. G.(1976), "Notes on Black Hole Evaporation",*Phys. Rev. D***14**(4): 870–892,Bibcode:1976PhRvD..1 doi:10.1103/PhysRevD.14.870

- Valtonen, M. J.; Lehto, H. J.; Nilsson, K.; Heidt, J.; Takalo, L. O.; Sillanpää, A.; Villforth, C.; Kidger, M.; et al. (2008), "A massive binary black-hole system in OJ 287 and a test of general relativity", *Nature* **452** (7189): 851–853, arXiv:0809.1280, Bibcode:2008Natur.452..851V, doi:10.1038/nature06896, PMID 18421348

- Wald, Robert M.(1975), "On Particle Creation by Black Holes",*Commun. Math. Phys.***45**(3): 9–34,Bibcode: doi:10.1007/BF01609863

- Wald, Robert M. (1984), *General Relativity*, University of Chicago Press, ISBN 0-226-87033-2

- Wald, Robert M. (1994), *Quantum field theory in curved spacetime and black hole thermodynamics*, University of Chicago Press, ISBN 0-226-87027-8

- Wald, Robert M. (2001),"The Thermodynamics of Black Holes",*Living Rev. Relativity***4**,Bibcode:2001LRR.....4.. doi:10.12942/lrr-2001-6, retrieved 2007-08-08

- Walsh, D.; Carswell, R. F.; Weymann, R. J. (1979), "0957 + 561 A, B: twin quasistellar objects or gravitational lens?", *Nature* **279** (5712): 381–4, Bibcode:1979Natur.279..381W, doi:10.1038/279381a0, PMID 16068158

- Wambsganss, Joachim (1998),"Gravitational Lensing in Astronomy",*Living Rev. Relativity***1**,arXiv:astro-ph/98 Bibcode:1998LRR.....1...12W, doi:10.12942/lrr-1998-12, retrieved 2007-07-20

- Weinberg, Steven (1972), *Gravitation and Cosmology*, John Wiley, ISBN 0-471-92567-5

- Weinberg, Steven (1995), *The Quantum Theory of Fields I: Foundations*, Cambridge University Press, ISBN 0-521-55001-7

- Weinberg, Steven (1996), *The Quantum Theory of Fields II: Modern Applications*, Cambridge University Press, ISBN 0-521-55002-5

- Weinberg, Steven (2000), *The Quantum Theory of Fields III: Supersymmetry*, Cambridge University Press, ISBN 0-521-66000-9

- Weisberg, Joel M.; Taylor, Joseph H. (2003), "The Relativistic Binary Pulsar B1913+16"", in Bailes, M.; Nice, D. J.; Thorsett, S. E., *Proceedings of "Radio Pulsars," Chania, Crete, August, 2002*, ASP Conference Series

- Weiss, Achim (2006), "Elements of the past: Big Bang Nucleosynthesis and observation", *Einstein Online* (Max Planck Institute for Gravitational Physics), retrieved 2007-02-24 External link in |work= (help)

- Wheeler, John A. (1990), *A Journey Into Gravity and Spacetime*, Scientific American Library, San Francisco: W. H. Freeman, ISBN 0-7167-6034-7

- Will, Clifford M. (1993), *Theory and experiment in gravitational physics*, Cambridge University Press, ISBN 0-521-43973-6

- Will, Clifford M. (2006), "The Confrontation between General Relativity and Experiment", *Living Rev. Relativity* **9**, arXiv:gr-qc/0510072, Bibcode:2006LRR.....9....3W, doi:10.12942/lrr-2006-3, retrieved 2007-06-12

- Zwiebach, Barton (2004), *A First Course in String Theory*, Cambridge University Press, ISBN 0-521-83143-1

8.12   Further reading

Popular books

- Geroch, R (1981), *General Relativity from A to B*, Chicago: University of Chicago Press, ISBN 0-226-28864-1

- Lieber, Lillian (2008), *The Einstein Theory of Relativity: A Trip to the Fourth Dimension*, Philadelphia: Paul Dry Books, Inc., ISBN 978-1-58988-044-3

- Wald, Robert M. (1992), *Space, Time, and Gravity: the Theory of the Big Bang and Black Holes*, Chicago: University of Chicago Press, ISBN 0-226-87029-4

- Wheeler, John; Ford, Kenneth (1998), *Geons, Black Holes, & Quantum Foam: a life in physics*, New York: W. W. Norton, ISBN 0-393-31991-1

Beginning undergraduate textbooks

- Callahan, James J. (2000), *The Geometry of Spacetime: an Introduction to Special and General Relativity*, New York: Springer, ISBN 0-387-98641-3

- Taylor, Edwin F.; Wheeler, John Archibald (2000), *Exploring Black Holes: Introduction to General Relativity*, Addison Wesley, ISBN 0-201-38423-X

Advanced undergraduate textbooks

- B. F. Schutz (2009), *A First Course in General Relativity (Second Edition)*, Cambridge University Press, ISBN 978-0-521-88705-2

- Cheng, Ta-Pei (2005), *Relativity, Gravitation and Cosmology: a Basic Introduction*, Oxford and New York: Oxford University Press, ISBN 0-19-852957-0

- Gron, O.; Hervik, S. (2007), *Einstein's General theory of Relativity*, Springer, ISBN 978-0-387-69199-2

- Hartle, James B. (2003), *Gravity: an Introduction to Einstein's General Relativity*, San Francisco: Addison-Wesley, ISBN 0-8053-8662-9

- Hughston, L. P. & Tod, K. P. (1991), *Introduction to General Relativity*, Cambridge: Cambridge University Press, ISBN 0-521-33943-X

- d'Inverno, Ray (1992), *Introducing Einstein's Relativity*, Oxford: Oxford University Press, ISBN 0-19-859686-3

- Ludyk, Günter (2013). *Einstein in Matrix Form* (1st ed.). Berlin: Springer. ISBN 9783642357978.

Graduate-level textbooks

- Carroll, Sean M. (2004), *Spacetime and Geometry: An Introduction to General Relativity*, San Francisco: Addison-Wesley, ISBN 0-8053-8732-3

- Grøn, Øyvind; Hervik, Sigbjørn (2007), *Einstein's General Theory of Relativity*, New York: Springer, ISBN 978-0-387-69199-2

- Landau, Lev D.; Lifshitz, Evgeny F. (1980), *The Classical Theory of Fields (4th ed.)*, London: Butterworth-Heinemann, ISBN 0-7506-2768-9

- Misner, Charles W.; Thorne, Kip. S.; Wheeler, John A. (1973). *Gravitation*, W. H. Freeman, ISBN 0-7167-0344-0

- Stephani, Hans (1990), *General Relativity: An Introduction to the Theory of the Gravitational Field*, Cambridge: Cambridge University Press, ISBN 0-521-37941-5

- Wald, Robert M. (1984). *General Relativity*, University of Chicago Press, ISBN 0-226-87033-2

8.13 External links

- Einstein Online – Articles on a variety of aspects of relativistic physics for a general audience; hosted by the Max Planck Institute for Gravitational Physics

- NCSA Spacetime Wrinkles – produced by the numerical relativity group at the NCSA, with an elementary introduction to general relativity

- **Courses**

- **Lectures**

- **Tutorials**

- Einstein's General Theory of Relativity on YouTube (lecture by Leonard Susskind recorded September 22, 2008 at Stanford University).

- Series of lectures on General Relativity given in 2006 at the Institut Henri Poincaré (introductory/advanced).

- General Relativity Tutorials by John Baez.

- Brown, Kevin. "Reflections on relativity". *Mathpages.com*. Retrieved May 29, 2005.

- Carroll, Sean M. "Lecture Notes on General Relativity". Retrieved January 5, 2014.

- Moor, Rafi. "Understanding General Relativity". Retrieved July 11, 2006.

- Waner, Stefan. "Introduction to Differential Geometry and General Relativity" (PDF). Retrieved 2015-04-05.

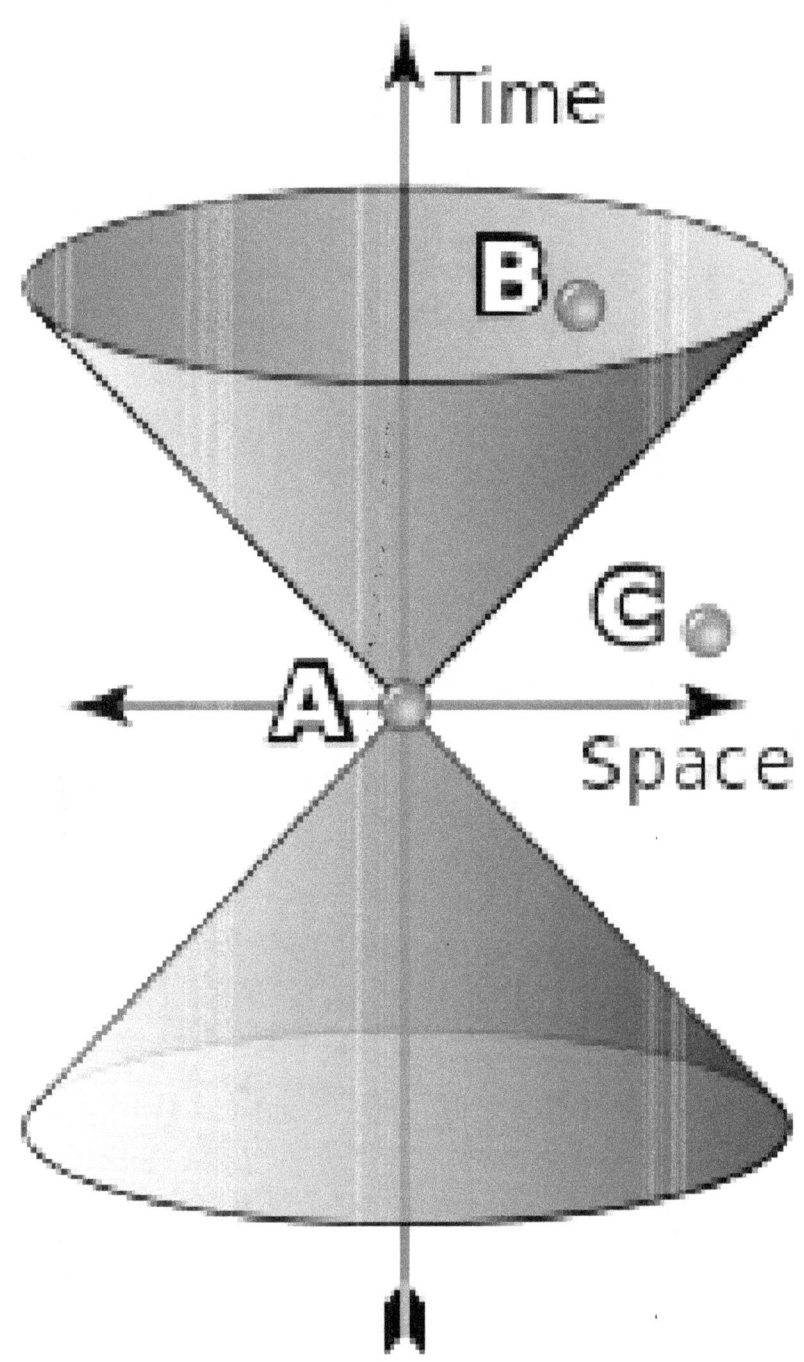

Light Cone

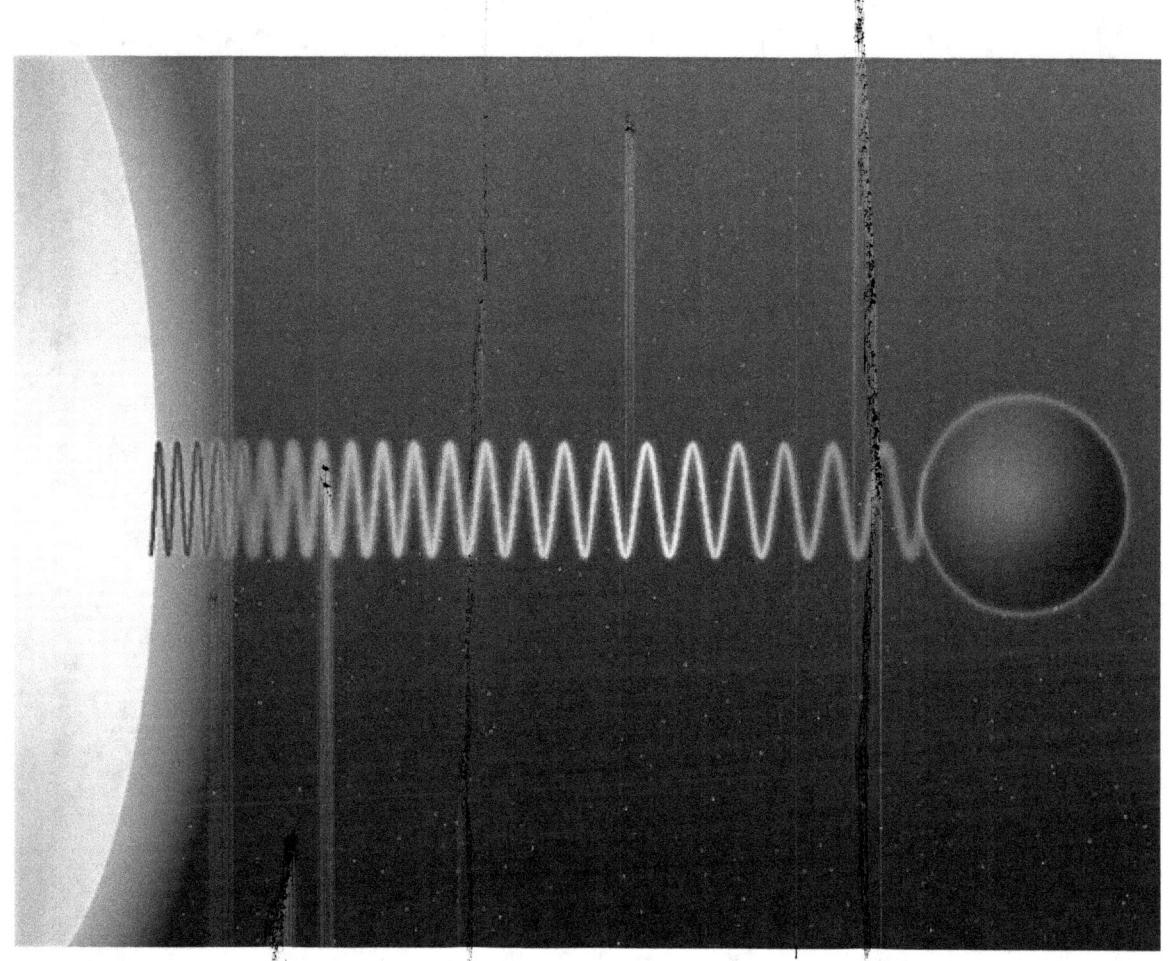

Schematic representation of the gravitational redshift of a light wave escaping from the surface of a massive body

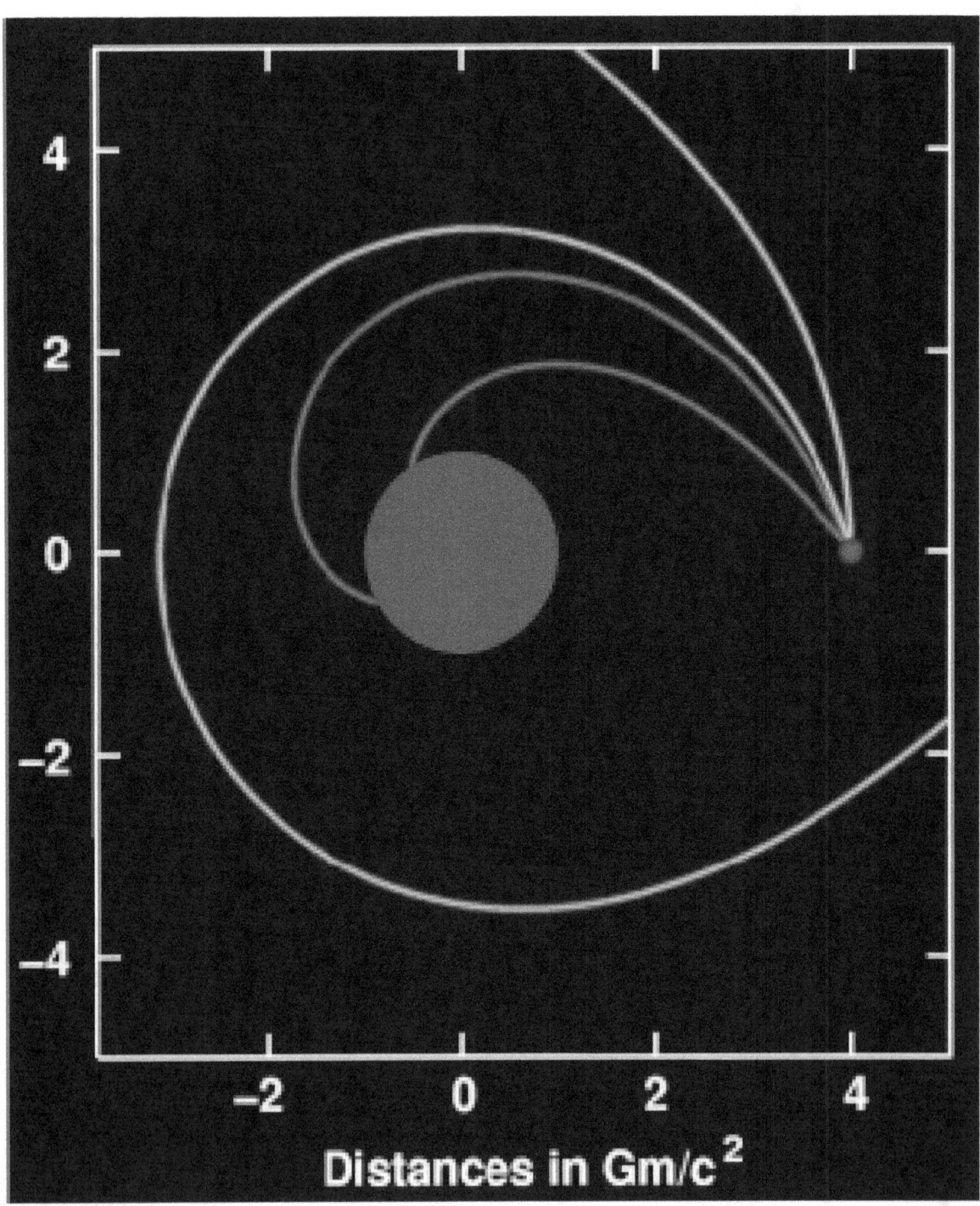

Deflection of light (sent out from the location shown in blue) near a compact body (shown in gray)

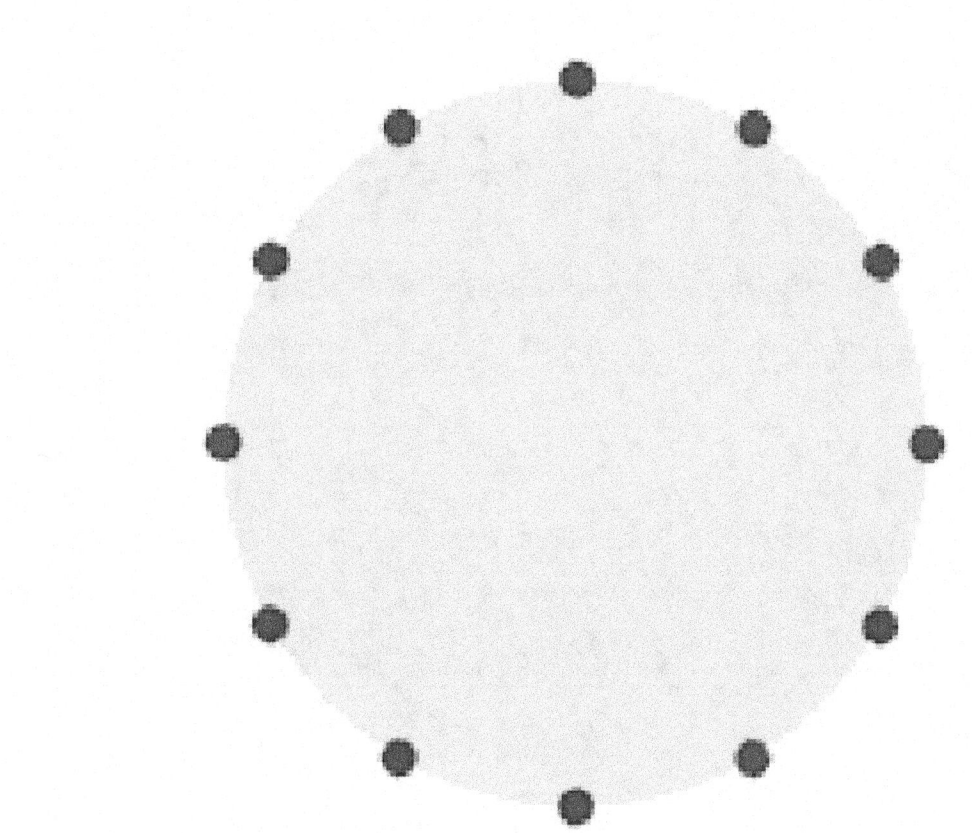

Ring of test particles influenced by gravitational wave

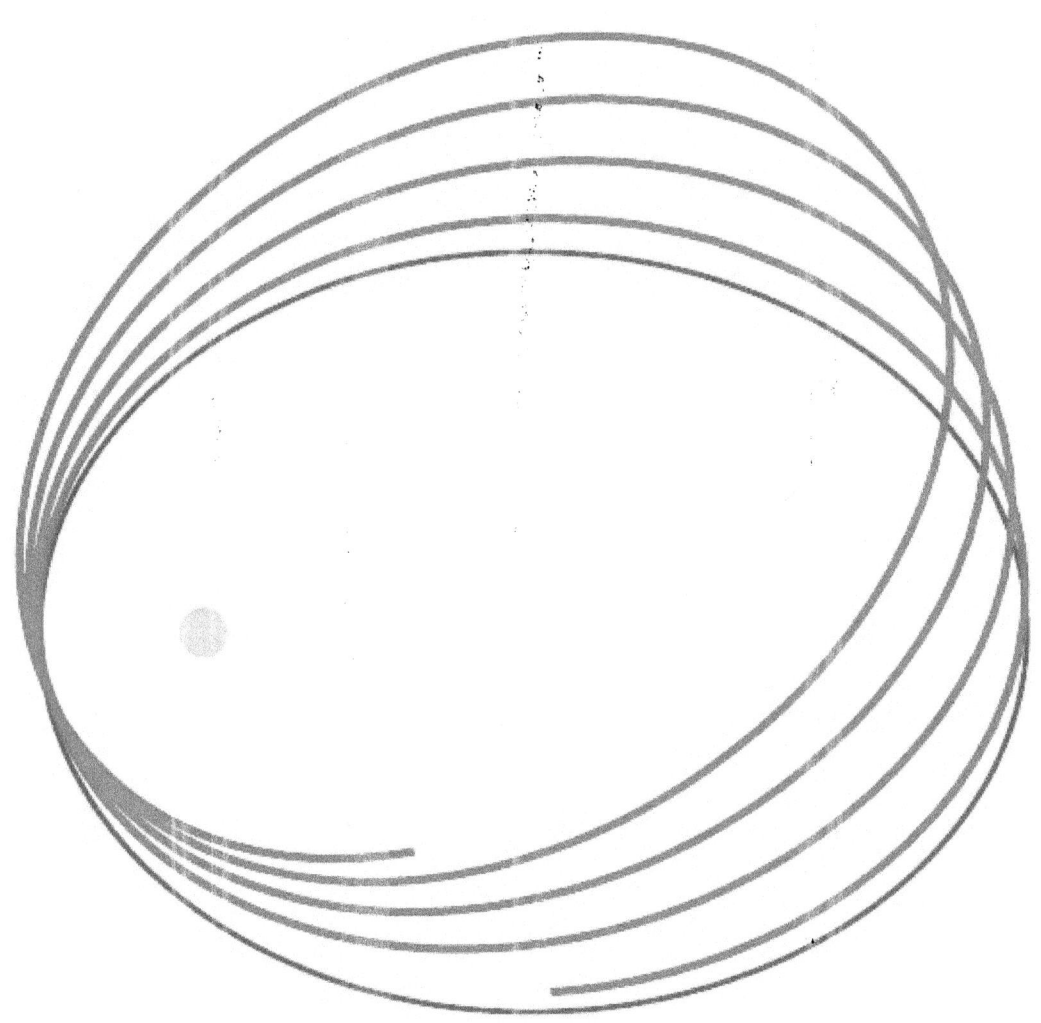

Newtonian (red) vs. Einsteinian orbit (blue) of a lone planet orbiting a star

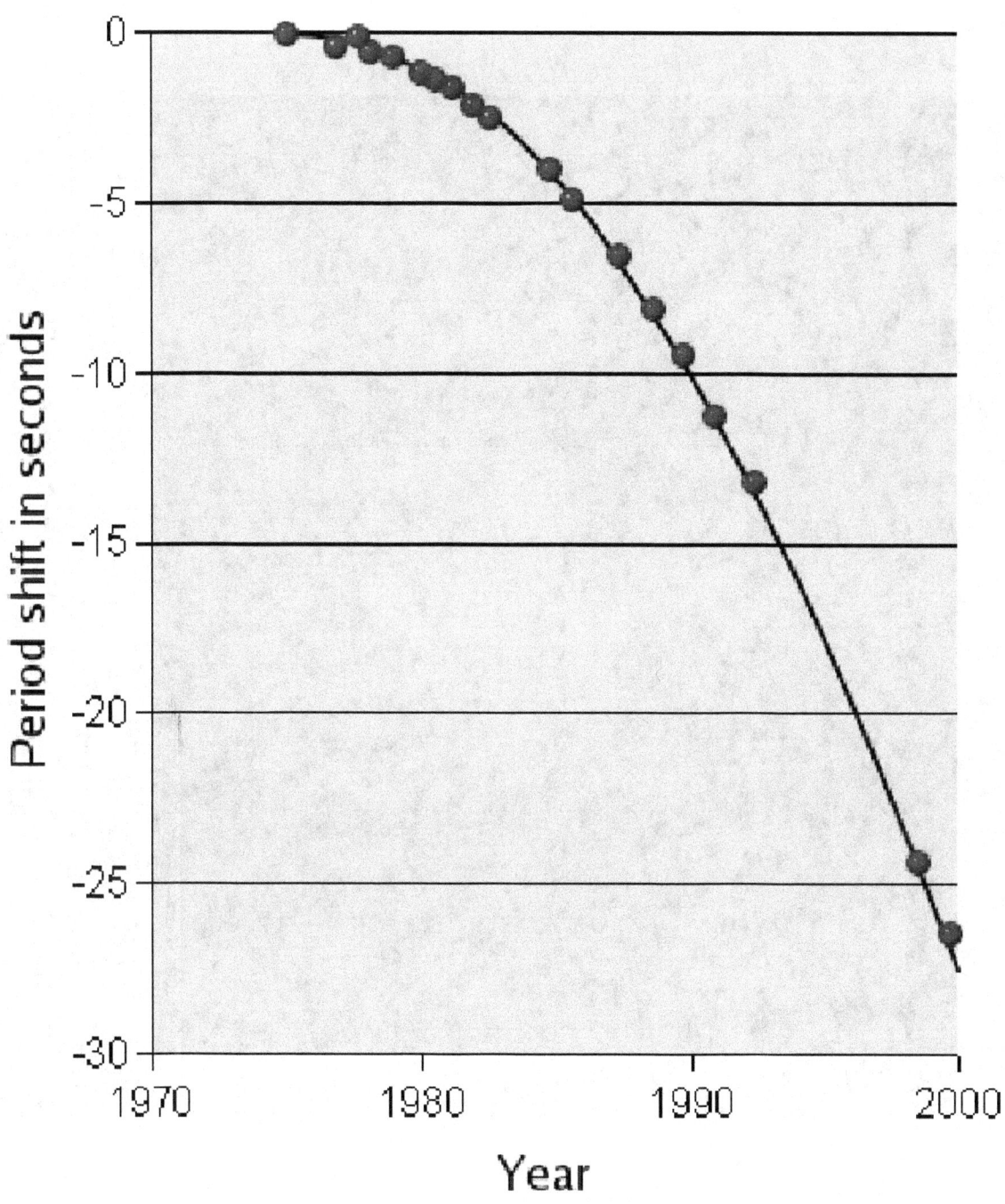

Orbital decay for PSR1913+16: time shift in seconds, tracked over three decades.[78]

Einstein cross: four images of the same astronomical object, produced by a gravitational lens

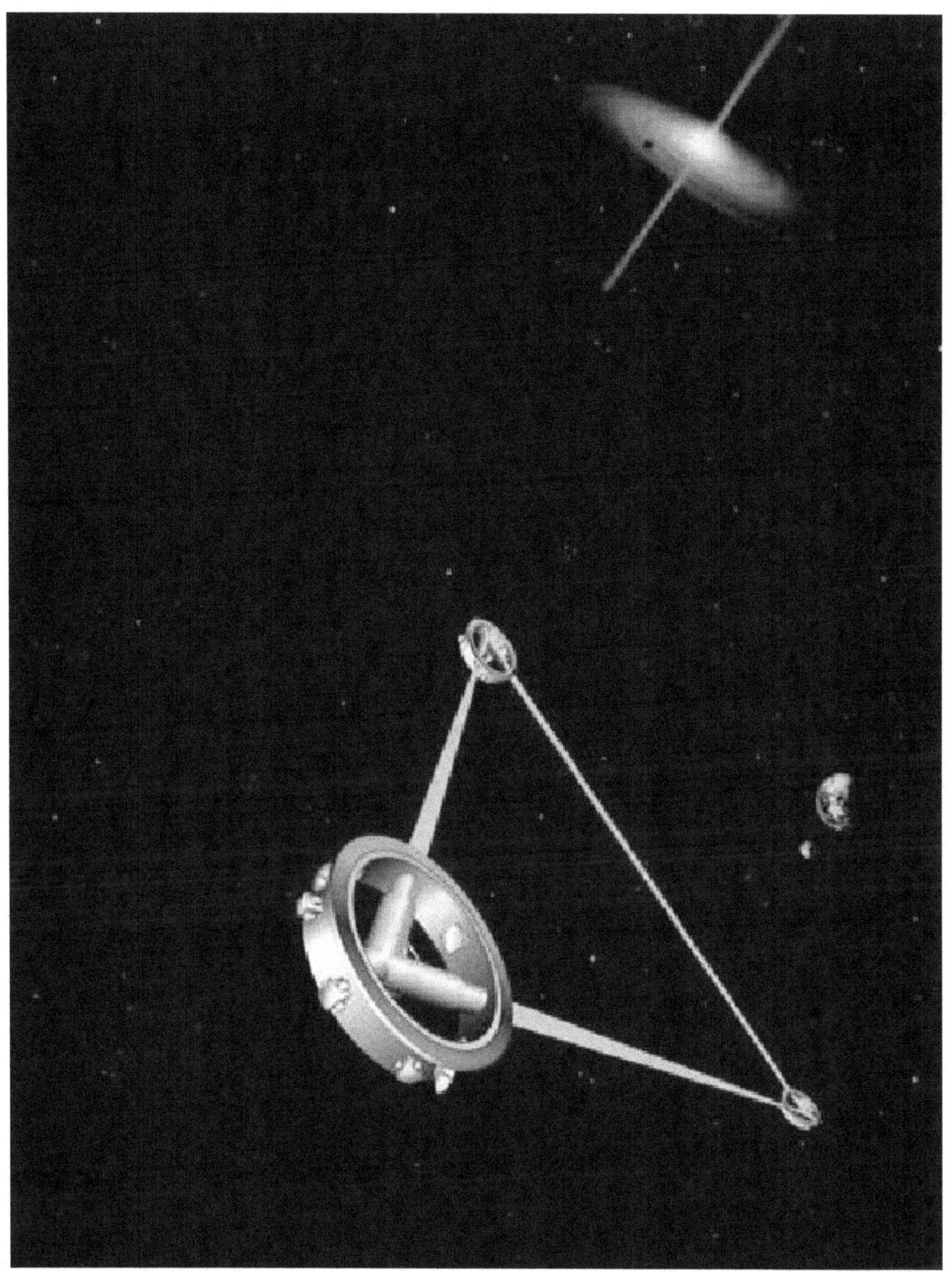

Artist's impression of the space-borne gravitational wave detector LISA

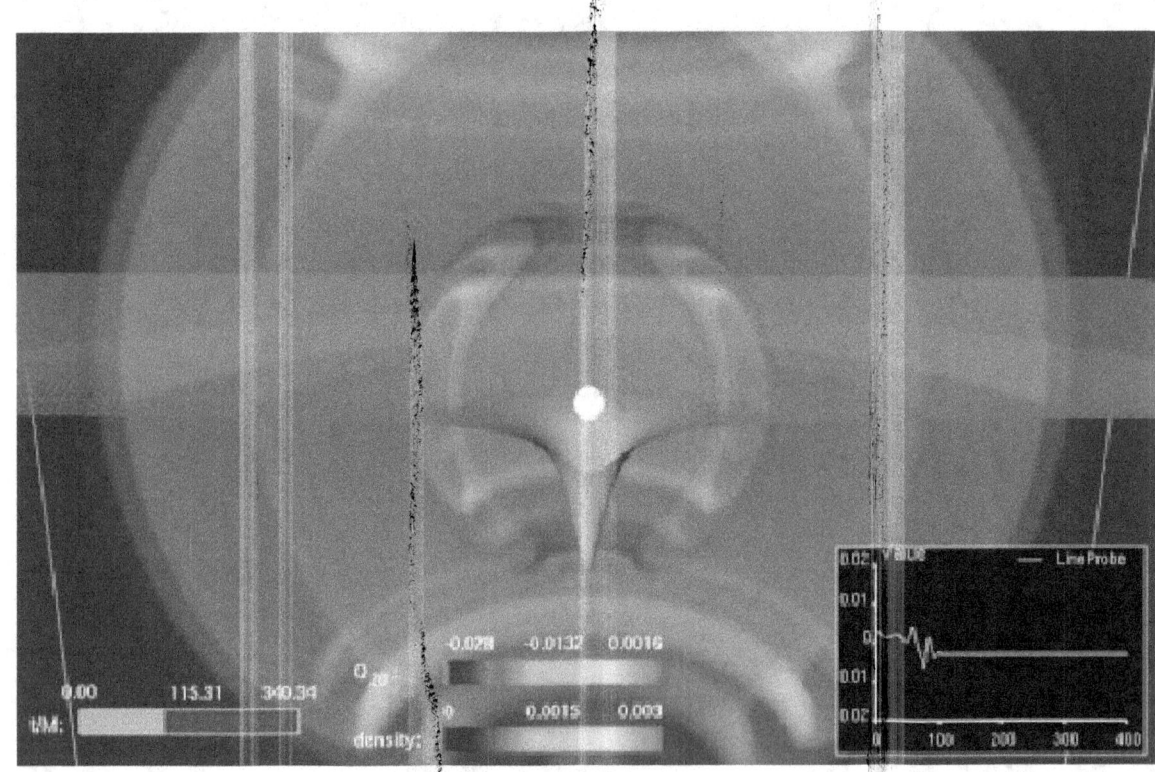

Simulation based on the equations of general relativity: a star collapsing to form a black hole while emitting gravitational waves

This blue horseshoe is a distant galaxy that has been magnified and warped into a nearly complete ring by the strong gravitational pull of the massive foreground luminous red galaxy.

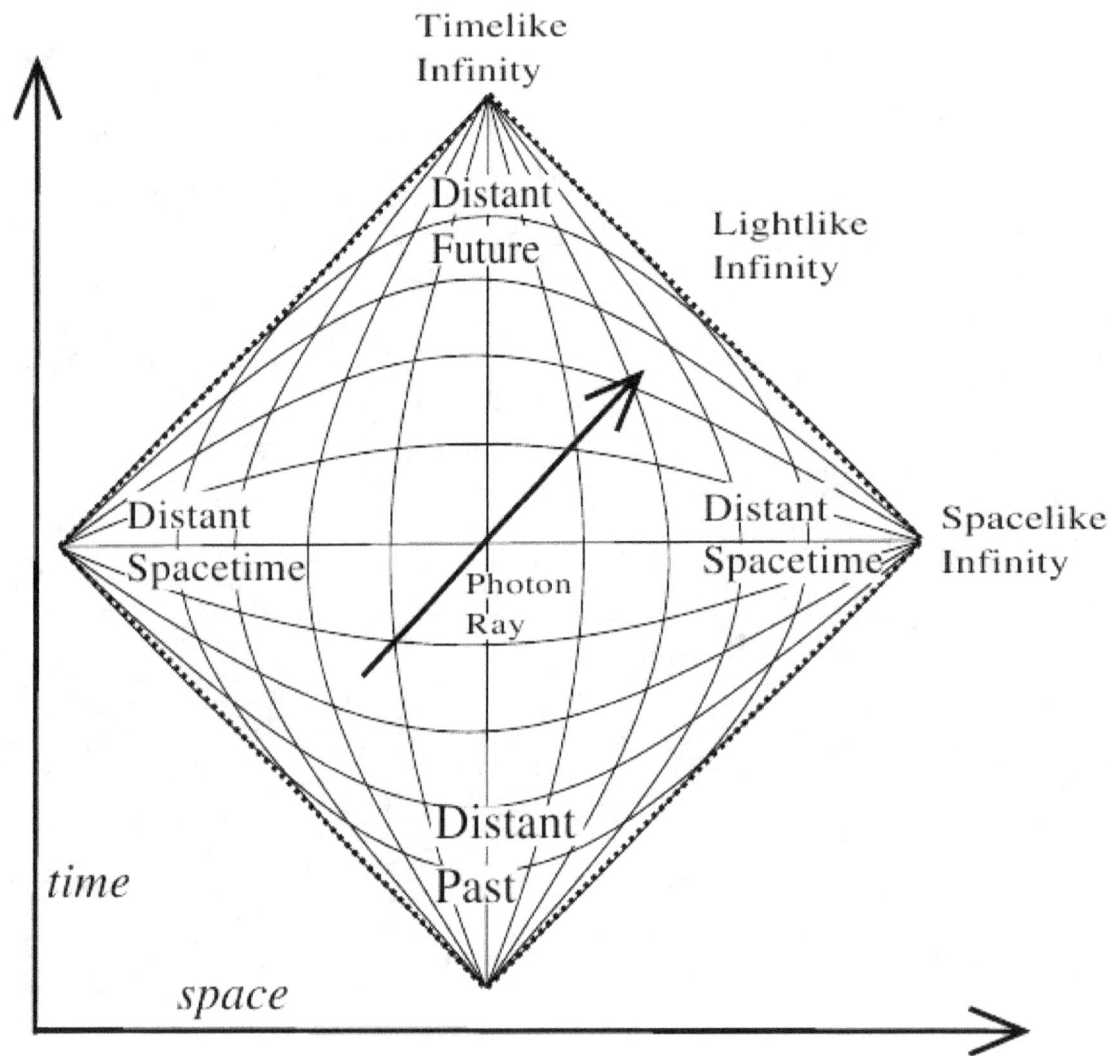

Penrose–Carter diagram of an infinite Minkowski universe

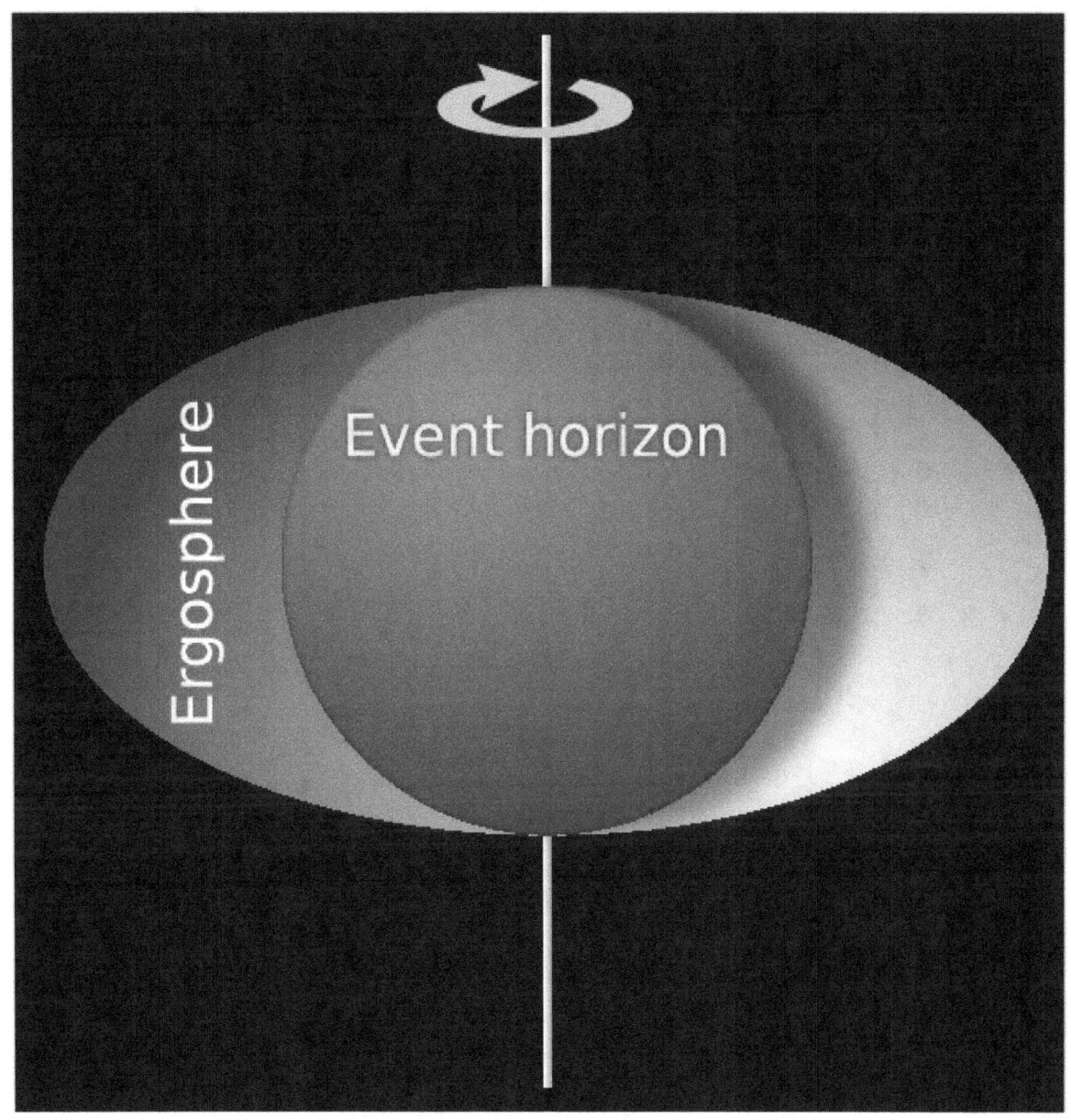

The ergosphere of a rotating black hole, which plays a key role when it comes to extracting energy from such a black hole

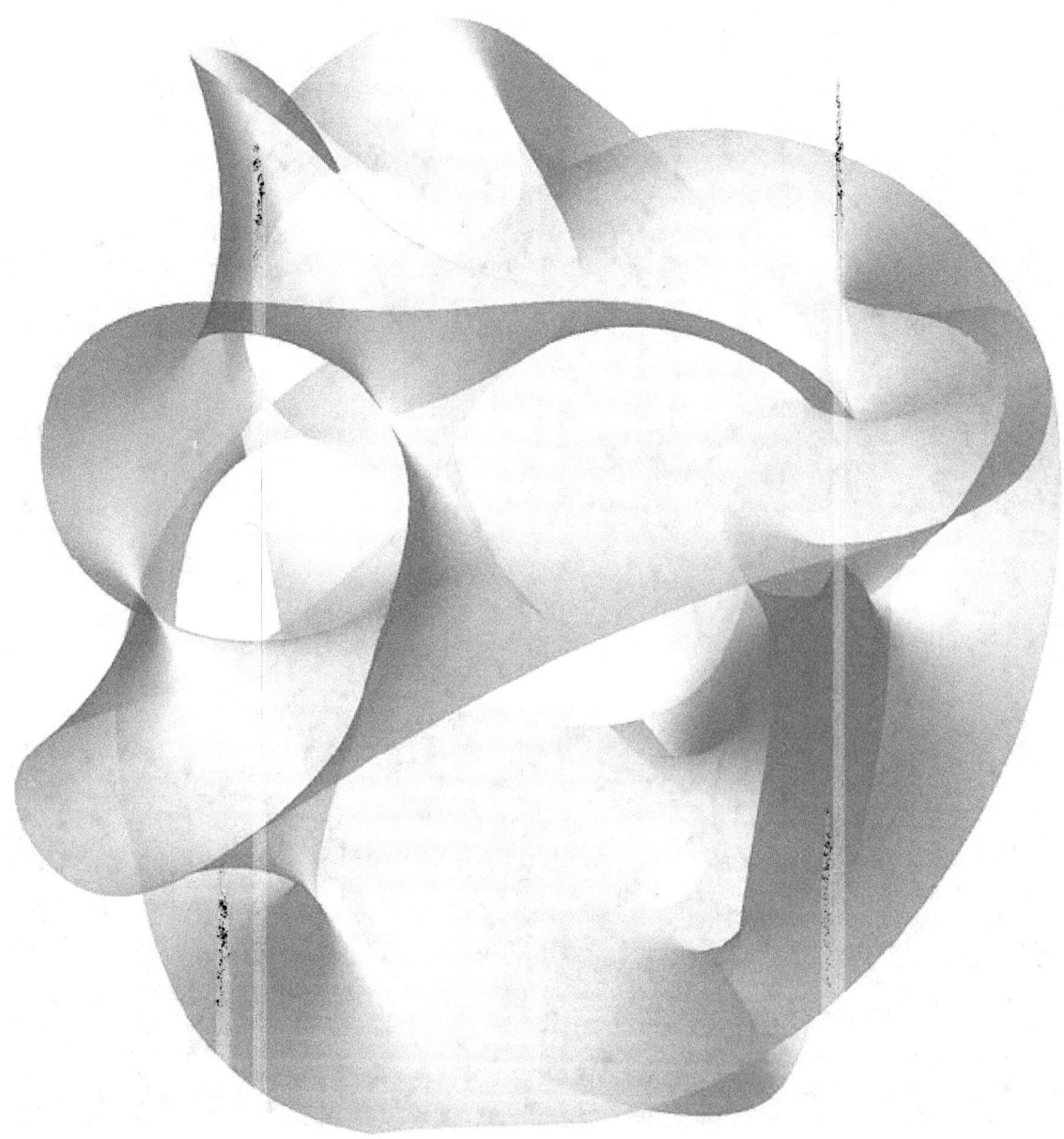

Projection of a Calabi–Yau manifold, one of the ways of compactifying the extra dimensions posited by string theory

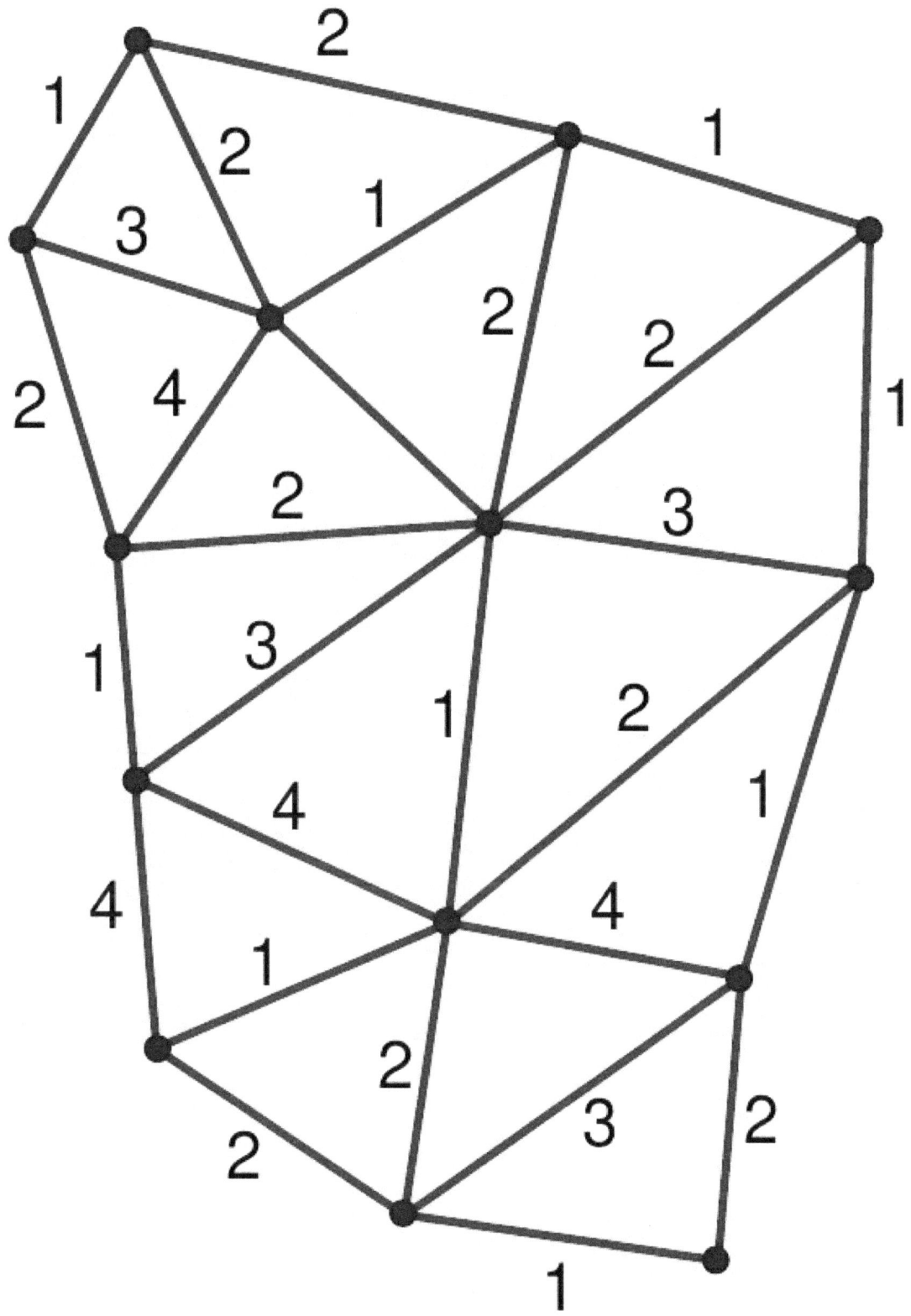

Simple spin network of the type used in loop quantum gravity

Chapter 9

Tests of general relativity

At its introduction in 1915, the general theory of relativity did not have a solid empirical foundation. It was known that it correctly accounted for the "anomalous" precession of the perihelion of Mercury and on philosophical grounds it was considered satisfying that it was able to unify Newton's law of universal gravitation with special relativity. That light appeared to bend in gravitational fields in line with the predictions of general relativity was found in 1919 but it was not until a program of precision tests was started in 1959 that the various predictions of general relativity were tested to any further degree of accuracy in the weak gravitational field limit, severely limiting possible deviations from the theory. Beginning in 1974, Hulse, Taylor and others have studied the behaviour of binary pulsars experiencing much stronger gravitational fields than found in our solar system. Both in the weak field limit (as in our solar system) and with the stronger fields present in systems of binary pulsars the predictions of general relativity have been extremely well tested locally.

The very strong gravitational fields that must be present close to black holes, especially those supermassive black holes which are thought to power active galactic nuclei and the more active quasars, belong to a field of intense active research. Observations of these quasars and active galactic nuclei are difficult, and interpretation of the observations is heavily dependent upon astrophysical models other than general relativity or competing fundamental theories of gravitation, but they are qualitatively consistent with the black hole concept as modelled in general relativity.

As a consequence of the equivalence principle, Lorentz invariance holds locally in freely falling reference frames. Experiments related to Lorentz invariance and thus special relativity (i.e., when gravitational effects can be neglected) are described in Tests of special relativity.

9.1 Classical tests

Albert Einstein proposed three tests of general relativity, subsequently called **the classical tests of general relativity**, in 1916:[1]

1. the perihelion precession of Mercury's orbit

2. the deflection of light by the Sun

3. the gravitational redshift of light

In the letter to the London Times on November 28, 1919, he described the theory of relativity and thanked his English colleagues for their understanding and testing of his work. He also mentioned three classical tests with comments:[2]

> "The chief attraction of the theory lies in its logical completeness. If a single one of the conclusions drawn from it proves wrong, it must be given up; to modify it without destroying the whole structure seems to be impossible."

9.1.1   Perihelion precession of Mercury

Transit of Mercury on November 8, 2006 with sunspots #921, 922, and 923

For more details on this topic, see Two-body problem in general relativity.

Under Newtonian physics, a two-body system consisting of a lone object orbiting a spherical mass would trace out an ellipse with the spherical mass at a focus. The point of closest approach, called the periapsis (or, as the central body in our Solar System is the sun, perihelion), is fixed. A number of effects in our solar system cause the perihelia of planets to precess (rotate) around the sun. The principal cause is the presence of other planets which perturb each other's orbit. Another (much less significant) effect is solar oblateness.

Mercury deviates from the precession predicted from these Newtonian effects. This anomalous rate of precession of the perihelion of Mercury's orbit was first recognized in 1859 as a problem in celestial mechanics, by Urbain Le Verrier. His re-analysis of available timed observations of transits of Mercury over the Sun's disk from 1697 to 1848 showed that the actual rate of the precession disagreed from that predicted from Newton's theory by 38" (arc seconds) per tropical century (later re-estimated at 43").[3] A number of *ad hoc* and ultimately unsuccessful solutions were proposed, but they tended to introduce more problems. In general relativity, this remaining precession, or change of orientation of the orbital ellipse within its orbital plane, is explained by gravitation being mediated by the curvature of spacetime. Einstein showed that general relativity[1] agrees closely with the observed amount of perihelion shift. This was a powerful factor motivating the adoption of general relativity.

Although earlier measurements of planetary orbits were made using conventional telescopes, more accurate measurements are now made with radar. The total observed precession of Mercury is 574.10±0.65 arc-seconds per century[4] relative to the inertial ICFR. This precession can be attributed to the following causes:

The correction by 42.98" is 3/2 multiple of classical prediction with PPN parameters $\gamma = \beta = 0$.[6]

Thus the effect can be fully explained by general relativity. More recent calculations based on more precise measurements have not materially changed the situation.

The other planets experience perihelion shifts as well, but, since they are farther from the sun and have longer periods, their shifts are lower, and could not be observed accurately until long after Mercury's. For example, the perihelion shift of Earth's orbit due to general relativity is of 3.84 seconds of arc per century, and Venus's is 8.62". Both values are in good agreement with observation.[7] The periapsis shift of binary pulsar systems have been measured, with PSR 1913+16 amounting to 4.2° per year.[8] These observations are consistent with general relativity.[9] It is also possible to measure periapsis shift in binary star systems which do not contain ultra-dense stars, but it is more difficult to model the classical effects precisely - for example, the alignment of the stars' spin to their orbital plane needs to be known and is hard to measure directly - so a few systems such as DI Herculis have been considered as problematic cases for general relativity.

9.1.2 Deflection of light by the Sun

For more details on this topic, see Kepler problem in general relativity.

Henry Cavendish in 1784 (in an unpublished manuscript) and Johann Georg von Soldner in 1801 (published in 1804) had pointed out that Newtonian gravity predicts that starlight will bend around a massive object.[10] The same value as Soldner's was calculated by Einstein in 1911 based on the equivalence principle alone. However, Einstein noted in 1915 in the process of completing general relativity, that his (and thus Soldner's) 1911-result is only half of the correct value. Einstein became the first to calculate the correct value for light bending.[11]

The first observation of light deflection was performed by noting the change in position of stars as they passed near the Sun on the celestial sphere. The observations were performed in May 1919 by Arthur Eddington and his collaborators during a total solar eclipse,[12] so that the stars near the Sun could be observed. Observations were made simultaneously in the cities of Sobral, Ceará, Brazil and in São Tomé and Príncipe on the west coast of Africa.[13] The result was considered spectacular news and made the front page of most major newspapers. It made Einstein and his theory of general relativity world-famous. When asked by his assistant what his reaction would have been if general relativity had not been confirmed by Eddington and Dyson in 1919, Einstein famously made the quip: "Then I would feel sorry for the dear Lord. The theory is correct anyway." [14]

The early accuracy, however, was poor. The results were argued by some[15] to have been plagued by systematic error and possibly confirmation bias, although modern reanalysis of the dataset[16] suggests that Eddington's analysis was accurate.[17][18] The measurement was repeated by a team from the Lick Observatory in the 1922 eclipse, with results that agreed with the 1919 results[18] and has been repeated several times since, most notably in 1953 by Yerkes Observatory astronomers[19] and in 1973 by a team from the University of Texas.[20] Considerable uncertainty remained in these measurements for almost fifty years, until observations started being made at radio frequencies. It was not until the 1960s that it was definitively accepted that the amount of deflection was the full value predicted by general relativity, and not half that number. The Einstein ring is an example of the deflection of light from distant galaxies by more nearby objects.

9.1.3 Gravitational redshift of light

Einstein predicted the gravitational redshift of light from the equivalence principle in 1907, but it is very difficult to measure astrophysically (see the discussion under *Equivalence Principle* below). Although it was measured by Walter

One of Eddington's photographs of the 1919 solar eclipse experiment, presented in his 1920 paper announcing its success

Sydney Adams in 1925, it was only conclusively tested when the Pound–Rebka experiment in 1959 measured the relative redshift of two sources situated at the top and bottom of Harvard University's Jefferson tower using an extremely sensitive

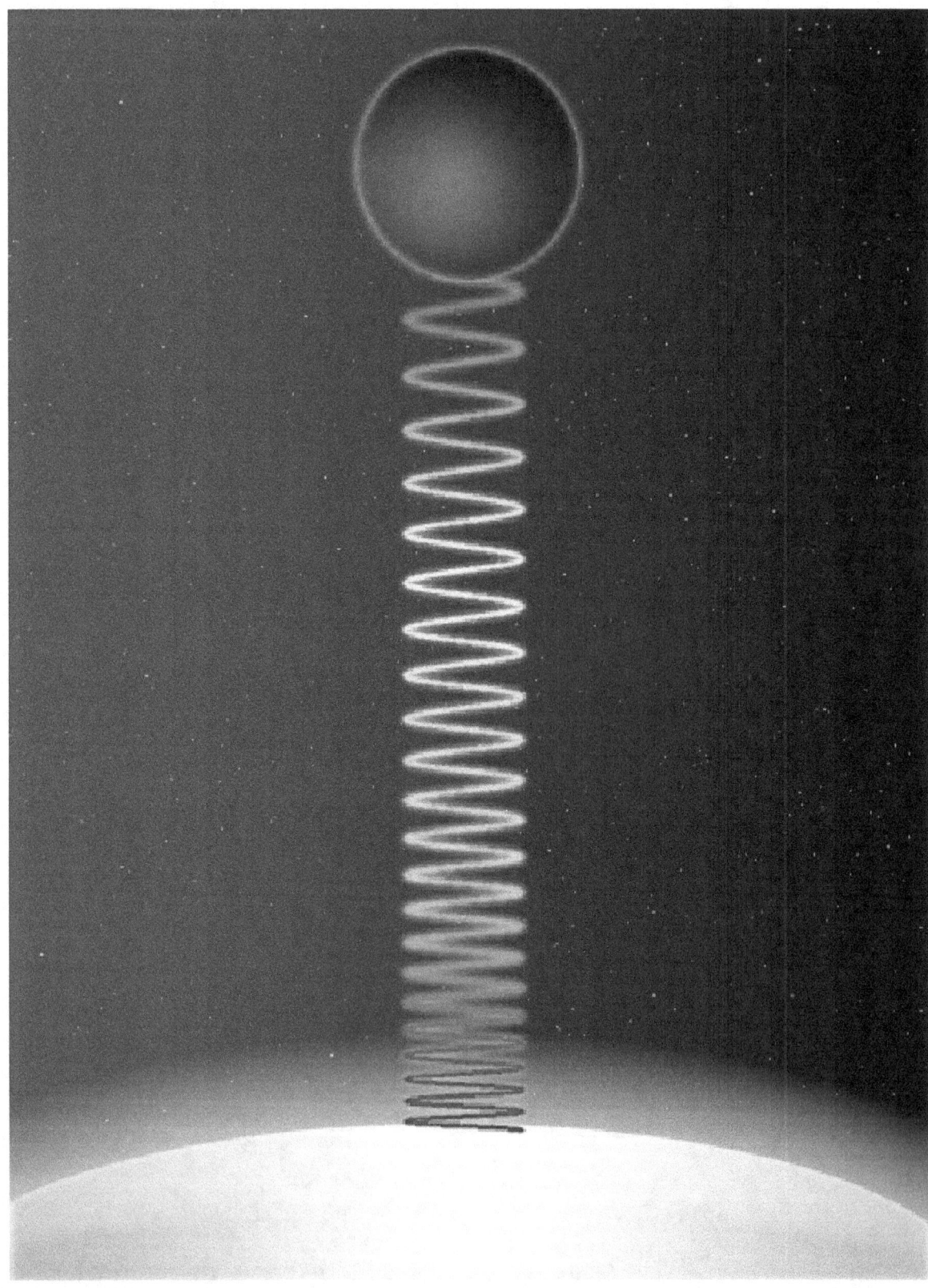

The gravitational redshift of a light wave as it moves upwards against a gravitational field (caused by the yellow star below).

phenomenon called the Mössbauer effect.[21][22] The result was in excellent agreement with general relativity. This was one of the first precision experiments testing general relativity.

9.2   Modern tests

The modern era of testing general relativity was ushered in largely at the impetus of Dicke and Schiff who laid out a framework for testing general relativity.[23][24][25] They emphasized the importance not only of the classical tests, but of null experiments, testing for effects which in principle could occur in a theory of gravitation, but do not occur in general relativity. Other important theoretical developments included the inception of alternative theories to general relativity, in particular, scalar-tensor theories such as the Brans–Dicke theory;[26] the parameterized post-Newtonian formalism in which deviations from general relativity can be quantified; and the framework of the equivalence principle.

Experimentally, new developments in space exploration, electronics and condensed matter physics have made additional precise experiments possible, such as the Pound–Rebka experiment, laser interferometry and lunar rangefinding.

9.2.1   Post-Newtonian tests of gravity

Early tests of general relativity were hampered by the lack of viable competitors to the theory: it was not clear what sorts of tests would distinguish it from its competitors. General relativity was the only known relativistic theory of gravity compatible with special relativity and observations. Moreover, it is an extremely simple and elegant theory. This changed with the introduction of Brans–Dicke theory in 1960. This theory is arguably simpler, as it contains no dimensionful constants, and is compatible with a version of Mach's principle and Dirac's large numbers hypothesis, two philosophical ideas which have been influential in the history of relativity. Ultimately, this led to the development of the parametrized post-Newtonian formalism by Nordtvedt and Will, which parametrizes, in terms of ten adjustable parameters, all the possible departures from Newton's law of universal gravitation to first order in the velocity of moving objects (*i.e.* to first order in v/c, where v is the velocity of an object and c is the speed of light). This approximation allows the possible deviations from general relativity, for slowly moving objects in weak gravitational fields, to be systematically analyzed. Much effort has been put into constraining the post-Newtonian parameters, and deviations from general relativity are at present severely limited.

The experiments testing gravitational lensing and light time delay limits the same post-Newtonian parameter, the so-called Eddington parameter γ, which is a straightforward parametrization of the amount of deflection of light by a gravitational source. It is equal to one for general relativity, and takes different values in other theories (such as Brans–Dicke theory). It is the best constrained of the ten post-Newtonian parameters, but there are other experiments designed to constrain the others. Precise observations of the perihelion shift of Mercury constrain other parameters, as do tests of the strong equivalence principle.

One of the goals of the mission BepiColombo is testing the general relativity theory by measuring the parameters gamma and beta of the parametrized post-Newtonian formalism with high accuracy.[27]

9.2.2   Gravitational lensing

One of the most important tests is gravitational lensing. It has been observed in distant astrophysical sources, but these are poorly controlled and it is uncertain how they constrain general relativity. The most precise tests are analogous to Eddington's 1919 experiment: they measure the deflection of radiation from a distant source by the sun. The sources that can be most precisely analyzed are distant radio sources. In particular, some quasars are very strong radio sources. The directional resolution of any telescope is in principle limited by diffraction; for radio telescopes this is also the practical limit. An important improvement in obtaining positional high accuracies (from milli-arcsecond to micro-arcsecond) was obtained by combining radio telescopes across the Earth. The technique is called very long baseline interferometry (VLBI). With this technique radio observations couple the phase information of the radio signal observed in telescopes separated over large distances. Recently, these telescopes have measured the deflection of radio waves by the Sun to extremely high precision, confirming the amount of deflection predicted by general relativity aspect to the 0.03% level.[28] At this level of precision systematic effects have to be carefully taken into account to determine the precise location of the telescopes on Earth. Some important effects are the Earth's nutation, rotation, atmospheric refraction, tectonic displacement and tidal waves. Another important effect is refraction of the radio waves by the solar corona. Fortunately, this effect has a characteristic spectrum, whereas gravitational distortion is independent of wavelength. Thus, careful analysis, using measurements at several frequencies, can subtract this source of error.

The entire sky is slightly distorted due to the gravitational deflection of light caused by the Sun (the anti-Sun direction excepted). This effect has been observed by the European Space Agency astrometric satellite Hipparcos. It measured the positions of about 10^5 stars. During the full mission about 3.5×10^6 relative positions have been determined, each to an accuracy of typically 3 milliarcseconds (the accuracy for an 8–9 magnitude star). Since the gravitation deflection perpendicular to the Earth-Sun direction is already 4.07 milliarcseconds, corrections are needed for practically all stars. Without systematic effects, the error in an individual observation of 3 milliarcseconds, could be reduced by the square root of the number of positions, leading to a precision of 0.0016 milliarcseconds. Systematic effects, however, limit the accuracy of the determination to 0.3% (Froeschlé, 1997).

In future, Gaia spacecraft will conduct a census of one billion stars in our Galaxy and measure their positions to an accuracy of 24 microarcseconds. Thus it will also provide stringent new tests of gravitational deflection of light caused by the Sun which was predicted by General relativity.[29]

9.2.3 Light travel time delay testing

Irwin I. Shapiro proposed another test, beyond the classical tests, which could be performed within the solar system. It is sometimes called the fourth "classical" test of general relativity. He predicted a relativistic time delay (Shapiro delay) in the round-trip travel time for radar signals reflecting off other planets.[30] The mere curvature of the path of a photon passing near the Sun is too small to have an observable delaying effect (when the round-trip time is compared to the time taken if the photon had followed a straight path), but general relativity predicts a time delay which becomes progressively larger when the photon passes nearer to the Sun due to the time dilation in the gravitational potential of the sun. Observing radar reflections from Mercury and Venus just before and after it will be eclipsed by the Sun gives agreement with general relativity theory at the 5% level.[31] More recently, the Cassini probe has undertaken a similar experiment which gave agreement with general relativity at the 0.002% level.[32] Very Long Baseline Interferometry has measured velocity-dependent (gravitomagnetic) corrections to the Shapiro time delay in the field of moving Jupiter [33][34] and Saturn.[35]

9.2.4 The equivalence principle

Main article: Equivalence principle

The equivalence principle, in its simplest form, asserts that the trajectories of falling bodies in a gravitational field should be independent of their mass and internal structure, provided they are small enough not to disturb the environment or be affected by tidal forces. This idea has been tested to incredible precision by Eötvös torsion balance experiments, which look for a differential acceleration between two test masses. Constraints on this, and on the existence of a composition-dependent fifth force or gravitational Yukawa interaction are very strong, and are discussed under fifth force and weak equivalence principle.

A version of the equivalence principle, called the strong equivalence principle, asserts that self-gravitation falling bodies, such as stars, planets or black holes (which are all held together by their gravitational attraction) should follow the same trajectories in a gravitational field, provided the same conditions are satisfied. This is called the Nordtvedt effect and is most precisely tested by the Lunar Laser Ranging Experiment.[36][37] Since 1969, it has continuously measured the distance from several rangefinding stations on Earth to reflectors on the Moon to approximately centimeter accuracy.[38] These have provided a strong constraint on several of the other post-Newtonian parameters.

Another part of the strong equivalence principle is the requirement that Newton's gravitational constant be constant in time, and have the same value everywhere in the universe. There are many independent observations limiting the possible variation of Newton's gravitational constant,[39] but one of the best comes from lunar rangefinding which suggests that the gravitational constant does not change by more than one part in 10^{11} per year. The constancy of the other constants is discussed in the Einstein equivalence principle section of the equivalence principle article.

Gravitational redshift

The first of the classical tests discussed above, the gravitational redshift, is a simple consequence of the Einstein equivalence principle and was predicted by Einstein in 1907. As such, it is not a test of general relativity in the same way as the post-Newtonian tests, because any theory of gravity obeying the equivalence principle should also incorporate the gravitational redshift. Nonetheless, confirming the existence of the effect was an important substantiation of relativistic gravity, since the absence of gravitational redshift would have strongly contradicted relativity. The first observation of the gravitational redshift was the measurement of the shift in the spectral lines from the white dwarf star Sirius B by Adams in 1925. Although this measurement, as well as later measurements of the spectral shift on other white dwarf stars, agreed with the prediction of relativity, it could be argued that the shift could possibly stem from some other cause, and hence experimental verification using a known terrestrial source was preferable.

Experimental verification of gravitational redshift using terrestrial sources took several decades, because it is difficult to find clocks (to measure time dilation) or sources of electromagnetic radiation (to measure redshift) with a frequency that is known well enough that the effect can be accurately measured. It was confirmed experimentally for the first time in 1960 using measurements of the change in wavelength of gamma-ray photons generated with the Mössbauer effect, which generates radiation with a very narrow line width. The experiment, performed by Pound and Rebka and later improved by Pound and Snyder, is called the Pound–Rebka experiment. The accuracy of the gamma-ray measurements was typically 1%. The blueshift of a falling photon can be found by assuming it has an equivalent mass based on its frequency $E = hf$ (where h is Planck's constant) along with $E = mc^2$, a result of special relativity. Such simple derivations ignore the fact that in general relativity the experiment compares clock rates, rather than energies. In other words, the "higher energy" of the photon after it falls can be equivalently ascribed to the slower running of clocks deeper in the gravitational potential well. To fully validate general relativity, it is important to also show that the rate of arrival of the photons is greater than the rate at which they are emitted. A very accurate gravitational redshift experiment, which deals with this issue, was performed in 1976,[40] where a hydrogen maser clock on a rocket was launched to a height of 10,000 km, and its rate compared with an identical clock on the ground. It tested the gravitational redshift to 0.007%.

Although the Global Positioning System (GPS) is not designed as a test of fundamental physics, it must account for the gravitational redshift in its timing system, and physicists have analyzed timing data from the GPS to confirm other tests. When the first satellite was launched, some engineers resisted the prediction that a noticeable gravitational time dilation would occur, so the first satellite was launched without the clock adjustment that was later built into subsequent satellites. It showed the predicted shift of 38 microseconds per day. This rate of discrepancy is sufficient to substantially impair function of GPS within hours if not accounted for. An excellent account of the role played by general relativity in the design of GPS can be found in Ashby 2003.

Other precision tests of general relativity,[41] not discussed here, are the Gravity Probe A satellite, launched in 1976, which showed gravity and velocity affect the ability to synchronize the rates of clocks orbiting a central mass; the Hafele–Keating experiment, which used atomic clocks in circumnavigating aircraft to test general relativity and special relativity together;[42][43] and the forthcoming Satellite Test of the Equivalence Principle.

9.2.5   Frame-dragging tests

Main article: Frame-dragging

Tests of the Lense–Thirring precession, consisting of small secular precessions of the orbit of a test particle in motion around a central rotating mass like, e.g., a planet or a star, have been performed with the LAGEOS satellites,[44] but many aspects of them remain controversial.[45] The same effect may have been detected in the data of the Mars Global Surveyor (MGS) spacecraft,[46] a former probe in orbit around Mars; also such a test raised a debate.[47][48] First attempts to detect the Sun's Lense–Thirring effect on the perihelia of the inner planets have been recently reported[49] as well. Frame dragging would cause the orbital plane of stars orbiting near a supermassive black hole to precess about the black hole spin axis. This effect should be detectable within the next few years via astrometric monitoring of stars at the center of the Milky Way galaxy.[50] By comparing the rate of orbital precession of two stars on different orbits, it is possible in

The LAGEOS-1 satellite. (D=60 cm)

principle to test the no-hair theorems of general relativity.[51]

The Gravity Probe B satellite, launched in 2004 and operated until 2005, detected frame-dragging and the geodetic effect. The experiment used four quartz spheres the size of ping pong balls coated with a superconductor. Data analysis continued through 2011 due to high noise levels and difficulties in modelling the noise accurately so that a useful signal could be found. Principal investigators at Stanford University reported on May 4, 2011, that they had accurately measured the framing effect relative to the distant star IM Pegasi, and the calculations proved to be in line with the prediction of Einstein's theory. The results, published in *Physical Review Letters* measured the geodetic effect with an error of about 0.2 percent. The results reported the frame dragging effect (caused by the Earth's rotation) added up to 37 milliarcseconds with an error of about 19 percent.[52] Investigator Francis Everitt explained that a milliarcsecond "is the width of a human hair seen at the distance of 10 miles".[53]

In January 2012, LARES satellite was launched on a Vega rocket[54] to measure Lense–Thirring effect with an accuracy of about 1%, according to its proponents.[55] This evaluation of the actual accuracy obtainable is a subject of debate.[56][57][58][59][60]

9.3   Strong field tests: Binary pulsars

Further information: Binary pulsar

Pulsars are rapidly rotating neutron stars which emit regular radio pulses as they rotate. As such they act as clocks

Artist's impression of the pulsar PSR J0348+0432 and its white dwarf companion.[61]

which allow very precise monitoring of their orbital motions. Observations of pulsars in orbit around other stars have all demonstrated substantial periapsis precessions that cannot be accounted for classically but can be accounted for by using general relativity. For example, the Hulse–Taylor binary pulsar PSR B1913+16 (a pair of neutron stars in which one is detected as a pulsar) has an observed precession of over 4° of arc per year (periastron shift per orbit only about 10^{-6}). This precession has been used to compute the masses of the components.

Similarly to the way in which atoms and molecules emit electromagnetic radiation, a gravitating mass that is in quadrupole type or higher order vibration, or is asymmetric and in rotation, can emit gravitational waves.[62] These gravitational waves are predicted to travel at the speed of light. For example, planets orbiting the Sun constantly lose energy via gravitational radiation, but this effect is so small that it is unlikely it will be observed in the near future (Earth radiates about 200 watts (see gravitational waves) of gravitational radiation).

Gravitational waves have been indirectly detected from the Hulse–Taylor binary (and other binary pulsars).[63] Precise timing of the pulses shows that the stars orbit only approximately according to Kepler's Laws: over time they gradually spiral towards each other, demonstrating an energy loss in close agreement with the predicted energy radiated by gravitational waves.[64][65] Thus, although the waves have not been directly measured, their effect is necessary to explain these orbits. For their discovery of the first binary pulsar and measuring its orbital decay due to gravitational-wave emission, Hulse and Taylor won the 1993 Nobel Prize in Physics.[66]

A "double pulsar" discovered in 2003, PSR J0737-3039, has a periastron precession of 16.90° per year; unlike the Hulse–Taylor binary, both neutron stars are detected as pulsars, allowing precision timing of both members of the system. Due to this, the tight orbit, the fact that the system is almost edge-on, and the very low transverse velocity of the system as seen from Earth, J0737−3039 provides by far the best system for strong-field tests of general relativity known so far. Several distinct relativistic effects are observed, including orbital decay as in the Hulse–Taylor system. After observing the system for two and a half years, four independent tests of general relativity were possible, the most precise (the Shapiro delay) confirming the general relativity prediction within 0.05%[67] (nevertheless the periastron shift per orbit is only about 0.0013% of a circle and thus it is not a higher-order relativity test).

In 2013, an international team of astronomers report new data from observing a pulsar-white dwarf system PSR J0348+0432, in which they have been able to measure a change in the orbital period 8 millionths of a second per year, and confirmed GR predictions in a regime of extreme gravitational fields never probed before;[68] but there are still some competing theories that would agree with these data.[69]

9.4 Direct detection of gravitational waves

As described above, binary pulsar observations have shown conclusively, although indirectly, that gravitational waves exist. A number of gravitational-wave detectors have been built with the intent of directly detecting the gravitational waves emanating from such astronomical events as the merger of two neutron stars or black holes. Currently, the most sensitive of these is the Laser Interferometer Gravitational-wave Observatory (LIGO), which was in operation from 2002 to 2010. To date, there has not been a single detection event by any of the existing detectors. Future detectors are being developed or planned, which will greatly improve the sensitivity of these experiments, such as the Advanced LIGO detector due to start operation in 2015, and the proposed evolved Laser Interferometer Space Antenna (eLISA). It is anticipated, for example, that Advanced LIGO will detect events possibly as often as daily, but there is great uncertainty in this (possibly a factor of 1000).[70]

General relativity predicts gravitational waves, as does any theory of gravitation that obeys special relativity and so has changes in the gravitational field propagate at a finite speed.[71] Continued failure to find waves as the detectors become more sensitive would tend to falsify both special and general relativity. However, because of the uncertainty in astrophysical event rates, a lack of detection is more likely to be initially attributed to a previous over-estimation of how many detectable gravitational-wave signals there should be — for example, it would not be surprising if Advanced LIGO operates for three years and detects nothing if detectable signals only occur on average once every ten years — and so update our understanding of star formation, stellar evolution or galaxy evolution. If, in the future, gravitational waves (of the predicted kind) were discovered, this would be evidence in favour of general relativity.

Once gravitational waves can be directly detected, it is possible to use them to learn about the Universe. This is gravitational-wave astronomy. Gravitational-wave astronomy can test general relativity by verifying that the observed waves are of the form predicted (for example, that they only have two transverse polarizations), and by checking that black holes are the objects described by solutions of the Einstein field equations.[72][73]

9.5 Cosmological tests

Tests of general relativity on the largest scales are not nearly so stringent as solar system tests.[74] The earliest such test was prediction and discovery of the expansion of the universe.[75] In 1922 Alexander Friedmann found that Einstein equations have non-stationary solutions (even in the presence of the cosmological constant).[76][77] In 1927 Georges Lemaître showed that static solutions of the Einstein equations, which are possible in the presence of the cosmological constant, are unstable, and therefore the static universe envisioned by Einstein could not exist (it must either expand or contract).[76] Lemaître made an explicit prediction that the universe should expand.[78] He also derived a redshift-distance relationship, which is now known as the Hubble Law.[78] Later, in 1931, Einstein himself agreed with the results of Friedmann and Lemaître.[76] The expansion of the universe discovered by Edwin Hubble in 1929[76] was then considered by many (and continues to be considered by some now) as a direct confirmation of general relativity.[79] In the 1930s, largely due to the work of E. A. Milne, it was realised that the linear relationship between redshift and distance derives from the general assumption of uniformity and isotropy rather than specifically from general relativity.[75] However the prediction of a non-static universe was non-trivial, indeed dramatic, and primarily motivated by general relativity.[80]

Some other cosmological tests include searches for primordial gravitational waves generated during cosmic inflation, which may be detected in the cosmic microwave background polarization[81] or by a proposed space-based gravitational-wave interferometer called the Big Bang Observer. Other tests at high redshift are constraints on other theories of gravity,[82][83] and the variation of the gravitational constant since big bang nucleosynthesis (it varied by no more than 40% since then).

9.6   See also

- Tests of special relativity

- Cooperstock's Energy Localization Hypothesis

- Square Kilometre Array

9.7   References

9.7.1   Notes

[1] Einstein, Albert (1916). "The Foundation of the General Theory of Relativity" (PDF). *Annalen der Physik* **49** (7): 769–822. Bibcode:1916AnP...354..769E. doi:10.1002/andp.19163540702. Retrieved 2006-09-03.

[2] Einstein, Albert (1919). "What Is The Theory Of Relativity?" (PDF). German History in Documents and Images. Retrieved 7 June 2013.

[3] U. Le Verrier (1859), (in French), "Lettre de M. Le Verrier à M. Faye sur la théorie de Mercure et sur le mouvement du périhélie de cette planète", Comptes rendus hebdomadaires des séances de l'Académie des sciences (Paris), vol. 49 (1859), pp.379–383.

[4] Clemence, G. M. (1947). "The Relativity Effect in Planetary Motions".*Reviews of Modern Physics***19**(4): 361–364.Bibcode: doi:10.1103/RevModPhys.19.361.

[5] Myles Standish, Jet Propulsion Laboratory (1998) http://classroom.sdmesa.edu/ssiegel/Physics%20197/labs/Mercury%20 pdf Precession.

[6] http://www.tat.physik.uni-tuebingen.de/~{}kokkotas/Teaching/Experimental_Gravity_files/Hajime_PPN.pdf - Perihelion shift of Mercury, page 11

[7] Biswas, Abhijit; Mani, Krishnan R. S. (2008). "Relativistic perihelion precession of orbits of Venus and the Earth". *Central European Journal of Physics*. v1 **6** (3): 754–758. arXiv:0802.0176. Bibcode:2008CEJPh...6..754B. doi:10.2478/s11534-008-0081-6.

[8] Matzner, Richard Alfred (2001). *Dictionary of geophysics, astrophysics, and astronomy*. CRC Press. p. 356. ISBN 0-8493-2891-8.

[9] Weisberg, J.M.; Taylor, J.H. (July 2005). "The Relativistic Binary Pulsar B1913+16: Thirty Years of Observations and Analysis". Written at San Francisco. In F.A. Rasio; I.H. Stairs. *ASP Conference Series* **328**. Aspen, Colorado, USA: Astronomical Society of the Pacific. p. 25. arXiv:astro-ph/0407149. Bibcode:2005ASPC..328...25W.

[10] Soldner, J. G. V. (1804). "On the deflection of a light ray from its rectilinear motion, by the attraction of a celestial body at which it nearly passes by". *Berliner Astronomisches Jahrbuch*: 161–172.

[11] Will, C.M. (2006). "The Confrontation between General Relativity and Experiment". *Living Rev. Relativity* **9**: 39. arXiv:gr-qc/0510072. Bibcode:2006LRR.....9....3W. doi:10.12942/lrr-2006-3.

[12] Dyson, F. W.; Eddington, A. S.; Davidson C. (1920). "A determination of the deflection of light by the Sun's gravitational field, from observations made at the total eclipse of 29 May 1919". *Philosophical Transactions of the Royal Society* **220A**: 291–333. doi:10.1098/rsta.1920.0009.

[13] Stanley, Matthew (2003). "'An Expedition to Heal the Wounds of War': The 1919 Eclipse and Eddington as Quaker Adventurer". *Isis* **94** (1): 57–89. doi:10.1086/376099. PMID 12725104.

[14] Rosenthal-Schneider, Ilse: Reality and Scientific Truth. Detroit: Wayne State University Press, 1980. p 74. See also Calaprice, Alice: The New Quotable Einstein. Princeton: Princeton University Press, 2005. p 227.)

[15] Harry Collins and Trevor Pinch. *The Golem*. ISBN 0-521-47736-0

[16] Daniel Kennefick (2007). "Not Only Because of Theory: Dyson, Eddington and the Competing Myths of the 1919 Eclipse Expedition". arXiv:0709.0685 [physics.hist-ph].

[17] Ball, Philip (2007). "Arthur Eddington was innocent!". *News@nature*. doi:10.1038/news070903-20.

[18] D. Kennefick, "Testing relativity from the 1919 eclipse- a question of bias", *Physics Today*, March 2009, pp. 37–42.

[19] van Biesbroeck, G.: The relativity shift at the 1952 February 25 eclipse of the Sun., *Astronomical Journal*, vol. 58, page 87, 1953.

[20] Texas Mauritanian Eclipse Team: Gravitational deflection of-light: solar eclipse of 30 June 1973 I. Description of procedures and final results., *Astronomical Journal*, vol. 81, page 452, 1976.

[21] Pound, R. V.; Rebka, Jr. G. A. (November 1, 1959). "Gravitational Red-Shift in Nuclear Resonance". *Physical Review Letters* **3** (9): 439–441. Bibcode:1959PhRvL...3..439P. doi:10.1103/PhysRevLett.3.439.

[22] Pound, R. V.; Rebka Jr. G. A. (April 1, 1960). "Apparent weight of photons". *Physical Review Letters* **4** (7): 337–341. Bibcode:1960PhRvL...4..337P. doi:10.1103/PhysRevLett.4.337.

[23] Dicke, R. H. (March 6, 1959). "New Research on Old Gravitation: Are the observed physical constants independent of the position, epoch, and velocity of the laboratory?". *Science* **129** (3349): 621–624. Bibcode:1959Sci...129..621D. doi:10.1126/science .129.3349.621.PMID17735811.

[24] Dicke, R. H. (1962). "Mach's Principle and Equivalence". *Evidence for gravitational theories: proceedings of course 20 of the International School of Physics "Enrico Fermi"* ed C. Møller.

[25] Schiff, L. I. (April 1, 1960). "On Experimental Tests of the General Theory of Relativity". *American Journal of Physics* **28** (4): 340–343. Bibcode:1960AmJPh..28..340S. doi:10.1119/1.1935800.

[26] Brans, C. H.; Dicke, R. H. (November 1, 1961). "Mach's Principle and a Relativistic Theory of Gravitation". *Physical Review* **124** (3): 925–935. Bibcode:1961PhRv..124..925B. doi:10.1103/PhysRev.124.925.

[27] Fact Sheet-BepiColombo

[28] Fomalont, E.B.; Kopeikin S.M.; Lanyi, G.; Benson, J. (July 2009). "Progress in Measurements of the Gravitational Bending of Radio Waves Using the VLBA". *Astrophysical Journal* **699** (2): 1395–1402. arXiv:0904.3992. Bibcode:2009ApJ...699.1395F. doi:10.1088/0004-637X/699/2/1395.

[29] Gaia overview

[30] Shapiro, I. I. (December 28, 1964). "Fourth test of general relativity".*Physical Review Letters***13**(26): 789–791.Bibcode:1964 doi:10.1103/PhysRevLett.13.789.

[31] Shapiro, I. I.; Ash M. E.; Ingalls R. P.; Smith W. B.; Campbell D. B.; Dyce R. B.; Jurgens R. F. & Pettengill G. H. (May 3, 1971). "Fourth Test of General Relativity: New Radar Result". *Physical Review Letters* **26** (18): 1132–1135. Bibcode:1971PhRvL..26 .1132S.doi:10.1103/PhysRevLett.26.1132.

[32] Bertotti B.; Iess L.; Tortora P. (2003). "A test of general relativity using radio links with the Cassini spacecraft". *Nature* **425** (6956): 374–376. Bibcode:2003Natur.425..374B. doi:10.1038/nature01997. PMID 14508481.

[33] Fomalont, E.B.; Kopeikin S.M. (November 2003). "The Measurement of the Light Deflection from Jupiter: Experimental Results". *Astrophysical Journal* **598** (1): 704–711. arXiv:astro-ph/0302294. Bibcode:2003ApJ...598..704F. doi:10.1086/378785.

[34] Kopeikin, S.M.; Fomalont E.B. (October 2007). "Gravimagnetism, causality, and aberration of gravity in the gravitational light-ray deflection experiments". *General Relativity and Gravitation* **39** (10): 1583–1624. arXiv:gr-qc/0510077. Bibcode:2007GRe Gr..39.1583K.doi:10.1007/s10714-007-0483-6.

[35] Fomalont, E.B.; Kopeikin, S. M.; Jones, D.; Honma, M.; Titov, O. (January 2010). "Recent VLBA/VERA/IVS tests of general relativity". *Proceedings of the International Astronomical Union, IAU Symposium* **261** (S261): 291–295. arXiv:0912.3421. Bibcode:2010IAUS..261..291F. doi:10.1017/S1743921309990536.

[36] Nordtvedt, Jr., K. (May 25, 1968). "Equivalence Principle for Massive Bodies. II. Theory". *Physical Review* **169** (5): 1017–1025. Bibcode:1968PhRv..169.1017N. doi:10.1103/PhysRev.169.1017.

[37] Nordtvedt, Jr., K. (June 25, 1968). "Testing Relativity with Laser Ranging to the Moon". *Physical Review* **170** (5): 1186–1187. Bibcode:1968PhRv..170.1186N. doi:10.1103/PhysRev.170.1186.

[38] Williams, J. G.; Turyshev, Slava G.; Boggs, Dale H. (December 29, 2004). "Progress in Lunar Laser Ranging Tests of Relativistic Gravity". *Physical Review Letters* **93** (5): 1017–1025. arXiv:gr-qc/0411113. Bibcode:2004PhRvL..93z1101W. doi:10.1103/PhysRevLett.93.261101.

[39] Uzan, J. P. (2003). "The fundamental constants and their variation: Observational status and theoretical motivations". *Reviews of Modern Physics* **75** (5): 403–. arXiv:hep-ph/0205340. Bibcode:2003RvMP...75..403U. doi:10.1103/RevModPhys.75.403.

[40] Vessot, R. F. C.; M. W. Levine; E. M. Mattison; E. L. Blomberg; T. E. Hoffman; G. U. Nystrom; B. F. Farrel; R. Decher et al. (December 29, 1980). "Test of Relativistic Gravitation with a Space-Borne Hydrogen Maser". *Physical Review Letters* **45** (26): 2081–2084. Bibcode:1980PhRvL..45.2081V. doi:10.1103/PhysRevLett.45.2081.

[41] "Gravitational Physics with Optical Clocks in Space" (PDF). *S. Schiller* (PDF). Heinrich Heine Universität Düsseldorf. 2007. Retrieved 19 March 2015.

[42] Hafele, J. C.; Keating, R. E. (July 14, 1972). "Around-the-World Atomic Clocks: Predicted Relativistic Time Gains". *Science* **177** (4044): 166–168. Bibcode:1972Sci...177..166H. doi:10.1126/science.177.4044.166. PMID 17779917.

[43] Hafele, J. C.; Keating, R. E. (July 14, 1972). "Around-the-World Atomic Clocks: Observed Relativistic Time Gains". *Science* **177** (4044): 168–170. Bibcode:1972Sci...177..168H. doi:10.1126/science.177.4044.168. PMID 17779918.

[44] Ciufolini I. & Pavlis E.C. (2004). "A confirmation of the general relativistic prediction of the Lense–Thirring effect". *Nature* **431** (7011): 958–960. Bibcode:2004Natur.431..958C. doi:10.1038/nature03007. PMID 15496915.

[45] Iorio L. (2009). "Conservative evaluation of the uncertainty in the LAGEOS-LAGEOS II Lense–Thirring test". *Central European Journal of Physics* **8** (1): 25. arXiv:0710.1022. Bibcode:2010CEJPh...8...25I. doi:10.2478/s11534-009-0060-6.

[46] Iorio L. (2006). "COMMENTS, REPLIES AND NOTES: A note on the evidence of the gravitomagnetic field of Mars". *Classical Quantum Gravity* **23** (17): 5451–5454. arXiv:gr-qc/0606092. Bibcode:2006CQGra..23.5451I. doi:10.1088/0264-9381/23/17/N01.

[47] Krogh K. (2007). "Comment on 'Evidence of the gravitomagnetic field of Mars'". *Classical Quantum Gravity* **24** (22): 5709–5715. Bibcode:2007CQGra..24.5709K. doi:10.1088/0264-9381/24/22/N01.

[48] Iorio L. (2009). "On the Lense–Thirring test with the Mars Global Surveyor in the gravitational field of Mars". *Central European Journal of Physics* **8** (3): 509. arXiv:gr-qc/0701146. Bibcode:2010CEJPh...8..509I. doi:10.2478/s11534-009-0117-6.

[49] Iorio L. (2008). "Advances in the Measurement of the Lense–Thirring Effect with Planetary Motions in the Field of the Sun". *Scholarly Research Exchange* **2008**: 1. arXiv:0807.0435. Bibcode:2008ScReE2008.5235I. doi:10.3814/2008/105235. 105235.

[50] Merritt, D.; Alexander, T.; Mikkola, S.; Will, C. (2010). "Testing Properties of the Galactic Center Black Hole Using Stellar Orbits". *Physical Review D* **81** (6): 062002. arXiv:0911.4718. Bibcode:2010PhRvD..81f2002M. doi:10.1103/PhysRevD.81.06

[51] Will, C. (2008). "Testing the General Relativistic "No-Hair" Theorems Using the Galactic Center Black Hole Sagittarius A*". *Astrophysical Journal Letters* **674** (1): L25–L28. arXiv:0711.1677. Bibcode:2008ApJ...674L..25W. doi:10.1086/528847.

[52] Everitt et al. (2011). "Gravity Probe B: Final Results of a Space Experiment to Test General Relativity". *Physical Review Letters* **106** (22): 221101. arXiv:1105.3456. Bibcode:2011PhRvL.106v1101E. doi:10.1103/PhysRevLett.106.221101. PMID 21702590.

[53] Ker Than. "Einstein Theories Confirmed by NASA Gravity Probe". News.nationalgeographic.com. Retrieved 2011-05-08.

[54] "Prepping satellite to test Albert Einstein".

[55] Ciufolini, I. et al. (2009). "Towards a One Percent Measurement of Frame Dragging by Spin with Satellite Laser Ranging to LAGEOS, LAGEOS 2 and LARES and GRACE Gravity Models". *Space Science Reviews* **148**: 71–104. Bibcode:2009SSRv..148 ...71C.doi:10.1007/s11214-009-9585-7.

[56] Iorio, L. (2009). "Will the recently approved LARES mission be able to measure the Lense–Thirring effect at 1%?". *General Relativity and Gravitation* **41** (8): 1717–1724. arXiv:0803.3278. Bibcode:2009GReGr..41.1717I. doi:10.1007/s10714-008-0742-1.

[57] Ciufolini, I.; Paolozzi A.; Pavlis E. C.; Ries J. C.; Koenig R.; Matzner R. A.; Sindoni G. & Neumayer H. (2009). "Towards a One Percent Measurement of Frame Dragging by Spin with Satellite Laser Ranging to LAGEOS, LAGEOS 2 and LARES and GRACE Gravity Models". *Space Science Reviews* **148**: 71–104. Bibcode:2009SSRv..148...71C. doi:10.1007/s11214-009-9585-7.

[58] Ciufolini, I.; Paolozzi A.; Pavlis E. C.; Ries J. C.; Koenig R.; Matzner R. A.; Sindoni G. & Neumayer H. (2010). "Gravitomagnetism and Its Measurement with Laser Ranging to the LAGEOS Satellites and GRACE Earth Gravity Models". *General Relativity and John Archibald Wheeler*. Astrophysics and Space Science Library **367**. SpringerLink. pp. 371–434. doi:10.1007/978-90-481-3735-0_17.

[59] Paolozzi, A.; Ciufolini I.; Vendittozzi C. (2011). "Engineering and scientific aspects of LARES satellite". *Acta Astronautica* **69** (3–4): 127–134. Bibcode:2011AcAau..69..127P. doi:10.1016/j.actaastro.2011.03.005. ISSN 0094-5765.

[60] Ries, J.C.; Ciufolini I.; Pavlis E.C.; Paolozzi A.; Koenig R.; Matzner R.A.; Sindoni G.; Neumayer H. (2011). "The Earth's frame-dragging via laser-ranged satellites: A Response to "Some considerations on the present-day results for the detection of frame-dragging after the final outcome of GP-B" by Iorio L.". *Europhysics Letters* **96** (3): 30002. Bibcode:2011EL.....9630002R. doi:10.1209/0295-5075/96/30002.

[61] "Einstein Was Right — So Far". *ESO Press Release*. Retrieved 30 April 2013.

[62] In general relativity, a perfectly spherical star (in vacuum) that expands or contracts while remaining perfectly spherical *cannot* emit any gravitational waves (similar to the lack of e/m radiation from a pulsating charge), as Birkhoff's theorem says that the geometry remains the same exterior to the star. More generally, a rotating system will only emit gravitational waves if it lacks the axial symmetry with respect to the axis of rotation.

[63] Stairs, Ingrid H. "Testing General Relativity with Pulsar Timing". *Living Reviews in Relativity* **6**. arXiv:astro-ph/0307536. Bibcode:2003LRR.....6....5S. doi:10.12942/lrr-2003-5.

[64] Weisberg, J. M.; Taylor, J. H.; Fowler, L. A. (October 1981). "Gravitational waves from an orbiting pulsar". *Scientific American* **245**: 74–82. Bibcode:1981SciAm.245...74W. doi:10.1038/scientificamerican1081-74.

[65] Weisberg, J. M.; Nice, D. J.; Taylor, J. H. (2010). "Timing Measurements of the Relativistic Binary Pulsar PSR B1913+16". *Astrophysical Journal* **722**: 1030–1034. arXiv:1011.0718v1. Bibcode:2010ApJ...722.1030W. doi:10.1088/0004-637X/722/2/10

[66] "Press Release: The Nobel Prize in Physics 1993". Nobel Prize. 13 October 1993. Retrieved 6 May 2014.

[67] Kramer, M. et al. (2006). "Tests of general relativity from timing the double pulsar". *Science* **314** (5796): 97–102. arXiv:astro-ph/0609417. Bibcode:2006Sci...314...97K. doi:10.1126/science.1132305. PMID 16973838.

[68] Antoniadis, John et al. (2013). "A Massive Pulsar in a Compact Relativistic Binary". *Science* (AAAS) **340** (6131). arXiv:1304. Bibcode:2013Sci...340..448A. doi:10.1126/science.1233232.

[69] "Massive double star is latest test for Einstein's gravity theory". *Ron Cowen*. Nature. 25 April 2013. Retrieved 7 May 2013.

[70] Abadie, J et al. (7 September 2010). "Predictions for the rates of compact binary coalescences observable by ground-based gravitational-wave detectors". *Classical and Quantum Gravity* **27** (17): 173001. arXiv:1003.2480. Bibcode:2010CQGra..27q3001A. doi:10.1088/0264-9381/27/17/173001.

[71] Schutz, Bernard F. "Gravitational waves on the back of an envelope". *American Journal of Physics* **52** (5): 412. Bibcode:1984AmJPh..52..412S. doi:10.1119/1.13627.

[72] Gair, Jonathan; Vallisneri, Michele; Larson, Shane L.; Baker, John G. "Testing General Relativity with Low-Frequency, Space-Based Gravitational-Wave Detectors". *Living Reviews in Relativity* **16**. arXiv:1212.5575. Bibcode:2013LRR....16....7G. doi:10.12942/lrr-2013-7.

[73] Yunes, Nicolás; Siemens, Xavier. "Gravitational-Wave Tests of General Relativity with Ground-Based Detectors and Pulsar-Timing Arrays". *Living Reviews in Relativity* **16**. arXiv:1304.3473. Bibcode:2013LRR....16....9Y. doi:10.12942/lrr-2013-9.

[74] Peebles, P. J. E. (December 2004). "Testing general relativity on the scales of cosmology". p. 106. arXiv:astro-ph/0410284. Bibcode:2005grg..conf..106P. doi:10.1142/9789812701688_0010. ISBN 978-981-256-424-5.

[75] Rudnicki, 1991, p. 28. *The Hubble Law was viewed by many as an observational confirmation of General Relativity in the early years*

[76] W.Pauli, 1958, pp.219–220

[77] Kragh, 2003, p. 152

[78] Kragh, 2003, p. 153

[79] Rudnicki, 1991, p. 28

[80] Chandrasekhar, 1980, p. 37

[81] Hand, Eric (2009). "Cosmology: The test of inflation". *Nature* **458**: 820–824. doi:10.1038/458820a.

[82] Reyes, Reinabelle et al. (2010). "Confirmation of general relativity on large scales from weak lensing and galaxy velocities". *Nature* **464**: 256–258. arXiv:1003.2185. Bibcode:2010Natur.464..256R. doi:10.1038/nature08857.

[83] Guzzo, L. et al. (2008). "A test of the nature of cosmic acceleration using galaxy redshift distortions". *Nature* **451**: 541–544. arXiv:0802.1944. Bibcode:2008Natur.451..541G. doi:10.1038/nature06555.

9.7.2   Other research papers

- Bertotti, B.; Iess, L.; Tortora, P. (2003). "A test of general relativity using radio links with the Cassini spacecraft". *Nature* **425** (6956): 374–6. Bibcode:2003Natur.425..374B. doi:10.1038/nature01997. PMID 14508481.

- Kopeikin, S.; Polnarev, A.; Schaefer, G.; Vlasov, I. (2007). "Gravimagnetic effect of the barycentric motion of the Sun and determination of the post-Newtonian parameter γ in the Cassini experiment". *Physics Letters A* **367** (4–5): 276. arXiv:gr-qc/0604060. Bibcode:2007PhLA..367..276K. doi:10.1016/j.physleta.2007.03.036.

- Brans, C.; Dicke, R. H. (1961). "Mach's principle and a relativistic theory of gravitation". *Phys. Rev.* **124** (3): 925–35. Bibcode:1961PhRv..124..925B. doi:10.1103/PhysRev.124.925.

- A. Einstein, "Über das Relativitätsprinzip und die aus demselben gezogene Folgerungen", *Jahrbuch der Radioaktivitaet und Elektronik* **4** (1907); translated "On the relativity principle and the conclusions drawn from it", in *The collected papers of Albert Einstein. Vol. 2 : The Swiss years: writings, 1900–1909* (Princeton University Press, Princeton, New Jersey, 1989), Anna Beck translator. Einstein proposes the gravitational redshift of light in this paper, discussed online at The Genesis of General Relativity.

- A. Einstein, "Über den Einfluß der Schwerkraft auf die Ausbreitung des Lichtes", *Annalen der Physik* **35** (1911); translated "On the Influence of Gravitation on the Propagation of Light" in *The collected papers of Albert Einstein. Vol. 3 : The Swiss years: writings, 1909–1911* (Princeton University Press, Princeton, New Jersey, 1994), Anna Beck translator, and in *The Principle of Relativity*, (Dover, 1924), pp 99–108, W. Perrett and G. B. Jeffery translators, ISBN 0-486-60081-5. The deflection of light by the sun is predicted from the principle of equivalence. Einstein's result is half the full value found using the general theory of relativity.

- Shapiro, S. S.; Davis, J. L.; Lebach, D. E.; Gregory J.S. (26 March 2004). "Measurement of the solar gravitational deflection of radio waves using geodetic very-long-baseline interferometry data, 1979–1999". *Physical Review Letters* (American Physical Society) **92** (121101): 121101. Bibcode:2004PhRvL..92l1101S. doi:10.1103/PhysRevLett.92.121101.PMID15089661.

- M. Froeschlé, F. Mignard and F. Arenou, "Determination of the PPN parameter γ with the Hipparcos data" Hipparcos Venice '97, ESA-SP-402 (1997).

- Will, Clifford M. (2006). "Was Einstein Right? Testing Relativity at the Centenary". *Annalen der Physik* **15**: 19–33. arXiv:gr-qc/0504086. Bibcode:2006AnP...518...19W. doi:10.1002/andp.200510170.

- Rudnicki, Conrad (1991). "What are the Empirical Bases of the Hubble Law" (PDF). *Apeiron* (9–10): 27–36. Retrieved 2009-06-23.

- Chandrasekhar, S. (1980). "The Role of General Relativity in Astronomy: Retrospect and Prospect" (PDF). *J. Astrophys. Astr.* **1** (1): 33–45. Bibcode:1980JApA....1...33C. doi:10.1007/BF02727948. Retrieved 2009-06-23.

- Kragh, Helge; Smith, Robert W. (2003). "Who discovered the expanding universe". *History of Science* **41**: 141–62. Bibcode:2003HisSc..41..141K. Retrieved 2013-02-15.

9.7.3 Textbooks

- S. M. Carroll, *Spacetime and Geometry: an Introduction to General Relativity*, Addison-Wesley, 2003. An introductory general relativity textbook.

- A. S. Eddington, *Space, Time and Gravitation*, Cambridge University Press, reprint of 1920 ed.

- A. Gefter. "Putting Einstein to the Test", *Sky and Telescope* July 2005, p. 38. A popular discussion of tests of general relativity.

- H. Ohanian and R. Ruffini, *Gravitation and Spacetime, 2nd Edition* Norton, New York, 1994, ISBN 0-393-96501-5. A general relativity textbook.

- Pauli, Wolfgang Ernst (1958). "Part IV. General Theory of Relativity". *Theory of Relativity*. Courier Dover Publications. ISBN 978-0-486-64152-2.

- C. M. Will, *Theory and Experiment in Gravitational Physics*, Cambridge University Press, Cambridge (1993). A standard technical reference.

- C. M. Will, *Was Einstein Right?: Putting General Relativity to the Test*, Basic Books (1993). This is a popular account of tests of general relativity.

- L. Iorio, *The Measurement of Gravitomagnetism: A Challenging Enterprise*, NOVA Science, Hauppauge (2007). It describes various theoretical and experimental/observational aspects of frame-dragging.

9.7.4 Living Reviews papers

- N. Ashby, "Relativity in the Global Positioning System", *Living Reviews in Relativity* (2003).

- C. M. Will, The Confrontation between General Relativity and Experiment, *Living Reviews in Relativity* (2006). An online, technical review, covering much of the material in *Theory and experiment in gravitational physics*. It is less comprehensive but more up to date.

9.8 External links

- the USENET Relativity FAQ experiments page

- Mathpages article on Mercury's perihelion shift (for amount of observed and GR shifts).

Chapter 10

Parameterized post-Newtonian formalism

Post-Newtonian formalism is a calculational tool that expresses Einstein's (nonlinear) equations of gravity in terms of the lowest-order deviations from Newton's law of universal gravitation. This allows approximations to Einstein's equations to be made in the case of weak fields. Higher order terms can be added to increase accuracy, but for strong fields sometimes it is preferable to solve the complete equations numerically. Some of these post-Newtonian approximations are expansions in a small parameter, which is the ratio of the velocity of the matter forming the gravitational field to the speed of light, which in this case is better called the speed of gravity. In the limit, when the fundamental speed of gravity becomes infinite, the post-Newtonian expansion reduces to Newton's law of gravity.

The **parameterized post-Newtonian formalism** or **PPN formalism** is a version of this formulation that explicitly details the parameters in which a general theory of gravity can differ from Newtonian gravity. It is used as a tool to compare Newtonian and Einsteinian gravity in the limit in which the gravitational field is weak and generated by objects moving slowly compared to the speed of light. In general, PPN formalism can be applied to all metric theories of gravitation in which all bodies satisfy the Einstein equivalence principle (EEP). The speed of light remains constant in PPN formalism and it assumes that the metric tensor is always symmetric.

10.1 History

The earliest parameterizations of the post-Newtonian approximation were performed by Sir Arthur Stanley Eddington in 1922. However, they dealt solely with the vacuum gravitational field outside an isolated spherical body. Dr. Ken Nordtvedt (1968, 1969) expanded this to include 7 parameters. Clifford Martin Will (1971) introduced a stressed, continuous matter description of celestial bodies.

The versions described here are based on Wei-Tou Ni (1972), Will and Nordtvedt (1972), Charles W. Misner et al. (1973) (see *Gravitation* (book)), and Will (1981, 1993) and have 10 parameters.

10.2 Beta-delta notation

Ten **post-Newtonian parameters** completely characterize the weak-field behavior of the theory. The formalism has been a valuable tool in tests of general relativity. In the notation of Will (1971), Ni (1972) and Misner et al. (1973) they have the following values:

$g_{\mu\nu}$ is the 4 by 4 symmetric metric tensor and indexes i and j go from 1 to 3.

In Einstein's theory, the values of these parameters are chosen (1) to fit Newton's Law of gravity in the limit of velocities and mass approaching zero, (2) to ensure conservation of energy, mass, momentum, and angular momentum, and (3) to make the equations independent of the reference frame. In this notation, general relativity has PPN parameters $\gamma = \beta = \beta_1 = \beta_2 = \beta_3 = \beta_4 = \Delta_1 = \Delta_2 = 1$ and $\zeta = \eta = 0$

10.3   Alpha-zeta notation

In the more recent notation of Will & Nordtvedt (1972) and Will (1981, 1993, 2006) a different set of ten PPN parameters is used.

$$\gamma = \gamma$$
$$\beta = \beta$$
$$\alpha_1 = 7\Delta_1 + \Delta_2 - 4\gamma - 4$$
$$\alpha_2 = \Delta_2 + \zeta - 1$$
$$\alpha_3 = 4\beta_1 - 2\gamma - 2 - \zeta$$
$$\zeta_1 = \zeta$$
$$\zeta_2 = 2\beta + 2\beta_2 - 3\gamma - 1$$
$$\zeta_3 = \beta_3 - 1$$
$$\zeta_4 = \beta_4 - \gamma$$
$$\xi \text{ is calculated from } 3\eta = 12\beta - 3\gamma - 9 + 10\xi - 3\alpha_1 + 2\alpha_2 - 2\zeta_1 - \zeta_2$$

The meaning of these is that α_1, α_2 and α_3 measure the extent of preferred frame effects. ζ_1, ζ_2, ζ_3, ζ_4 and α_3 measure the failure of conservation of energy, momentum and angular momentum.

In this notation, general relativity has PPN parameters

$$\gamma = \beta = 1 \text{ and } \alpha_1 = \alpha_2 = \alpha_3 = \zeta_1 = \zeta_2 = \zeta_3 = \zeta_4 = \xi = 0$$

The mathematical relationship between the metric, metric potentials and PPN parameters for this notation is:

$$g_{00} = -1 + 2U - 2\beta U^2 - 2\xi\Phi_W + (2\gamma + 2 + \alpha_3 + \zeta_1 - 2\xi)\Phi_1 + 2(3\gamma - 2\beta + 1 + \zeta_2 + \xi)\Phi_2$$
$$+ 2(1 + \zeta_3)\Phi_3 + 2(3\gamma + 3\zeta_4 - 2\xi)\Phi_4 - (\zeta_1 - 2\xi)A - (\alpha_1 - \alpha_2 - \alpha_3)w^2 U$$
$$- \alpha_2 w^i w^j U_{ij} + (2\alpha_3 - \alpha_1)w^i V_i + O(\epsilon^3)$$

$$g_{0i} = -\tfrac{1}{2}(4\gamma + 3 + \alpha_1 - \alpha_2 + \zeta_1 - 2\xi)V_i - \tfrac{1}{2}(1 + \alpha_2 - \zeta_1 + 2\xi)W_i - \tfrac{1}{2}(\alpha_1 - 2\alpha_2)w^i U - \alpha_2 w^j U_{ij} + O(\epsilon^{\frac{5}{2}})$$

$$g_{ij} = (1 + 2\gamma U)\delta_{ij} + O(\epsilon^2)$$

where repeated indexes are summed. ϵ is on the order of potentials such as U, the square magnitude of the coordinate velocities of matter, etc. w^i is the velocity vector of the PPN coordinate system relative to the mean rest-frame of the universe. $w^2 = \delta_{ij}w^i w^j$ is the square magnitude of that velocity. $\delta_{ij} = 1$ if and only if $i = j$, 0 otherwise.

There are ten metric potentials, U, U_{ij}, Φ_W, A, Φ_1, Φ_2, Φ_3, Φ_4, V_i and W_i, one for each PPN parameter to ensure a unique solution. 10 linear equations in 10 unknowns are solved by inverting a 10 by 10 matrix. These metric potentials have forms such as:

$$U(\mathbf{x}, t) = \int \frac{\rho(\mathbf{x}', t)}{|\mathbf{x} - \mathbf{x}'|} d^3 x'$$

which is simply another way of writing the Newtonian gravitational potential,

$$U_{ij} = \int \frac{\rho(\mathbf{x}', t)(x - x')_i(x - x')_j}{|\mathbf{x} - \mathbf{x}'|^3} d^3 x'$$

$$\Phi_W = \int \frac{\rho(\mathbf{x}', t)\rho(\mathbf{x}'', t)(x - x')_i}{|\mathbf{x} - \mathbf{x}'|^3} \left(\frac{(x' - x'')^i}{|\mathbf{x} - \mathbf{x}'|} - \frac{(x - x'')^i}{|\mathbf{x}' - \mathbf{x}''|} \right) d^3 x' d^3 x''$$

$$A = \int \frac{\rho(\mathbf{x}',t)\,(\mathbf{v}(\mathbf{x}',t)\cdot(\mathbf{x}-\mathbf{x}'))^2}{|\mathbf{x}-\mathbf{x}'|^3}\,d^3x'$$

$$\Phi_1 = \int \frac{\rho(\mathbf{x}',t)\mathbf{v}(\mathbf{x}',t)^2}{|\mathbf{x}-\mathbf{x}'|}\,d^3x'$$

$$\Phi_2 = \int \frac{\rho(\mathbf{x}',t)U(\mathbf{x}',t)}{|\mathbf{x}-\mathbf{x}'|}\,d^3x'$$

$$\Phi_3 = \int \frac{\rho(\mathbf{x}',t)\Pi(\mathbf{x}',t)}{|\mathbf{x}-\mathbf{x}'|}\,d^3x'$$

$$\Phi_4 = \int \frac{p(\mathbf{x}',t)}{|\mathbf{x}-\mathbf{x}'|}\,d^3x'$$

$$V_i = \int \frac{\rho(\mathbf{x}',t)v(\mathbf{x}',t)_i}{|\mathbf{x}-\mathbf{x}'|}\,d^3x'$$

$$W_i = \int \frac{\rho(\mathbf{x}',t)\,(\mathbf{v}(\mathbf{x}',t)\cdot(\mathbf{x}-\mathbf{x}'))\,(x-x')_i}{|\mathbf{x}-\mathbf{x}'|^3}\,d^3x'$$

where ρ is the density of rest mass, Π is the internal energy per unit rest mass, p is the pressure as measured in a local freely falling frame momentarily comoving with the matter, and $\mathbf{v}$ is the coordinate velocity of the matter.

Stress-energy tensor for a perfect fluid takes form

$$T^{00} = \rho(1 + \Pi + \mathbf{v}^2 + 2U)$$

$$T^{0i} = \rho(1 + \Pi + \mathbf{v}^2 + 2U + p/\rho)v^i$$

$$T^{ij} = \rho(1 + \Pi + \mathbf{v}^2 + 2U + p/\rho)v^i v^j + p\delta^{ij}(1 - 2\gamma U)$$

10.4   How to apply PPN

Examples of the process of applying PPN formalism to alternative theories of gravity can be found in Will (1981, 1993). It is a nine step process:

- Step 1: Identify the variables, which may include: (a) dynamical gravitational variables such as the metric $g_{\mu\nu}$, scalar field ϕ , vector field K_μ , tensor field $B_{\mu\nu}$ and so on; (b) prior-geometrical variables such as a flat background metric $\eta_{\mu\nu}$, cosmic time function t , and so on; (c) matter and non-gravitational field variables.

- Step 2: Set the cosmological boundary conditions. Assume a homogeneous isotropic cosmology, with isotropic coordinates in the rest frame of the universe. A complete cosmological solution may or may not be needed. Call the results $g_{\mu\nu}^{(0)} = \mathrm{diag}(-c_0, c_1, c_1, c_1)$, ϕ_0 , $K_\mu^{(0)}$, $B_{\mu\nu}^{(0)}$.

- Step 3: Get new variables from $h_{\mu\nu} = g_{\mu\nu} - g_{\mu\nu}^{(0)}$, with $\phi - \phi_0$, $K_\mu - K_\mu^{(0)}$ or $B_{\mu\nu} - B_{\mu\nu}^{(0)}$ if needed.

- Step 4: Substitute these forms into the field equations, keeping only such terms as are necessary to obtain a final consistent solution for $h_{\mu\nu}$. Substitute the perfect fluid stress tensor for the matter sources.

- Step 5: Solve for h_{00} to $O(2)$. Assuming this tends to zero far from the system, one obtains the form $h_{00} = 2\alpha U$ where U is the Newtonian gravitational potential and α may be a complicated function including the gravitational "constant" G . The Newtonian metric has the form $g_{00} = -c_0 + 2\alpha U$, $g_{0j} = 0$, $g_{ij} = \delta_{ij}c_1$. Work in units where the gravitational "constant" measured today far from gravitating matter is unity so set $G_{\text{today}} = \alpha/c_0 c_1 = 1$

- Step 6: From linearized versions of the field equations solve for h_{ij} to $O(2)$ and h_{0j} to $O(3)$.

- Step 7: Solve for h_{00} to $O(4)$. This is the messiest step, involving all the nonlinearities in the field equations. The stress–energy tensor must also be expanded to sufficient order.

- Step 8: Convert to local quasi-Cartesian coordinates and to standard PPN gauge.

- Step 9: By comparing the result for $g_{\mu\nu}$ with the equations presented in PPN with alpha-zeta parameters, read off the PPN parameter values.

10.5   Comparisons between theories of gravity

Main article: Alternatives to general relativity § PPN parameters for a range of theories

A table comparing PPN parameters for 23 theories of gravity can be found in Alternatives to general relativity#PPN parameters for a range of theories.

Most metric theories of gravity can be lumped into categories. Scalar theories of gravitation include conformally flat theories and stratified theories with time-orthogonal space slices.

In conformally flat theories such as Nordström's theory of gravitation the metric is given by $\mathbf{g} = f\eta$ and for this metric $\gamma = -1$, which violently disagrees with observations. In stratified theories such as Yilmaz theory of gravitation the metric is given by $\mathbf{g} = f_1 \mathbf{dt} \otimes \mathbf{dt} + f_2\eta$ and for this metric $\alpha_1 = -4(\gamma + 1)$, which also disagrees violently with observations.

Another class of theories is the quasilinear theories such as Whitehead's theory of gravitation. For these $\xi = \beta$. The relative magnitudes of the harmonics of the Earth's tides depend on ξ and α_2 , and measurements show that quasilinear theories disagree with observations of Earth's tides.

Another class of metric theories is the bimetric theory. For all of these α_2 is non-zero. From the precession of the solar spin we know that $\alpha_2 < 4 \times 10^{-7}$, and that effectively rules out bimetric theories.

Another class of metric theories is the scalar tensor theories, such as Brans–Dicke theory. For all of these, $\gamma = \frac{1+\omega}{2+\omega}$. The limit of $\gamma - 1 < 2.3 \times 10^{-5}$ means that ω would have to be very large, so these theories are looking less and less likely as experimental accuracy improves.

The final main class of metric theories is the vector-tensor theories. For all of these the gravitational "constant" varies with time and α_2 is non-zero. Lunar laser ranging experiments tightly constrain the variation of the gravitational "constant" with time and $\alpha_2 < 4 \times 10^{-7}$, so these theories are also looking unlikely.

There are some metric theories of gravity that do not fit into the above categories, but they have similar problems.

10.6   Accuracy from experimental tests

Bounds on the PPN parameters Will (2006)

† Will, C.M., *Is momentum conserved? A test in the binary system PSR 1913 + 16*, Astrophysical Journal, Part 2 - Letters (ISSN 0004-637X), vol. 393, no. 2, July 10, 1992, p. L59-L61.

‡ Based on $6\zeta_4 = 3\alpha_3 + 2\zeta_1 - 3\zeta_3$ from Will (1976, 2006). It is theoretically possible for an alternative model of gravity to bypass this bound, in which case the bound is $|\zeta_4| < 0.4$ from Ni (1972).

10.7   References

- Eddington, A. S. (1922) The Mathematical Theory of Relativity, Cambridge University Press.

- Misner, C. W., Thorne, K. S. & Wheeler, J. A. (1973) Gravitation, W. H. Freeman and Co.

- Nordtvedt Jr, K. (1968) Equivalence principle for massive bodies II: Theory, Phys. Rev. 169, 1017-1025.

- Nordtvedt Jr, K. (1969) Equivalence principle for massive bodies including rotational energy and radiation pressure, Phys. Rev. 180, 1293-1298.

- Will, C. M. (1971) Theoretical frameworks for testing relativistic gravity II: Parameterized post-Newtonian hydro-dynamics and the Nordtvedt effect, Astrophys. J. 163, 611-628.

- Will, C. M. (1976) Active mass in relativistic gravity: Theoretical interpretation of the Kreuzer experiment, Astrophys. J., 204, 224-234.

- Will, C. M. (1981, 1993) Theory and Experiment in Gravitational Physics, Cambridge University Press. ISBN 0-521-43973-6.

- Will, C. M., (2006) The Confrontation between General Relativity and Experiment, http://relativity.livingreviews.org/Articles/lrr-2006-3/

- Will, C. M., and Nordtvedt Jr., K (1972) Conservation laws and preferred frames in relativistic gravity I, The Astrophysical Journal 177, 757.

10.8   See also

- Alternatives to general relativity#PPN parameters for a range of theories

- Linearized gravity

- Peskin-Takeuchi parameter The same thing as PPN, but for electroweak theory instead of gravitation

- Tests of general relativity

Chapter 11

Linearized gravity

Linearized gravity is an approximation scheme in general relativity in which the nonlinear contributions from the spacetime metric are ignored, simplifying the study of many problems while still producing useful approximate results.

11.1 The method

In linearized gravity the metric tensor, g, of spacetime is treated as a sum of an exact solution of Einstein's equations (often Minkowski spacetime) and a perturbation h.

$$g = \eta + h$$

where η is the nondynamical background metric that is being perturbed about, and h represents the deviation of the true metric (g) from flat spacetime.

The perturbation is treated using the methods of perturbation theory, "linearized" by ignoring all terms of order higher than one (quadratic in h, cubic in h etc...) in the perturbation.

11.2 Applications

The Einstein field equations (EFE), being nonlinear in the metric, are difficult to solve exactly and the above perturbation scheme allows linearised Einstein field equations to be obtained. These equations are linear in the metric, and the sum of two solutions of the linearized EFE is also a solution. The idea of 'ignoring the nonlinear part' is thus encapsulated in this linearization procedure.

The method is used to derive the Newtonian limit, including the first corrections, much like for a derivation of the existence of gravitational waves that led, after quantization, to gravitons. This is why the conceptual approach of linearized gravity is the canonical one in particle physics, string theory, and more generally quantum field theory where classical (bosonic) fields are expressed as coherent states of particles.

This approximation is also known as the weak-field approximation as it is only valid for h very small.

11.2.1 Weak-field approximation

In a weak-field approximation, the gauge symmetry is associated with diffeomorphisms with small "displacements" (diffeomorphisms with large displacements obviously violate the weak field approximation), which has the exact form (for infinitesimal transformations)

$$\delta_\xi h = \delta_\xi g - \delta_\xi \eta = \mathcal{L}_{\vec\xi} g = \mathcal{L}_{\vec\xi} \eta + \mathcal{L}_{\vec\xi} h = \left[\xi_{\nu;\mu} + \xi_{\mu;\nu} + \xi^\alpha h_{\mu\nu;\alpha} + \xi^\alpha_{;\mu} h_{\alpha\nu} + \xi^\alpha_{;\nu} h_{\mu\alpha} \right] dx^\mu \otimes dx^\nu$$

Where $\mathcal{L}$ is the Lie derivative and we used the fact that η does not transform (by definition). Note that we are raising and lowering the indices with respect to η and not g and taking the covariant derivatives (Levi-Civita connection) with respect to η. This is the standard practice in linearized gravity. The way of thinking in linearized gravity is this: the background metric η is the metric and h is a field propagating over the spacetime with this metric.

In the weak field limit, this gauge transformation simplifies to

$$\delta_\xi h_{\mu\nu} \approx \left(\mathcal{L}_{\vec\xi} \eta \right)_{\mu\nu} = \xi_{\nu;\mu} + \xi_{\mu;\nu}$$

The weak-field approximation is useful in finding the values of certain constants, for example in the Einstein field equations and in the Schwarzschild metric.

11.3   Linearised Einstein field equations

The **linearised Einstein field equations** (linearised EFE) are an approximation to Einstein's field equations that is valid for a weak gravitational field and is used to simplify many problems in general relativity and to discuss the phenomena of gravitational radiation. The approximation can also be used to derive Newtonian gravity as the weak-field approximation of Einsteinian gravity.

The equations are obtained by assuming the spacetime metric is only slightly different from some baseline metric (usually a Minkowski metric). Then the difference in the metrics can be considered as a field on the baseline metric, whose behaviour is approximated by a set of linear equations.

11.3.1   Derivation for the Minkowski metric

Starting with the metric for a spacetime in the form

$$g_{ab} = \eta_{ab} + h_{ab}$$

where η_{ab} is the Minkowski metric and h_{ab} — sometimes written as $\epsilon \, \gamma_{ab}$ — is the deviation of g_{ab} from it. h must be negligible compared to η : $|h_{\mu\nu}| \ll 1$ (and similarly for all derivatives of h). Then one ignores all products of h (or its derivatives) with h or its derivatives (equivalent to ignoring all terms of higher order than 1 in ϵ). It is further assumed in this approximation scheme that all indices of h and its derivatives are raised and lowered with η .

The metric h is clearly symmetric, since g and η are. The consistency condition $g_{ab} g^{bc} = \delta_a{}^c$ shows that

$$g^{ab} = \eta^{ab} - h^{ab}$$

The Christoffel symbols can be calculated as

$$2\Gamma^a_{bc} = (h^a{}_{b,c} + h^a{}_{c,b} - h_{bc,}{}^a)$$

where $h_{bc,}{}^a \overset{\text{def}}{=} \eta^{ar} h_{bc,r}$, and this is used to calculate the Riemann tensor:

$$2R^a{}_{bcd} = 2(\Gamma^a_{bd,c} - \Gamma^a_{bc,d}) = \eta^{ae}(h_{eb,dc} + h_{ed,bc} - h_{bd,ec} - h_{eb,cd} - h_{ec,bd} + h_{bc,ed}) =$$

$$= \eta^{ae}(h_{ed,bc} - h_{bd,ec} - h_{ec,bd} + h_{bc,ed}) = h^a_{d,bc} - h_{bd,~c}^{~~a} + h_{bc,~d}^{~~a} - h^a_{~c,bd}$$

Using $R_{bd} = \delta^c_{~a}R^a_{~bcd}$ gives

$$2R_{bd} = h^r_{d,br} + h^r_{b,dr} - h_{,bd} - h_{bd,rs}\eta^{rs}$$

For Ricci scalar we have:

$$R = R_{bd}\eta^{bd} = h^{ab}_{,ab} - \Box h$$

Then the linearized Einstein equations are

$$8\pi T_{bd} = R_{bd} - R_{ac}\eta^{ac}\eta_{bd}/2$$

or

$$8\pi T_{bd} = (h^r_{d,br} + h^r_{b,dr} - h_{,bd} - h_{bd,r}^{~~r} - h^r_{s,r}{}^s\eta_{bd})/2 + (h_{,a}^{~a}\eta_{bd} + h_{ac,r}^{~~~r}\eta^{ac}\eta_{bd})/4$$

Or, equivalently:

$$8\pi(T_{bd} - T_{ac}\eta^{ac}\eta_{bd}/2) = R_{bd}$$
$$16\pi(T_{bd} - T_{ac}\eta^{ac}\eta_{bd}/2) = h^r_{d,br} + h^r_{b,dr} - h_{,bd} - h_{bd,rs}\eta^{rs}$$

11.4   With a coordinate condition

If one uses the Lorentz invariant harmonic coordinate condition

$$h_{\alpha\beta,\gamma}\eta^{\beta\gamma} = \frac{1}{2}h_{\beta\gamma,\alpha}\eta^{\beta\gamma} .$$

then the last form above of the linearized Einstein equation simplifies to

$$16\pi(T_{bd} - T_{ac}\eta^{ac}\eta_{bd}/2) = -h_{bd,rs}\eta^{rs} .$$

To solve it, this can be rewritten as

$$\Delta h_{bd} = \frac{-16\pi G}{c^4}(T_{bd} - T_{ac}\eta^{ac}\eta_{bd}/2) + \frac{\partial^2 h_{bd}}{c^2\partial t^2}$$

where Δ is the Laplacian on a spatial slice. If the stress-energy changes slowly (velocities are low compared to c), then this gives

$$h_{bd}(r) = \frac{-1}{4\pi}\int \left(\frac{-16\pi G}{c^4}(T_{bd}(s) - T_{ac}(s)\eta^{ac}\eta_{bd}/2) + \frac{\partial^2 h_{bd}(s)}{c^2\partial t^2}\right)\frac{1}{|r-s|}d^3s$$

as a generalization of the Newtonian formula for gravitational potential. This is solved iteratively by first replacing the second time derivative by zero and then inserting the h so obtained repeatedly until convergence.

11.5   Applications

The linearised EFE are used primarily in the theory of gravitational radiation, where the gravitational field far from the source is approximated by these equations.

11.6   See also

- Correspondence principle

- Gravitoelectromagnetism

- Lanczos tensor

- Parameterized post-Newtonian formalism

- Quasinormal mode

11.7   References

- Stephani, Hans (1990). *General Relativity: An Introduction to the Theory of the Gravitational Field,*. Cambridge: Cambridge University Press. ISBN 0-521-37941-5.

- Adler, Ronald; Bazin, Maurice' & Schiffer, Menahem (1965). *Introduction to General Relativity*. New York: McGraw-Hill. ISBN 0-07-000423-4.

Chapter 12

ADM formalism

The **ADM formalism**, named for its authors Richard Arnowitt, Stanley Deser and Charles W. Misner, is a Hamiltonian formulation of general relativity that plays an important role in quantum gravity and numerical relativity. It was first published in 1959.[2]

The comprehensive review of the formalism that the authors published in 1962[3] has been reprinted in the journal *General Relativity and Gravitation*,[4] while the original papers can be found in the archives of *Physical Review*.[2][5][6][7][8][9][10][11][12]

12.1 Overview

The formalism supposes that spacetime is foliated into a family of spacelike surfaces Σ_t, labeled by their time coordinate t, and with coordinates on each slice given by x^i. The dynamic variables of this theory are taken to be the metric tensor of three dimensional spatial slices $\gamma_{ij}(t, x^k)$ and their conjugate momenta $\pi^{ij}(t, x^k)$. Using these variables it is possible to define a Hamiltonian, and thereby write the equations of motion for general relativity in the form of Hamilton's equations.

In addition to the twelve variables γ_{ij} and π^{ij}, there are four Lagrange multipliers: the lapse function, N, and components of shift vector field, N_i. These describe how each of the "leaves" Σ_t of the foliation of spacetime are welded together. The equations of motion for these variables can be freely specified; this freedom corresponds to the freedom to specify how to lay out the coordinate system in space and time.

12.2 Derivation

12.2.1 Notation

Most references adopt notation in which four dimensional tensors are written in abstract index notation, and that Greek indices are spacetime indices taking values (0, 1, 2, 3) and Latin indices are spatial indices taking values (1, 2, 3). In the derivation here, a superscript (4) is prepended to quantities that typically have both a three-dimensional and a four-dimensional version, such as the metric tensor for three-dimensional slices g_{ij} and the metric tensor for the full four-dimensional spacetime $^{(4)}g_{\mu\nu}$.

The text here uses Einstein notation in which summation over repeated indices is assumed.

Two types of derivatives are used: Partial derivatives are denoted either by the operator ∂_i or by subscripts preceded by a comma. Covariant derivatives are denoted either by the operator ∇_i or by subscripts preceded by a semicolon.

The absolute value of the determinant of the matrix of metric tensor coefficients is represented by g (with no indices). Other tensor symbols written without indices represent the trace of the corresponding tensor such as $\pi = g^{ij}\pi_{ij}$.

Richard Arnowitt, Stanley Deser and Charles Misner at the ADM-50: A Celebration
of Current GR Innovation *conference held in November 2009[1] to honor the 50th anniversary of their paper.*

12.2.2 Lagrangian formulation

The starting point for the ADM formulation is the Lagrangian

$$\mathcal{L} = {}^{(4)}R\sqrt{{}^{(4)}g}$$

which is a product of the square root of the determinant of the four-dimensional metric tensor for the full spacetime and its Ricci scalar. This is the Lagrangian from the Einstein–Hilbert action.

The desired outcome of the derivation is to define an embedding of three-dimensional spatial slices in the four-dimensional spacetime. The metric of the three-dimensional slices

$$g_{ij} = {}^{(4)}g_{ij}$$

will be the generalized coordinates for a Hamiltonian formulation. The conjugate momenta can then be computed

$$\pi^{ij} = \sqrt{{}^{(4)}g}\left({}^{(4)}\Gamma^0_{pq} - g_{pq}\,{}^{(4)}\Gamma^0_{rs}g^{rs}\right)g^{ip}g^{jq}$$

using standard techniques and definitions. The symbols ${}^{(4)}\Gamma^0_{ij}$ are Christoffel symbols associated with the metric of the full four-dimensional spacetime. The lapse

$$N = \left(-{}^{(4)}g^{00}\right)^{-1/2}$$

and the shift vector

$$N_i = {}^{(4)}g_{0i}$$

are the remaining elements of the four-metric tensor.

Having identified the quantities for the formulation, the next step is to rewrite the Lagrangian in terms of these variables. The new expression for the Lagrangian

$$\mathcal{L} = -g_{ij}\partial_t\pi^{ij} - NH - N_iP^i - 2\partial_i(\pi^{ij}N_j - \tfrac{1}{2}\pi N^i + \nabla^i N\sqrt{g})$$

is conveniently written in terms of the two new quantities

$$H = -\sqrt{g}\left[{}^{(3)}R + g^{-1}\left(\tfrac{1}{2}\pi^2 - \pi^{ij}\pi_{ij}\right)\right]$$

and

$$P^i = -2\pi^{ij}{}_{;j}$$

which are known as the Hamiltonian constraint and the momentum constraint respectively. Note also that the lapse and the shift appear in the Hamiltonian as Lagrange multipliers.

12.2.3 Equations of motion

Although the variables in the Lagrangian represent the metric tensor on three-dimensional spaces embedded in the four-dimensional spacetime, it is possible and desirable to use the usual procedures from Lagrangian mechanics to derive "equations of motion" that describe the time evolution of both the metric g_{ij} and its conjugate momentum π^{ij}. The result

$$\partial_t g_{ij} = 2Ng^{-1/2}(\pi_{ij} - \tfrac{1}{2}\pi g_{ij}) + N_{i;j} + N_{j;i}$$

and

$$\partial_t \pi^{ij} = -N\sqrt{g}(R^{ij} - \frac{1}{2}Rg^{ij}) + \frac{1}{2}Ng^{-1/2}g^{ij}(\pi^{mn}\pi_{mn} - \frac{1}{2}\pi^2) - 2Ng^{-1/2}(\pi^{in}\pi_n{}^j - \frac{1}{2}\pi\pi^{ij})$$

$$-\sqrt{g}(\nabla^i\nabla^j N - g^{ij}\nabla^n\nabla_n N) + \nabla_n(\pi^{ij}N^n) - N^i{}_{;n}\pi^{nj} - N^j{}_{;n}\pi^{ni}$$

is a non-linear set of partial differential equations.

Taking variations with respect to the lapse and shift provide constraint equations

$$H = 0$$

and

$$P^i = 0$$

and the lapse and shift themselves can be freely specified, reflecting the fact that coordinate systems can be freely specified in both space and time.

12.3   Application to quantum gravity

Using the ADM formulation, it is possible to attempt to construct a quantum theory of gravity, in the same way that one constructs the Schrödinger equation corresponding to a given Hamiltonian in quantum mechanics. That is, replace the canonical momenta $\pi^{ij}(t, x^k)$ and the spatial metric functions by linear functional differential operators

$$\hat{g}_{ij}(t, x^k) \rightarrow g_{ij}(t, x^k)$$

$$\hat{\pi}^{ij}(t, x^k) \rightarrow -i\frac{\delta}{\delta g_{ij}(t, x^k)}$$

More precisely, the replacing of classical variables by operators is restricted by commutation relations. The hats represents operators in quantum theory. This leads to the Wheeler–DeWitt equation.

12.4   Application to numerical solutions of the Einstein equations

There are relatively few exact solutions to the Einstein field equations. In order to find other solutions, there is an active field of study known as numerical relativity in which supercomputers are used to find approximate solutions to the equations. In order to construct such solutions numerically, most researchers start with a formulation of the Einstein equations closely related to the ADM formulation. The most common approaches start with an initial value problem based on the ADM formalism.

In Hamiltonian formulations, the basic point is replacement of set of second order equations by another first order set of equations. We may get this second set of equations by Hamiltonian formulation in an easy way. Of course this is very useful for numerical physics, because the reduction of order of differential equations must be done, if we want to prepare equations for a computer.

12.5   ADM energy

ADM energy is a special way to define the energy in general relativity which is only applicable to some special geometries of spacetime that asymptotically approach a well-defined metric tensor at infinity — for example a spacetime that asymptotically approaches Minkowski space. The ADM energy in these cases is defined as a function of the deviation of the metric tensor from its prescribed asymptotic form. In other words, the ADM energy is computed as the strength of the gravitational field at infinity.

If the required asymptotic form is time-independent (such as the Minkowski space itself), then it respects the time-translational symmetry. Noether's theorem then implies that the ADM energy is conserved. According to general relativity, the conservation law for the total energy does not hold in more general, time-dependent backgrounds – for example, it is completely violated in physical cosmology. Cosmic inflation in particular is able to produce energy (and mass) from "nothing" because the vacuum energy density is roughly constant, but the volume of the Universe grows exponentially.

12.6 See also

- Canonical coordinates
- Canonical gravity
- Hamiltonian mechanics
- Hamilton–Jacobi–Einstein equation
- Wheeler–DeWitt equation
- Peres metric

12.7 References

[1] ADM-50: A Celebration of Current GR Innovation

[2] Arnowitt, R.; Deser, S.; Misner, C. (1959). "Dynamical Structure and Definition of Energy in General Relativity". *Physical Review* **116** (5): 1322–1330. Bibcode:1959PhRv..116.1322A. doi:10.1103/PhysRev.116.1322.

[3] Chapter 7 (pp. 227–265) of Louis Witten (ed.), *Gravitation: An introduction to current research*, Wiley: New York, 1962.

[4] Arnowitt, R.; Deser, S.; Misner, C. (2008). "Republication of: The dynamics of general relativity". *General Relativity and Gravitation* **40** (9): 1997–2027. arXiv:gr-qc/0405109. Bibcode:2008GReGr..40.1997A. doi:10.1007/s10714-008-0661-1.

[5] Arnowitt, R.; Deser, S. (1959). "Quantum Theory of Gravitation: General Formulation and Linearized Theory". *Physical Review* **113** (2): 745–750. Bibcode:1959PhRv..113..745A. doi:10.1103/PhysRev.113.745.

[6] Arnowitt, R.; Deser, S.; Misner, C. (1960). "Canonical Variables for General Relativity". *Physical Review* **117** (6): 1595–1602. Bibcode:1960PhRv..117.1595A. doi:10.1103/PhysRev.117.1595.

[7] Arnowitt, R.; Deser, S.; Misner, C. (1960). "Finite Self-Energy of Classical Point Particles". *Physical Review Letters* **4** (7): 375–377. Bibcode:1960PhRvL...4..375A. doi:10.1103/PhysRevLett.4.375.

[8] Arnowitt, R.; Deser, S.; Misner, C. (1960). "Energy and the Criteria for Radiation in General Relativity". *Physical Review* **118** (4): 1100–1104. Bibcode:1960PhRv..118.1100A. doi:10.1103/PhysRev.118.1100.

[9] Arnowitt, R.; Deser, S.; Misner, C. (1960). "Gravitational-Electromagnetic Coupling and the Classical Self-Energy Problem". *Physical Review* **120**: 313–320. Bibcode:1960PhRv..120..313A. doi:10.1103/PhysRev.120.313.

[10] Arnowitt, R.; Deser, S.; Misner, C. (1960). "Interior Schwarzschild Solutions and Interpretation of Source Terms". *Physical Review* **120**: 321–324. Bibcode:1960PhRv..120..321A. doi:10.1103/PhysRev.120.321.

[11] Arnowitt, R.; Deser, S.; Misner, C. (1961). "Wave Zone in General Relativity". *Physical Review* **121** (5): 1556–1566. Bibcode:1961PhRv..121.1556A. doi:10.1103/PhysRev.121.1556.

[12] Arnowitt, R.; Deser, S.; Misner, C. (1961). "Coordinate Invariance and Energy Expressions in General Relativity". *Physical Review* **122** (3): 997–1006. Bibcode:1961PhRv..122..997A. doi:10.1103/PhysRev.122.997.

- Kiefer, Claus (2007). *Quantum Gravity*. Oxford, New York: Oxford University Press. ISBN 978-0-19-921252-1.

Chapter 13

Alternatives to general relativity

Alternatives to general relativity are physical theories that attempt to describe the phenomena of gravitation in competition to Einstein's theory of general relativity.

There have been many different attempts at constructing an ideal theory of gravity. These attempts can be split into four broad categories:

- Straightforward alternatives to general relativity (GR), such as the Cartan, Brans–Dicke and Rosen bimetric theories.

- Those that attempt to construct a quantized gravity theory such as loop quantum gravity.

- Those that attempt to unify gravity and other forces such as Kaluza–Klein.

- Those that attempt to do several at once, such as M-theory.

This article deals only with straightforward alternatives to GR. For quantized gravity theories, see the article quantum gravity. For the unification of gravity and other forces, see the article classical unified field theories. For those theories that attempt to do several at once, see the article theory of everything.

13.1 Motivations

Motivations for developing new theories of gravity have changed over the years, with the first one to explain planetary orbits (Newton) and more complicated orbits (e.g. Lagrange). Then came unsuccessful attempts to combine gravity and either wave or corpuscular theories of gravity. The whole landscape of physics was changed with the discovery of Lorentz transformations, and this led to attempts to reconcile it with gravity. At the same time, experimental physicists started testing the foundations of gravity and relativity – Lorentz invariance, the gravitational deflection of light, the Eötvös experiment. These considerations led to and past the development of general relativity.

After that, motivations differ. Two major concerns were the development of quantum theory and the discovery of the strong and weak nuclear forces. Attempts to quantize and unify gravity are outside the scope of this article, and so far none has been completely successful.

After general relativity (GR), attempts were made either to improve on theories developed before GR, or to improve GR itself. Many different strategies were attempted, for example the addition of spin to GR, combining a GR-like metric with a space-time that is static with respect to the expansion of the universe, getting extra freedom by adding another parameter. At least one theory was motivated by the desire to develop an alternative to GR that is completely free from singularities.

Experimental tests improved along with the theories. Many of the different strategies that were developed soon after GR were abandoned, and there was a push to develop more general forms of the theories that survived, so that a theory would be ready the moment any test showed a disagreement with GR.

By the 1980s, the increasing accuracy of experimental tests had all led to confirmation of GR, no competitors were left except for those that included GR as a special case. Further, shortly after that, theorists switched to string theory which was starting to look promising, but has since lost popularity. In the mid-1980s a few experiments were suggesting that gravity was being modified by the addition of a fifth force (or, in one case, of a fifth, sixth and seventh force) acting on the scale of meters. Subsequent experiments eliminated these.

Motivations for the more recent alternative theories are almost all cosmological, associated with or replacing such constructs as "inflation", "dark matter" and "dark energy". Investigation of the Pioneer anomaly has caused renewed public interest in alternatives to General Relativity.

13.2 Notation in this article

Main article: Mathematics of general relativity

c is the speed of light, G is the gravitational constant. "Geometric variables" are not used.

Latin indexes go from 1 to 3, Greek indexes go from 0 to 3. The Einstein summation convention is used.

$\eta_{\mu\nu}$ is the Minkowski metric. $g_{\mu\nu}$ is a tensor, usually the metric tensor. These have signature $(-,+,+,+)$.

Partial differentiation is written $\partial_\mu \phi$ or $\phi_{,\mu}$. Covariant differentiation is written $\nabla_\mu \phi$ or $\phi_{;\mu}$.

13.3 Classification of theories

Theories of gravity can be classified, loosely, into several categories. Most of the theories described here have:

- an 'action' (see the principle of least action, a variational principle based on the concept of action)

- a Lagrangian density

- a metric

If a theory has a Lagrangian density for gravity, say L, then the gravitational part of the action S is the integral of that.

$$S = \int L\sqrt{-g}\, \mathrm{d}^4 x$$

In this equation it is usual, though not essential, to have $g = -1$ at spatial infinity when using Cartesian coordinates. For example the Einstein–Hilbert action uses

$$L \propto R$$

where R is the scalar curvature, a measure of the curvature of space.

Almost every theory described in this article has an action. It is the only known way to guarantee that the necessary conservation laws of energy, momentum and angular momentum are incorporated automatically; although it is easy to construct an action where those conservation laws are violated. The original 1983 version of MOND did not have an action.

A few theories have an action but not a Lagrangian density. A good example is Whitehead (1922), the action there is termed non-local.

A theory of gravity is a "metric theory" if and only if it can be given a mathematical representation in which two conditions hold:

Condition 1: There exists a symmetric metric tensor $g_{\mu\nu}$ of signature (−, +, +, +), which governs proper-length and proper-time measurements in the usual manner of special and general relativity:

$$d\tau^2 = -g_{\mu\nu}\, dx^\mu\, dx^\nu$$

where there is a summation over indices μ and ν.

Condition 2: Stressed matter and fields being acted upon by gravity respond in accordance with the equation:

$$0 = \nabla_\nu T^{\mu\nu} = T^{\mu\nu}{}_{,\nu} + \Gamma^\mu_{\sigma\nu} T^{\sigma\nu} + \Gamma^\nu_{\sigma\nu} T^{\mu\sigma}$$

where $T^{\mu\nu}$ is the stress–energy tensor for all matter and non-gravitational fields, and where ∇_ν is the covariant derivative with respect to the metric and $\Gamma^\alpha_{\sigma\nu}$ is the Christoffel symbol. The stress–energy tensor should also satisfy an energy condition.

Metric theories include (from simplest to most complex):

- Scalar field theories (includes Conformally flat theories & Stratified theories with conformally flat space slices)
  - Bergman
  - Coleman
  - Einstein (1912)
  - Einstein–Fokker theory
  - Lee–Lightman–Ni
  - Littlewood
  - Ni
  - Nordström's theory of gravitation (first metric theory of gravity to be developed)
  - Page–Tupper
  - Papapetrou
  - Rosen (1971)
  - Whitrow–Morduch
  - Yilmaz theory of gravitation (attempted to eliminate event horizons from the theory.)
- Quasilinear theories (includes Linear fixed gauge)
  - Bollini–Giambiagi–Tiomno
  - Deser–Laurent
  - Whitehead's theory of gravity (intended to use only retarded potentials)
- Tensor theories
  - Einstein's GR
  - Fourth order gravity (allows the Lagrangian to depend on second-order contractions of the Riemann curvature tensor)
    - f(R) gravity (allows the Lagrangian to depend on higher powers of the Ricci scalar)
  - Gauss–Bonnet gravity
  - Lovelock theory of gravity (allows the Lagrangian to depend on higher-order contractions of the Riemann curvature tensor)
- Scalar-tensor theories

- Bekenstein
- Bergmann-Wagoner
- Brans–Dicke theory (the most well-known alternative to GR, intended to be better at applying Mach's principle)
- Jordan
- Nordtvedt
- Thiry
- Chameleon
- Pressuron

- Vector-tensor theories
  - Hellings–Nordtvedt
  - Will–Nordtvedt

- Bimetric theories
  - Lightman–Lee
  - Rastall
  - Rosen (1975)

- Other metric theories

(see section Modern theories below)

Non-metric theories include

- Belinfante–Swihart
- Einstein–Cartan theory (intended to handle spin-orbital angular momentum interchange)
- Kustaanheimo (1967)
- Teleparallelism
- Gauge theory gravity

A word here about Mach's principle is appropriate because a few of these theories rely on Mach's principle (e.g. Whitehead (1922)), and many mention it in passing (e.g. Einstein–Grossmann (1913), Brans–Dicke (1961)). Mach's principle can be thought of a half-way-house between Newton and Einstein. It goes this way:[1]

- Newton: Absolute space and time.
- Mach: The reference frame comes from the distribution of matter in the universe.
- Einstein: There is no reference frame.

So far, all the experimental evidence points to Mach's principle being wrong, but it has not entirely been ruled out.

13.4   Early theories, 1686 to 1916

Main articles: History of gravitational theory and History of general relativity

Newton (1686)

In Newton's (1686) theory (rewritten using more modern mathematics) the density of mass ρ generates a scalar field, the gravitational potential ϕ in joules per kilogram, by

$$\frac{\partial^2 \phi}{\partial x^j \partial x^j} = 4\pi G \rho \,.$$

Using the Nabla operator ∇ for the gradient and divergence (partial derivatives), this can be conveniently written as:

$$\nabla^2 \phi = 4\pi G \rho \,.$$

This scalar field governs the motion of a free-falling particle by:

$$\frac{d^2 x^j}{dt^2} = -\frac{\partial \phi}{\partial x^j} \,.$$

At distance, r, from an isolated mass, M, the scalar field is

$$\phi = -GM/r \,.$$

The theory of Newton, and Lagrange's improvement on the calculation (applying the variational principle), completely fails to take into account relativistic effects of course, and so can be rejected as a viable theory of gravity. Even so, Newton's theory is thought to be exactly correct in the limit of weak gravitational fields and low speeds and all other theories of gravity need to reproduce Newton's theory in the appropriate limits.

Mechanical explanations (1650–1900)

To explain Newton's theory, some mechanical explanations of gravitation (incl. Le Sage's theory) were created between 1650 and 1900, but they were overthrown because most of them lead to an unacceptable amount of drag, which is not observed. Other models are violating the energy conservation law and are incompatible with modern thermodynamics.

Electrostatic models (1870–1900)

At the end of the 19th century, many tried to combine Newton's force law with the established laws of electrodynamics, like those of Weber, Carl Friedrich Gauss, Bernhard Riemann and James Clerk Maxwell. Those models were used to explain the perihelion advance of Mercury. In 1890, Lévy succeeded in doing so by combining the laws of Weber and Riemann, whereby the speed of gravity is equal to the speed of light in his theory. And in another attempt, Paul Gerber (1898) even succeeded in deriving the correct formula for the Perihelion shift (which was identical to that formula later used by Einstein). However, because the basic laws of Weber and others were wrong (for example, Weber's law was superseded by Maxwell's theory), those hypothesis were rejected.[2] In 1900, Hendrik Lorentz tried to explain gravity on the basis of his Lorentz ether theory and the Maxwell equations. He assumed, like Ottaviano Fabrizio Mossotti and Johann Karl Friedrich Zöllner, that the attraction of opposite charged particles is stronger than the repulsion of equal charged particles. The resulting net force is exactly what is known as universal gravitation, in which the speed of gravity is that of light. But Lorentz calculated that the value for the perihelion advance of Mercury was much too low.[3]

Lorentz-invariant models (1905–1910)

Based on the principle of relativity, Henri Poincaré (1905, 1906), Hermann Minkowski (1908), and Arnold Sommerfeld (1910) tried to modify Newton's theory and to establish a Lorentz invariant gravitational law, in which the speed of gravity is that of light. However, as in Lorentz's model, the value for the perihelion advance of Mercury was much too low.[4]

Einstein (1908, 1912)

Einstein's two part publication in 1912 (and before in 1908) is really only important for historical reasons. By then he knew of the gravitational redshift and the deflection of light. He had realized that Lorentz transformations are not generally applicable, but retained them. The theory states that the speed of light is constant in free space but varies in the presence of matter. The theory was only expected to hold when the source of the gravitational field is stationary. It includes the principle of least action:

$$\delta \int d\tau = 0$$

$$d\tau^2 = -\eta_{\mu\nu} dx^\mu dx^\nu$$

where $\eta_{\mu\nu}$ is the Minkowski metric, and there is a summation from 1 to 4 over indices μ and ν.

Einstein and Grossmann (1913) includes Riemannian geometry and tensor calculus.

$$\delta \int d\tau = 0$$

$$d\tau^2 = -g_{\mu\nu} dx^\mu dx^\nu$$

The equations of electrodynamics exactly match those of GR. The equation

$$T^{\mu\nu} = \rho \frac{dx^\mu}{d\tau} \frac{dx^\nu}{d\tau}$$

is not in GR. It expresses the stress–energy tensor as a function of the matter density.

Abraham (1912)

While this was going on, Abraham was developing an alternative model of gravity in which the speed of light depends on the gravitational field strength and so is variable almost everywhere. Abraham's 1914 review of gravitation models is said to be excellent, but his own model was poor.

Nordström (1912)

The first approach of Nordström (1912) was to retain the Minkowski metric and a constant value of c but to let mass depend on the gravitational field strength ϕ. Allowing this field strength to satisfy

$$\Box \phi = \rho$$

where ρ is rest mass energy and $\Box$ is the d'Alembertian,

$$m = m_0 \exp(\phi/c^2)$$

and

$$-\frac{\partial \phi}{\partial x^{\mu}} = \dot{u}_{\mu} + \frac{u_{\mu}}{c^2 \phi}$$

where u is the four-velocity and the dot is a differential with respect to time.

The second approach of Nordström (1913) is remembered as the first logically consistent relativistic field theory of gravitation ever formulated. From (note, notation of Pais (1982) not Nordström):

$$\delta \int \psi d\tau = 0$$

$$d\tau^2 = -\eta_{\mu\nu} dx^{\mu} dx^{\nu}$$

where ψ is a scalar field.

$$-\frac{\partial T^{\mu\nu}}{\partial x^{\nu}} = T \frac{1}{\psi} \frac{\partial \psi}{\partial x_{\mu}}$$

This theory is Lorentz invariant, satisfies the conservation laws, correctly reduces to the Newtonian limit and satisfies the weak equivalence principle.

Einstein and Fokker (1914)

This theory is Einstein's first treatment of gravitation in which general covariance is strictly obeyed. Writing:

$$\delta \int ds = 0$$

$$ds^2 = g_{\mu\nu} dx^{\mu} dx^{\nu}$$

$$g_{\mu\nu} = \psi^2 \eta_{\mu\nu}$$

they relate Einstein-Grossmann (1913) to Nordström (1913). They also state:

$$T \propto R.$$

That is, the trace of the stress energy tensor is proportional to the curvature of space.

Einstein (1916, 1917)

This theory is what we now know of as General Relativity. Discarding the Minkowski metric entirely, Einstein gets:

$$\delta \int ds = 0$$

$$ds^2 = g_{\mu\nu} dx^{\mu} dx^{\nu}$$

$$R_{\mu\nu} = \frac{8\pi G}{c^4} \left(T_{\mu\nu} - \frac{1}{2} g_{\mu\nu} T \right)$$

which can also be written

$$T^{\mu\nu} = \frac{c^4}{8\pi G} \left(R^{\mu\nu} - \frac{1}{2} g^{\mu\nu} R \right) .$$

Five days before Einstein presented the last equation above, Hilbert had submitted a paper containing an almost identical equation. See relativity priority dispute. Hilbert was the first to correctly state the Einstein–Hilbert action for GR, which is:

$$S = \frac{c^4}{16\pi G} \int R\sqrt{-g} d^4 x + S_m$$

where G is Newton's gravitational constant, $R = R_\mu{}^\mu$ is the Ricci curvature of space, $g = \det(g_{\mu\nu})$ and S_m is the action due to mass.

GR is a tensor theory, the equations all contain tensors. Nordström's theories, on the other hand, are scalar theories because the gravitational field is a scalar. Later in this article you will see scalar-tensor theories that contain a scalar field in addition to the tensors of GR, and other variants containing vector fields as well have been developed recently.

13.5 Theories from 1917 to the 1980s

This section includes alternatives to GR published after GR but before the observations of galaxy rotation that led to the hypothesis of "dark matter".

Those considered here include (see Will (1981),[5] Lang (2002)[6]):

Listed by date (the hyperlinks take you further down this article)

Whitehead (1922), Cartan (1922, 1923), Fierz & Pauli (1939), Birkhov (1943), Milne (1948), Thiry (1948), Papapetrou (1954a, 1954b), Littlewood (1953), Jordan (1955), Bergman (1956), Belinfante & Swihart (1957), Yilmaz (1958, 1973), Brans & Dicke (1961), Whitrow & Morduch (1960, 1965), Kustaanheimo (1966), Kustaanheimo & Nuotio (1967), Deser & Laurent (1968), Page & Tupper (1968), Bergmann (1968), Bollini-Giambiagi-Tiomno (1970), Nordtveldt (1970), Wagoner (1970), Rosen (1971, 1975, 1975), Wei-Tou Ni (1972, 1973), Will & Nordtveldt (1972), Hellings & Nordtveldt (1973), Lightman & Lee (1973), Lee, Lightman & Ni (1974), Bekenstein (1977), Barker (1978), Rastall (1979)

These theories are presented here without a cosmological constant or added scalar or vector potential unless specifically noted, for the simple reason that the need for one or both of these was not recognised before the supernova observations by the Supernova Cosmology Project and High-Z Supernova Search Team. How to add a cosmological constant or quintessence to a theory is discussed under Modern Theories (see also here).

13.5.1 Scalar field theories

See also: Scalar theories of gravitation

The scalar field theories of Nordström (1912, 1913) have already been discussed. Those of Littlewood (1953), Bergman (1956), Yilmaz (1958), Whitrow and Morduch (1960, 1965) and Page and Tupper (1968) follow the general formula give by Page and Tupper.

According to Page and Tupper (1968), who discuss all these except Nordström (1913), the general scalar field theory comes from the principle of least action:

$$\delta \int f\left(\frac{\phi}{c^2}\right) ds = 0$$

where the scalar field is,

$$\phi = GM/r$$

and c may or may not depend on ϕ.

In Nordström (1912),

$$f(\phi/c^2) = \exp(-\phi/c^2), \qquad c = c_\infty$$

In Littlewood (1953) and Bergmann (1956),

$$f(\phi/c^2) = \exp(-\phi/c^2 - (\phi/c^2)^2/2), \qquad c = c_\infty$$

In Whitrow and Morduch (1960),

$$f(\phi/c^2) = 1, \qquad c^2 = c_\infty^2 - 2\phi$$

In Whitrow and Morduch (1965),

$$f(\phi/c^2) = \exp(-\phi/c^2), \qquad c^2 = c_\infty^2 - 2\phi$$

In Page and Tupper (1968),

$$f(\phi/c^2) = \phi/c^2 + \alpha(\phi/c^2)^2, \qquad c_\infty^2/c^2 = 1 + 4(\phi/c_\infty^2) + (15 + 2\alpha)(\phi/c_\infty^2)^2$$

Page and Tupper (1968) matches Yilmaz (1958) (see also Yilmaz theory of gravitation) to second order when $\alpha = -7/2$.

The gravitational deflection of light has to be zero when c is constant. Given that variable c and zero deflection of light are both in conflict with experiment, the prospect for a successful scalar theory of gravity looks very unlikely. Further, if the parameters of a scalar theory are adjusted so that the deflection of light is correct then the gravitational redshift is likely to be wrong.

Ni (1972) summarised some theories and also created two more. In the first, a pre-existing special relativity space-time and universal time coordinate acts with matter and non-gravitational fields to generate a scalar field. This scalar field acts together with all the rest to generate the metric.

The action is:

$$S = \frac{1}{16\pi G} \int d^4x \sqrt{-g} L_\phi + S_m$$

$$L_\phi = \phi R - 2g^{\mu\nu}\partial_\mu\phi\partial_\nu\phi$$

Misner et al. (1973) gives this without the ϕR term. S_m is the matter action.

$$\Box\phi = 4\pi T^{\mu\nu} \left[\eta_{\mu\nu}e^{-2\phi} + \left(e^{2\phi} + e^{-2\phi}\right)\partial_\mu t\partial_\nu t\right]$$

t is the universal time coordinate. This theory is self-consistent and complete. But the motion of the solar system through the universe leads to serious disagreement with experiment.

In the second theory of Ni (1972) there are two arbitrary functions $f(\phi)$ and $k(\phi)$ that are related to the metric by:

$$ds^2 = e^{-2f(\phi)}dt^2 - e^{2f(\phi)}\left[dx^2 + dy^2 + dz^2\right]$$

$$\eta^{\mu\nu}\partial_\mu\partial_\nu\phi = 4\pi\rho^* k(\phi)$$

Ni (1972) quotes Rosen (1971) as having two scalar fields ϕ and ψ that are related to the metric by:

$$ds^2 = \phi^2 dt^2 - \psi^2\left[dx^2 + dy^2 + dz^2\right]$$

In Papapetrou (1954a) the gravitational part of the Lagrangian is:

$$L_\phi = e^\phi \left(\tfrac{1}{2}e^{-\phi}\partial_\alpha\phi\partial_\alpha\phi + \tfrac{3}{2}e^\phi\partial_0\phi\partial_0\phi\right)$$

In Papapetrou (1954b) there is a second scalar field χ. The gravitational part of the Lagrangian is now:

$$L_\phi = e^{\frac{1}{2}(3\phi+\chi)} \left(-\tfrac{1}{2}e^{-\phi}\partial_\alpha\phi\partial_\alpha\phi - e^{-\phi}\partial_\alpha\phi\partial_\chi\phi + \tfrac{3}{2}e^{-\chi}\partial_0\phi\partial_0\phi\right)$$

13.5.2 Bimetric theories

See also: Bimetric theory

Bimetric theories contain both the normal tensor metric and the Minkowski metric (or a metric of constant curvature), and may contain other scalar or vector fields.

Rosen (1973, 1975) Bimetric Theory The action is:

$$S = \frac{1}{64\pi G}\int d^4x\sqrt{-\eta}\,\eta^{\mu\nu}g^{\alpha\beta}g^{\gamma\delta}(g_{\alpha\gamma|\mu}g_{\alpha\delta|\nu} - \tfrac{1}{2}g_{\alpha\beta|\mu}g_{\gamma\delta|\nu}) + S_m$$

where the vertical line "|" denotes covariant derivative with respect to η. The field equations may be written in the form:

$$\Box_\eta g_{\mu\nu} - g^{\alpha\beta}\eta^{\gamma\delta}g_{\mu\alpha|\gamma}g_{\nu\beta|\delta} = -16\pi G\sqrt{g/\eta}(T_{\mu\nu} - \tfrac{1}{2}g_{\mu\nu}T)$$

Lightman-Lee (1973) developed a metric theory based on the non-metric theory of Belinfante and Swihart (1957a, 1957b). The result is known as BSLL theory. Given a tensor field $B_{\mu\nu}$, $B = B_{\mu\nu}\eta^{\mu\nu}$, and two constants a and f the action is:

$$S = \frac{1}{16\pi G}\int d^4x\sqrt{-\eta}(aB^{\mu\nu|\alpha}B_{\mu\nu|\alpha} + fB_{,\alpha}B^{,\alpha}) + S_m$$

and the stress–energy tensor comes from:

$$a\Box_\eta B^{\mu\nu} + f\eta^{\mu\nu}\Box_\eta B = -4\pi G\sqrt{g/\eta}T^{\alpha\beta}(\partial g_{\alpha\beta}/\partial B_{\mu\nu})$$

In Rastall (1979), the metric is an algebraic function of the Minkowski metric and a Vector field.[5] The Action is:

$$S = \frac{1}{16\pi G} \int d^4 x \sqrt{-g} F(N) K^{\mu\nu} K_{\mu\nu} + S_m$$

where

$$F(N) = -N/(2+N) \text{ and } N = g^{\mu\nu} K_\mu K_\nu$$

(see Will (1981) for the field equation for $T^{\mu\nu}$ and K_μ).

13.5.3   Quasilinear theories

In Whitehead (1922), the physical metric g is constructed (by Synge) algebraically from the Minkowski metric η and matter variables, so it doesn't even have a scalar field. The construction is:

$$g_{\mu\nu}(x^\alpha) = \eta_{\mu\nu} - 2 \int_{\Sigma^-} \frac{y_\mu^- y_\nu^-}{(w^-)^3} [\sqrt{-g}\rho u^\alpha d\Sigma_\alpha]^-$$

where the superscript (-) indicates quantities evaluated along the past η light cone of the field point x^α and

$$(y^\mu)^- = x^\mu - (x^\mu)^- \ , (y^\mu)^-(y_\mu)^- = 0,$$
$$w^- = (y^\mu)^-(u_\mu)^- \ , (u_\mu) = dx^\mu/d\sigma,$$
$$d\sigma^2 = \eta_{\mu\nu} dx^\mu dx^\nu$$

Nevertheless the metric construction (from a non-metric theory) using the "length contraction" ansatz[7] is criticised.[8]

Deser and Laurent (1968) and Bollini-Giambiagi-Tiomno (1970) are Linear Fixed Gauge (LFG) theories. Taking an approach from quantum field theory, combine a Minkowski spacetime with the gauge invariant action of a spin-two tensor field (i.e. graviton) $h_{\mu\nu}$ to define

$$g_{\mu\nu} = \eta_{\mu\nu} + h_{\mu\nu}$$

The action is:

$$S = \frac{1}{16\pi G} \int d^4 x \sqrt{-\eta} [2h^{\mu\nu}_{|\nu} h^{|\lambda}_{\mu\lambda} - 2h^{\mu\nu}_{|\nu} h^\lambda_{\lambda|\mu} + h^\nu_{\nu|\mu} h^{\lambda|\mu}_\lambda - h^{\mu\nu|\lambda} h_{\mu\nu|\lambda}] + S_m$$

The Bianchi identity associated with this partial gauge invariance is wrong. LFG theories seek to remedy this by breaking the gauge invariance of the gravitational action through the introduction of auxiliary gravitational fields that couple to $h_{\mu\nu}$.

A cosmological constant can be introduced into a quasilinear theory by the simple expedient of changing the Minkowski background to a de Sitter or anti-de Sitter spacetime, as suggested by G. Temple in 1923. Temple's suggestions on how to do this were criticized by C. B. Rayner in 1955.[9]

13.5.4   Tensor theories

Einstein's general relativity is the simplest plausible theory of gravity that can be based on just one symmetric tensor field (the metric tensor). Others include: Gauss–Bonnet gravity, f(R) gravity, and Lovelock theory of gravity.

13.5.5 Scalar-tensor theories

See also: Scalar-tensor theory, Brans–Dicke theory, Dilaton, Chameleon_particle and Pressuron

These all contain at least one free parameter, as opposed to GR which has no free parameters.

Although not normally considered a Scalar-Tensor theory of gravity, the 5 by 5 metric of Kaluza–Klein reduces to a 4 by 4 metric and a single scalar. So if the 5th element is treated as a scalar gravitational field instead of an electromagnetic field then Kaluza–Klein can be considered the progenitor of Scalar-Tensor theories of gravity. This was recognised by Thiry (1948).

Scalar-Tensor theories include Thiry (1948), Jordan (1955), Brans and Dicke (1961), Bergman (1968), Nordtveldt (1970), Wagoner (1970), Bekenstein (1977) and Barker (1978).

The action S is based on the integral of the Lagrangian L_ϕ .

$$S = \frac{1}{16\pi G} \int d^4x \sqrt{-g} L_\phi + S_m$$

$$L_\phi = \phi R - \frac{\omega(\phi)}{\phi} g^{\mu\nu} \partial_\mu \phi \partial_\nu \phi + 2\phi\lambda(\phi)$$

$$S_m = \int d^4x \sqrt{g} G_N L_m$$

$$T^{\mu\nu} \overset{\text{def}}{=} \frac{2}{\sqrt{g}} \frac{\delta S_m}{\delta g_{\mu\nu}}$$

where $\omega(\phi)$ is a different dimensionless function for each different scalar-tensor theory. The function $\lambda(\phi)$ plays the same role as the cosmological constant in GR. G_N is a dimensionless normalization constant that fixes the present-day value of G . An arbitrary potential can be added for the scalar.

The full version is retained in Bergman (1968) and Wagoner (1970). Special cases are:

Nordtvedt (1970), $\lambda = 0$

Since λ was thought to be zero at the time anyway, this would not have been considered a significant difference. The role of the cosmological constant in more modern work is discussed under Cosmological constant.

Brans–Dicke (1961), ω is constant

Bekenstein (1977) Variable Mass Theory Starting with parameters r and q , found from a cosmological solution, $\phi = [1 - qf(\phi)]f(\phi)^{-r}$ determines function f then

$$\omega(\phi) = -\frac{3}{2} - \frac{1}{4}f(\phi)[(1 - 6q)qf(\phi) - 1][r + (1 - r)qf(\phi)]^{-2}$$

Barker (1978) Constant G Theory

$$\omega(\phi) = (4 - 3\phi)/(2\phi - 2)$$

Adjustment of $\omega(\phi)$ allows Scalar Tensor Theories to tend to GR in the limit of $\omega \to \infty$ in the current epoch. However, there could be significant differences from GR in the early universe.

So long as GR is confirmed by experiment, general Scalar-Tensor theories (including Brans–Dicke) can never be ruled out entirely, but as experiments continue to confirm GR more precisely and the parameters have to be fine-tuned so that the predictions more closely match those of GR.

13.5.6   Vector-tensor theories

Before we start, Will (2001) has said: "Many alternative metric theories developed during the 1970s and 1980s could be viewed as "straw-man" theories, invented to prove that such theories exist or to illustrate particular properties. Few of these could be regarded as well-motivated theories from the point of view, say, of field theory or particle physics. Examples are the vector-tensor theories studied by Will, Nordtvedt and Hellings."

Hellings and Nordtvedt (1973) and Will and Nordtvedt (1972) are both vector-tensor theories. In addition to the metric tensor there is a timelike vector field K_μ . The gravitational action is:

$$S = \frac{1}{16\pi G} \int d^4x \sqrt{-g}[R + \omega K_\mu K^\mu R + \eta K^\mu K^\nu R_{\mu\nu} - \epsilon F_{\mu\nu}F^{\mu\nu} + \tau K_{\mu;\nu}K^{\mu;\nu}] + S_m$$

where ω , η , ϵ and τ are constants and

$$F_{\mu\nu} = K_{\nu;\mu} - K_{\mu;\nu}$$

See Will (1981) for the field equations for $T^{\mu\nu}$ and K_μ .

Will and Nordtvedt (1972) is a special case where

$$\omega = \eta = \epsilon = 0 \; ; \tau = 1$$

Hellings and Nordtvedt (1973) is a special case where

$$\tau = 0 \; ; \epsilon = 1 \; ; \eta = -2\omega$$

These vector-tensor theories are semi-conservative, which means that they satisfy the laws of conservation of momentum and angular momentum but can have preferred frame effects. When $\omega = \eta = \epsilon = \tau = 0$ they reduce to GR so, so long as GR is confirmed by experiment, general vector-tensor theories can never be ruled out.

13.5.7   Other metric theories

Others metric theories have been proposed: that of Bekenstein (2004) is discussed under Modern Theories.

13.5.8   Non-metric theories

See also: Einstein–Cartan theory and Cartan connection

Cartan's theory is particularly interesting both because it is a non-metric theory and because it is so old. The status of Cartan's theory is uncertain. Will (1981) claims that all non-metric theories are eliminated by Einstein's Equivalence Principle (EEP). Will (2001) tempers that by explaining experimental criteria for testing non-metric theories against EEP. Misner et al. (1973) claims that Cartan's theory is the only non-metric theory to survive all experimental tests up to that date and Turyshev (2006) lists Cartan's theory among the few that have survived all experimental tests up to that date. The following is a quick sketch of Cartan's theory as restated by Trautman (1972).

Cartan (1922, 1923) suggested a simple generalization of Einstein's theory of gravitation. He proposed a model of space time with a metric tensor and a linear "connection" compatible with the metric but not necessarily symmetric. The torsion tensor of the connection is related to the density of intrinsic angular momentum. Independently of Cartan, similar ideas were put forward by Sciama, by Kibble in the years 1958 to 1966, culminating in a 1976 review by Hehl et al.

The original description is in terms of differential forms, but for the present article that is replaced by the more familiar language of tensors (risking loss of accuracy). As in GR, the Lagrangian is made up of a massless and a mass part. The Lagrangian for the massless part is:

$$L = \frac{1}{32\pi G}\Omega_\nu^\mu g^{\nu\xi} x^\eta x^\zeta \varepsilon_{\xi\mu\eta\zeta}$$

$$\Omega_\nu^\mu = d\omega_\nu^\mu + \omega_\xi^\eta$$

$$\nabla x^\mu = -\omega_\nu^\mu x^\nu$$

The ω_ν^μ is the linear connection. $\varepsilon_{\xi\mu\eta\zeta}$ is the completely antisymmetric pseudo-tensor (Levi-Civita symbol) with $\varepsilon_{0123} = \sqrt{-g}$, and $g^{\nu\xi}$ is the metric tensor as usual. By assuming that the linear connection is metric, it is possible to remove the unwanted freedom inherent in the non-metric theory. The stress–energy tensor is calculated from:

$$T^{\mu\nu} = \frac{1}{16\pi G}(g^{\mu\nu}\eta_\eta^\xi - g^{\xi\mu}\eta_\eta^\nu - g^{\xi\nu}\eta_\eta^\mu)\Omega_\xi^\eta$$

The space curvature is not Riemannian, but on a Riemannian space-time the Lagrangian would reduce to the Lagrangian of GR.

Some equations of the non-metric theory of Belinfante and Swihart (1957a, 1957b) have already been discussed in the section on bimetric theories.

A distinctively non-metric theory is given by gauge theory gravity, which replaces the metric in its field equations with a pair of gauge fields in flat spacetime. On the one hand, the theory is quite conservative because it is substantially equivalent to Einstein–Cartan theory (or general relativity in the limit of vanishing spin), differing mostly in the nature of its global solutions. On the other hand, it is radical because it replaces differential geometry with geometric algebra.

13.6 Testing of alternatives to general relativity

Main article: Tests of general relativity

Any putative alternative to general relativity would need to meet a variety of tests for it to become accepted. For in-depth coverage of these tests, see Misner et al. (1973) Ch.39, Will (1981) Table 2.1, and Ni (1972). Most such tests can be categorized as in the following subsections.

13.6.1 Self-consistency

Self-consistency among non-metric theories includes eliminating theories allowing tachyons, ghost poles and higher order poles, and those that have problems with behaviour at infinity.

Among metric theories, self-consistency is best illustrated by describing several theories that fail this test. The classic example is the spin-two field theory of Fierz and Pauli (1939); the field equations imply that gravitating bodies move in straight lines, whereas the equations of motion insist that gravity deflects bodies away from straight line motion. Yilmaz (1971, 1973) contains a tensor gravitational field used to construct a metric; it is mathematically inconsistent because the functional dependence of the metric on the tensor field is not well defined.

13.6.2 Completeness

To be complete, a theory of gravity must be capable of analysing the outcome of every experiment of interest. It must therefore mesh with electromagnetism and all other physics. For instance, any theory that cannot predict from first principles the movement of planets or the behaviour of atomic clocks is incomplete.

Many early theories are incomplete in that it is unclear whether the density ρ used by the theory should be calculated from the stress–energy tensor T as $\rho = T_{\mu\nu}u^\mu u^\nu$ or as $\rho = T_{\mu\nu}\delta^{\mu\nu}$, where u is the four-velocity, and δ is the Kronecker delta.

The theories of Thirry (1948) and Jordan (1955) are incomplete unless Jordan's parameter η is set to -1, in which case they match the theory of Brans–Dicke (1961) and so are worthy of further consideration.

Milne (1948) is incomplete because it makes no gravitational red-shift prediction.

The theories of Whitrow and Morduch (1960, 1965), Kustaanheimo (1966) and Kustaanheimo and Nuotio (1967) are either incomplete or inconsistent. The incorporation of Maxwell's equations is incomplete unless it is assumed that they are imposed on the flat background space-time, and when that is done they are inconsistent, because they predict zero gravitational redshift when the wave version of light (Maxwell theory) is used, and nonzero redshift when the particle version (photon) is used. Another more obvious example is Newtonian gravity with Maxwell's equations: light as photons is deflected by gravitational fields (by twice that of GR) but light as waves is not.

13.6.3   Classical tests

Main article: Tests of general relativity

There are three "classical" tests (dating back to the 1910s or earlier) of the ability of gravity theories to handle relativistic effects; they are:

- gravitational redshift

- gravitational lensing (generally tested around the Sun)

- anomalous perihelion advance of the planets (see Tests of General Relativity)

Each theory should reproduce the observed results in these areas, which have to date always aligned with the predictions of general relativity.

In 1964, Irwin I. Shapiro found a fourth test, called the Shapiro delay. It is usually regarded as a "classical" test as well.

13.6.4   Agreement with Newtonian mechanics and special relativity

As an example of disagreement with Newtonian experiments, Birkhoff (1943) theory predicts relativistic effects fairly reliably but demands that sound waves travel at the speed of light. This was the consequence of an assumption made to simplify handling the collision of masses.

13.6.5   The Einstein equivalence principle (EEP)

Main article: Equivalence principle

The EEP has three components.

The first is the uniqueness of free fall, also known as the Weak Equivalence Principle (WEP). This is satisfied if inertial mass is equal to gravitational mass. η is a parameter used to test the maximum allowable violation of the WEP. The first tests of the WEP were done by Eötvös before 1900 and limited η to less than 5×10^{-9}. Modern tests have reduced that to less than 5×10^{-13}.

The second is Lorentz invariance. In the absence of gravitational effects the speed of light is constant. The test parameter for this is δ. The first tests of Lorentz invariance were done by Michelson and Morley before 1890 and limited δ to less than 5×10^{-3}. Modern tests have reduced this to less than 1×10^{-21}.

The third is local position invariance, which includes spatial and temporal invariance. The outcome of any local non-gravitational experiment is independent of where and when it is performed. Spatial local position invariance is tested using gravitational redshift measurements. The test parameter for this is α. Upper limits on this found by Pound and Rebka in 1960 limited α to less than 0.1. Modern tests have reduced this to less than 1×10^{-4}.

Schiff's conjecture states that any complete, self-consistent theory of gravity that embodies the WEP necessarily embodies EEP. This is likely to be true if the theory has full energy conservation.

Metric theories satisfy the Einstein Equivalence Principle. Extremely few non-metric theories satisfy this. For example, the non-metric theory of Belinfante & Swihart (1957) is eliminated by the *THεμ* formalism for testing EEP. Gauge theory gravity is a notable exception, where the strong equivalence principle is essentially the minimal coupling of the gauge covariant derivative.

13.6.6 Parametric post-Newtonian (PPN) formalism

Main article: Parameterized post-Newtonian formalism

See also Tests of general relativity, Misner et al. (1973) and Will (1981) for more information.

Work on developing a standardized rather than ad-hoc set of tests for evaluating alternative gravitation models began with Eddington in 1922 and resulted in a standard set of PPN numbers in Nordtvedt and Will (1972) and Will and Nordtvedt (1972). Each parameter measures a different aspect of how much a theory departs from Newtonian gravity. Because we are talking about deviation from Newtonian theory here, these only measure weak-field effects. The effects of strong gravitational fields are examined later.

These ten are called : γ , β , η , α_1 , α_2 , α_3 , ζ_1 , ζ_2 , ζ_3 , ζ_4

γ is a measure of space curvature, being zero for Newtonian gravity and one for GR.

β is a measure of nonlinearity in the addition of gravitational fields, one for GR.

η is a check for preferred location effects.

α_1 , α_2 , α_3 measure the extent and nature of "preferred-frame effects". Any theory of gravity with at least one α nonzero is called a preferred-frame theory.

ζ_1 , ζ_2 , ζ_3 , ζ_4 , α_3 measure the extent and nature of breakdowns in global conservation laws. A theory of gravity possesses 4 conservation laws for energy-momentum and 6 for angular momentum only if all five are zero.

13.6.7 Strong gravity and gravitational waves

Main article: Tests of general relativity

PPN is only a measure of weak field effects. Strong gravity effects can be seen in compact objects such as white dwarfs, neutron stars, and black holes. Experimental tests such as the stability of white dwarfs, spin-down rate of pulsars, orbits of binary pulsars and the existence of a black hole horizon can be used as tests of alternative to GR.

GR predicts that gravitational waves travel at the speed of light. Many alternatives to GR say that gravitational waves travel faster than light. If true, this could result in failure of causality.

13.6.8 Cosmological tests

Many of these have been developed recently. For those theories that aim to replace dark matter, the galaxy rotation curve, the Tully-Fisher relation, the faster rotation rate of dwarf galaxies, and the gravitational lensing due to galactic clusters act as constraints.

For those theories that aim to replace inflation, the size of ripples in the spectrum of the cosmic microwave background radiation is the strictest test.

For those theories that incorporate or aim to replace dark energy, the supernova brightness results and the age of the universe can be used as tests.

Another test is the flatness of the universe. With GR, the combination of baryonic matter, dark matter and dark energy add up to make the universe exactly flat. As the accuracy of experimental tests improve, alternatives to GR that aim to replace dark matter or dark energy will have to explain why.

13.7   Results of testing theories

13.7.1   PPN parameters for a range of theories

(See Will (1981) and Ni (1972) for more details. Misner et al. (1973) gives a table for translating parameters from the notation of Ni to that of Will)

General Relativity is now more than 90 years old, during which one alternative theory of gravity after another has failed to agree with ever more accurate observations. One illustrative example is Parameterized post-Newtonian formalism (PPN).

The following table lists PPN values for a large number of theories. If the value in a cell matches that in the column heading then the full formula is too complicated to include here.

† The theory is incomplete, and ζ_4 can take one of two values. The value closest to zero is listed.

All experimental tests agree with GR so far, and so PPN analysis immediately eliminates all the scalar field theories in the table.

A full list of PPN parameters is not available for Whitehead (1922), Deser-Laurent (1968), Bollini-Giambiagi-Tiomino (1970), but in these three cases $\beta = \xi$, which is in strong conflict with GR and experimental results. In particular, these theories predict incorrect amplitudes for the Earth's tides. (A minor modification of Whitehead's theory avoids this problem. However, the modification predicts the Nordtvedt effect, which has been experimentally constrained.)

13.7.2   Theories that fail other tests

The stratified theories of Ni (1973), Lee Lightman and Ni (1974) are non-starters because they all fail to explain the perihelion advance of Mercury.

The bimetric theories of Lightman and Lee (1973), Rosen (1975), Rastall (1979) all fail some of the tests associated with strong gravitational fields.

The scalar-tensor theories include GR as a special case, but only agree with the PPN values of GR when they are equal to GR to within experimental error. As experimental tests get more accurate, the deviation of the scalar-tensor theories from GR is being squashed to zero.

The same is true of vector-tensor theories, the deviation of the vector-tensor theories from GR is being squashed to zero. Further, vector-tensor theories are semi-conservative; they have a nonzero value for α_2 which can have a measurable effect on the Earth's tides.

Non-metric theories, such as Belinfante and Swihart (1957a, 1957b), usually fail to agree with experimental tests of Einstein's equivalence principle.

And that leaves, as a likely valid alternative to GR, nothing except possibly Cartan (1922).

That was the situation until cosmological discoveries pushed the development of modern alternatives.

13.8 Modern theories 1980s to present

This section includes alternatives to GR published after the observations of galaxy rotation that led to the hypothesis of "dark matter".

There is no known reliable list of comparison of these theories.

Those considered here include: Beckenstein (2004), Moffat (1995), Moffat (2002), Moffat (2005a, b).

These theories are presented with a cosmological constant or added scalar or vector potential.

13.8.1 Motivations

Motivations for the more recent alternatives to GR are almost all cosmological, associated with or replacing such constructs as "inflation", "dark matter" and "dark energy". The basic idea is that gravity agrees with GR at the present epoch but may have been quite different in the early universe.

There was a slow dawning realisation in the physics world that there were several problems inherent in the then big bang scenario, two of these were the horizon problem and the observation that at early times when quarks were first forming there was not enough space on the universe to contain even one quark. Inflation theory was developed to overcome these. Another alternative was constructing an alternative to GR in which the speed of light was larger in the early universe.

The discovery of unexpected rotation curves for galaxies took everyone by surprise. Could there be more mass in the universe than we are aware of, or is the theory of gravity itself wrong? The consensus now is that the missing mass is "cold dark matter", but that consensus was only reached after trying alternatives to general relativity and some physicists still believe that alternative models of gravity might hold the answer.

The discovery of the accelerated expansion of the universe by the supernova surveys led to the rapid reinstatement of Einstein's cosmological constant, and quintessence arrived as an alternative to the cosmological constant. At least one new alternative to GR attempted to explain the supernova surveys' results in a completely different way.

Another observation that sparked recent interest in alternatives to General Relativity is the Pioneer anomaly. It was quickly discovered that alternatives to GR could explain this anomaly. This is now believed to be accounted for by non-uniform thermal radiation.

13.8.2 Cosmological constant and quintessence

(also see Cosmological constant, Einstein–Hilbert action, Quintessence (physics))

The cosmological constant Λ is a very old idea, going back to Einstein in 1917. The success of the Friedmann model of the universe in which $\Lambda = 0$ led to the general acceptance that it is zero, but the use of a non-zero value came back with a vengeance when data from supernovae indicated that the expansion of the universe is accelerating

First, let's see how it influences the equations of Newtonian gravity and General Relativity.

In Newtonian gravity, the addition of the cosmological constant changes the Newton-Poisson equation from:

$$\nabla^2 \phi = 4\pi \rho \, G;$$

to

$$\nabla^2 \phi - \Lambda \phi = 4\pi \rho \, G;$$

In GR, it changes the Einstein–Hilbert action from

$$S = \frac{1}{16\pi G} \int R\sqrt{-g} \, d^4 x + S_m$$

to

$$S = \frac{1}{16\pi G} \int (R - 2\Lambda)\sqrt{-g}\, d^4x + S_m$$

which changes the field equation

$$T^{\mu\nu} = \frac{1}{8\pi G}\left(R^{\mu\nu} - \frac{1}{2}g^{\mu\nu}R\right)$$

to

$$T^{\mu\nu} = \frac{1}{8\pi G}\left(R^{\mu\nu} - \frac{1}{2}g^{\mu\nu}R + g^{\mu\nu}\Lambda\right)$$

In alternative theories of gravity, a cosmological constant can be added to the action in exactly the same way.

The cosmological constant is not the only way to get an accelerated expansion of the universe in alternatives to GR. We've already seen how the scalar potential $\lambda(\phi)$ can be added to scalar tensor theories. This can also be done in every alternative the GR that contains a scalar field ϕ by adding the term $\lambda(\phi)$ inside the Lagrangian for the gravitational part of the action, the L_ϕ part of

$$S = \frac{1}{16\pi G} \int d^4x\, \sqrt{-g}L_\phi + S_m$$

Because $\lambda(\phi)$ is an arbitrary function of the scalar field, it can be set to give an acceleration that is large in the early universe and small at the present epoch. This is known as quintessence.

A similar method can be used in alternatives to GR that use vector fields, including Rastall (1979) and vector-tensor theories. A term proportional to

$$K^\mu K^\nu g_{\mu\nu}$$

is added to the Lagrangian for the gravitational part of the action.

13.8.3   Relativistic MOND

(see Modified Newtonian dynamics, Tensor-vector-scalar gravity, and Bekenstein (2004) for more details).

The original theory of MOND by Milgrom was developed in 1983 as an alternative to "dark matter". Departures from Newton's law of gravitation are governed by an acceleration scale, not a distance scale. MOND successfully explains the Tully-Fisher observation that the luminosity of a galaxy should scale as the fourth power of the rotation speed. It also explains why the rotation discrepancy in dwarf galaxies is particularly large.

There were several problems with MOND in the beginning.

1. It did not include relativistic effects

2. It violated the conservation of energy, momentum and angular momentum

3. It was inconsistent in that it gives different galactic orbits for gas and for stars

4. It did not state how to calculate gravitational lensing from galaxy clusters.

By 1984, problems 2 and 3 had been solved by introducing a Lagrangian (AQUAL). A relativistic version of this based on scalar-tensor theory was rejected because it allowed waves in the scalar field to propagate faster than light. The Lagrangian of the non-relativistic form is:

$$L = -\frac{a_0^2}{8\pi G} f \left[\frac{|\nabla \phi|^2}{a_0^2} \right] - \rho \phi$$

The relativistic version of this has:

$$L = -\frac{a_0^2}{8\pi G} \tilde{f} \left(l_0^2 g^{\mu\nu} \partial_\mu \phi \partial_\nu \phi \right)$$

with a nonstandard mass action. Here f and $\tilde{f}$ are arbitrary functions selected to give Newtonian and MOND behaviour in the correct limits, and $l_0 = c^2/a_0$ is the MOND length scale.

By 1988, a second scalar field (PCC) fixed problems with the earlier scalar-tensor version but is in conflict with the perihelion precession of Mercury and gravitational lensing by galaxies and clusters.

By 1997, MOND had been successfully incorporated in a stratified relativistic theory [Sanders], but as this is a preferred frame theory it has problems of its own.

Bekenstein (2004) introduced a tensor-vector-scalar model (TeVeS). This has two scalar fields ϕ and σ and vector field U_α. The action is split into parts for gravity, scalars, vector and mass.

$$S = S_g + S_s + S_v + S_m$$

The gravity part is the same as in GR.

$$S_s = -\frac{1}{2} \int \left[\sigma^2 h^{\alpha\beta} \phi_{,\alpha} \phi_{,\beta} + \frac{1}{2} G l_0^{-2} \sigma^4 F(kG\sigma^2) \right] \sqrt{-g} \, d^4x$$

$$S_v = -\frac{K}{32\pi G} \int \left[g^{\alpha\beta} g^{\mu\nu} U_{[\alpha,\mu]} U_{[\beta,\nu]} - \frac{2\lambda}{K} \left(g^{\mu\nu} U_\mu U_\nu + 1 \right) \right] \sqrt{-g} \, d^4x$$

$$S_m = \int L \left(\tilde{g}_{\mu\nu}, f^\alpha, f^\alpha_{|\mu}, \cdots \right) \sqrt{-g} \, d^4x$$

where

$$h^{\alpha\beta} = g^{\alpha\beta} - U^\alpha U^\beta$$

$$\tilde{g}^{\alpha\beta} = e^{2\phi} g^{\alpha\beta} + 2U^\alpha U^\beta \sinh(2\phi)$$

k, K are constants, square brackets in indices $U_{[\alpha,\mu]}$ represent anti-symmetrization, λ is a Lagrange multiplier (calculated elsewhere), and L is a Lagrangian translated from flat spacetime onto the metric $\tilde{g}^{\alpha\beta}$. Note that G need not equal the observed gravitational constant G_{Newton}.

F is an arbitrary function, and

$$F(\mu) = \frac{3}{4} \frac{\mu^2 (\mu - 2)^2}{1 - \mu}$$

is given as an example with the right asymptotic behaviour; note how it becomes undefined when $\mu = 1$

The PPN parameters of this theory are calculated in.[10] which shows that all its parameters are equal to GR's, except for

$$\alpha_1 = \frac{4G}{K}\left((2K-1)e^{-4\phi_0} - e^{4\phi_0} + 8\right) - 8$$

$$\alpha_2 = \frac{6G}{2-K} - \frac{2G(K+4)e^{4\phi_0}}{(2-K)^2} - 1$$

both of which expressed in geometric units where $c = G_{Newtonian} = 1$; so

$$G^{-1} = \frac{2}{2-K} + \frac{k}{4\pi}.$$

The parameter ϕ_0 measures the value of the scalar field ϕ at infinity, and is given by

$$\frac{K}{2-K} = e^{-4\phi_0} - 1.$$

Milgrom[11] proposed a "bimetric MOND" or "BIMOND" theory, with action

$$S - S_M - \hat{S}_M = -\frac{c^4}{16\pi G} \int \left[\beta g^{1/2} R + \alpha \hat{g}^{1/2}\hat{R} - 2(g\hat{g})^{1/4} f(\kappa) l_0^{-2} \mathcal{M}\left(l_0^m \Upsilon^{(m)}\right)\right] d^4x$$

with S_M and $\hat{S}_M$ the (noninteracting) matter actions attached to the two metrics, Υ a tensor derived from the difference in the metrics' connections, $\kappa = (g/\hat{g})^{\frac{1}{4}}$ the ratio between the two metric traces, and α, β are free parameters. $\mathcal{M}$ is a function which depends on some contractions of the Υ tensors.

Assuming that $\mathcal{M}$ depends only on the scalar contraction of Υ, Milgrom obtained as a nonrelativistic limit his bi-potential version of MOND with action

$$S - S_M = -\frac{1}{8\pi G} \int \left[\beta\left(\nabla\phi\right)^2 + \alpha\left(\nabla\hat{\phi}\right)^2 - a_0^2 \mathcal{M}\left(\left(\nabla\phi - \nabla\hat{\phi}\right)^2 a_0^{-2}\right)\right] d^4x$$

$$S_M = \rho(v^2/2 - \phi)$$

Here $\mathcal{M}(z)$ should scale as $z^{-1/4}$ in the deep-MOND limit and as z in the Newtonian limit.

13.8.4 Moffat's theories

J. W. Moffat (1995) developed a non-symmetric gravitation theory (NGT). This is not a metric theory. It was first claimed that it does not contain a black hole horizon, but Burko and Ori (1995) have found that NGT can contain black holes. Later, Moffat claimed that it has also been applied to explain rotation curves of galaxies without invoking "dark matter". Damour, Deser & MaCarthy (1993) have criticised NGT, saying that it has unacceptable asymptotic behaviour.

The mathematics is not difficult but is intertwined so the following is only a brief sketch. Starting with a non-symmetric tensor $g_{\mu\nu}$, the Lagrangian density is split into

$$L = L_R + L_M$$

where L_M is the same as for matter in GR.

$$L_R = \sqrt{-g}\left[R(W) - 2\lambda - \frac{1}{4}\mu^2 g^{\mu\nu} g_{[\mu\nu]} \right] - \frac{1}{6}g^{\mu\nu}W_\mu W_\nu$$

where $R(W)$ is a curvature term analogous to but not equal to the Ricci curvature in GR, λ and μ^2 are cosmological constants, $g_{[\nu\mu]}$ is the antisymmetric part of $g_{\nu\mu}$. W_μ is a connection, and is a bit difficult to explain because it's defined recursively. However, $W_\mu \approx -2g^{\nu}_{[\mu\nu]}$

Moffat's (2002) theory is a scalar-tensor bimetric gravity theory (BGT) and is one of the many theories of gravity in which the speed of light is faster in the early universe. These theories were motivated partly be the desire to avoid the "horizon problem" without invoking inflation. It has a variable G. The theory also attempts to explain the dimming of supernovae from a perspective other than the acceleration of the universe and so runs the risk of predicting an age for the universe that is too small.

Moffat's (2005a) metric-skew-tensor-gravity (MSTG) theory is able to predict rotation curves for galaxies without either dark matter or MOND, and claims that it can also explain gravitational lensing of galaxy clusters without dark matter. It has variable G, increasing to a final constant value about a million years after the big bang.

The theory seems to contain an asymmetric tensor $A_{\mu\nu}$ field and a source current J_μ vector. The action is split into:

$$S = S_G + S_F + S_{FM} + S_M$$

Both the gravity and mass terms match those of GR with cosmological constant. The skew field action and the skew field matter coupling are:

$$S_F = \int d^4x \sqrt{-g}\left(\frac{1}{12}F_{\mu\nu\rho}F^{\mu\nu\rho} - \frac{1}{4}\mu^2 A_{\mu\nu}A^{\mu\nu} \right)$$

$$S_{FM} = \int d^4x\, \epsilon^{\alpha\beta\mu\nu} A_{\alpha\beta}\partial_\mu J_\nu$$

where

$$F_{\mu\nu\rho} = \partial_\mu A_{\nu\rho} + \partial_\rho A_{\mu\nu}$$

and $\epsilon^{\alpha\beta\mu\nu}$ is the Levi-Civita symbol. The skew field coupling is a Pauli coupling and is gauge invariant for any source current. The source current looks like a matter fermion field associated with baryon and lepton number.

Moffat (2005b) Scalar-tensor-vector gravity (SVTG) theory.

The theory contains a tensor, vector and three scalar fields. But the equations are quite straightforward. The action is split into: $S = S_G + S_K + S_S + S_M$ with terms for gravity, vector field K_μ, scalar fields G, ω & μ, and mass. S_G is the standard gravity term with the exception that G is moved inside the integral.

$$S_K = -\int d^4x \sqrt{-g}\,\omega\left(\frac{1}{4}B_{\mu\nu}B^{\mu\nu} + V(K) \right)$$

where $B_{\mu\nu} = \partial_\mu K_\nu - \partial_\nu K_\mu$

$$S_S = -\int d^4x \sqrt{-g}\frac{1}{G^3}\left(\frac{1}{2}g^{\mu\nu}\nabla_\mu G \nabla_\nu G - V(G) \right)$$
$$+ \frac{1}{G}\left(\frac{1}{2}g^{\mu\nu}\nabla_\mu\omega\nabla_\nu\omega - V(\omega) \right) + \frac{1}{\mu^2 G}\left(\frac{1}{2}g^{\mu\nu}\nabla_\mu\mu\nabla_\nu\mu - V(\mu) \right)$$

The potential function for the vector field is chosen to be:

$$V(K) = -\frac{1}{2}\mu^2 \phi^\mu \phi_\mu - \frac{1}{4}g(\phi^\mu \phi_\mu)^2$$

where g is a coupling constant. The functions assumed for the scalar potentials are not stated.

13.9   Footnotes

[1] this isn't exactly the way Mach originally stated it, see other variants in Mach principle

[2] Zenneck, J. (1903). "Gravitation". *Encyklopädie der mathematischen Wissenschaften mit Einschluss ihrer Anwendungen* (in German) **5**: 25–67. doi:10.1007/978-3-663-16016-8_2.

[3] Lorentz, H.A. (1900). "Considerations on Gravitation". *Proc. Acad. Amsterdam* **2**: 559–574.

[4] Walter, S. (2007). Renn, J., ed. "Breaking in the 4-vectors: the four-dimensional movement in gravitation, 1905–1910". *The Genesis of General Relativity* (Berlin: Springer) **3**: 193–252.

[5] A later edition is Will (1993). See also Ni (1972)

[6] Although an important source for this article, the presentations of Turyshev (2006) and Lang (2002) contain many errors of fact

[7] http://arxiv.org/pdf/0704.1574v2.pdf - Retarded electric and magnetic fields of a moving charge: Feynman's derivation of Liénard-Wiechert potentials revisited

[8] http://gsjournal.net/Science-Journals/Research%20Papers-Relativity%20Theory/Download/4217 - On the Multiple Interpretations of Gravity

[9] Gary Gibbons; Will (2008). "On the Multiple Deaths of Whitehead's Theory of Gravity". *Stud.Hist.Philos.Mod.Phys.* **39**: 41–61. arXiv:gr-qc/0611006. doi:10.1016/j.shpsb.2007.04.004. Cf. Ronny Desmet and Michel Weber (edited by), Whitehead. The Algebra of Metaphysics. Applied Process Metaphysics Summer Institute Memorandum, Louvain-la-Neuve, Éditions Chromatika, 2010.

[10] Sagi, Eva (July 2009). "Preferred frame parameters in the tensor-vector-scalar theory of gravity and its generalization". arXiv:0905.4001. Bibcode:2009PhRvD..80d4032S. doi:10.1103/PhysRevD.80.044032.

[11] Milgrom, M (2009). "Bimetric MOND gravity". *Physical Review D* **80** (12). arXiv:0912.0790. Bibcode:2009PhRvD..80l3536M. doi:10.1103/PhysRevD.80.123536.

13.10   References

- Barker, B. M. (1978). "General scalar-tensor theory of gravity with constant G". *The Astrophysical Journal* **219**: 5. Bibcode:1978ApJ...219....5B. doi:10.1086/155749.

- Bekenstein, Jacob (1977). "Are particle rest masses variable? Theory and constraints from solar system experiments". *Physical Review D* **15** (6): 1458–1468. Bibcode:1977PhRvD..15.1458B. doi:10.1103/PhysRevD.15.1458.

- Bekenstein, J. D. (2004) Revised gravitation theory for the modified Newtonian dynamics paradigm. Phys. Rev. D 70, 083509

- Belinfante, F. J. and Swihart, J. C. (1957a) Phenomenological linear theory of gravitation Part I, Ann. Phys. 1, 168

- Belinfante, F. J. and Swihart, J. C. (1957b) Phenomenological linear theory of gravitation Part II, Ann. Phys. 2, 196

- Bergman, O. (1956) Scalar field theory as a theory of gravitation, Amer. J. Phys. 24, 39

- Bergmann, P. G. (1968) Comments on the scalar-tensor theory, Int. J. Theor. Phys. 1, 25-36

- Birkhoff, G. D. (1943) Matter, electricity and gravitation in flat space-time. Proc. Nat Acad. Sci. U.S. 29, 231-239

- Bollini, C. G., Giambiagi, J. J., and Tiomno, J. (1970) A linear theory of gravitation, Nuovo Com. Lett. 3, 65-70

- Burko, L.M. and Ori, A. (1995) On the Formation of Black Holes in Nonsymmetric Gravity, Phys. Rev. Lett. 75, 2455-2459

- Brans, C. and Dicke, R. H. (1961) Mach's principle and a relativistic theory of gravitation. Phys. Rev. 124, 925-935

- Carroll, Sean. Video lecture discussion on the possibilities and constraints to revision of the General Theory of Relativity. Dark Energy or Worse: Was Einstein Wrong?

- Cartan, É. (1922) Sur une généralisation de la notion de courbure de Riemann et les espaces à torsion. Acad. Sci. Paris, Comptes Rend. 174, 593-595

- Cartan, É. (1923) Sur les variétés à connexion affine et la théorie de la relativité généralisée. Annales Scientifiques de l'École Normale Superieure Sér. 3, 40, 325-412. http://archive.numdam.org/article/ASENS_1923_3_40__325_0.pdf

- Damour; Deser; McCarthy (1993). "Nonsymmetric Gravity has Unacceptable Global Asymptotics". arXiv:gr-qc/9312030 [gr-qc].

- Deser, S. and Laurent, B. E. (1968) Gravitation without self-interaction, Annals of Physics 50, 76-101

- Einstein, A. (1912a) Lichtgeschwindigkeit und Statik des Gravitationsfeldes. Annalen der Physik 38, 355-369

- Einstein, A. (1912b) Zur Theorie des statischen Gravitationsfeldes. Annalen der Physik 38, 443

- Einstein, A. and Grossmann, M. (1913), Z. Math Physik 62, 225

- Einstein, A. and Fokker, A. D. (1914) Die Nordströmsche Gravitationstheorie vom Standpunkt des absoluten Differentkalküls. Annalen der Physik 44, 321-328

- Einstein, A. (1916) Annalen der Physik 49, 769

- Einstein, A. (1917) Über die Spezielle und die Allgemeinen Relativatätstheorie, Gemeinverständlich, Vieweg, Braunschweig

- Fierz, M. and Pauli, W. (1939) On relativistic wave equations for particles of arbitrary spin in an electromagnetic field. Proc. Royal Soc. London 173, 211-232

- J.Foukzon, S.A.Podosenov, A.A.Potapov, E.Menkova, Bimetric Theory of Gravitational-Inertial Field in Riemannian and in Finsler-Lagrange Approximation 2010.http://arxiv.org/abs/1007.3290

- Hellings, Ronald; Nordtvedt, Kenneth (1973). "Vector-Metric Theory of Gravity". *Physical Review D* 7 (12): 3593–3602. Bibcode:1973PhRvD...7.3593H. doi:10.1103/PhysRevD.7.3593.

- Jordan, P.(1955) Schwerkraft und Weltall, Vieweg, Braunschweig

- Kustaanheimo, P. (1966) Route dependence of the gravitational redshift. Phys. Lett. 23, 75-77

- Kustaanheimo, P. E. and Nuotio, V. S. (1967) Publ. Astron. Obs. Helsinki No. 128

- Lang, R. (2002) Experimental foundations of general relativity,http://www.mppmu.mpg.de/~{}rlang/talks/melbou ppt

- Lee, D.; Lightman, A.; Ni, W. (1974). "Conservation laws and variational principles in metric theories of gravity". *Physical Review D* 10 (6): 1685-1700. Bibcode:1974PhRvD..10.1685L. doi:10.1103/PhysRevD.10.1685.

- Lightman, Alan; Lee, David (1973). "New Two-Metric Theory of Gravity with Prior Geometry". *Physical Review D* **8** (10): 3293–3302. Bibcode:1973PhRvD...8.3293L. doi:10.1103/PhysRevD.8.3293.

- Littlewood, D. E. (1953) Proceedings of the Cambridge Philosophical Society 49, 90-96

- Milne E. A. (1948) Kinematic Relativity, Clarendon Press, Oxford

- Misner, C. W., Thorne, K. S. and Wheeler, J. A. (1973) Gravitation, W. H. Freeman & Co.

- Moffat (1995). "Nonsymmetric Gravitational Theory". *Physics Letters B* **355** (3–4): 447–452. arXiv:gr-qc/9411006. Bibcode:1995PhLB..355..447M. doi:10.1016/0370-2693(95)00670-G.

- Moffat (2003). "Bimetric Gravity Theory, Varying Speed of Light and the Dimming of Supernovae". *International Journal of Modern Physics D [Gravitation; Astrophysics and Cosmology]* **12** (2): 281. arXiv:gr-qc/0202012. Bibcode:2003IJMPD..12..281M. doi:10.1142/S0218271803002366.

- Moffat (2005). "Gravitational Theory, Galaxy Rotation Curves and Cosmology without Dark Matter". *Journal of Cosmology and Astroparticle Physics* **2005** (5): 003–003. arXiv:astro-ph/0412195. Bibcode:2005JCAP...05..003M. doi:10.1088/1475-7516/2005/05/003.

- Moffat (2006). "Scalar-Tensor-Vector Gravity Theory". *Journal of Cosmology and Astroparticle Physics* **2006** (3): 004–004. arXiv:gr-qc/0506021. Bibcode:2006JCAP...03..004M. doi:10.1088/1475-7516/2006/03/004.

- Newton, I. (1686) *Philosophiæ Naturalis Principia Mathematica*

- Ni, Wei-Tou (1972). "Theoretical Frameworks for Testing Relativistic Gravity.IV. a Compendium of Metric Theories of Gravity and Their POST Newtonian Limits". *The Astrophysical Journal* **176**: 769. Bibcode:1972ApJ...176. .769N.doi:10.1086/151677.

- Ni, Wei-Tou (1973). "A New Theory of Gravity".*Physical Review D***7**(10): 2880–2883.Bibcode:1973PhRvD...7. doi:10.1103/PhysRevD.7.2880.

- Nordtvedt, Jr., K. (1970) Post-Newtonian metric for a general class of scalar-tensor gravitational theories with observational consequences, The Astrophysical Journal 161, 1059

- Nordtvedt, Jr., K. and Will C. M. (1972) Conservation laws and preferred frames in relativistic gravity II, The Astrophysical Journal 177, 775

- Nordström, G. (1912), Relativitätsprinzip und Gravitation. *Phys. Zeitschr.* 13, 1126

- Nordström, G. (1913), Zur Theorie der Gravitation vom Standpunkt des Relativitätsprinzips, *Annalen der Physik* 42, 533

- Pais, A. (1982) *Subtle is the Lord*, Clarendon Press

- Page, C. and Tupper, B. O. J. (1968) Scalar gravitational theories with variable velocity of light, *Mon. Not. R. Astr. Soc.* 138, 67-72

- Papapetrou, A. (1954a) Zs Phys., 139, 518

- Papapetrou, A. (1954b) Math. Nach., 12, 129 & Math. Nach., 12, 143

- Poincaré, H. (1908) *Science and Method*

- Rastall, P. (1979) The Newtonian theory of gravitation and its generalization, *Canadian Journal of Physics* 57, 944-973

- Rosen, N. (1971) Theory of gravitation, *Physical Review D* 3, 2317

- Rosen, N. (1973) A bimetric theory of gravitation, *General Relativity and Gravitation* 4, 435-447.

- Rosen, N. (1975) A bimetric theory of gravitation II, *General Relativity and Gravitation* 6, 259-268

- Seljak, Uros, et al. (2010) Study Validates General Relativity on Cosmic Scale, abstract appears in physorg.com

- Thiry, Y. (1948) Les équations de la théorie unitaire de Kaluza, *Comptes Rendus Acad. Sci* (Paris) 226, 216

- Trautman, A. (1972) On the Einstein–Cartan equations I, Bulletin de l'Academie Polonaise des Sciences 20, 185-190

- Turyshev, S. G. (2006) Testing gravity in the solar system,http://star-www.st-and.ac.uk/~{}hz4/workshop/worksh turyshev.pdf

- Wagoner, Robert V. (1970). "Scalar-Tensor Theory and Gravitational Waves". *Physical Review D* **1** (12): 3209–3216. Bibcode:1970PhRvD...1.3209W. doi:10.1103/PhysRevD.1.3209.

- Whitehead, A.N. (1922) *The Principles of Relativity*, Cambridge Univ. Press

- Whitrow, G. J. and Morduch, G. E. (1960) General relativity and Lorentz-invariant theories of gravitations, *Nature* 188, 790-794

- Whitrow, G. J. and Morduch, G. E. (1965) Relativistic theories of gravitation, *Vistas in Astronomy* 6, 1-67

- Will, C. M. (1981, 1993) *Theory and Experiment in Gravitational Physics*, Cambridge Univ. Press

- Will, C. M. (2006) The Confrontation between General Relativity and Experiment, *Living Rev. Relativity* 9 (3), http://www.livingreviews.org/lrr-2006-3

- Will, C. M. and Nordtvedt Jr., K. (1972) Conservation laws and preferred frames in relativistic gravity I, *The Astrophysical Journal* 177, 757

- Yilmaz, H. (1958) New approach to general relativity, *Phys. Rev.* 111, 1417

- Yilmaz, H. (1973) New approach to relativity and gravitation, *Annals of Physics* 81, 179-200

Chapter 14

Quantum gravity

Quantum gravity (**QG**) is a field of theoretical physics that seeks to describe the force of gravity according to the principles of quantum mechanics.

The current understanding of gravity is based on Albert Einstein's general theory of relativity, which is formulated within the framework of classical physics. On the other hand, the nongravitational forces are described within the framework of quantum mechanics, a radically different formalism for describing physical phenomena based on probability.[1] The necessity of a quantum mechanical description of gravity follows from the fact that one cannot consistently couple a classical system to a quantum one.[2]

Although a quantum theory of gravity is needed in order to reconcile general relativity with the principles of quantum mechanics, difficulties arise when one attempts to apply the usual prescriptions of quantum field theory to the force of gravity.[3] From a technical point of view, the problem is that the theory one gets in this way is not renormalizable and therefore cannot be used to make meaningful physical predictions. As a result, theorists have taken up more radical approaches to the problem of quantum gravity, the most popular approaches being string theory and loop quantum gravity.[4] A recent development is the theory of causal fermion systems which gives quantum mechanics, general relativity, and quantum field theory as limiting cases.[5][6][7][8][9][10]

Strictly speaking, the aim of quantum gravity is only to describe the quantum behavior of the gravitational field and should not be confused with the objective of unifying all fundamental interactions into a single mathematical framework. While any substantial improvement into the present understanding of gravity would aid further work towards unification, study of quantum gravity is a field in its own right with various branches having different approaches to unification. Although some quantum gravity theories, such as string theory, try to unify gravity with the other fundamental forces, others, such as loop quantum gravity, make no such attempt; instead, they make an effort to quantize the gravitational field while it is kept separate from the other forces. A theory of quantum gravity that is also a grand unification of all known interactions is sometimes referred to as a theory of everything (TOE).

One of the difficulties of quantum gravity is that quantum gravitational effects are only expected to become apparent near the Planck scale, a scale far smaller in distance (equivalently, far larger in energy) than what is currently accessible at high energy particle accelerators. As a result, quantum gravity is a mainly theoretical enterprise, although there are speculations about how quantum gravity effects might be observed in existing experiments.[11]

14.1 Overview

Much of the difficulty in meshing these theories at all energy scales comes from the different assumptions that these theories make on how the universe works. Quantum field theory depends on particle fields embedded in the flat space-time of special relativity. General relativity models gravity as a curvature within space-time that changes as a gravitational mass moves. Historically, the most obvious way of combining the two (such as treating gravity as simply another particle field) ran quickly into what is known as the renormalization problem. In the old-fashioned understanding of renormalization, gravity particles would attract each other and adding together all of the interactions results in many infinite values which

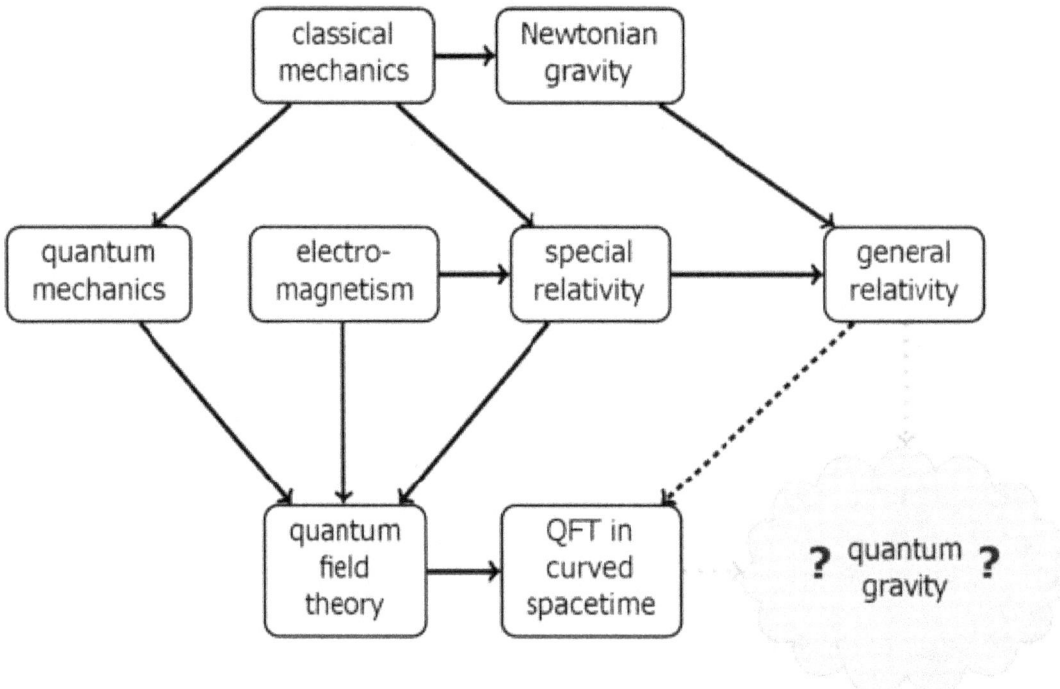

Diagram showing where quantum gravity sits in the hierarchy of physics theories

cannot easily be cancelled out mathematically to yield sensible, finite results. This is in contrast with quantum electrodynamics where, given that the series still do not converge, the interactions sometimes evaluate to infinite results, but those are few enough in number to be removable via renormalization.

14.1.1 Effective field theories

Quantum gravity can be treated as an effective field theory. Effective quantum field theories come with some high-energy cutoff, beyond which we do not expect that the theory provides a good description of nature. The "infinities" then become large but finite quantities depending on this finite cutoff scale, and correspond to processes that involve very high energies near the fundamental cutoff. These quantities can then be absorbed into an infinite collection of coupling constants, and at energies well below the fundamental cutoff of the theory, to any desired precision; only a finite number of these coupling constants need to be measured in order to make legitimate quantum-mechanical predictions. This same logic works just as well for the highly successful theory of low-energy pions as for quantum gravity. Indeed, the first quantum-mechanical corrections to graviton-scattering and Newton's law of gravitation have been explicitly computed[12] (although they are so astronomically small that we may never be able to measure them). In fact, gravity is in many ways a much better quantum field theory than the Standard Model, since it appears to be valid all the way up to its cutoff at the Planck scale.

While confirming that quantum mechanics and gravity are indeed consistent at reasonable energies, it is clear that near or above the fundamental cutoff of our effective quantum theory of gravity (the cutoff is generally assumed to be of the order of the Planck scale), a new model of nature will be needed. Specifically, the problem of combining quantum mechanics and gravity becomes an issue only at very high energies, and may well require a totally new kind of model.

14.1.2   Quantum gravity theory for the highest energy scales

The general approach to deriving a quantum gravity theory that is valid at even the highest energy scales is to assume that such a theory will be simple and elegant and, accordingly, to study symmetries and other clues offered by current theories that might suggest ways to combine them into a comprehensive, unified theory. One problem with this approach is that it is unknown whether quantum gravity will actually conform to a simple and elegant theory, as it should resolve the dual conundrums of special relativity with regard to the uniformity of acceleration and gravity, and general relativity with regard to spacetime curvature.

Such a theory is required in order to understand problems involving the combination of very high energy and very small dimensions of space, such as the behavior of black holes, and the origin of the universe.

14.2   Quantum mechanics and general relativity

14.2.1   The graviton

Main article: Graviton

At present, one of the deepest problems in theoretical physics is harmonizing the theory of general relativity, which describes gravitation, and applications to large-scale structures (stars, planets, galaxies), with quantum mechanics, which describes the other three fundamental forces acting on the atomic scale. This problem must be put in the proper context, however. In particular, contrary to the popular claim that quantum mechanics and general relativity are fundamentally incompatible, one can demonstrate that the structure of general relativity essentially follows inevitably from the quantum mechanics of interacting theoretical spin-2 massless particles (called gravitons).[13][14][15][16][17]

While there is no concrete proof of the existence of gravitons, quantized theories of matter may necessitate their existence. Supporting this theory is the observation that all fundamental forces except gravity have one or more known messenger particles, leading researchers to believe that at least one most likely does exist; they have dubbed this hypothetical particle the *graviton*. The predicted find would result in the classification of the graviton as a "force particle" similar to the photon of the electromagnetic field. Many of the accepted notions of a unified theory of physics since the 1970s assume, and to some degree depend upon, the existence of the graviton. These include string theory, superstring theory, M-theory, and loop quantum gravity. Detection of gravitons is thus vital to the validation of various lines of research to unify quantum mechanics and relativity theory.

14.2.2   The dilaton

Main article: Dilaton

The dilaton made its first appearance in Kaluza–Klein theory, a five-dimensional theory that combined gravitation and electromagnetism. Generally, it appears in string theory. More recently, it has appeared in the lower-dimensional many-bodied gravity problem[18] based on the field theoretic approach of Roman Jackiw. The impetus arose from the fact that complete analytical solutions for the metric of a covariant N-body system have proven elusive in general relativity. To simplify the problem, the number of dimensions was lowered to $(1+1)$, i.e. one spatial dimension and one temporal dimension. This model problem, known as $R=T$ theory[19] (as opposed to the general $G=T$ theory) was amenable to exact solutions in terms of a generalization of the Lambert W function. It was also found that the field equation governing the dilaton (derived from differential geometry) was the Schrödinger equation and consequently amenable to quantization.[20]

Thus, one had a theory which combined gravity, quantization and even the electromagnetic interaction, promising ingredients of a fundamental physical theory. It is worth noting that the outcome revealed a previously unknown and already existing *natural link* between general relativity and quantum mechanics. However, this theory needs to be generalized in $(2+1)$ or $(3+1)$ dimensions although, in principle, the field equations are amenable to such generalization as shown with the inclusion of a one-graviton process[21] and yielding the correct Newtonian limit in d dimensions if a dilaton is included. However, it is not yet clear what the fully generalized field equation governing the dilaton in (3+1) dimensions should be. This is further complicated by the fact that gravitons can propagate in *(3+1)* dimensions and consequently that would imply gravitons and dilatons exist in the real world. Moreover, detection of the dilaton is expected to be even more elusive than the graviton. However, since this approach allows for the combination of gravitational, electromagnetic and quantum effects, their coupling could potentially lead to a means of vindicating the theory, through cosmology and perhaps even *experimentally*.

14.2.3 Nonrenormalizability of gravity

Further information: Renormalization

General relativity, like electromagnetism, is a classical field theory. One might expect that, as with electromagnetism, there should be a corresponding quantum field theory.

However, gravity is perturbatively nonrenormalizable.[22][23] For a quantum field theory to be well-defined according to this understanding of the subject, it must be asymptotically free or asymptotically safe. The theory must be characterized by a choice of *finitely many* parameters, which could, in principle, be set by experiment. For example, in quantum electrodynamics, these parameters are the charge and mass of the electron, as measured at a particular energy scale.

On the other hand, in quantizing gravity, there are *infinitely many independent parameters* (counterterm coefficients) needed to define the theory. For a given choice of those parameters, one could make sense of the theory, but since we can never do infinitely many experiments to fix the values of every parameter, we do not have a meaningful physical theory:

- At low energies, the logic of the renormalization group tells us that, despite the unknown choices of these infinitely many parameters, quantum gravity will reduce to the usual Einstein theory of general relativity.

- On the other hand, if we could probe very high energies where quantum effects take over, then *every one* of the infinitely many unknown parameters would begin to matter, and we could make no predictions at all.

As explained below, there is a way around this problem by treating QG as an effective field theory.

Any meaningful theory of quantum gravity that makes sense and is predictive at all energy scales must have some deep principle that reduces the infinitely many unknown parameters to a finite number that can then be measured.

- One possibility is that normal perturbation theory is not a reliable guide to the renormalizability of the theory, and that there really *is* a UV fixed point for gravity. Since this is a question of non-perturbative quantum field theory, it is difficult to find a reliable answer, but some people still pursue this option.

- Another possibility is that there are new symmetry principles that constrain the parameters and reduce them to a finite set. This is the route taken by string theory, where all of the excitations of the string essentially manifest themselves as new symmetries.

14.2.4 QG as an effective field theory

Main article: Effective field theory

In an effective field theory, all but the first few of the infinite set of parameters in a non-renormalizable theory are suppressed by huge energy scales and hence can be neglected when computing low-energy effects. Thus, at least in the low-energy regime, the model is indeed a predictive quantum field theory.[12] (A very similar situation occurs for the very similar effective field theory of low-energy pions.) Furthermore, many theorists agree that even the Standard Model should really be regarded as an effective field theory as well, with "nonrenormalizable" interactions suppressed by large energy scales and whose effects have consequently not been observed experimentally.

Recent work[12] has shown that by treating general relativity as an effective field theory, one can actually make legitimate predictions for quantum gravity, at least for low-energy phenomena. An example is the well-known calculation of the tiny first-order quantum-mechanical correction to the classical Newtonian gravitational potential between two masses.

14.2.5   Spacetime background dependence

Main article: Background independence

A fundamental lesson of general relativity is that there is no fixed spacetime background, as found in Newtonian mechanics and special relativity; the spacetime geometry is dynamic. While easy to grasp in principle, this is the hardest idea to understand about general relativity, and its consequences are profound and not fully explored, even at the classical level. To a certain extent, general relativity can be seen to be a relational theory,[24] in which the only physically relevant information is the relationship between different events in space-time.

On the other hand, quantum mechanics has depended since its inception on a fixed background (non-dynamic) structure. In the case of quantum mechanics, it is time that is given and not dynamic, just as in Newtonian classical mechanics. In relativistic quantum field theory, just as in classical field theory, Minkowski spacetime is the fixed background of the theory.

String theory

String theory can be seen as a generalization of quantum field theory where instead of point particles, string-like objects propagate in a fixed spacetime background, although the interactions among closed strings give rise to space-time in a dynamical way. Although string theory had its origins in the study of quark confinement and not of quantum gravity, it was soon discovered that the string spectrum contains the graviton, and that "condensation" of certain vibration modes of strings is equivalent to a modification of the original background. In this sense, string perturbation theory exhibits exactly the features one would expect of a perturbation theory that may exhibit a strong dependence on asymptotics (as seen, for example, in the AdS/CFT correspondence) which is a weak form of background dependence.

Background independent theories

Loop quantum gravity is the fruit of an effort to formulate a background-independent quantum theory.

Topological quantum field theory provided an example of background-independent quantum theory, but with no local degrees of freedom, and only finitely many degrees of freedom globally. This is inadequate to describe gravity in 3+1 dimensions, which has local degrees of freedom according to general relativity. In 2+1 dimensions, however, gravity is a topological field theory, and it has been successfully quantized in several different ways, including spin networks.

14.2.6   Semi-classical quantum gravity

Quantum field theory on curved (non-Minkowskian) backgrounds, while not a full quantum theory of gravity, has shown many promising early results. In an analogous way to the development of quantum electrodynamics in the early part of the 20th century (when physicists considered quantum mechanics in classical electromagnetic fields), the consideration of quantum field theory on a curved background has led to predictions such as black hole radiation.

Phenomena such as the Unruh effect, in which particles exist in certain accelerating frames but not in stationary ones, do not pose any difficulty when considered on a curved background (the Unruh effect occurs even in flat Minkowskian backgrounds). The vacuum state is the state with the least energy (and may or may not contain particles). See Quantum field theory in curved spacetime for a more complete discussion.

14.2.7 Points of tension

There are other points of tension between quantum mechanics and general relativity.

- First, classical general relativity breaks down at singularities, and quantum mechanics becomes inconsistent with general relativity in the neighborhood of singularities (however, no one is certain that classical general relativity applies near singularities in the first place).

- Second, it is not clear how to determine the gravitational field of a particle, since under the Heisenberg uncertainty principle of quantum mechanics its location and velocity cannot be known with certainty. The resolution of these points may come from a better understanding of general relativity.[25]

- Third, there is the problem of time in quantum gravity. Time has a different meaning in quantum mechanics and general relativity and hence there are subtle issues to resolve when trying to formulate a theory which combines the two.[26]

14.3 Candidate theories

There are a number of proposed quantum gravity theories.[27] Currently, there is still no complete and consistent quantum theory of gravity, and the candidate models still need to overcome major formal and conceptual problems. They also face the common problem that, as yet, there is no way to put quantum gravity predictions to experimental tests, although there is hope for this to change as future data from cosmological observations and particle physics experiments becomes available.[28][29]

14.3.1 String theory

Main article: String theory

One suggested starting point is ordinary quantum field theories which, after all, are successful in describing the other three basic fundamental forces in the context of the standard model of elementary particle physics. However, while this leads to an acceptable effective (quantum) field theory of gravity at low energies,[30] gravity turns out to be much more problematic at higher energies. Where, for ordinary field theories such as quantum electrodynamics, a technique known as renormalization is an integral part of deriving predictions which take into account higher-energy contributions,[31] gravity turns out to be nonrenormalizable: at high energies, applying the recipes of ordinary quantum field theory yields models that are devoid of all predictive power.[32]

One attempt to overcome these limitations is to replace ordinary quantum field theory, which is based on the classical concept of a point particle, with a quantum theory of one-dimensional extended objects: string theory.[33] At the energies reached in current experiments, these strings are indistinguishable from point-like particles, but, crucially, different modes of oscillation of one and the same type of fundamental string appear as particles with different (electric and other) charges. In this way, string theory promises to be a unified description of all particles and interactions.[34] The theory is successful in that one mode will always correspond to a graviton, the messenger particle of gravity; however, the price to pay are unusual features such as six extra dimensions of space in addition to the usual three for space and one for time.[35]

In what is called the second superstring revolution, it was conjectured that both string theory and a unification of general relativity and supersymmetry known as supergravity[36] form part of a hypothesized eleven-dimensional model known as M-theory, which would constitute a uniquely defined and consistent theory of quantum gravity.[37][38] As presently understood, however, string theory admits a very large number (10^{500} by some estimates) of consistent vacua, comprising the so-called "string landscape". Sorting through this large family of solutions remains a major challenge.

14.3.2 Loop quantum gravity

Main article: Loop quantum gravity

Loop quantum gravity is based first of all on the idea to take seriously the insight of general relativity that spacetime is a dynamical field and therefore is a quantum object. The second idea is that the quantum discreteness that determines the particle-like behavior of other field theories (for instance, the photons of the electromagnetic field) also affects the structure of space.

The main result of loop quantum gravity is the derivation of a granular structure of space at the Planck length. This is derived as follows. In the case of electromagnetism, the quantum operator representing the energy of each frequency of the field has discrete spectrum. Therefore the energy of each frequency is quantized, and the quanta are the photons. In the case of gravity, the operators representing the area and the volume of each surface or space region have discrete spectrum. Therefore area and volume of any portion of space are quantized, and the quanta are elementary quanta of space. It follows that spacetime has an elementary quantum granular structure at the Planck scale, which cuts-off the ultraviolet infinities of quantum field theory.

The quantum state of spacetime is described in the theory by means of a mathematical structure called spin networks. Spin networks were initially introduced by Roger Penrose in abstract form, and later shown by Carlo Rovelli and Lee Smolin to derive naturally from a non perturbative quantization of general relativity. Spin networks do not represent quantum states of a field in spacetime: they represent directly quantum states of spacetime.

The theory is based on the reformulation of general relativity known as Ashtekar variables, which represent geometric gravity using mathematical analogues of electric and magnetic fields.[39][40] In the quantum theory space is represented by a network structure called a spin network, evolving over time in discrete steps.[41][42][43][44]

The dynamics of the theory is today constructed in several versions. One version starts with the canonical quantization of general relativity. The analogue of the Schrödinger equation is a Wheeler–DeWitt equation, which can be defined in the theory.[45] In the covariant, or spinfoam formulation of the theory, the quantum dynamics is obtained via a sum over discrete versions of spacetime, called spinfoams. These represent histories of spin networks.

14.3.3   Scale Relativity

Main article: Scale relativity

Most quantum gravity theories assume quantum laws as a starting point. However, in the framework of scale relativity, this is not needed.[46] The theory is an extension of special and general relativity, including the relativity of scale transformations. It thus takes a geometrical approach to the problem, where quantum phenomena became a manifestation of the fractality of spacetime. This is similar to the geometrical interpretation of gravitation in general relativity, where gravitation become a manifestation of spacetime curvature instead of a force. Although much remains to be developed, validated predictions have already been obtained in physics, astrophysics and cosmology.

14.3.4   Other approaches

There are a number of other approaches to quantum gravity. The approaches differ depending on which features of general relativity and quantum theory are accepted unchanged, and which features are modified.[47][48] Examples include:

- Acoustic metric and other analog models of gravity

- Asymptotic safety in quantum gravity

- Euclidean quantum gravity

- Causal dynamical triangulation[49]

- Causal fermion systems,[5][6][7][8][9][10] giving quantum mechanics, general relativity and quantum field theory as limiting cases.

- Causal sets[50]

- Covariant Feynman path integral approach

- Group field theory[51]

- Wheeler-DeWitt equation

- Geometrodynamics

- Hořava–Lifshitz gravity

- MacDowell–Mansouri action

- Noncommutative geometry.

- Path-integral based models of quantum cosmology[52]

- Regge calculus

- String-nets giving rise to gapless helicity ±2 excitations with no other gapless excitations[53]

- Superfluid vacuum theory a.k.a. theory of BEC vacuum

- Supergravity

- Twistor theory[54]

- Canonical quantum gravity

- E8 Theory

14.4 Weinberg–Witten theorem

In quantum field theory, the Weinberg–Witten theorem places some constraints on theories of composite gravity/emergent gravity. However, recent developments attempt to show that if locality is only approximate and the holographic principle is correct, the Weinberg–Witten theorem would not be valid.

14.5 Experimental tests

As was emphasized above, quantum gravitational effects are extremely weak and therefore difficult to test. For this reason, the possibility of experimentally testing quantum gravity had not received much attention prior to the late 1990s. However, in the past decade, physicists have realized that evidence for quantum gravitational effects can guide the development of the theory. Since theoretical development has been slow, the field of phenomenological quantum gravity, which studies the possibility of experimental tests, has obtained increased attention.[55][56]

The most widely pursued possibilities for quantum gravity phenomenology include violations of Lorentz invariance, imprints of quantum gravitational effects in the cosmic microwave background (in particular its polarization), and decoherence induced by fluctuations in the space-time foam.

The BICEP2 experiment detected what was initially thought to be primordial B-mode polarization caused by gravitational waves in the early universe. If truly primordial, these waves were born as quantum fluctuations in gravity itself. Cosmologist Ken Olum (Tufts University) stated: "I think this is the only observational evidence that we have that actually shows that gravity is quantized....It's probably the only evidence of this that we will ever have."[57]

14.6 See also

14.7 References

[1] Griffiths, David J. (2004). *Introduction to Quantum Mechanics*. Pearson Prentice Hall. OCLC 803860989.

[2] Wald, Robert M. (1984). *General Relativity*. University of Chicago Press. p. 382. OCLC 471881415.

[3] Zee, Anthony (2010). *Quantum Field Theory in a Nutshell* (2nd ed.). Princeton University Press. p. 172. OCLC 659549695.

[4] Penrose, Roger (2007). *The road to reality : a complete guide to the laws of the universe*. Vintage. p. 1017. OCLC 716437154.

[5] F. Finster, J. Kleiner, Causal Fermion Systems as a Candidate for a Unified Physical Theory, http://arxiv.org/abs/1502.03587

[6] F. Finster, The Principle of the Fermionic Projector, hep-th/0001048, hep-th/0202059, hep-th/0210121, AMS/IP Studies in Advanced Mathematics, vol. **35**, American Mathematical Society, Providence, RI, 2006.

[7] F. Finster, A formulation of quantum field theory realizing a sea of interacting Dirac particles, arXiv:0911.2102 [hep-th], Lett. Math. Phys. **97** (2011), no. 2, 165–183.

[8] F. Finster, An action principle for an interacting fermion system and its analysis in the continuum limit, arXiv:0908.1542 [math-ph] (2009).

[9] F. Finster, The continuum limit of a fermion system involving neutrinos: Weak and gravitational interactions, arXiv:1211.3351 [math-ph] (2012).

[10] F. Finster, Perturbative quantum field theory in the framework of the fermionic projector, arXiv:1310.4121 [math-ph], J. Math. Phys. **55** (2014), no. 4, 042301.

[11] Quantum effects in the early universe might have an observable effect on the structure of the present universe, for example, or gravity might play a role in the unification of the other forces. Cf. the text by Wald cited above.

[12] Donoghue (1995). "Introduction to the Effective Field Theory Description of Gravity". arXiv:gr-qc/9512024. (verify against ISBN 9789810229085)

[13] Kraichnan, R. H. (1955). "Special-Relativistic Derivation of Generally Covariant Gravitation Theory". *Physical Review* **98** (4): 1118–1122. Bibcode:1955PhRv...98.1118K. doi:10.1103/PhysRev.98.1118.

[14] Gupta, S. N. (1954). "Gravitation and Electromagnetism". *Physical Review* **96** (6): 1683–1685. Bibcode:1954PhRv...96.1683G. doi:10.1103/PhysRev.96.1683.

[15] Gupta, S. N.(1957). "Einstein's and Other Theories of Gravitation".*Reviews of Modern Physics* **29**(3): 334–336. Bibcode:1957 doi:10.1103/RevModPhys.29.334.

[16] Gupta, S. N. (1962). "Quantum Theory of Gravitation". *Recent Developments in General Relativity*. Pergamon Press. pp. 251–258.

[17] Deser, S. (1970). "Self-Interaction and Gauge Invariance". *General Relativity and Gravitation* 1: 9–18. arXiv:gr-qc/0411023. Bibcode:1970GReGr...1....9D. doi:10.1007/BF00759198.

[18] Ohta, Tadayuki; Mann, Robert (1996). "Canonical reduction of two-dimensional gravity for particle dynamics". *Classical and Quantum Gravity* **13** (9): 2585–2602. arXiv:gr-qc/9605004. Bibcode:1996CQGra..13.2585O. doi:10.1088/0264-9381/13/9/022.

[19] Sikkema, A E; Mann, R B (1991). "Gravitation and cosmology in (1+1) dimensions". *Classical and Quantum Gravity* **8**: 219–235. Bibcode:1991CQGra...8..219S. doi:10.1088/0264-9381/8/1/022.

[20] Farrugia; Mann; Scott (2007). "N-body Gravity and the Schroedinger Equation". *Classical and Quantum Gravity* **24** (18): 4647–4659. arXiv:gr-qc/0611144. Bibcode:2007CQGra..24.4647F. doi:10.1088/0264-9381/24/18/006.

[21] Mann, R B; Ohta, T (1997). "Exact solution for the metric and the motion of two bodies in (1+1)-dimensional gravity". *Physical Review D* **55** (8): 4723–4747. arXiv:gr-qc/9611008. Bibcode:1997PhRvD..55.4723M. doi:10.1103/PhysRevD.55.4723.

[22] Feynman, R. P.; Morinigo, F. B.; Wagner, W. G.; Hatfield, B. (1995). *Feynman lectures on gravitation*. Addison-Wesley. ISBN 0-201-62734-5.

[23] Hamber, H. W. (2009). *Quantum Gravitation - The Feynman Path Integral Approach*. Springer Publishing. ISBN 978-3-540-85292-6.

[24] Smolin, Lee (2001). *Three Roads to Quantum Gravity*. Basic Books. pp. 20–25. ISBN 0-465-07835-4 Pages 220–226 are annotated references and guide for further reading.

[25] Hunter Monroe (2005). "Singularity-Free Collapse through Local Inflation". arXiv:astro-ph/0506506.

[26] Edward Anderson (2010). "The Problem of Time in Quantum Gravity". arXiv:1009.2157 [gr-qc]. (also published as chapter 4 of ISBN 9781611229578)

[27] A timeline and overview can be found in Rovelli, Carlo (2000). "Notes for a brief history of quantum gravity". arXiv:gr-qc/0006061. (verify against ISBN 9789812777386)

[28] Ashtekar, Abhay (2007). "Loop Quantum Gravity: Four Recent Advances and a Dozen Frequently Asked Questions". *11th Marcel Grossmann Meeting on Recent Developments in Theoretical and Experimental General Relativity*. p. 126. arXiv:0705.2222. Bibcode:2008mgm..conf..126A. doi:10.1142/9789812834300_0008.

[29] Schwarz, John H. (2007). "String Theory: Progress and Problems". *Progress of Theoretical Physics Supplement* **170**: 214–226. arXiv:hep-th/0702219. Bibcode:2007PThPS.170..214S. doi:10.1143/PTPS.170.214.

[30] Donoghue, John F. (editor) (1995). "Introduction to the Effective Field Theory Description of Gravity". In Cornet, Fernando. *Effective Theories: Proceedings of the Advanced School, Almunecar, Spain, 26 June–1 July 1995*. Singapore: World Scientific. arXiv:gr-qc/9512024. ISBN 981-02-2908-9.

[31] Weinberg, Steven (1996). "Chapters 17–18". *The Quantum Theory of Fields II: Modern Applications*. Cambridge University Press. ISBN 0-521-55002-5.

[32] Goroff, Marc H.; Sagnotti, Augusto; Sagnotti, Augusto (1985). "Quantum gravity at two loops". *Physics Letters B* **160**: 81–86. Bibcode:1985PhLB..160...81G. doi:10.1016/0370-2693(85)91470-4.

[33] An accessible introduction at the undergraduate level can be found in Zwiebach, Barton (2004). *A First Course in String Theory*. Cambridge University Press. ISBN 0-521-83143-1., and more complete overviews in Polchinski, Joseph (1998). *String Theory Vol. I: An Introduction to the Bosonic String*. Cambridge University Press. ISBN 0-521-63303-6. and Polchinski, Joseph (1998b). *String Theory Vol. II: Superstring Theory and Beyond*. Cambridge University Press. ISBN 0-521-63304-4.

[34] Ibanez, L. E. (2000). "The second string (phenomenology) revolution". *Classical & Quantum Gravity* **17** (5): 1117–1128. arXiv:hep-ph/9911499. Bibcode:2000CQGra..17.1117I. doi:10.1088/0264-9381/17/5/321.

[35] For the graviton as part of the string spectrum, e.g. Green, Schwarz & Witten 1987, sec. 2.3 and 5.3; for the extra dimensions, ibid sec. 4.2.

[36] Weinberg, Steven (2000). "Chapter 31". *The Quantum Theory of Fields II: Modern Applications*. Cambridge University Press. ISBN 0-521-55002-5.

[37] Townsend, Paul K. (1996). *Four Lectures on M-Theory*. ICTP Series in Theoretical Physics. p. 385. arXiv:hep-th/9612121. Bibcode:1997hepcbconf..385T.

[38] Duff, Michael (1996). "M-Theory (the Theory Formerly Known as Strings)". *International Journal of Modern Physics A* **11** (32): 5623–5642. arXiv:hep-th/9608117. Bibcode:1996IJMPA..11.5623D. doi:10.1142/S0217751X96002583.

[39] Ashtekar, Abhay (1986). "New variables for classical and quantum gravity". *Physical Review Letters* **57** (18): 2244–2247. Bibcode:1986PhRvL..57.2244A. doi:10.1103/PhysRevLett.57.2244. PMID 10033673.

[40] Ashtekar, Abhay (1987). "New Hamiltonian formulation of general relativity". *Physical Review D* **36** (6): 1587–1602. Bibcode:1987PhRvD..36.1587A. doi:10.1103/PhysRevD.36.1587.

[41] Thiemann, Thomas (2006). "Loop Quantum Gravity: An Inside View". *Approaches to Fundamental Physics*. Lecture Notes in Physics **721**: 185. arXiv:hep-th/0608210. Bibcode:2007LNP...721..185T. doi:10.1007/978-3-540-71117-9_10. ISBN 978-3-540-71115-5.

[42] Rovelli, Carlo (1998). "Loop Quantum Gravity". *Living Reviews in Relativity* **1**. Retrieved 2008-03-13.

[43] Ashtekar, Abhay; Lewandowski, Jerzy (2004). "Background Independent Quantum Gravity: A Status Report". *Classical & Quantum Gravity* **21** (15): R53–R152. arXiv:gr-qc/0404018. Bibcode:2004CQGra..21R..53A. doi:10.1088/0264-9381/21/15/

[44] Thiemann, Thomas (2003). "Lectures on Loop Quantum Gravity". *Lecture Notes in Physics*. Lecture Notes in Physics **631**: 41–135. arXiv:gr-qc/0210094. Bibcode:2003LNP...631...41T. doi:10.1007/978-3-540-45230-0_3. ISBN 978-3-540-40810-9.

[45] Rovelli, Carlo (2004). *Quantum Gravity*. Cambridge University Press. ISBN 0521715962.

[46] Nottale, L. (2011). *Scale Relativity and Fractal Space-Time: A New Approach to Unifying Relativity and Quantum Mechanics*. World Scientific Publishing Company. ISBN 1848166508.;p. 458

[47] Isham, Christopher J. (1994). "Prima facie questions in quantum gravity". In Ehlers, Jürgen; Friedrich, Helmut. *Canonical Gravity: From Classical to Quantum*. Springer. arXiv:gr-qc/9310031. ISBN 3-540-58339-4.

[48] Sorkin, Rafael D. (1997). "Forks in the Road, on the Way to Quantum Gravity". *International Journal of Theoretical Physics* **36** (12): 2759–2781. arXiv:gr-qc/9706002. Bibcode:1997IJTP...36.2759S. doi:10.1007/BF02435709.

[49] Loll, Renate (1998). "Discrete Approaches to Quantum Gravity in Four Dimensions". *Living Reviews in Relativity* **1**: 13. arXiv:gr-qc/9805049. Bibcode:1998LRR.....1...13L. doi:10.12942/lrr-1998-13. Retrieved 2008-03-09.

[50] Sorkin, Rafael D. (2005). "Causal Sets: Discrete Gravity". In Gomberoff, Andres; Marolf, Donald. *Lectures on Quantum Gravity*. Springer. arXiv:gr-qc/0309009. ISBN 0-387-23995-2.

[51] See Daniele Oriti and references therein.

[52] Hawking, Stephen W. (1987). "Quantum cosmology". In Hawking, Stephen W.; Israel, Werner. *300 Years of Gravitation*. Cambridge University Press. pp. 631–651. ISBN 0-521-37976-8.

[53] Wen 2006

[54] See ch. 33 in Penrose 2004 and references therein.

[55] Hossenfelder, Sabine (2011). "Experimental Search for Quantum Gravity". In V. R. Frignanni. *Classical and Quantum Gravity: Theory, Analysis and Applications*. Chapter 5: Nova Publishers. ISBN 978-1-61122-957-8.

[56] Hossenfelder, Sabine (2010-10-17). V. R. Frignanni, ed. "Experimental Search for Quantum Gravity". *Classical and Quantum Gravity: Theory, Analysis and Applications* (Nova Publishers) **5** (2011). arXiv:1010.3420. Bibcode:2010arXiv1010.3420H. |chapter= ignored (help)

[57] Camille Carlisle. "First Direct Evidence of Big Bang Inflation". SkyandTelescope.com. Retrieved March 18, 2014.

14.8   Further reading

- Ahluwalia, D. V. (2002). "Interface of Gravitational and Quantum Realms". *Modern Physics Letters A* **17** (15–17): 1135. arXiv:gr-qc/0205121. Bibcode:2002MPLA...17.1135A. doi:10.1142/S021773230200765X.

- Ashtekar, Abhay (2005). "The winding road to quantum gravity" (PDF). *Current Science* **89**: 2064–2074.

- Carlip, Steven (2001). "Quantum Gravity: a Progress Report". *Reports on Progress in Physics* **64** (8): 885–942. arXiv:gr-qc/0108040. Bibcode:2001RPPh...64..885C. doi:10.1088/0034-4885/64/8/301.

- Herbert W. Hamber (2009). *Quantum Gravitation*. Springer Publishing. doi:10.1007/978-3-540-85293-3. ISBN 978-3-540-85292-6.

- Kiefer, Claus (2007). *Quantum Gravity*. Oxford University Press. ISBN 0-19-921252-X.

- Kiefer, Claus (2005). "Quantum Gravity: General Introduction and Recent Developments". *Annalen der Physik* **15**: 129–148. arXiv:gr-qc/0508120. Bibcode:2006AnP...518..129K. doi:10.1002/andp.200510175.

- Lämmerzahl, Claus, ed. (2003). *Quantum Gravity: From Theory to Experimental Search*. Lecture Notes in Physics. Springer. ISBN 3-540-40810-X.

- Rovelli, Carlo (2004). *Quantum Gravity*. Cambridge University Press. ISBN 0-521-83733-2.

- Trifonov, Vladimir (2008). "GR-friendly description of quantum systems". *International Journal of Theoretical Physics* **47** (2): 492–510. arXiv:math-ph/0702095. Bibcode:2008IJTP...47..492T. doi:10.1007/s10773-007-9474-3.

Gravity Probe B (GP-B) has measured spacetime curvature near Earth to test related models in application of Einstein's general theory of relativity.

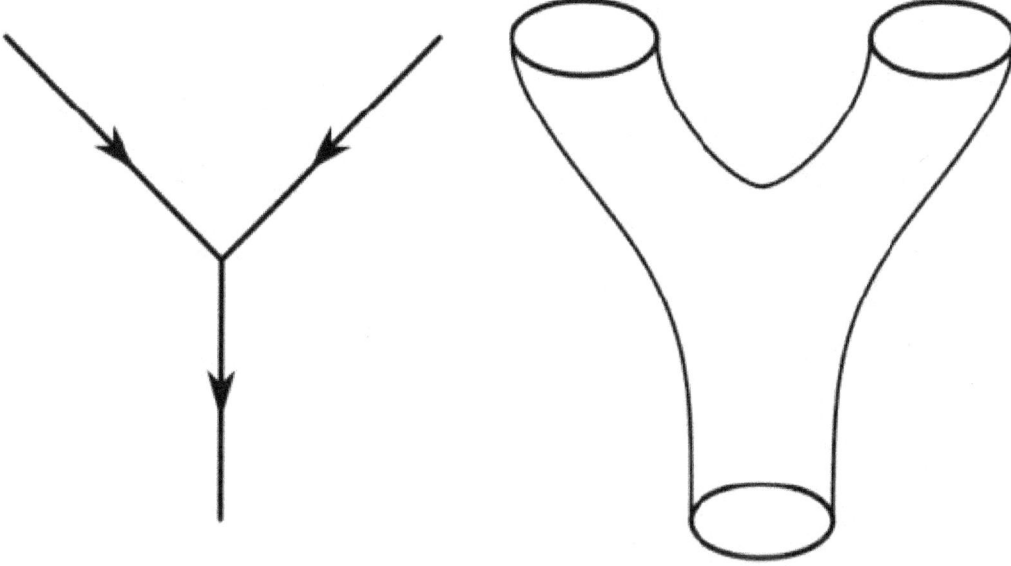

Interaction in the subatomic world: world lines of point-like particles in the Standard Model or a world sheet swept up by closed strings in string theory

Projection of a Calabi–Yau manifold, one of the ways of compactifying the extra dimensions posited by string theory

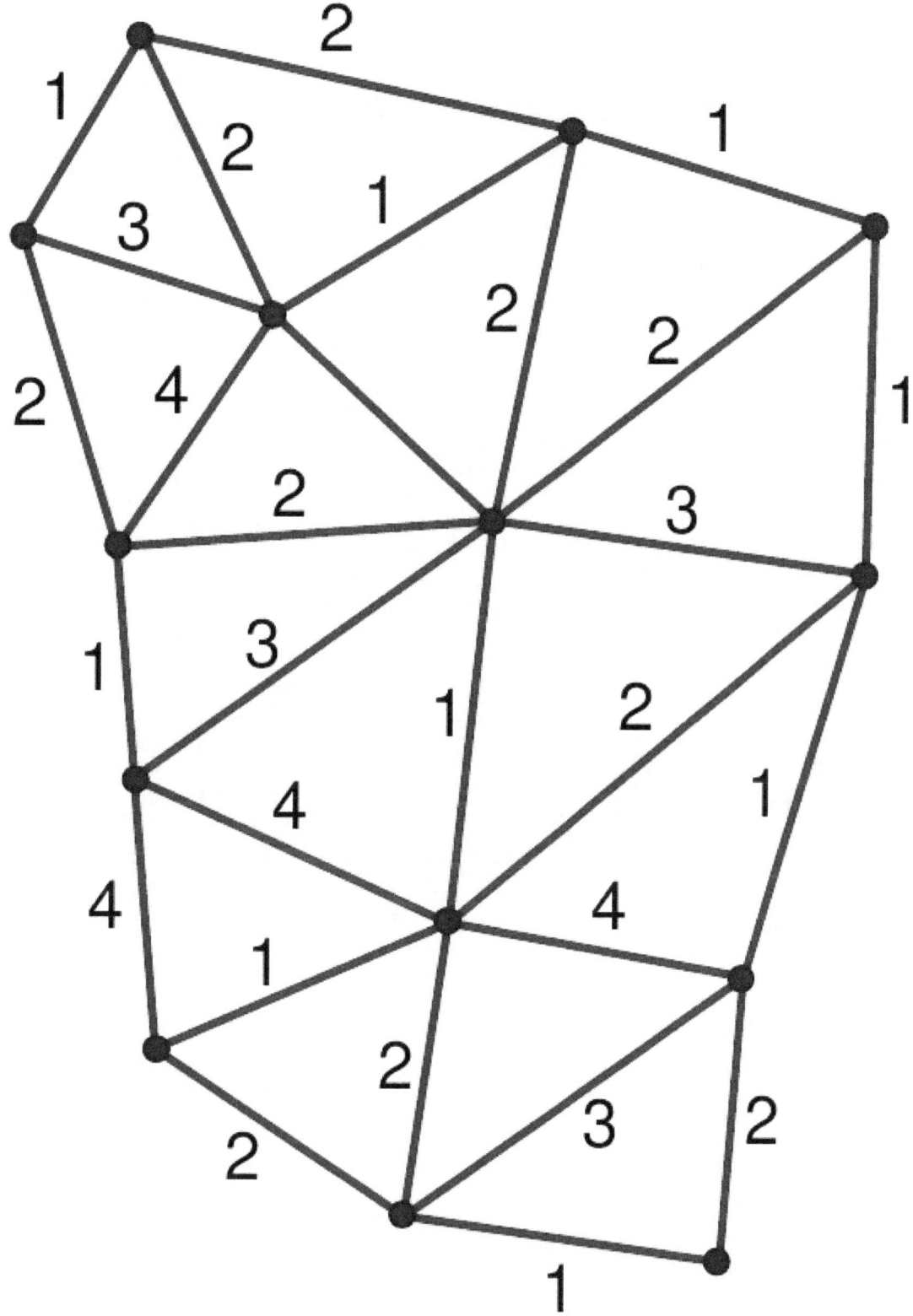

Simple spin network of the type used in loop quantum gravity

Schrödinger's flower. Morphogenesis of a flower-like structure, solution of a growth process equation that takes the form of a Schrödinger equation under fractal conditions.

Chapter 15

Antimatter gravity measurement

Antimatter gravity measurement pertains to testing whether or not antimatter is subject to gravity in the same way as typical matter. A current and prevalent hypothesis, based on Einstein's equivalence principle, states that gravity affects matter and antimatter exactly the same. Yet, there is no direct evidence to support this hypothesis.

Additionally, the hypothesis of dark matter and dark energy means that 95% of the mass of the observable universe cannot currently be detected; and can only be inferred from calculations and observable indirect effects. Therefore, researchers are unable to account for 95% of the universe's gravity. This means that gravitational effects on antimatter are actually unknown. Direct experiments are needed. Recycling atoms and antimatter interferometry using antimatter hydrogen atoms is a proposed solution.[1][2][3][4]

15.1 Experiments

- ALPHA is studying tapped antihydrogen

- ATRAP is developing techniques for cold antihydrogen

- AEGIS proposes to diffract antihydrogen

15.2 References

[1] Hamilton, P. et al. (2014). "Antimatter Interferometry for Gravity Measurements". *Physical Review Letters* **112**: 121102. arXiv:1308.1079. Bibcode:2014PhRvL.11211102H. doi:10.1103/PhysRevLett.112.121102. Lay summary.

[2] Regenfus, C. (26 September 2004). *Measurement of antimatter gravity with an (anti) matter wave interferometer* (PDF). *SPSC Villars Meetings 2004* (CERN). Retrieved 3 April 2014.

[3] Amole, C. et al. (2013). "Description and first application of a new technique to measure the gravitational mass of antihydrogen". *Nature Communications* **4**: 1785. Bibcode:2013NatCo...4E1785A. doi:10.1038/ncomms2787. PMC 3644108. PMID 23653197.

[4] "ALPHA probes antimatter gravity". ALPHA. 14 January 2014. Retrieved 2 April 2014.

Chapter 16

Bi-scalar tensor vector gravity

Not to be confused with Scalar–tensor–vector gravity or Tensor–vector–scalar gravity.

Bi-scalar tensor vector gravity theory (**BSTV**)[1] is an extension of the *tensor–vector–scalar gravity* theory (TeVeS).[2] TeVeS is a relativistic generalization of Mordehai Milgrom's Modified Newtonian Dynamics MOND paradigm proposed by Jacob Bekenstein.[3] BSTV was proposed by R.H.Sanders. BSTV makes TeVeS more flexible by making a non-dynamical scalar field in TeVeS into a dynamical one.[4]

16.1   References

[1] Sanders, R.H. (2005-07-01) A tensor-vector-scalar framework for modified dynamics and cosmic dark matter *Cornell University Library*, retrieved July 11, 2010

[2] Jacob D. Bekenstein (2004), "Relativistic gravitation theory for the modified Newtonian dynamics paradigm", *Phys. Rev. D* **70**: 083509, arXiv:astro-ph/0403694, Bibcode:2004PhRvD..70h3509B, doi:10.1103/PhysRevD.70.083509

[3] M. Milgrom (1983), "A modification of the Newtonian dynamics as a possible alternative to the hidden mass hypothesis", *Astrophys. J.* **270**: 365–370, Bibcode:1983ApJ...270..365M, doi:10.1086/161130

[4] Zhao, HongSheng; Bacon, David J.; Taylor, Andy N.; Horne, Keith (2005-12-16) Testing Bekenstein's Relativistic MOND gravity with Lensing Data *Cornell University Library*, retrieved July 11, 2010

Chapter 17

Bimetric gravity

Bimetric gravity or **bigravity** refers to a class of modified mathematical theories of gravity (or gravitation) in which two metric tensors are used instead of one. The second metric may be introduced at high energies, with the implication that the speed of light could be energy-dependent. If the two metrics are dynamical and interact then there will be two graviton modes, one massive and one massless, and thus bimetric theories are closely related to massive gravity.[1]

There are several different bimetric theories, such as those attributed to Nathan Rosen (1909–1995)[2][3] or Mordehai Milgrom with Modified Newtonian Dynamics (MOND). More recently, developments in massive gravity have also led to new consistent theories of bimetric gravity. Though none has been shown to account for physical observations more accurately or more consistently than the theory of general relativity, Rosen's theory has been shown to be inconsistent with observations of the Hulse–Taylor binary pulsar.[2] Some of these theories lead to cosmic acceleration at late times and are therefore alternatives to dark energy.[4][5]

17.1   Rosen's bigravity (1940)

In general relativity (GR), it is assumed that the distance between two points in spacetime is given by the metric tensor. Einstein's field equation is then used to calculate the form of the metric based on the distribution of energy and momentum.

Rosen (1940) has proposed that at each point of space-time, there is a Euclidean metric tensor γ_{ij} in addition to the Riemannian metric tensor g_{ij}. Thus at each point of space-time there are two metrics:

1. $ds^2 = g_{ij}dx^i dx^j$
2. $d\sigma^2 = \gamma_{ij}dx^i dx^j$

The first metric tensor, g_{ij}, describes the geometry of space-time and thus the gravitational field. The second metric tensor, γ_{ij}, refers to the flat space-time and describes the inertial forces. The Christoffel symbols formed from g_{ij} and γ_{ij} are denoted by $\{^i_{jk}\}$ and Γ^i_{jk} respectively. The quantities Δ are defined such that

$$\Delta^i_{jk} = \{^i_{jk}\} - \Gamma^i_{jk} \qquad (1)$$

Two kinds of covariant differentiation then arise: g-differentiation based on g_{ij} (denoted by a semicolon), and 3-differentiation based on γ_{ij} (denoted by a slash). Ordinary partial derivatives are represented by a comma. Let $R^\lambda_{ij\sigma}$ and $P^\lambda_{ij\sigma}$ be the Riemann curvature tensors calculated from g_{ij} and γ_{ij}, respectively. In the above approach the curvature tensor $P^\lambda_{ij\sigma}$ is zero, since γ_{ij} is the flat space-time metric.

From (1) one finds that though $\{:\}$ and Γ are not tensors, but Δ is a tensor having the same form as $\{:\}$ except that the ordinary partial derivative is replaced by 3-covariant derivative. A straightforward calculation yields the Riemann curvature tensor

$$R^h_{ijk} = -\Delta^h_{ij/k} + \Delta^h_{ik/j} + \Delta^h_{mj}\Delta^m_{ik} - \Delta^h_{mk}\Delta^m_{ij}$$

Each term on right hand side is a tensor. It is seen that from GR one can go to the new formulation just by replacing $\{:\}$ by Δ, ordinary differentiation by 3-covariant differentiation, $\sqrt{-g}$ by $\sqrt{\frac{g}{\gamma}}$, integration measure d^4x by $\sqrt{-\gamma}d^4x$, where $g = det(g_{ij})$, $\beta = det(\gamma_{ij})$ and $d^4x = dx^1 dx^2 dx^3 dx^4$. Having once introduced γ_{ij} into the theory, one has a great number of new tensors and scalars at one's disposal. One can set up other field equations other than Einstein's. It is possible that some of these will be more satisfactory for the description of nature.

The geodesic equation in bimetric relativity (BR) takes the form

$$\frac{d^2x}{ds^2} + \Gamma^i_{jk}\frac{dx^j}{ds}\frac{dx^k}{ds} + \Delta^i_{jk}\frac{dx^j}{ds}\frac{dx^k}{ds} = 0 \qquad (2)$$

It is seen from equations (1) and (2) that Γ can be regarded as describing the inertial field because it vanishes by a suitable coordinate transformation.

The quantity Δ, being a tensor, is independent of any coordinate system and hence may be regarded as describing the permanent gravitational field.

Rosen (1973) has found BR satisfying the covariance and equivalence principle. In 1966, Rosen showed that the introduction of the space metric into the framework of general relativity not only enables one to get the energy momentum density tensor of the gravitational field, but also enables one to obtain this tensor from a variational principle. The field equations of BR derived from the variational principle are

$$K^i_j = N^i_j - \frac{1}{2}\delta^i_j N = -8\pi\kappa T^i_j \qquad (3)$$

where

$$N^i_j = \frac{1}{2}\gamma^{\alpha\beta}(g^{hi}g_{hj/\alpha})_{/\beta}$$

or

$$N^i_j = \frac{1}{2}\gamma^{\alpha\beta}\Big\{(g^{hi}g_{hj,\alpha})_{,\beta} - (g^{hi}g_{mj}\Gamma^m_{h\alpha})_{,\beta} - \gamma^{\alpha\beta}(\Gamma^i_{j\alpha})_{,\beta} + \Gamma^i_{\lambda\beta}[g^{h\lambda}g_{hj,\alpha} - g^{h\lambda}g_{mj}\Gamma^m_{h\alpha} - \Gamma^\lambda_{j\alpha}]$$

$$-\Gamma^\lambda_{j\beta}[g^{hi}g_{h\lambda,\alpha} - g^{hi}g_{m\lambda}\Gamma^m_{h\alpha} - \Gamma^i_{\lambda\alpha}] + \Gamma^\lambda_{\alpha\beta}[g^{hi}g_{hj,\lambda} - g^{hi}g_{mj}\Gamma^m_{h\lambda} - \Gamma^i_{j\lambda}]\Big\}$$

$$N = g^{ij}N_{ij}, \qquad \kappa = \sqrt{\frac{g}{\gamma}}.$$

and T^i_j is the energy-momentum tensor.

The variational principle also leads to the relation

$$T^i_{j;i} = 0.$$

Hence from (3)

$$K^i_{j;i} = 0,$$

which implies that in a BR, a test particle in a gravitational field moves on a geodesic with respect to g_{ij}.

It is found that the BR and GR theories differ in the following cases:

- propagation of electromagnetic waves
- the external field of a high density star
- the behaviour of intense gravitational waves propagating through a strong static gravitational field.

The predictions of gravitational radiation in Rosen's theory have been shown to be in conflict with observations of the Hulse–Taylor binary pulsar.[2]

17.2 Massive bigravity

Since 2010 there has been renewed interest in bigravity after the development by Claudia de Rham, Gregory Gabadadze, and Andrew Tolley (dRGT) of a healthy theory of massive gravity.[6] Massive gravity is a bimetric theory in the sense that nontrivial interaction terms for the metric $g_{\mu\nu}$ can only be written down with the help of a second metric, as the only nonderivative term that can be written using one metric is a cosmological constant. In the dRGT theory, a nondynamical "reference metric" $f_{\mu\nu}$ is introduced, and the interaction terms are built out of the matrix square root of $g^{-1}f$.

In dRGT massive gravity, the reference metric must be specified by hand. One can give the reference metric an Einstein-Hilbert term, in which case $f_{\mu\nu}$ is not chosen but instead evolves dynamically in response to $g_{\mu\nu}$ and possibly matter. This *massive bigravity* was introduced by Fawad Hassan and Rachel Rosen[1] as an extension of dRGT massive gravity.

The dRGT theory is crucial to developing a theory with two dynamical metrics because general bimetric theories are plagued by the Boulware-Deser ghost, a possible sixth polarization for a massive graviton.[7] The dRGT potential is constructed specifically to render this ghost nondynamical, and as long as the kinetic term for the second metric is of the Einstein-Hilbert form, the resulting theory remains ghost-free.[1]

The action for the ghost-free massive bigravity is given by[8]

$$S = -\frac{M_g^2}{2}\int d^4x\sqrt{-g}R(g) - \frac{M_f^2}{2}\int d^4x\sqrt{-f}R(f) + m^2 M_g^2 \int d^4x\sqrt{-g}\sum_{n=0}^{4}\beta_n e_n(\mathbb{X}) + \int d^4x\sqrt{-g}\mathcal{L}_\mathrm{m}(g,\Phi_i).$$

As in standard general relativity, the metric $g_{\mu\nu}$ has an Einstein-Hilbert kinetic term proportional to the Ricci scalar $R(g)$ and a minimal coupling to the matter Lagrangian $\mathcal{L}_\mathrm{m}$, with Φ_i representing all of the matter fields, such as those of the Standard Model. An Einstein-Hilbert term is also given for $f_{\mu\nu}$. Each metric has its own Planck mass, M_g and M_f. The interaction potential is the same as in dRGT massive gravity. The β_i are dimensionless coupling constants and m (or specifically $\beta_i^{1/2}m$) is related to the mass of the massive graviton. This theory propagates seven degrees of freedom, corresponding to a massless graviton and a massive graviton (although the massive and massless states do not align with either of the metrics).

The interaction potential is built out of the elementary symmetric polynomials e_n of the eigenvalues of the matrices $\mathbb{K} = \mathbb{I} - \sqrt{g^{-1}f}$ or $\mathbb{X} = \sqrt{g^{-1}f}$, parametrized by dimensionless coupling constants α_i or β_i, respectively. Here $\sqrt{g^{-1}f}$ is the matrix square root of the matrix $g^{-1}f$. Written in index notation, $\mathbb{X}$ is defined by the relation

$$X^\mu{}_\alpha X^\alpha{}_\nu = g^{\mu\alpha}f_{\nu\alpha}.$$

The e_n can be written directly in terms of $\mathbb{X}$ as

$e_0(\mathbb{X}) = 1,$

$e_1(\mathbb{X}) = [\mathbb{X}],$

$e_2(\mathbb{X}) = \dfrac{1}{2}\left([\mathbb{X}]^2 - [\mathbb{X}^2]\right),$

$e_3(\mathbb{X}) = \dfrac{1}{6}\left([\mathbb{X}]^3 - 3[\mathbb{X}][\mathbb{X}^2] + 2[\mathbb{X}^3]\right),$

$e_4(\mathbb{X}) = \det\mathbb{X}.$

where brackets indicate a trace, $[\mathbb{X}] \equiv X^a{}_\mu$. It is the particular antisymmetric combination of terms in each of the e_n which is responsible for rendering the Boulware-Deser ghost nondynamical.

17.3   See also

- Accelerating universe

- Alternatives to general relativity

- Bimetric theory

- DGP model

- Scalar–tensor theory

17.4   References

[1] Hassan, S.F.; Rosen, Rachel A. (2012). "Bimetric Gravity from Ghost-free Massive Gravity". *JHEP* **1202**: 126. arXiv:1109.3515. Bibcode:2012JHEP...02..126H. doi:10.1007/JHEP02(2012)126.

[2] *The New Physics*, Paul Davies, 1992, 526 pages, web: Books-Google-ak.

[3] "Nathan Rosen — The Man and His Life-Work", Technion.ac.il, 2011, web: Technion-rosen.

[4] Akrami, Yashar; Koivisto, Tomi S.; Sandstad, Marit (2013). "Accelerated expansion from ghost-free bigravity: a statistical analysis with improved generality".*JHEP***1303**: 099.arXiv:1209.0457.Bibcode:2013JHEP...03..099A.doi:10.1007/JHEP03(2013)

[5] Akrami, Yashar; Hassan, S.F.; Könnig, Frank; Schmidt-May, Angnis; Solomon, Adam R. (2015). "Bimetric gravity is cosmologically viable". arXiv:1503.07521. Bibcode:2015arXiv150307521A.

[6] de Rham, Claudia; Gabadadze, Gregory; Tolley, Andrew J. (2010). "Resummation of Massive Gravity". *Phys.Rev.Lett.* **106**: 231101. arXiv:1011.1232. Bibcode:2011PhRvL.106w1101D. doi:10.1103/PhysRevLett.106.231101.

[7]Boulware, David G.; Deser, Stanley (1972). "Can gravitation have a finite range?".*Phys.Rev.***D6**: 3368–3382.Bibcode:1972Ph doi:10.1103/PhysRevD.6.3368.

[8] Hassan, S.F.; Rosen, Rachel A. (2011). "On Non-Linear Actions for Massive Gravity". *JHEP* **1107**: 009. arXiv:1103.6055. Bibcode:2011JHEP...07..009H. doi:10.1007/JHEP07(2011)009.

Chapter 18

Composite gravity

In theoretical physics, **composite gravity** refers to models that attempted to derive general relativity in a framework where the graviton is constructed as a composite bound state of more elementary particles, usually fermions. A theorem by Steven Weinberg and Edward Witten shows that this is not possible in Lorentz covariant theories: massless particles with spin greater than one are forbidden. The AdS/CFT correspondence may be viewed as a loophole in their argument. However, in this case not only the graviton is emergent: a whole spacetime dimension is emergent, too.[1]

18.1   See also

- Weinberg–Witten theorem

18.2   References

[1] "Probing composite gravity in colliders". scitation.aip.org. Retrieved 2008-07-08.

Chapter 19

Conformal gravity

Conformal gravity are gravity theories that are invariant under conformal transformations in the Riemannian geometry sense; more accurately, they are invariant under Weyl transformations $g_{ab} \to \Omega^2(x)g_{ab}$ where g_{ab} is the metric tensor and $\Omega(x)$ is a function on spacetime.

19.1 Weyl-squared theories

The simplest theory in this category has the square of the Weyl tensor as the Lagrangian

$$S = \int \mathrm{d}^4x \sqrt{-g}\, C_{abcd}C^{abcd},$$

where C_{abcd} is the Weyl tensor. This is to be contrasted with the usual Einstein–Hilbert action where the Lagrangian is just the Ricci scalar. The equation of motion upon varying the metric is called the Bach equation,

$$2\nabla_a\nabla_d C^a{}_{bc}{}^d + C^a{}_{bc}{}^d R_{ad} = 0,$$

where R_{ab} is the Ricci tensor. Conformally flat metrics are solutions of this equation.

Since these theories lead to fourth order equations for the fluctuations around a fixed background, they are not manifestly unitary. It has therefore been generally believed that they could not be consistently quantized. This is now disputed.[1]

19.2 Four-derivative theories

Conformal gravity is an example of a 4-derivative theory. This means that each term in the wave equation can contain up to 4 derivatives. There are pros and cons of 4-derivative theories. The pros are that the quantized version of the theory is more convergent and renormalisable. The cons are that there may be issues with causality. A simpler example of a 4-derivative wave equation is the scalar 4-derivative wave equation:

$$\Box^2 \Phi = 0$$

The solution for this in a central field of force is:

$$\Phi(r) = 1 - \frac{2m}{r} + ar + br^2$$

The first two terms are the same as a normal wave equation. Because this equation is a simpler approximation to conformal gravity, m corresponds to the mass of the central source. The last two terms are unique to 4-derivative wave equations. It has been suggested that small values be assigned to them to account for the galactic acceleration constant (also known as dark matter) and the dark energy constant.[2] The solution equivalent to the Schwarzschild solution in general relativity for a spherical source for conformal gravity has a metric with:

$$\phi(r) = g^{00} = (1 - 6bc)^{\frac{1}{2}} - \frac{2b}{r} + cr + \frac{d}{3}r^2$$

to show the difference between general relativity. 6bc is very small so can be ignored. The problem is that now c is the total mass-energy of the source, b is the integral of density times distance to source squared. So this is a completely different potential to general relativity and not just a small modification.

The main issue with conformal gravity theories, as well as any theory with higher derivatives, is the typical presence of ghosts, which point to instabilities of the quantum version of the theory, although there might be a solution to the ghost problem.[3]

An alternative approach is to consider the gravitational constant as a symmetry broken scalar field in which case you would consider a small correction to Newtonian gravity like this (where we consider ε to be a small correction:

$$\Box\Phi + \varepsilon^2\Box^2\Phi = 0$$

in which case the general solution is the same as the Newtonian case except there can be an additional term:

$$\Phi = 1 - \frac{2m}{r}(1 + \alpha\sin(r/\varepsilon + \beta))$$

where there is can be an additional component varying sinusoidally over space. The wavelength of this variation could be quite large such as an atomic width. Thus there appears to be several stable potentials around a gravitational force in this model.

19.3   Conformal unification to the Standard Model

By adding a suitable gravitational term to the standard model action with gravitational coupling, the theory develops a local conformal (Weyl) invariance in the unitary gauge for the local SU(2). The gauge is fixed by requiring the Higgs scalar to be a constant. This mechanism generates the masses for the vector bosons and matter fields with no physical degrees of freedom for the Higgs.[4] [5]

19.4   See also

- Conformal supergravity

19.5   Notes

[1] Mannheim, Philip D. (2007-07-16). "PASCOS-07, Imperial College London, July 2007" **0707**. p. 2283. arXiv:0707.2283. Bibcode:2007arXiv0707.2283M. |contribution= ignored (help)

[2] Mannheim, Philip D. (2005-08-01). "Alternatives to Dark Matter and Dark Energy". *Prog.Part.Nucl.Phys.* **56** (2): 340. arXiv:astro-ph/0505266. Bibcode:2006PrPNP..56..340M. doi:10.1016/j.ppnp.2005.08.001.

[3] Mannheim, Philip D. (2006-09-06). "Solution to the ghost problem in fourth order derivative theories". *Found. Phys.* **37** (4–5): 532. arXiv:hep-th/0608154. Bibcode:2007FoPh...37..532M. doi:10.1007/s10701-007-9119-7.

[4] Montag, J. Lee (1992). "Spontaneously Broken Conformal Symmetry and the Standard Model"

[5] Pawlowski, M.; Raczka, R. (1994). "A Unified Conformal Model for Fundamental Interactions without Dynamical Higgs Field". *Foundations of Physics* **24** (9): 1305–1327. arXiv:hep-th/9407137. Bibcode:1994FoPh...24.1305P. doi:10.1007/BF02148570

19.6 References

- E.S. Fradkin and A.A. Tseytlin (1985). "Conformal Supergravity". *Phys. Rept.* **119**(4–5): 233–362. Bibcode:1985 doi:10.1016/0370-1573(85)90138-3.

- Conformal gravity on arxiv.org

19.7 External links

- Conformal gravity: New light on dark matter and dark energy talk at the SETI Institute

- Falsification of Mannheim's conformal gravity at CERN

- Mannheim's rebuttal of above at arXiv.

Chapter 20

Entropic gravity

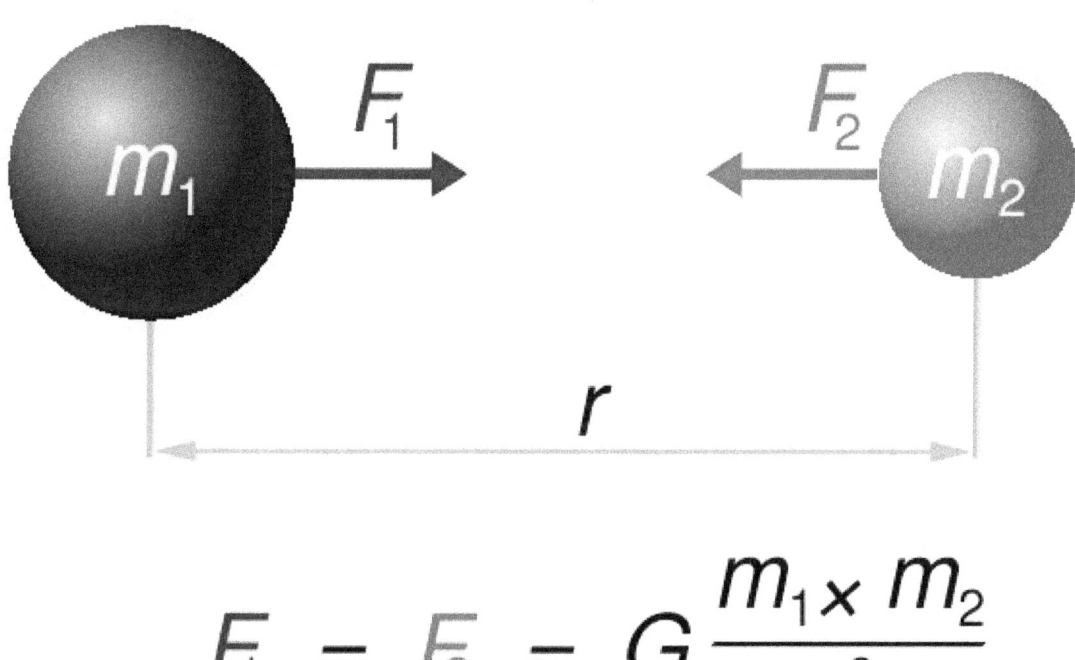

$$F_1 = F_2 = G\frac{m_1 \times m_2}{r^2}$$

Verlinde's statistical description of gravity as an entropic force leads to the correct inverse square distance law of attraction between classical bodies.

Entropic gravity is a theory in modern physics that describes gravity as an entropic force—not a fundamental interaction mediated by a quantum field theory and a gauge particle (like photons for the electromagnetic force, and gluons for the strong nuclear force), but a probabilistic consequence of physical systems' tendency to increase their entropy. The proposal has been intensely contested in the physics community but it has also sparked a new line of research into thermodynamic properties of gravity.

20.1 Origin

The probabilistic description of gravity has a history that goes back at least to research on black hole thermodynamics by Bekenstein and Hawking in the mid-1970s. These studies suggest a deep connection between gravity and thermodynamics, which describes the behavior of heat. In 1995 Jacobson demonstrated that the Einstein field equations describing relativistic gravitation can be derived by combining general thermodynamic considerations with the equivalence principle.[1] Subsequently, other physicists, most notably Thanu Padmanabhan, began to explore links between gravity and entropy.[2][3]

20.2 Erik Verlinde's theory

In 2009, Erik Verlinde disclosed a conceptual model that describes gravity as an entropic force.[4] On 6 January 2010 he published a preprint of a 29-page paper titled *On the Origin of Gravity and the Laws of Newton*.[5] The paper was published in the Journal of High Energy Physics in April 2011.[6] Reversing the logic of over 300 years, it argued (similar to Jacobson's result) that gravity is a consequence of the "information associated with the positions of material bodies". This model combines the thermodynamic approach to gravity with Gerardus 't Hooft's holographic principle. It implies that gravity is not a fundamental interaction, but an emergent phenomenon which arises from the statistical behavior of microscopic degrees of freedom encoded on a holographic screen. The paper drew a variety of responses from the scientific community. Andrew Strominger, a string theorist at Harvard said "Some people have said it can't be right, others that it's right and we already knew it — that it's right and profound, right and trivial."[7]

In July 2011 Verlinde presented the further development of his ideas in a contribution to the Strings 2011 conference, including an explanation for the origin of dark matter.[8]

Verlinde's article also attracted a large amount of media exposure,[9][10] and led to immediate follow-up work in cosmology,[the dark energy hypothesis,[13] cosmological acceleration,[14][15] cosmological inflation,[16] and loop quantum gravity.[17] Also, a specific microscopic model has been proposed that indeed leads to entropic gravity emerging at large scales.[18]

20.3 Criticism and experimental tests

Entropic gravity, as proposed by Verlinde in his original article, reproduces Einstein field equations and, in a Newtonian approximation, a 1/r potential for gravitational forces. Since it does not make new physical predictions, it can not be falsified with existing experimental methods, at this time, any more than Newtonian gravity and general relativity.

Even so, entropic gravity in its current form has been severely challenged on formal grounds. Matt Visser, professor of mathematics at Victoria University of Wellington, NZ in "Conservative Entropic Forces"[19] has shown that the attempt to model conservative forces in the general Newtonian case (i.e. for arbitrary potentials and an unlimited number of discrete masses) leads to unphysical requirements for the required entropy and involves an unnatural number of temperature baths of differing temperatures. Visser concludes:

> There is no reasonable doubt concerning the physical reality of entropic forces, and no reasonable doubt that classical (and semi-classical) general relativity is closely related to thermodynamics [52–55]. Based on the work of Jacobson [1–6], Thanu Padmanabhan [7–12], and others, there are also good reasons to suspect a thermodynamic interpretation of the fully relativistic Einstein equations might be possible. Whether the specific proposals of Verlinde [26] are anywhere near as fundamental is yet to be seen — the rather baroque construction needed to accurately reproduce n-body Newtonian gravity in a Verlinde-like setting certainly gives one pause.

For the derivation of Einstein's equations from an entropic gravity perspective, Tower Wang shows in [20] that the inclusion of energy-momentum conservation and cosmological homogeneity and isotropy requirements severely restrict a wide class of potential modifications of entropic gravity, some of which have been used to generalize entropic gravity beyond the singular case of an entropic model of Einstein's equations. Wang asserts that

As indicated by our results, the modified entropic gravity models of form (2), if not killed, should live in a very narrow room to assure the energy-momentum conservation and to accommodate a homogeneous isotropic universe.

20.3.1 Entropic gravity and quantum coherence

Another way of criticism of the entropic gravity is a reason that entropic processes should break quantum coherence. Recent experiments with ultra-cold neutrons in the gravitational field of Earth show that neutrons lie on discrete levels exactly as predicted by Schrödinger equation considering the gravitation to be a conservative potential field without any decoherent factors. Archil Kobakhidze argues that this result disproves entropic gravity.[21][22] Luboš Motl gives popular explanations of this problem in his blog.[23][24]

20.4   See also

- Entropic elasticity of an ideal chain

- Entropic force

- Gravitation

- Induced gravity

20.5   References

[1] Jacobson, Theodore (4 April 1995). "Thermodynamics of Spacetime: The Einstein Equation of State". *Phys. Rev. Lett.* **75** (7): 1260–1263. arXiv:gr-qc/9504004. Bibcode:1995PhRvL..75.1260J. doi:10.1103/PhysRevLett.75.1260.

[2] Padmanabhan, Thanu (26 November 2009). "Thermodynamical Aspects of Gravity: New insights". *Rep. Prog. Phys.* **73** (4): 6901. arXiv:0911.5004. Bibcode:2010RPPh...73d6901P. doi:10.1088/0034-4885/73/4/046901.

[3] Mok, H.M. (13 August 2004). "Further Explanation to the Cosmological Constant Problem by Discrete Space-time Through Modified Holographic Principle". arXiv:physics/0408060 [physics.gen-ph].

[4] van Calmthout, Martijn (12 December 2009). "Is Einstein een beetje achterhaald?". *de Volkskrant* (in Dutch). Retrieved 6 September 2010.

[5] Verlinde, Eric (6 January 2010). "Title: On the Origin of Gravity and the Laws of Newton". arXiv:1001.0785 [hep-th].

[6] E.P. Verlinde. "On the Origin of Gravity and the Laws of Newton". *JHEP*. arXiv:1001.0785. Bibcode:2011JHEP...04..029V. doi:10.1007/JHEP04(2011)029.

[7] Overbye, Dennis (12 July 2010). "A Scientist Takes On Gravity". *The New York Times*. Retrieved 6 September 2010.

[8] E. Verlinde, The Hidden Phase Space of our Universe, Strings 2011, Uppsala, 1 July 2011.

[9] The entropy force: a new direction for gravity, New Scientist, 20 January 2010, issue 2744

[10] Gravity is an entropic form of holographic information, *Wired Magazine*, 20 January 2010

[11] Fu-Wen Shu; Yungui Gong (2010). "Equipartition of energy and the first law of thermodynamics at the apparent horizon". arXiv:1001.3237 [gr-qc].

[12] Rong-Gen Cai; Li-Ming Cao; Nobuyoshi Ohta (2010). "Friedmann Equations from Entropic Force", *Phys. Rev.* **D 81** (6). arXiv:1001.3470. Bibcode:2010PhRvD..81f1501C. doi:10.1103/PhysRevD.81.061501.

[13] It from Bit: How to get rid of dark energy, Johannes Koelman, 2010

[14] Easson; Frampton; Smoot (2010). "Entropic Accelerating Universe". *Phys. Lett. B* **696** (3): 273–277. arXiv:1002.4278. Bibcode:2011PhLB..696..273E. doi:10.1016/j.physletb.2010.12.025.

[15] Yi-Fu Cai; Jie Liu; Hong Li (2010). "Entropic cosmology: a unified model of inflation and late-time acceleration". *Phys. Lett. B* **690** (3): 213–219. arXiv:1003.4526. Bibcode:2010PhLB..690..213C. doi:10.1016/j.physletb.2010.05.033.

[16] Yi Wang (2010). "Towards a Holographic Description of Inflation and Generation of Fluctuations from Thermodynamics". arXiv:1001.4786 [hep-th].

[17] Lee Smolin (2010). "Newtonian gravity in loop quantum gravity". arXiv:1001.3668 [gr-qc].

[18] Jarmo Mäkelä (2010). "Notes Concerning "On the Origin of Gravity and the Laws of Newton" by E. Verlinde (arXiv:1001.0785)". arXiv:1001.3808 [gr-qc].

[19] Visser, Matt. "Conservative entropic forces". arXiv:1108.5240.

[20] Wang, Tower. "Modified entropic gravity revisited". arXiv:1211.5722.

[21] Kobakhidze, Archil. "Gravity is not an entropic force". arXiv:1009.5414.

[22] Kobakhidze, Archil. "Once more: gravity is not an entropic force". arXiv:1108.4161.

[23] Motl, Luboš. "Why gravity can't be entropic". *The Reference Frame*. Retrieved 10 March 2015.

[24] Motl, Luboš. "Once more: gravity is not an entropic force". *The Reference Frame*. Retrieved 29 April 2015.

20.6 Further reading

- It from bit - Entropic gravity for pedestrians, J. Koelman

- Gravity: the inside story, T Padmanabhan

- Experiments Show Gravity Is Not an Emergent Phenomenon

Chapter 21

Extended theories of gravity

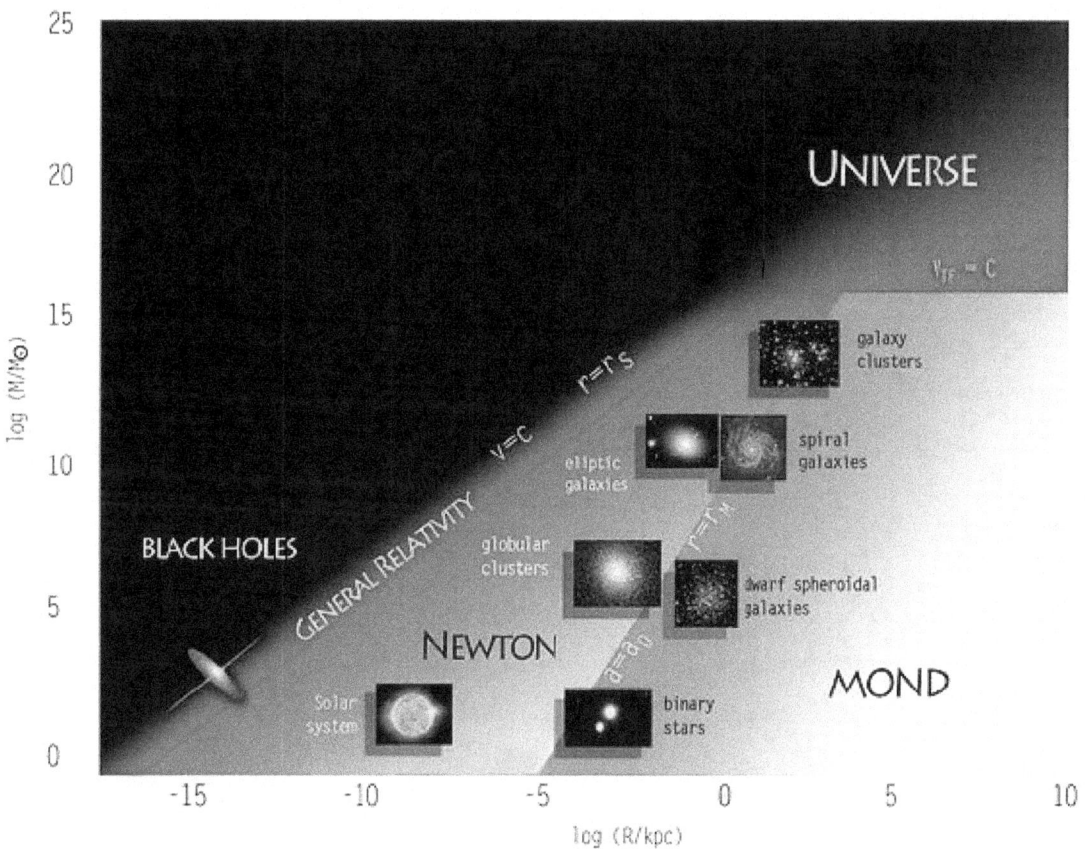

Distribution of astronomical systems in the phase space diagram or gravity, plotted by X. Hernández

Extended theories of gravity are alternative theories of gravity developed from the exact starting points investigated first by Einstein and Hilbert. These are theories describing gravity, which are metric theory, "a linear connection" or related affine theories, or metric-affine gravitation theory. Rather than trying to discover correct calculations for the matter side of the Einstein field equations; which include inflation, dark energy, dark matter, large-scale structure, and possibly quantum gravity; it is proposed, instead, to change the gravitational side of the equation.[1][2]

21.1 Proposed theories

21.1.1 Hernández et al.

One such theory is also an extension to general relativity and Newton's Universal gravity law ($f(\chi) = \chi^{\frac{3}{2}}$), first proposed in 2010 by the Mexican astronomers Xavier Hernández Doring, Sergio Mendoza Ramos et al., researchers at the Astronomy Institute, at the National Autonomous University of Mexico.[3][4] This theory is in accordance with observations of kinematics of the solar system, extended binary stars,[5] and all types of galaxies and galactic groups and clouds.[6] It also reproduces the gravitational lensing effect with out the need of postulating dark matter.[7]

There is some evidence that it could also explain the dark energy phenomena[8][9] and give a nice solution to the initial conditions problem.[10]

These results can be classified as a metric f(R) gravity theory, more properly an f(R,T) theory, derived from an action principle. This approach to solve the dark matter problem takes into account the Tully–Fisher relation as an empirical law that applies always at scales larger than the Milgrom radius.[11]

21.2 See also

- Modified Newtonian Dynamics
- Alternatives to general relativity

21.3 References

[1] Capozziello, S.; De Laurentis, M. (2011). "Extended Theories of Gravity". *Physics Reports* **509** (4–5): 167–321. arXiv:1108.6266. Bibcode:2011PhR...509..167C. doi:10.1016/j.physrep.2011.09.003.

[2] Capozziello, S.; Francaviglia, M. (2008). "Extended theories of gravity and their cosmological and astrophysical applications". *General Relativity and Gravitation* **40** (2–3): 357–420. arXiv:0706.1146. Bibcode:2008GReGr..40..357C. doi:10.1007/s10714-007-0551-y.

[3] Mendoza, S.; Hernandez, X.; Hidalgo, J. C.; Bernal, T. (2011). "A natural approach to extended Newtonian gravity: Tests and predictions across astrophysical scales". *Monthly Notices of the Royal Astronomical Society* **411** (411): 226–234. arXiv:1006.5037. Bibcode:2011MNRAS.411..226M. doi:10.1111/j.1365-2966.2010.17685.x.

[4] Hidalgo, J. C.; Mendoza, S.; Hernandez, X.; Bernal, T.; Jimenez, M. A.; Allen, C. (2012). "Non-relativistic Extended Gravity and its applications across different astrophysical scales". *AIP Conference Proceedings*. AIP Conference Proceedings **1458**: 427–430. arXiv:1202.4189. Bibcode:2012AIPC.1458..427H. doi:10.1063/1.4734451.

[5] Hernandez, X.; Jiménez, M. A.; Allen, C. (2012). "Wide binaries as a critical test of Classical Gravity". *European Physical Journal C* **72** (2): 1884. arXiv:1105.1873. Bibcode:2012EPJC...72.1884H. doi:10.1140/epjc/s10052-012-1884-6.

[6] Hernandez, X. (2012). "A Phase Space Diagram for Gravity". *Entropy* **14** (12): 848. arXiv:1203.4248. Bibcode:2012Entrp..14..8 doi:10.3390/e14050848.

[7] Mendoza, S.; Bernal, T.; Hernandez, X.; Hidalgo, J. C.; Torres, L. A. (2013). "Gravitational lensing with f(χ)=χ$^{3/2}$ gravity in accordance with astrophysical observations". *Monthly Notices of the Royal Astronomical Society* **433** (3): 1802–1812. arXiv:1208.6241. Bibcode:2013MNRAS.433.1802M. doi:10.1093/mnras/stt752.

[8] Mendoza, S. (2012). "Extending Cosmology: The Metric Approach". In Olmo, G. J. *Open Questions in Cosmology*. INTECH. pp. 133–156. arXiv:1208.3408. doi:10.5772/53878. ISBN 978-953-51-0880-1.

[9] Carranza, D. A.; Mendoza, S.; Torres, L. A. (2012). "A cosmological dust model with extended f(χ) gravity". *European Physical Journal C* **73**: 2282. arXiv:1208.2502. Bibcode:2013EPJC...73.2282C. doi:10.1140/epjc/s10052-013-2282-4.

[10] Hernandez, X.; Jimenez. M. A. (2013). "A first linear cosmological structure formation scenario under extended gravity". arXiv:1307.0777 [astro-ph.CO].

[11] Capozziello, S.; De Laurentis, M. (2013). "Extended Gravity: State of the Art and Perspectives". In Rosquist, K.; Jantzen, R. T.; Ruffini, R. *Proceedings of the Thirteenth Marcel Grossman Meeting on General Relativity*. World Scientific. arXiv:1307.4523.

21.4   Further reading

- Wands, D. (1994). "Extended gravity theories and the Einstein–Hilbert action". *Classical and Quantum Gravity* **11** (1): 269. arXiv:gr-qc/9307034. Bibcode:1994CQGra..11..269W. doi:10.1088/0264-9381/11/1/025.

- Allemandi, G.; Capone, M.; Capozziello, S.; Francaviglia, M. (2006). "Conformal aspects of the Palatini approach in Extended Theories of Gravity". *General Relativity and Gravitation* **38** (1): 33. arXiv:hep-th/0409198. Bibcode:2006GReGr..38...33A. doi:10.1007/s10714-005-0208-7.

21.5   External links

- Sergio Mendoza's web page

21.6   News

- El universal.com (Spanish).

- La jornada.mx (Spanish).

- La crónica.com (Spanish).

Chapter 22

Gauge gravitation theory

In quantum field theory, **gauge gravitation theory** is the effort to extend Yang–Mills theory, which provides a universal description of the fundamental interactions, to describe gravity. It should not be confused with the related but distinct gauge theory gravity.

The first gauge model of gravity was suggested by R. Utiyama in 1956[1] just two years after birth of the gauge theory itself.[2] However, the initial attempts to construct the gauge theory of gravity by analogy with the gauge models of internal symmetries encountered a problem of treating general covariant transformations and establishing the gauge status of a pseudo-Riemannian metric (a tetrad field).

In order to overcome this drawback, representing tetrad fields as gauge fields of the translation group was attempted.[3] Infinitesimal generators of general covariant transformations were considered as those of the translation gauge group, and a tetrad (coframe) field was identified with the translation part of an affine connection on a world manifold X. Any such connection is a sum $K = \Gamma + \Theta$ of a linear world connection Γ and a soldering form $\Theta = \Theta_\mu^a dx^\mu \otimes \vartheta_a$ where $\vartheta_a = \vartheta_a^\lambda \partial_\lambda$ is a non-holonomic frame. For instance, if K is the Cartan connection, then $\Theta = \theta = dx^\mu \otimes \partial_\mu$ is the canonical soldering form on X. There are different physical interpretations of the translation part Θ of affine connections. In gauge theory of dislocations, a field Θ describes a distortion.[4] At the same time, given a linear frame ϑ_a, the decomposition $\theta = \vartheta^a \otimes \vartheta_a$ motivates many authors to treat a coframe ϑ^a as a translation gauge field.[5]

Difficulties of constructing gauge gravitation theory by analogy with the Yang-Mills one result from the gauge transformations in these theories belonging to different classes. In the case of internal symmetries, the gauge transformations are just vertical automorphisms of a principal bundle $P \to X$ leaving its base X fixed. On the other hand, gravitation theory is built on the principal bundle FX of the tangent frames to X. It belongs to the category of natural bundles $T \to X$ for which diffeomorphisms of the base X canonically give rise to automorphisms of T.[6] These automorphisms are called general covariant transformations. General covariant transformations are sufficient in order to restate Einstein's General Relativity and metric-affine gravitation theory as the gauge ones.

In terms of gauge theory on natural bundles, gauge fields are linear connections on a world manifold X, defined as principal connections on the linear frame bundle FX, and a metric (tetrad) gravitational field plays the role of a Higgs field responsible for spontaneous symmetry breaking of general covariant transformations.[7]

Spontaneous symmetry breaking is a quantum effect when the vacuum is not invariant under the transformation group. In classical gauge theory, spontaneous symmetry breaking occurs if the structure group G of a principal bundle $P \to X$ is reducible to a closed subgroup H, i.e., there exists a principal subbundle of P with the structure group H.[8] By virtue of the well-known theorem, there exists one-to-one correspondence between the reduced principal subbundles of P with the structure group H and the global sections of the quotient bundle $P/H \to X$. These sections are treated as classical Higgs fields.

The idea of the pseudo-Riemannian metric as a Higgs field appeared while constructing non-linear (induced) representations of the general linear group $GL(4, \mathbb{R})$, of which the Lorentz group is a Cartan subgroup.[9] The geometric equivalence principle postulating the existence of a reference frame in which Lorentz invariants are defined on the whole world manifold is the theoretical justification of that the structure group $GL(4, \mathbb{R})$ of the linear frame bundle FX is reduced to the Lorentz group. Then the very definition of a pseudo-Riemannian metric on a manifold X as a global section of the quotient bundle $FX/O(1,3) \to X$ leads to its physical interpretation as a Higgs field. The physical reason for world symmetry breaking is the existence of Dirac fermion matter, whose symmetry group is the universal two-sheeted covering $SL(2,\mathbb{C})$ of the restricted Lorentz group, $SO^+(1,3)$.[10]

22.1   See also

- Ashtekar variables

- Metric-affine gravitation theory

- Einstein–Cartan theory

- Spontaneous symmetry breaking

- Teleparallelism

- Reduction of the structure group

- Higgs field (classical)

- General covariant transformations

- Equivalence principle (geometric)

- Affine gauge theory

22.2   Notes

[1] R Utiyama, Invariant theoretical interpretation of interaction, Physical Review **101** (1956) 1597. doi:10.1103/PhysRev.101.1597

[2] Blagojević, Milutin; Hehl, Friedrich W. (2013). *Gauge Theories of Gravitation: A Reader with Commentaries*. World Scientific. ISBN 978-184-8167-26-1.

[3] F.Hehl, J. McCrea, E. Mielke, Y. Ne'eman, Metric-affine gauge theory of gravity: field equations, Noether identities, world spinors, and breaking of dilaton invariance, Physics Reports **258** (1995) 1. doi:10.1016/0370-1573(94)00111-F

[4] C.Malyshev, The dislocation stress functions from the double curl $T(3)$-gauge equations: Linearity and look beyond, Annals of Physics **286** (2000) 249. doi:10.1006/aphy.2000.6088

[5] M. Blagojević, Gravitation and Gauge Symmetries (IOP Publishing, Bristol, 2002).

[6] I. Kolář, P. W. Michor, J. Slovák, Natural Operations in Differential Geometry (Springer-Verlag, Berlin, Heidelberg, 1993).

[7] D.Ivanenko, G.Sardanashvily, The gauge treatment of gravity, Physics Reports **94** (1983) 1. doi:10.1016/0370-1573(83)90046-7

[8] L. Nikolova, V. Rizov, Geometrical approach to the reduction of gauge theories with spontaneous broken symmetries, Reports on Mathematical Physics **20** (1984) 287. doi:10.1016/0034-4877(84)90039-9

[9] M. Leclerk, The Higgs sector of gravitational gauge theories, Annals of Physics **321** (2006) 708. doi:10.1016/j.aop.2005.08.009

[10] G. Sardanashvily, O.Zakharov, Gauge Gravitation Theory (World Scientific, Singapore, 1992).

22.3 References

- I. Kirsch, A Higgs mechanism for gravity, Phys. Rev. **D72** (2005) 024001; arXiv: hep-th/0503024.

- G. Sardanashvily, Classical gauge gravitation theory, Int. J. Geom. Methods Mod. Phys. **8** (2011) 1869-1895; arXiv: 1110.1176.

- Yu. Obukhov, Poincaré gauge gravity: selected topics, Int. J. Geom. Methods Mod. Phys. **3** (2006) 95-138; arXiv: gr-qc/0601090.

Chapter 23

Gauge theory gravity

Gauge theory gravity (GTG) is a theory of gravitation cast in the mathematical language of geometric algebra. To those familiar with general relativity, it is highly reminiscent of the tetrad formalism although there are significant conceptual differences. Most notably, the background in GTG is flat, Minkowski spacetime. The equivalence principle is not assumed, but instead follows from the fact that the gauge covariant derivative is minimally coupled. As in general relativity, equations structurally identical to the Einstein field equations are derivable from a variational principle. A spin tensor can also be supported in a manner similar to Einstein–Cartan–Sciama–Kibble theory. GTG was first proposed by Lasenby, Doran, and Gull in 1998[1] as a fulfillment of partial results presented in 1993.[2] The theory has not been widely adopted by the rest of the physics community, who have mostly opted for differential geometry approaches like that of the related gauge gravitation theory.

23.1   Mathematical foundation

The foundation of GTG comes from two principles. First, *position-gauge invariance* demands that arbitrary local displacements of fields not affect the physical content of the field equations. Second, *rotation-gauge invariance* demands that arbitrary local rotations of fields not affect the physical content of the field equations. These principles lead to the introduction of a new pair of linear functions, the position-gauge field and the rotation-gauge field. A displacement by some arbitrary function f

$$x \mapsto x' = f(x)$$

gives rise to the position-gauge field defined by the mapping on its adjoint,

$$\bar{\mathsf{h}}(a,x) \mapsto \bar{\mathsf{h}}'(a,x) = \bar{\mathsf{h}}(f^{-1}(a),f(x)),$$

which is linear in its first argument and a is a constant vector. Similarly, a rotation by some arbitrary rotor R gives rise to the rotation-gauge field

$$\bar{\ }(a,x) \mapsto \bar{\ }'(a,x) = R\bar{\ }(a,x)R^{\dagger} - 2a \cdot \nabla RR^{\dagger}.$$

We can define two different covariant directional derivatives

$$a \cdot D = a \cdot \bar{\mathsf{h}}(\nabla) + \frac{1}{2} \ (\mathsf{h}(a))$$

$$a \cdot \mathcal{D} = a \cdot \bar{\mathsf{h}}(\nabla) + (\mathsf{h}(a))$$

or with the specification of a coordinate system

$$D_\mu = \partial_\mu + \frac{1}{2}\Omega_\mu$$

$$\mathcal{D}_\mu = \partial_\mu + \Omega_\mu \times,$$

where × denotes the commutator product.

The first of these derivatives is better suited for dealing directly with spinors whereas the second is better suited for observables. The GTG analog of the Riemann tensor is built from the commutation rules of these derivatives.

$$[D_\mu, D_\nu]\psi = \frac{1}{2}\mathsf{R}_{\mu\nu}\psi$$

$$\mathcal{R}(a \wedge b) = \mathsf{R}(\mathsf{h}(a \wedge b))$$

23.2 Field equations

The field equations are derived by postulating the Einstein–Hilbert action governs the evolution of the gauge fields, i.e.

$$S = \int \left[\frac{1}{2\kappa}(\mathcal{R} - 2\Lambda) + \mathcal{L}_{\mathrm{M}} \right] (\det \mathsf{h})^{-1}\, \mathrm{d}^4 x.$$

Minimizing variation of the action with respect to the two gauge fields results in the field equations

$$\mathcal{G}(a) - \Lambda a = \kappa \mathcal{T}(a)$$

$$\mathcal{D} \wedge \bar{\mathsf{h}}(a) = \kappa \mathcal{S} \cdot \bar{\mathsf{h}}(a),$$

where $\mathcal{T}$ is the covariant energy–momentum tensor and $\mathcal{S}$ is the covariant spin tensor. Importantly, these equations do not give an evolving curvature of spacetime but rather merely give the evolution of the gauge fields within the flat spacetime. Moreover, the existence of the spin tensor does *not* endow spacetime with torsion.

23.3 Relation to general relativity

For those more familiar with general relativity, it is possible to define a metric tensor from the position-gauge field in a manner similar to tetrads. In the tetrad formalism, a set of four vectors $\{e_{(a)}{}^\mu\}$ are introduced. The Greek index μ is raised or lowered by multiplying and contracting with the spacetime's metric tensor. The parenthetical Latin index (a) is a label for each of the four tetrads, which is raised and lowered as if it were multiplied and contracted with a separate Minkowski metric tensor. GTG, roughly, reverses the roles of these indices. The metric is implicitly assumed to be Minkowski in the selection of the spacetime algebra. The information contained in the other set of indices gets subsumed by the behavior of the gauge fields.

We can make the associations

$$g_\mu = \mathsf{h}^{-1}(e_\mu)$$

$$g^\mu = \bar{\mathsf{h}}(e^\mu)$$

for a covariant vector and contravariant vector in a curved spacetime, where now the unit vectors $\{e_\mu\}$ are the chosen coordinate basis. These can define the metric using the rule

$$g_{\mu\nu} = g_\mu \cdot g_\nu.$$

Following this procedure, it is possible to show that for the most part the observable predictions of GTG agree with Einstein–Cartan–Sciama–Kibble theory for non-vanishing spin and reduce to general relativity for vanishing spin. GTG does, however, make different predictions about global solutions. For example, in the study of a point mass, the choice of a "Newtonian gauge" yields a solution similar to the Schwarzschild metric in Gullstrand–Painlevé coordinates. General relativity permits an extension known as the Kruskal–Szekeres coordinates. GTG, on the other hand, forbids any such extension.

23.4   References

[1] Lasenby, Anthony; Chris Doran; Stephen Gull (1998), "Gravity, gauge theories and geometric algebra", *Philosophical Transactions of the Royal Society A* **356**: 487–582, arXiv:gr-qc/0405033, Bibcode:1998RSPTA.356..487L, doi:10.1098/rsta.1998.0178

[2] Doran, Chris; Anthony Lasenby; Stephen Gull (1993), F. Brackx, R. Delanghe, H. Serras, eds., "Gravity as a gauge theory in the spacetime algebra", *Third International Conference on Clifford Algebras and their Applications in Mathematical Physics*

23.5   External links

- David Hestenes: Spacetime calculus for gravitation theory – an account of the mathematical formalism explicitly directed to GTG

Chapter 24

Gauge vector–tensor gravity

Gauge vector–tensor gravity[1] (GVT) is a relativistic generalization of Mordehai Milgrom's Modified Newtonian dynamics (MOND) paradigm[2] where gauge fields cause the MOND behavior. The former covariant realizations of MOND such as the Bekenestein's Tensor–vector–scalar gravity and the Moffat's Scalar–tensor–vector gravity attribute MONDian behavior to some scalar fields. GVT is the first example wherein the MONDian behavior is mapped to the gauge vector fields. The main features of GVT can be summarized as follows:

- As it is derived from the action principle, GVT respects conservation laws;

- In the weak-field approximation of the spherically symmetric, static solution, GVT reproduces the MOND acceleration formula;

- It can accommodate gravitational lensing.

- It is in total agreement with the Einstein–Hilbert action in the strong and Newtonian gravities.

Its dynamical degrees of freedom are:

- Two gauge fields: $B_\mu, \hat{B}_\mu$;

- A metric, $g_{\mu\nu}$.

24.1   Details

The physical geometry, as seen by particles, represents the Finsler geometry–Randers type:

$$ds = \sqrt{-g_{\mu\nu}dx^\mu dx^\nu} + (B_\mu + \hat{B}_\mu)dx^\mu$$

This implies that the orbit of a particle with mass m can be derived from the following effective action:

$$S = m \int d\tau (\tfrac{1}{2}\dot{x}^\mu \dot{x}^\nu g_{\mu\nu} + (B_\mu + \hat{B}_\mu)\dot{x}^\mu) .$$

The geometrical quantities are Riemannian. GVT, thus, is a bi-geometric gravity.

24.1.1   Action

The metric's action coincides to that of the Einstein–Hilbert gravity:

$$S_{\text{Grav}} = \tfrac{1}{16\pi G} \int d^4 x \sqrt{-g} R$$

where R is the Ricci scalar constructed out from the metric. The action of the gauge fields follow:

$$S_B = -\frac{1}{16\pi G \kappa l^2} \int d^4x \sqrt{-g}\, L(\tfrac{l^2}{4} B_{\mu\nu} B^{\mu\nu})$$

and

$$S_{\tilde{B}} = -\frac{1}{16\pi G \tilde{\kappa} \tilde{l}^2} \int d^4x \sqrt{-g}\, L(\tfrac{\tilde{l}^2}{4} \tilde{B}_{\mu\nu} \tilde{B}^{\mu\nu})$$

where L has the following MOND asymptotic behaviors

$$L(x) = \begin{cases} x & , \quad \text{for } x \gg 1 \\ \frac{2}{3}|x|^{\frac{3}{2}} & , \quad \text{for } x \leq 1 \end{cases} ,$$

and $\kappa, \tilde{\kappa}$ represent the coupling constants of the theory while $l, \tilde{l}$ are the parameters of the theory and $l < \tilde{l}$.

24.1.2   Coupling to the matter

Metric couples to the energy-momentum tensor. The matter current is the source field of both gauge fields. The matter current is

$$J^\mu = \rho u^\mu$$

where ρ is the density and u^μ represents the four velocity.

24.2   Regimes of the GVT theory

GVT accommodates the Newtonian and MOND regime of gravity; but it admits the post-MONDian regime.

24.2.1   Strong and Newtonian regimes

The strong and Newtonian regime of the theory is defined to be where holds:

$$\begin{aligned} L(\tfrac{l^2}{4} B_{\mu\nu} B^{\mu\nu}) &= \tfrac{l^2}{4} B_{\mu\nu} B^{\mu\nu} . \\ L(\tfrac{\tilde{l}^2}{4} \tilde{B}_{\mu\nu} \tilde{B}^{\mu\nu}) &= \tfrac{\tilde{l}^2}{4} \tilde{B}_{\mu\nu} \tilde{B}^{\mu\nu} . \end{aligned}$$

The consistency between the gravitoelectromagnetism approximation to the GVT theory and that predicted and measured by the Einstein–Hilbert gravity demands that

$$\kappa + \tilde{\kappa} = 0$$

which results to $B_\mu + \tilde{B}_\mu = 0$. So the theory coincides to the Einstein–Hilbert gravity in its Newtonian and strong regimes.

24.2.2   MOND regime

The MOND regime of the theory is defined to be

$$\begin{aligned} L(\tfrac{l^2}{4} B_{\mu\nu} B^{\mu\nu}) &= \left| \tfrac{l^2}{4} B_{\mu\nu} B^{\mu\nu} \right|^{\frac{3}{2}} . \\ L(\tfrac{\tilde{l}^2}{4} \tilde{B}_{\mu\nu} \tilde{B}^{\mu\nu}) &= \tfrac{\tilde{l}^2}{4} \tilde{B}_{\mu\nu} \tilde{B}^{\mu\nu} . \end{aligned}$$

So the action for the B_μ field becomes aquadratic. For the static mass distribution, the theory then converts to the AQUAL model of gravity[3] with the critical acceleration of

$$a_0 = \frac{4\sqrt{2}\kappa c^2}{l}$$

So the GVT theory is capable of reproducing the flat rotational velocity curves of galaxies. The current observations do not fix κ which is supposedly of order one.

24.2.3 Post-MONDian regime

The post-MONDian regime of the theory is defined where both of the actions of the B_μ, $\dot{B}_\mu$ are aquadratic. The MOND type behavior is suppressed in this regime due to the contribution of the second gauge field.

24.3 See also

- Modified newtonian dynamics
- Scalar–tensor–vector gravity
- General theory of relativity
- Law of universal gravitation
- Pioneer anomaly
- Nonsymmetric gravitational theory
- Dark matter
- Dark energy
- Dark fluid
- Tensor
- Vector
- Scalar – scalar field

24.4 References

[1] Exirifard, Qasem (27 August 2013). "GravitoMagnetic force in modified Newtonian dynamics". *Journal of Cosmology and Astroparticle Physics* **2013** (08): 046–046. doi:10.1088/1475-7516/2013/08/046.

[2] Milgrom, M. (1 July 1983). "A modification of the Newtonian dynamics as a possible alternative to the hidden mass hypothesis". *The Astrophysical Journal* **270**: 365. doi:10.1086/161130.

[3] Bekenstein, J.; Milgrom, M. (1 November 1984). "Does the missing mass problem signal the breakdown of Newtonian gravity?". *The Astrophysical Journal* **286**: 7. doi:10.1086/162570.

Chapter 25

Gauss's law for gravity

This article is about Gauss's law concerning the gravitational field. For analogous laws concerning different fields, see Gauss's law and Gauss's law for magnetism. For Gauss's theorem, a mathematical theorem relevant to all of these laws, see Divergence theorem.

In physics, **Gauss's law for gravity**, also known as **Gauss's flux theorem for gravity**, is a law of physics that is essentially equivalent to Newton's law of universal gravitation. It is named after Carl Friedrich Gauss. Although Gauss's law for gravity is equivalent to Newton's law, there are many situations where Gauss's law for gravity offers a more convenient and simple way to do a calculation than Newton's law.

The form of Gauss's law for gravity is mathematically similar to Gauss's law for electrostatics, one of Maxwell's equations. Gauss's law for gravity has the same mathematical relation to Newton's law that Gauss's law for electricity bears to Coulomb's law. This is because both Newton's law and Coulomb's law describe inverse-square interaction in a 3-dimensional space.

25.1 Qualitative statement of the law

Main article: Gravitational field

The gravitational field **g** (also called gravitational acceleration) is a vector field – a vector at each point of space (and time). It is defined so that the gravitational force experienced by a particle is equal to the mass of the particle multiplied by the gravitational field at that point.

Gravitational flux is a surface integral of the gravitational field over a closed surface, analogous to how magnetic flux is a surface integral of the magnetic field.

Gauss's law for gravity states:

> *The gravitational flux through any closed surface is proportional to the enclosed mass.*

25.2 Integral form

The integral form of Gauss's law for gravity states:

where

$$\oiint$$

∂V (also written $\oint_{\partial V}$) denotes a surface integral over a closed surface,

∂V is any closed surface (the *boundary* of a closed volume V),

$d\mathbf{A}$ is a vector, whose magnitude is the area of an infinitesimal piece of the surface ∂V, and whose direction is the outward-pointing surface normal (see surface integral for more details),

$\mathbf{g}$ is the gravitational field,

G is the universal gravitational constant, and

M is the total mass enclosed within the surface ∂V.

The left-hand side of this equation is called the flux of the gravitational field. Note that according to the law it is always negative (or zero), and never positive. This can be contrasted with Gauss's law for electricity, where the flux can be either positive or negative. The difference is because *charge* can be either positive or negative, while *mass* can only be positive.

25.3 Differential form

The differential form of Gauss's law for gravity states

where $\nabla\cdot$ denotes divergence, G is the universal gravitational constant, and ρ is the mass density at each point.

25.3.1 Relation to the integral form

The two forms of Gauss's law for gravity are mathematically equivalent. The divergence theorem states:

$$\oint_{\partial V} \mathbf{g} \cdot d\mathbf{A} = \int_V \nabla \cdot \mathbf{g} \, dV$$

where V is a closed region bounded by a simple closed oriented surface ∂V and dV is an infinitesimal piece of the volume V (see volume integral for more details). The gravitational field $\mathbf{g}$ must be a continuously differentiable vector field defined on a neighborhood of V.

Given also that

$$M = \int_V \rho \, dV$$

we can apply the divergence theorem to the integral form of Gauss's law for gravity, which becomes:

$$\int_V \nabla \cdot \mathbf{g} \, dV = -4\pi G \int_V \rho \, dV$$

which can be rewritten:

$$\int_V (\nabla \cdot \mathbf{g}) \, dV = \int_V (-4\pi G \rho) \, dV.$$

This has to hold simultaneously for every possible volume V; the only way this can happen is if the integrands are equal. Hence we arrive at

$$\nabla \cdot \mathbf{g} = -4\pi G \rho,$$

which is the differential form of Gauss's law for gravity.

It is possible to derive the integral form from the differential form using the reverse of this method.

Although the two forms are equivalent, one or the other might be more convenient to use in a particular computation.

25.4   Relation to Newton's law

25.4.1   Deriving Gauss's law from Newton's law

Gauss's law for gravity can be derived from Newton's law of universal gravitation, which states that the gravitational field due to a point mass is:

$$\mathbf{g}(\mathbf{r}) = -GM \frac{\mathbf{e_r}}{r^2}$$

where

> $\mathbf{e_r}$ is the radial unit vector,
>
> r is the radius, $|\mathbf{r}|$.
>
> M is the mass of the particle, which is assumed to be a point mass located at the origin.

A proof using vector calculus is shown in the box below. It is mathematically identical to the proof of Gauss's law (in electrostatics) starting from Coulomb's law.[1]

25.4.2   Deriving Newton's law from Gauss's law and irrotationality

It is impossible to mathematically prove Newton's law from Gauss's law *alone*, because Gauss's law specifies the divergence of $\mathbf{g}$ but does not contain any information regarding the curl of $\mathbf{g}$ (see Helmholtz decomposition). In addition to Gauss's law, the assumption is used that $\mathbf{g}$ is irrotational (has zero curl), as gravity is a conservative force:

$$\nabla \times \mathbf{g} = 0$$

Even these are not enough: Boundary conditions on $\mathbf{g}$ are also necessary to prove Newton's law, such as the assumption that the field is zero infinitely far from a mass.

The proof of Newton's law from these assumptions is as follows:

25.5   Poisson's equation and gravitational potential

Since the gravitational field has zero curl (equivalently, gravity is a conservative force) as mentioned above, it can be written as the gradient of a scalar potential, called the gravitational potential:

$$\mathbf{g} = -\nabla \phi.$$

Then the differential form of Gauss's law for gravity becomes Poisson's equation:

$$\nabla^2 \phi = 4\pi G\rho.$$

This provides an alternate means of calculating the gravitational potential and gravitational field. Although computing $\mathbf{g}$ via Poisson's equation is mathematically equivalent to computing $\mathbf{g}$ directly from Gauss's law, one or the other approach may be an easier computation in a given situation.

In radially symmetric systems, the gravitational potential is a function of only one variable (namely, $r = |\mathbf{r}|$), and Poisson's equation becomes (see Del in cylindrical and spherical coordinates):

$$\frac{1}{r^2}\frac{\partial}{\partial r}\left(r^2 \frac{\partial \phi}{\partial r}\right) = 4\pi G\rho(r)$$

while the gravitational field is:

$$\mathbf{g}(\mathbf{r}) = -\mathbf{e_r}\frac{\partial \phi}{\partial r}.$$

When solving the equation it should be taken into account that in the case of finite densities $\partial\phi/\partial r$ has to be continuous at boundaries (discontinuities of the density), and zero for $r = 0$.

25.6   Applications

Gauss's law can be used to easily derive the gravitational field in certain cases where a direct application of Newton's law would be more difficult (but not impossible). See the article Gaussian surface for more details on how these derivations are done. Three such applications are as follows:

25.6.1   Bouguer plate

Main article: Bouguer plate

We can conclude (by using a "Gaussian pillbox") that for an infinite, flat plate (Bouguer plate) of any finite thickness, the gravitational field outside the plate is perpendicular to the plate, towards it, with magnitude $2\pi G$ times the mass per unit area, independent of the distance to the plate[2] (see also gravity anomalies).

More generally, for a mass distribution with the density depending on one Cartesian coordinate z only, gravity for any z is $2\pi G$ times (the mass per unit area above z, minus the mass per unit area below z).

In particular, a combination of two equal parallel infinite plates does not produce any gravity inside.

25.6.2 Cylindrically symmetric mass distribution

In the case of an infinite cylindrically symmetric mass distribution we can conclude (by using a cylindrical Gaussian surface) that the field strength at a distance r from the center is inward with a magnitude of $2G/r$ times the total mass per unit length at a smaller distance (from the axis), regardless of any masses at a larger distance.

For example, inside an infinite hollow cylinder, the field is zero.

25.6.3 Spherically symmetric mass distribution

Main article: Shell theorem

In the case of a spherically symmetric mass distribution we can conclude (by using a spherical Gaussian surface) that the field strength at a distance r from the center is inward with a magnitude of G/r^2 times only the total mass within a smaller distance than r. All the mass at a greater distance than r from the center can be ignored.

For example, a hollow sphere does not produce any net gravity inside. The gravitational field inside is the same as if the hollow sphere were not there (i.e. the resultant field is that of any masses inside and outside the sphere only).

Although this follows in one or two lines of algebra from Gauss's law for gravity, it took Isaac Newton several pages of cumbersome calculus to derive it directly using his law of gravity; see the article shell theorem for this direct derivation.

25.7 Derivation from Lagrangian

Main article: Lagrangian (field theory)

The Lagrangian density for Newtonian gravity is

$$\mathcal{L}(\vec{x}, t) = -\rho(\vec{x}, t)\phi(\vec{x}, t) - \frac{1}{8\pi G}(\nabla\phi(\vec{x}, t))^2$$

Applying Hamilton's principle to this Lagrangian, the result is Gauss's law for gravity:

$$4\pi G\rho(\vec{x}, t) = \nabla^2\phi(\vec{x}, t).$$

See Lagrangian (field theory) for details.

25.8 In fiction

In Arthur C. Clarke's science fiction novel, *2010: Odyssey Two*, while investigating the alien Monolith orbiting Jupiter, the *Leonov* 's chief scientist, Vasili Orlov, has engineer Curnow park one of the revived *Discovery* 's space pods a short distance from the Monolith's two-kilometer-long surface, recalling Bauguer's Anomaly, derived from Gauss's law. He remarks, "I've just remembered an exercise from one of my college astronomy courses - the gravitational attraction of an infinite flat plate. I never thought I'd have a chance of using it in real life." [3]

25.9 See also

- Carl Friedrich Gauss

- Divergence theorem

- Gauss's law for electricity

- Gauss's law for magnetism

- Vector calculus

- Integral

- Flux

- Gaussian surface

25.10   References

[1] See, for example, Griffiths, David J. (1998). *Introduction to Electrodynamics* (3rd ed.). Prentice Hall. p. 50. ISBN 0-13-805326-X.

[2] The mechanics problem solver, by Fogiel, pp 535–536

[3] Clarke, Arthur C. (1982) *2010: Odyssey Two*. Del Rey ISBN 0-345-41397-0

- For usage of the term "Gauss's law for gravity" see, for example, this article.

Chapter 26

Gauss–Bonnet gravity

In general relativity, **Gauss–Bonnet gravity**, also referred to as **Einstein–Gauss–Bonnet gravity**,[1] is a modification of the Einstein–Hilbert action to include the Gauss–Bonnet term (named after Carl Friedrich Gauss and Pierre Ossian Bonnet) $G = R^2 - 4R^{\mu\nu}R_{\mu\nu} + R^{\mu\nu\rho\sigma}R_{\mu\nu\rho\sigma}$

$$\int d^D x \sqrt{-g}\, G$$

This term is only nontrivial in 4+1D or greater, and as such, only applies to extra dimensional models. In 3+1D and lower, it reduces to a topological surface term. This follows from the generalized Gauss–Bonnet theorem on a 4D manifold

$$\frac{1}{8\pi^2} \int d^4 x \sqrt{-g}\, G = \chi(M)$$

Despite being quadratic in the Riemann tensor (and Ricci tensor), terms containing more than 2 partial derivatives of the metric cancel out, making the Euler–Lagrange equations second order quasilinear partial differential equations in the metric. Consequently, there are no additional dynamical degrees of freedom, as in say f(R) gravity.

More generally, we may consider

$$\int d^D x \sqrt{-g}\, f(G)$$

term for some function f. Nonlinearities in f render this coupling nontrivial even in 3+1D. However, fourth order terms reappear with the nonlinearities.

26.1 See also

- Einstein–Hilbert action
- f(R) gravity
- Lovelock gravity

26.2 References

[1] Lovelock, David (1971), "The Einstein tensor and its generalizations", *J. Math. Phys.* **12** (3): 498, Bibcode:1971JMP....12..498L, doi:10.1063/1.1665613

Chapter 27

Gravitational field

In physics, a **gravitational field** is a model used to explain the influence that a massive body extends into the space around itself, producing a force on another massive body.[1] Thus, a gravitational field is used to explain gravitational phenomena, and is measured in newtons per kilogram (N/kg). In its original concept, gravity was a force between point masses. Following Newton, Laplace attempted to model gravity as some kind of radiation field or fluid, and since the 19th century explanations for gravity have usually been taught in terms of a field model, rather than a point attraction.

In a field model, rather than two particles attracting each other, the particles distort spacetime via their mass, and this distortion is what is perceived and measured as a "force". In such a model one states that matter moves in certain ways in response to the curvature of spacetime,[2] and that there is either *no gravitational force*,[3] or that gravity is a fictitious force.[4]

27.1 Classical mechanics

In classical mechanics as in physics, a gravitational field is a physical quantity.[5] A gravitational field can be defined using Newton's law of universal gravitation. Determined in this way, the gravitational field **g** around a single particle of mass M is a vector field consisting at every point of a vector pointing directly towards the particle. The magnitude of the field at every point is calculated applying the universal law, and represents the force per unit mass on any object at that point in space. Because the force field is conservative, there is a scalar potential energy per unit mass, Φ, at each point in space associated with the force fields; this is called gravitational potential.[6] The gravitational field equation is[7]

$$\mathbf{g} = \frac{\mathbf{F}}{m} = -\frac{\mathrm{d}^2 \mathbf{R}}{\mathrm{d}t^2} = -GM \frac{\hat{\mathbf{R}}}{|\mathbf{R}|^2} = -\nabla\Phi,$$

where **F** is the gravitational force, m is the mass of the test particle, **R** is the position of the test particle, $\hat{\mathbf{R}}$ is a unit vector in the direction of **R**, t is time, G is the gravitational constant, and ∇ is the del operator.

This includes Newton's law of gravitation, and the relation between gravitational potential and field acceleration. Note that $\mathrm{d}^2\mathbf{R}/\mathrm{d}t^2$ and $\mathbf{F}/m$ are both equal to the gravitational acceleration **g** (equivalent to the inertial acceleration, so same mathematical form, but also defined as gravitational force per unit mass[8]). The negative signs are inserted since the force acts antiparallel to the displacement. The equivalent field equation in terms of mass density ρ of the attracting mass are:

$$-\nabla \cdot \mathbf{g} = \nabla^2\Phi = 4\pi G\rho$$

which contains Gauss' law for gravity, and Poisson's equation for gravity. Newton's and Gauss' law are mathematically equivalent, and are related by the divergence theorem. Poisson's equation is obtained by taking the divergence of both sides of the previous equation. These classical equations are differential equations of motion for a test particle in the presence of a gravitational field, i.e. setting up and solving these equations allows the motion of a test mass to be determined and described.

The field around multiple particles is simply the vector sum of the fields around each individual particle. An object in such a field will experience a force that equals the vector sum of the forces it would feel in these individual fields. This is mathematically:[9]

$$\mathbf{g}_j^{(net)} = \sum_{i \neq j} \mathbf{g}_i = \frac{1}{m_j} \sum_{i \neq j} \mathbf{F}_i = -G \sum_{i \neq j} m_i \frac{\hat{\mathbf{R}}_{ij}}{|\mathbf{R}_i - \mathbf{R}_j|^2} = -\sum_{i \neq j} \nabla \Phi_i$$

i.e. the gravitational field on mass mj is the sum of all gravitational fields due to all other masses mi, except the mass mj itself. The unit vector $\hat{\mathbf{R}}_{ij}$ is in the direction of $\mathbf{R}i - \mathbf{R}j$.

27.2 General relativity

See also: Gravitational acceleration § General relativity and Gravitational potential § General relativity

In general relativity the gravitational field is determined by solving the Einstein field equations,[10]

$$\mathbf{G} = \frac{8\pi G}{c^4} \mathbf{T}.$$

Here $\mathbf{T}$ is the stress–energy tensor, $\mathbf{G}$ is the Einstein tensor, and c is the speed of light.

These equations are dependent on the distribution of matter and energy in a region of space, unlike Newtonian gravity, which is dependent only on the distribution of matter. The fields themselves in general relativity represent the curvature of spacetime. General relativity states that being in a region of curved space is equivalent to accelerating up the gradient of the field. By Newton's second law, this will cause an object to experience a fictitious force if it is held still with respect to the field. This is why a person will feel himself pulled down by the force of gravity while standing still on the Earth's surface. In general the gravitational fields predicted by general relativity differ in their effects only slightly from those predicted by classical mechanics, but there are a number of easily verifiable differences, one of the most well known being the bending of light in such fields.

27.3 See also

- Classical mechanics

- Gravitation

- Gravitational potential

- Newton's law of universal gravitation

- Newton's laws of motion

- Potential energy

- Speed of gravity

- Tests of general relativity

- Defining equation (physics)

27.4 Notes

[1] Richard Feynman (1970). *The Feynman Lectures on Physics Vol I*. Addison Wesley Longman. ISBN 978-0-201-02115-8.

[2] Geroch, Robert (1981). *General relativity from A to B*. University of Chicago Press. p. 181. ISBN 0-226-28864-1., Chapter 7, page 181

[3] Grøn, Øyvind; Hervik, Sigbjørn (2007). *Einstein's general theory of relativity: with modern applications in cosmology*. Springer Japan. p. 256. ISBN 0-387-69199-5., Chapter 10, page 256

[4] J. Foster, J. D. Nightingale, J. Foster, J. D. Nightingale; J. Foster, J. D. Nightingale, J. Foster, J. D. Nightingale (2006). *A short course in general relativity* (3 ed.). Springer Science & Business. p. 55. ISBN 0-387-26078-1., Chapter 2, page 55

[5] Richard Feynman (1970). *The Feynman Lectures on Physics Vol II*. Addison Wesley Longman. ISBN 978-0-201-02115-8. A "field" is any physical quantity which takes on different values at different points in space.

[6] Dynamics and Relativity, J.R. Forshaw, A.G. Smith, Wiley, 2009, ISBN 978-0-470-01460-8

[7] Encyclopaedia of Physics, R.G. Lerner, G.L. Trigg, 2nd Edition, VHC Publishers, Hans Warlimont, Springer, 2005

[8] Essential Principles of Physics, P.M. Whelan, M.J. Hodgeson, 2nd Edition, 1978, John Murray, ISBN 0-7195-3382-1

[9] Classical Mechanics (2nd Edition), T.W.B. Kibble, European Physics Series, Mc Graw Hill (UK), 1973, ISBN 0-07-084018-0.

[10] Gravitation, J.A. Wheeler, C. Misner, K.S. Thorne, W.H. Freeman & Co, 1973, ISBN 0-7167-0344-0

Chapter 28

Higher-dimensional Einstein gravity

Higher-dimensional Einstein gravity is any of various physical theories that attempt to generalise to higher dimensions various results of the well established theory of standard (four-dimensional) Einstein gravity, that is, general relativity. This attempt at generalisation has been strongly influenced in recent decades by string theory.

At present, this work can probably be most fairly described as extended theoretical speculation. Currently, it has no *direct* observational and experimental support, in contrast to four-dimensional general relativity. However, this theoretical work has led to the possibility of proving the existence of extra dimensions. This is best demonstrated by the proof of Harvey Reall and Roberto Emparan that there is a 'black ring' solution in 5 dimensions. If such a 'black ring' could be produced in a particle accelerator such as the Large Hadron Collider, this would provide the evidence that higher dimensions exist.

28.1   Exact solutions

The higher-dimensional generalization of the Kerr metric was discovered by Myers and Perry.[1] Like the Kerr metric, the Myers-Perry metric has spherical horizon topology. The construction involves making a Kerr-Schild ansatz; by a similar method, the solution has been generalized to include a cosmological constant. The **black ring** is a solution of five-dimensional general relativity. It inherits its name from the fact that its event horizon is topologically $S^1 \times S^2$. This is in contrast to other known black hole solutions in five dimensions which have horizon topology S^3.

28.2   Black hole uniqueness

In four dimensions, Hawking proved that the topology of the event horizon of a non-rotating black hole must be spherical. Because the proof uses the Gauss–Bonnet theorem, it does not generalize to higher dimensions. The discovery of black ring solutions in five dimensions shows that other topologies are allowed in higher dimensions, but it is unclear precisely which topologies are allowed. It has been shown that the horizon must be of positive Yamabe type, meaning that it must admit a metric of positive scalar curvature.

28.3   See also

- General relativity

- Kaluza–Klein theory

- Gauss–Bonnet gravity

28.4 References

[1] Robert C. Myers, M.J. Perry (1986). "Black Holes in Higher Dimensional Space-Times". *Annals Phys.* **172**: 304–347. Bibcode:1986AnPhy.172..304M. doi:10.1016/0003-4916(86)90186-7.

Chapter 29

Higher-dimensional supergravity

Higher-dimensional supergravity is the supersymmetric generalization of general relativity in higher dimensions. Supergravity can be formulated in any number of dimensions up to eleven. This article focuses upon supergravity (SUGRA) in greater than four dimensions.

29.1 Supermultiplets

Fields related by supersymmetry transformations form a supermultiplet; the one that contains a graviton is called the supergravity multiplet.

The name of a supergravity theory generally includes the number of dimensions of spacetime that it inhabits, and also the number $\mathcal{N}$ of gravitinos that it has. Sometimes one also includes the choices of supermultiplets in the name of theory. For example, an $\mathcal{N} = 2$, $(9 + 1)$-dimensional supergravity enjoys 9 spatial dimensions, one time and 2 gravitinos. While the field content of different supergravity theories varies considerably, all supergravity theories contain at least one gravitino and they all contain a single graviton. Thus every supergravity theory contains a single supergravity supermultiplet. It is still not known whether one can construct theories with multiple gravitons that are not equivalent to multiple decoupled theories with a single graviton in each. In maximal supergravity theories (see below), all fields are related by supersymmetry transformations so that there is only one supermultiplet: the supergravity multiplet.

29.2 Gauged supergravity versus Yang–Mills supergravity

Often an abuse of nomenclature is used when "gauge supergravity" refers to a supergravity theory in which fields in the theory are charged with respect to vector fields in the theory. However, when the distinction is important, the following is the correct nomenclature. If a global (i.e. rigid) R-symmetry is gauged, the gravitino is charged with respect to some vector fields, and the theory is called gauged supergravity. When other global (rigid) symmetries (e.g., if the theory is a non-linear sigma model) of the theory are gauged such that some (non-gravitino) fields are charged with respect to vectors, it is known as a Yang–Mills–Einstein supergravity theory. Of course, one can imagine having a "gauged Yang–Mills–Einstein" theory using a combination of the above gaugings.

29.3 Counting gravitinos

Gravitinos are fermions, which means that according to the spin-statistics theorem they must have an odd number of spinorial indices. In fact the gravitino field has one spinor and one vector index, which means that gravitinos transform as a tensor product of a spinorial representation and the vector representation of the Lorentz group. This is a Rarita-Schwinger spinor.

While there is only one vector representation for each Lorentz group, in general there are several different spinorial representations. Technically these are really representations of the double cover of the Lorentz group called a spin group.

The canonical example of a spinorial representation is the Dirac spinor, which exists in every number of space-time dimensions. However the Dirac spinor representation is not always irreducible. When calculating the number $\mathcal{N}$, one always counts the number of *real* irreducible representations. The spinors with spins less than 3/2 that exist in each number of dimensions will be classified in the following subsection.

29.4 A classification of spinors

The available spinor representations depends on k: The maximal compact subgroup of the little group of the Lorentz group that preserves the momentum of a massless particle is Spin($d - 1$) × Spin($d - k - 1$), where k is equal to the number d of spatial dimensions minus the number $d - k$ of time dimensions. (See helicity (particle physics)) For example, in our world, this is $3 - 1 = 2$. Due to the mod 8 Bott periodicity of the homotopy groups of the Lorentz group, really we only need to consider k modulo 8.

For any value of k there is a Dirac representation, which is always of real dimension $2^{1+\lfloor \frac{2d-k}{2} \rfloor}$ where $\lfloor x \rfloor$ is the greatest integer less than or equal to x. When $-2 \leq k \leq 2$ (mod 8) there is a real Majorana spinor representation, whose dimension is half that of the Dirac representation. When k is even there is a Weyl spinor representation, whose real dimension is again half that of the Dirac spinor. Finally when k is divisible by eight, that is, when k is zero modulo eight, there is a Majorana-Weyl spinor, whose real dimension is one quarter that of the Dirac spinor.

Occasionally one also considers symplectic Majorana spinor which exist when $3 \leq k \leq 5$, which have half has many components as Dirac spinors. When $k=4$ these may also be Weyl, yielding Weyl symplectic Majorana spinors which have one quarter as many components as Dirac spinors.

29.5 Choosing chiralities

Spinors in n-dimensions are representations (really modules) not only of the n-dimensional Lorentz group, but also of a Lie algebra called the n-dimensional Clifford algebra. The most commonly used basis of the complex $2^{\lfloor n \rfloor}$-dimensional representation of the Clifford algebra, the representation that acts on the Dirac spinors, consists of the gamma matrices.

When n is even the product of all of the gamma matrices, which is often referred to as Γ_5 as it was first considered in the case $n = 4$, is not itself a member of the Clifford algebra. However, being a product of elements of the Clifford algebra, it is in the algebra's universal cover and so has an action on the Dirac spinors.

In particular, the Dirac spinors may be decomposed into eigenspaces of Γ_5 with eigenvalues equal to $\pm(-1)^{-k/2}$, where k is the number of spatial minus temporal dimensions in the spacetime. The spinors in these two eigenspaces each form projective representations of the Lorentz group, known as Weyl spinors. The eigenvalue under Γ_5 is known as the chirality of the spinor, which can be left or right-handed.

A particle that transforms as a single Weyl spinor is said to be chiral. The CPT theorem, which is required by Lorentz invariance in Minkowski space, implies that when there is a single time direction such particles have antiparticles of the opposite chirality.

Recall that the eigenvalues of Γ_5, whose eigenspaces are the two chiralities, are $\pm(-1)^{-k/2}$. In particular, when k is equal to two modulo four the two eigenvalues are complex conjugate and so the two chiralities of Weyl representations are complex conjugate representations.

Complex conjugation in quantum theories corresponds to time inversion. Therefore the CPT theorem implies that when the number of Minkowski dimensions is divisible by four (so that k is equal to 2 modulo 4) there be an equal number of left-handed and right-handed supercharges. On the other hand, if the dimension is equal to 2 modulo 4, there can be different numbers of left and right-handed supercharges, and so often one labels the theory by a doublet $\mathcal{N} = (\mathcal{N}_L, \mathcal{N}_R)$ where $\mathcal{N}_L$ and $\mathcal{N}_R$ are the number of left-handed and right-handed supercharges respectively.

29.6   Counting supersymmetries

All supergravity theories are invariant under transformations in the super-Poincaré algebra, although individual configurations are not in general invariant under every transformation in this group. The super-Poincaré group is generated by the Super-Poincaré algebra, which is a Lie superalgebra. A Lie superalgebra is a $\mathbf{Z}_2$ graded algebra in which the elements of degree zero are called bosonic and those of degree one are called fermionic. A commutator, that is an antisymmetric bracket satisfying the Jacobi identity is defined between each pair of generators of fixed degree except for pairs of fermionic generators, for which instead one defines a symmetric bracket called an anticommutator.

The fermionic generators are also called supercharges. Any configuration which is invariant under any of the supercharges is said to be BPS, and often nonrenormalization theorems demonstrate that such states are particularly easily treated because they are unaffected by many quantum corrections.

The supercharges transform as spinors, and the number of irreducible spinors of these fermionic generators is equal to the number of gravitinos $\mathcal{N}$ defined above. Often $\mathcal{N}$ is defined to be the number of fermionic generators, instead of the number of gravitinos, because this definition extends to supersymmetric theories without gravity.

Sometimes it is convenient to characterize theories not by the number $\mathcal{N}$ of irreducible representations of gravitinos or supercharges, but instead by the total Q of their dimensions. This is because some features of the theory have the same Q-dependence in any number of dimensions. For example, one is often only interested in theories in which all particles have spin less than or equal to two. This requires that Q not exceed 32, except possibly in special cases in which the supersymmetry is realized in an unconventional, nonlinear fashion with products of bosonic generators in the anticommutators of the fermionic generators.

29.7   Examples

29.7.1   Why less than 32 SUSYs?

The supergravity theories that have attracted the most interest contain no spins higher than two. This means, in particular, that they do not contain any fields that transform as symmetric tensors of rank higher than two under Lorentz transformations. The consistency of interacting higher spin field theories is, however, presently a field of very active interest.

The supercharges in every super-Poincaré algebra are generated by a multiplicative basis of m fundamental supercharges, and an additive basis of the supercharges (this definition of supercharges is a bit more broad than that given above) is given by a product of any subset of these m fundamental supercharges. The number of subsets of m elements is 2^m, thus the space of supercharges is 2^m-dimensional.

The fields in a supersymmetric theory form representations of the super-Poincaré algebra. It can be shown that when m is greater than 5 there are no representations that contain only fields of spin less than or equal to two. Thus we are interested in the case in which m is less than or equal to 5, which means that the maximal number of supercharges is 32. A supergravity theory with precisely 32 supersymmetries is known as a maximal supergravity.

Above we saw that the number of supercharges in a spinor depends on the dimension and the signature of spacetime. The supercharges occur in spinors. Thus the above limit on the number of supercharges cannot be satisfied in a spacetime of arbitrary dimension. Below we will describe some of the cases in which it is satisfied.

29.7.2   A 12-dimensional two-time theory

The highest dimension in which spinors exist with only 32 supercharges is 12. If there are 11 spatial directions and 1 time direction then there will be Weyl and Majorana spinors which both are of dimension 64, and so are too large. However, some authors have considered nonlinear actions of the supersymmetry in which higher spin fields may not appear.

If instead one considers 10 spatial direction and a second temporal dimension then there is a Majorana-Weyl spinor, which as desired has only 32 components. For an overview of two-time theories by one of their main proponents, Itzhak

Bars, see his paper Two-Time Physics and Two-Time Physics on arxiv.org. He considered 12-dimensional supergravity in Supergravity, p-brane duality and hidden space and time dimensions.

It was widely, but not universally, thought that two-time theories may have problems. For example, there could be causality problems (disconnect between cause and effect) and unitarity problems (negative probability, ghosts). Also, the Hamiltonian-based approach to quantum mechanics may have to be modified in the presence of a second Hamiltonian for the other time. However in Two-Time Physics it was demonstrated that such potential problems are solved with an appropriate gauge symmetry.

Some other two time theories describe low-energy behavior, such as Cumrun Vafa's F-theory that is also formulated with the help of 12 dimensions. F-theory itself however is not a two-time theory. One can understand 2 of the 12-dimensions of F-theory as a bookkeeping device; they should not be confused with the other 10 spacetime coordinates. These two dimensions are somehow dual to each other and should not be treated independently.

29.7.3   11-dimensional maximal SUGRA

This maximal supergravity is the classical limit of M-theory. There is, classically, only one 11-dimensional supergravity theory. Like all maximal supergravities, it contains a single supermultiplet, the supergravity supermultiplet. This supermultiplet contains the graviton, a Majorana gravitino and a 3-form gauge field often called the C-field.

It contains two p-brane solutions, a 2-brane and a 5-brane, which are electrically and magnetically charged, respectively, with respect to the C-field. This means that 2-brane and 5-brane charge are the violations of the Bianchi identities for the dual C-field and original C-field respectively. The supergravity 2-brane and 5-brane are the long-wavelength limits (see also the historical survey above) of the M2-brane and M5-brane in M-theory.

29.7.4   10d SUGRA theories

Type IIA SUGRA: $N = (1, 1)$

This maximal supergravity is the classical limit of type IIA string theory. The field content of the supergravity super-multiplet consists of a graviton, a Majorana gravitino, a Kalb-Ramond field, odd-dimensional Ramond-Ramond gauge potentials, a dilaton and a dilatino.

The Bianchi identities of the Ramond-Ramond gauge potentials C_{2k-1} can be violated by adding sources ρ, which are called D$(8 - 2k)$-branes

$$ddC_{2k-1} = \rho.$$

In the democratic formulation of type IIA supergravity there exist Ramond-Ramond gauge potentials for $0 < k < 6$, which leads to D0-branes (also called D-particles), D2-branes, D4-branes, D6-branes and, if one includes the case $k = -1$, D8-branes. In addition there are fundamental strings and their electromagnetic duals, which are called NS5-branes.

Although obviously there are no -1-form gauge connections, the corresponding 0-form field strength, G_0 may exist. This field strength is called the **Romans mass** and when it is not equal to zero the supergravity theory is called massive IIA supergravity or Romans IIA supergravity. From the above Bianchi identity we see that a D8-brane is a domain wall between zones of differing G_0, thus in the presence of a D8-brane at least part of the spacetime will be described by the Romans theory.

IIA SUGRA from 11d SUGRA

IIA SUGRA is the dimensional reduction of 11-dimensional supergravity on a circle. This means that 11d supergravity on the spacetime $M^{10} \times S^1$ is equivalent to IIA supergravity on the 10-manifold M^{10} where one eliminates modes with masses proportional to the inverse radius of the circle S^1.

In particular the field and brane content of IIA supergravity can be derived via this dimensional reduction procedure. The field G_0 however does not arise from the dimensional reduction, massive IIA is not known to be the dimensional reduction of any higher-dimensional theory. The 1-form Ramond-Ramond potential C_1 is the usual 1-form connection that arises from the Kaluza–Klein procedure, it arises from the components of the 11-d metric that contain one index along the compactified circle. The IIA 3-form gauge potential C_3 is the reduction of the 11d 3-form gauge potential components with indices that do not lie along the circle, while the IIA Kalb-Ramond 2-form B-field consists of those components of the 11-dimensional 3-form with one index along the circle. The higher forms in IIA are not independent degrees of freedom, but are obtained from the lower forms using Hodge duality.

Similarly the IIA branes descend from the 11-dimension branes and geometry. The IIA D0-brane is a Kaluza–Klein momentum mode along the compactified circle. The IIA fundamental string is an 11-dimensional membrane which wraps the compactified circle. The IIA D2-brane is an 11-dimensional membrane that does not wrap the compactified circle. The IIA D4-brane is an 11-dimensional 5-brane that wraps the compactified circle. The IIA NS5-brane is an 11-dimensional 5-brane that does not wrap the compactified circle. The IIA D6-brane is a Kaluza–Klein monopole, that is, a topological defect in the compact circle fibration. The lift of the IIA D8-brane to 11-dimensions is not known, as one side of the IIA geometry as a nontrivial Romans mass, and an 11-dimensional original of the Romans mass is unknown.

Type IIB SUGRA: $N = (2, 0)$

This maximal supergravity is the classical limit of type IIB string theory. The field content of the supergravity supermultiplet consists of a graviton, a Weyl gravitino, a Kalb-Ramond field, even-dimensional Ramond-Ramond gauge potentials, a dilaton and a dilatino.

The Ramond-Ramond fields are sourced by odd-dimensional D($2k + 1$)-branes, which host supersymmetric $U(1)$ gauge theories. As in IIA supergravity, the fundamental string is an electric source for the Kalb-Ramond B-field and the NS5-brane is a magnetic source. Unlike that of the IIA theory, the NS5-brane hosts a worldvolume $U(1)$ supersymmetric gauge theory with $\mathcal{N} = (1, 1)$ supersymmetry, although some of this supersymmetry may be broken depending on the geometry of the spacetime and the other branes that are present.

This theory enjoys an SL(2, $\mathbf{R}$) symmetry known as S-duality that interchanges the Kalb-Ramond field and the RR 2-form and also mixes the dilaton and the RR 0-form axion.

Type I gauged SUGRA: $N = (1, 0)$

These are the classical limits of type I string theory and the two heterotic string theories. There is a single Majorana-Weyl spinor of supercharges, which in 10 dimensions contains 16 supercharges. As 16 is less than 32, the maximal number of supercharges, type I is not a maximal supergravity theory.

In particular this implies that there is more than one variety of supermultiplet. In fact, there are two. As usual, there is a supergravity supermultiplet. This is smaller than the supergravity supermultiplet in type II, it contains only the graviton, a Majorana-Weyl gravitino, a 2-form gauge potential, the dilaton and a dilatino. Whether this 2-form is considered to be a Kalb-Ramond field or Ramond-Ramond field depends on whether one considers the supergravity theory to be a classical limit of a heterotic string theory or type I string theory. There is also a vector supermultiplet, which contains a one-form gauge potential called a gluon and also a Majorana-Weyl gluino.

Unlike type IIA and IIB supergravities, for which the classical theory is unique, as a classical theory $\mathcal{N} = 1$ supergravity is consistent with a single supergravity supermultiplet and any number of vector multiplets. It is also consistent without the supergravity supermultiplet, but then it would contain no graviton and so would not be a supergravity theory. While one may add multiple supergravity supermultiplets, it is not known if they may consistently interact. One is free not only to determine the number, if any, of vector supermultiplets, but also there is some freedom in determining their couplings. They must describe a classical super Yang–Mills gauge theory, but the choice of gauge group is arbitrary. In addition one is free to make some choices of gravitational couplings in the classical theory.

While there are many varieties of classical $\mathcal{N} = 1$ supergravities, not all of these varieties are the classical limits of quantum theories. Generically the quantum versions of these theories suffer from various anomalies, as can be seen already at 1-loop in the hexagon Feynman diagrams. In 1984 and 1985 Michael Green and John H. Schwarz have shown that if one includes precisely 496 vector supermultiplets and chooses certain couplings of the 2-form and the metric then the gravitational anomalies cancel. This is called the Green-Schwarz anomaly cancellation mechanism.

In addition, anomaly cancellation requires one to cancel the gauge anomalies. This fixes the gauge symmetry algebra to be either $\mathfrak{so}(32)$, $\mathfrak{e}_8 \oplus \mathfrak{e}_8$, $\mathfrak{e}_8 \oplus 248\mathfrak{u}(1)$ or $496\mathfrak{u}(1)$. However, only the first two Lie algebras can be gotten from superstring theory. Quantum theories with at least 8 supercharges tend to have continuous moduli spaces of vacua. In compactifications of these theories, which have 16 supercharges, there exist degenerate vacua with different values of various Wilson loops. Such Wilson loops may be used to break the gauge symmetries to various subgroups. In particular the above gauge symmetries may be broken to obtain not only the standard model gauge symmetry but also symmetry groups such as SO(10) and SU(5) that are popular in GUT theories.

29.7.5   9d SUGRA theories

In 9-dimensional Minkowski space the only irreducible spinor representation is the Majorana spinor, which has 16 components. Thus supercharges inhabit Majorana spinors of which there are at most two.

Maximal 9d SUGRA from 10d

In particular, if there are two Majorana spinors then one obtains the 9-dimensional maximal supergravity theory. Recall that in 10 dimensions there were two inequivalent maximal supergravity theories, IIA and IIB. The dimensional reduction of either IIA or IIB on a circle is the unique 9-dimensional supergravity. In other words, IIA or IIB on the product of a 9-dimensional space M^9 and a circle is equivalent to the 9-dimension theory on M^9, with Kaluza–Klein modes if one does not take the limit in which the circle shrinks to zero.

T-duality

More generally one could consider the 10-dimensional theory on a nontrivial circle bundle over M^9. Dimensional reduction still leads to a 9-dimensional theory on M^9, but with a 1-form gauge potential equal to the connection of the circle bundle and a 2-form field strength which is equal to the Chern class of the old circle bundle. One may then lift this theory to the other 10-dimensional theory, in which case one finds that the 1-form gauge potential lifts to the Kalb-Ramond field. Similarly, the connection of the fibration of the circle in the second 10-dimensional theory is the integral of the Kalb-Ramond field of the original theory over the compactified circle.

This transformation between the two 10-dimensional theories is known as T-duality. While T-duality in supergravity involves dimensional reduction and so loses information, in the full quantum string theory the extra information is stored in string winding modes and so T-duality is a duality between the two 10-dimensional theories. The above construction can be used to obtain the relation between the circle bundle's connection and dual Kalb-Ramond field even in the full quantum theory.

$N = 1$ Gauged SUGRA

Main article: Gauged supergravity

As was the case in the parent 10-dimensional theory, 9-dimensional N=1 supergravity contains a single supergravity multiplet and an arbitrary number of vector multiplets. These vector multiplets may be coupled so as to admit arbitrary gauge theories, although not all possibilities have quantum completions. Unlike the 10-dimensional theory, as was described in the previous subsection, the supergravity multiplet itself contains a vector and so there will always be at least a U(1) gauge symmetry, even in the N=2 case.

29.8   The mathematics

The Lagrangian for 11D supergravity found by brute force by Cremmer, Julia and Scherk[1] is:

$$
\begin{aligned}
L \;=\; & +\tfrac{1}{2\kappa^2}eR - \tfrac{1}{2}e\overline{\psi}_M \Gamma^{MNP}D_N[\tfrac{1}{2}(\omega - \varpi)]\psi_P \\
& +\tfrac{1}{48}eF^2_{MNPQ} + \tfrac{\sqrt{2}\kappa}{384}e(\overline{\psi}_M \Gamma^{MNPQRS}\psi_S \\
& +12\overline{\psi}^N \Gamma^{PQ}\psi^R)(F + \overline{F})_{NPQR} + \tfrac{\sqrt{2}\kappa}{3456}\varepsilon^{M_1 \ldots M_{11}} F_{M_1 \ldots M_4} F_{M_5 \ldots M_8} A_{M_9 M_{10} M_{11}}
\end{aligned}
$$

which contains the three types of field:

$$
e^A_M \, , \, \psi_M \, , \, A_{MNP}
$$

The symmetry of this supergravity theory is given by the supergroup OSp(1|32) which gives the subgroups O(1) for the bosonic symmetry and Sp(32) for the fermion symmetry. This is because spinors need 32 components in 11 dimensions. 11D supergravity can be compactified down to 4 dimensions which then has OSp(8|4) symemtry. (We still have $8 \times 4 = 32$ so there are still the same number of components.) Spinors need 4 components in 4 dimensions. This gives O(8) for the gauge group which is too small to contain the Standard Model gauge group U(1) × SU(2) × SU(3) which would need at least O(10).

29.8.1   Maximal SUGRA from Freund-Rubin

29.9   Notes and references

[1] E. Cremmer, B. Julia, J. Scherk: Phys. Lett. 76B, 409 (1978).

Chapter 30

Hoyle–Narlikar theory of gravity

The **Hoyle–Narlikar theory of gravity**[1] is a Machian theory of gravity proposed by Fred Hoyle and Jayant Narlikar that fits into the quasi steady state model of the universe.[2] The gravitational constant G is arbitrary and is determined by the mean density of matter in the universe. The theory was inspired by the Wheeler–Feynman absorber theory for electrodynamics.[3]

Currently the theory does not fit into WMAP data.[4] Narlikar and his followers are working on adding mini bangs with various creation fields to explain the anisotropy of the universe.[5][6]

30.1   See also

- Self-creation cosmology

- Mach's principle

- Brans–Dicke theory

- Non-standard cosmology

- Wheeler–Feynman absorber theory

30.2   Notes

[1] "Cosmology: Math Plus Mach Equals Far-Out Gravity". Time. Jun 26, 1964. Retrieved 7 August 2010.

[2] F.Hoyle and J.V.Narlikar(1964). "A New Theory of Gravitation".*Proceedings of the Royal Society A*.Bibcode:1964RSPSA.282. doi:10.1098/rspa.1964.0227.

[3] Hoyle, Narlikar (1995). "Cosmology and action-at-a-distance electrodynamics". *Reviews of Modern Physics* **67** (1): 113. Bibcode:1995RvMP...67..113H. doi:10.1103/RevModPhys.67.113.

[4] Edward L. Wright. "Errors in the Steady State and Quasi-SS Models". Retrieved 7 August 2010.

[5] J.V. Narlikar, R.G. Vishwakarma, Amir Hajian, Tarun Souradeep, G. Burbidge, F. Hoyle (2002). "Inhomogeneities in the Microwave Background Radiation interpreted within the framework of the Quasi-Steady State Cosmology". *Astrophysical Journal* **585**: 1–11. arXiv:astro-ph/0211036. Bibcode:2003ApJ...585....1N. doi:10.1086/345928.

[6] J. V. Narlikar and N. C. Rana (1983). "Cosmic microwave background spectrum in the Hoyle-Narlikar cosmology". *Physics Letters A* **99** (2-3): 75–76. Bibcode:1983PhLA...99...75N. doi:10.1016/0375-9601(83)90927-1.

Chapter 31

Induced gravity

Induced gravity (or **emergent gravity**) is an idea in quantum gravity that space-time curvature and its dynamics emerge as a mean field approximation of underlying microscopic degrees of freedom, similar to the fluid mechanics approximation of Bose–Einstein condensates. The concept was originally proposed by Andrei Sakharov in 1967.

Sakharov observed that many condensed matter systems give rise to emergent phenomena that are identical to general relativity. For example, crystal defects can look like curvature and torsion in an Einstein–Cartan spacetime. This allows one to create a theory of gravity with torsion from a World Crystal model of spacetime[1] in which the lattice spacing is of the order of a Planck length. Sakharov's idea was to start with an arbitrary background pseudo-Riemannian manifold (in modern treatments, possibly with torsion) and introduce quantum fields (matter) on it but not introduce any gravitational dynamics explicitly. This gives rise to an effective action which to one-loop order contains the Einstein–Hilbert action with a cosmological constant. In other words, general relativity arises as an emergent property of matter fields and is not put in by hand. On the other hand, such models typically predict huge cosmological constants.

Some argue that the particular models proposed by Sakharov and others have been proven impossible by the Weinberg–Witten theorem. However, models with emergent gravity are possible as long as other things, such as spacetime dimensions, emerge together with gravity. Developments in AdS/CFT correspondence after 1997 suggest that the microphysical degrees of freedom in induced gravity might be radically different. The bulk space-time arises as an emergent phenomenon of the quantum degrees of freedom that live in the boundary of the space-time.

31.1   See also

- Entropic gravity

- Causal dynamical triangulation

31.2   References

[1] H. Kleinert (1987). "Gravity as Theory of Defects in a Crystal with Only Second-Gradient Elasticity". *Annalen der Physik* **44**: 117. Bibcode:1987AnP...499..117K. doi:10.1002/andp.19874990206.

31.3   External links

- Carlos Barcelo, Stefano Liberati, Matt Visser, *Living Rev.Rel.* 8:12, 2005.

- D. Berenstein, *Emergent Gravity from CFT*, online lecture.

- C. J. Hogan *Quantum Indeterminacy of Emergent Spacetime*, preprint

- A.D. Sakharov, *Vacuum Quantum Fluctuations in Curved Space and the Theory of Gravitation*, 1967.

- Matt Visser, *Sakharov's induced gravity: a modern perspective*, 2002.

- H. Kleinert, *Multivalued Fields in Condensed Matter, Electrodynamics, and Gravitation*, 2008.

Chapter 32

Loop quantum gravity

Loop quantum gravity (**LQG**) is a theory that attempts to describe the quantum properties of the universe and gravity. It is also a theory of quantum spacetime because, according to general relativity, gravity is a manifestation of the geometry of spacetime. LQG is an attempt to merge quantum mechanics and general relativity. The main output of the theory is a physical picture of space where space is granular. The granularity is a direct consequence of the quantization. It has the same nature as the granularity of the photons in the quantum theory of electromagnetism and the discrete levels of the energy of the atoms. Here, it is space itself that is discrete. In other words, there is a minimum distance possible to travel through it.

More precisely, space can be viewed as an extremely fine fabric or network "woven" of finite loops. These networks of loops are called spin networks. The evolution of a spin network over time is called a spin foam. The predicted size of this structure is the Planck length, which is approximately 10^{-35} meters. According to the theory, there is no meaning to distance at scales smaller than the Planck scale. Therefore, LQG predicts that not just matter, but space itself, has an atomic structure.

Today LQG is a vast area of research, developing in several directions, which involves about 30 research groups worldwide.[1] They all share the basic physical assumptions and the mathematical description of quantum space. The full development of the theory is being pursued in two directions: the more traditional canonical loop quantum gravity, and the newer covariant loop quantum gravity, more commonly called spin foam theory.

Research into the physical consequences of the theory is proceeding in several directions. Among these, the most well-developed is the application of LQG to cosmology, called loop quantum cosmology (LQC). LQC applies LQG ideas to the study of the early universe and the physics of the Big Bang. Its most spectacular consequence is that the evolution of the universe can be continued beyond the Big Bang. The Big Bang appears thus to be replaced by a sort of cosmic Big Bounce.

32.1 History

Main article: History of loop quantum gravity

In 1986, Abhay Ashtekar reformulated Einstein's general relativity in a language closer to that of the rest of fundamental physics. Shortly after, Ted Jacobson and Lee Smolin realized that the formal equation of quantum gravity, called the Wheeler–DeWitt equation, admitted solutions labelled by loops, when rewritten in the new Ashtekar variables, and Carlo Rovelli and Lee Smolin defined a nonperturbative and background-independent quantum theory of gravity in terms of these loop solutions. Jorge Pullin and Jerzy Lewandowski understood that the intersections of the loops are essential for the consistency of the theory, and the theory should be formulated in terms of intersecting loops, or graphs.

In 1994, Rovelli and Smolin showed that the quantum operators of the theory associated to area and volume have a discrete spectrum. That is, geometry is quantized. This result defines an explicit basis of states of quantum geometry,

258

which turned out to be labelled by Roger Penrose's spin networks, which are graphs labelled by spins.

The canonical version of the dynamics was put on firm ground by Thomas Thiemann, who defined an anomaly-free Hamiltonian operator, showing the existence of a mathematically consistent background-independent theory. The covariant or spinfoam version of the dynamics developed during several decades, and crystallized in 2008, from the joint work of research groups in France, Canada, UK, Poland, and Germany, lead to the definition of a family of transition amplitudes, which in the classical limit can be shown to be related to a family of truncations of general relativity.[2] The finiteness of these amplitudes was proven in 2011.[3][4] It requires the existence of a positive cosmological constant, and this is consistent with observed acceleration in the expansion of the Universe.

32.2   General covariance and background independence

Main articles: General covariance, background-independent and diffeomorphism

In theoretical physics, general covariance is the invariance of the form of physical laws under arbitrary differentiable coordinate transformations. The essential idea is that coordinates are only artifices used in describing nature, and hence should play no role in the formulation of fundamental physical laws. A more significant requirement is the principle of general relativity that states that the laws of physics take the same form in all reference systems. This is a generalization of the principle of special relativity which states that the laws of physics take the same form in all inertial frames.

In mathematics, a diffeomorphism is an isomorphism in the category of smooth manifolds. It is an invertible function that maps one differentiable manifold to another, such that both the function and its inverse are smooth. These are the defining symmetry transformations of General Relativity since the theory is formulated only in terms of a differentiable manifold.

In general relativity, general covariance is intimately related to "diffeomorphism invariance". This symmetry is one of the defining features of the theory. However, it is a common misunderstanding that "diffeomorphism invariance" refers to the invariance of the physical predictions of a theory under arbitrary coordinate transformations; this is untrue and in fact every physical theory is invariant under coordinate transformations this way. Diffeomorphisms, as mathematicians define them, correspond to something much more radical; intuitively a way they can be envisaged is as simultaneously dragging all the physical fields (including the gravitational field) over the bare differentiable manifold while staying in the same coordinate system. Diffeomorphisms are the true symmetry transformations of general relativity, and come about from the assertion that the formulation of the theory is based on a bare differentiable manifold, but not on any prior geometry — the theory is background-independent (this is a profound shift, as all physical theories before general relativity had as part of their formulation a prior geometry). What is preserved under such transformations are the coincidences between the values the gravitational field take at such and such a "place" and the values the matter fields take there. From these relationships one can form a notion of matter being located with respect to the gravitational field, or vice versa. This is what Einstein discovered: that physical entities are located with respect to one another only and not with respect to the spacetime manifold. As Carlo Rovelli puts it: "No more fields on spacetime: just fields on fields.".[5] This is the true meaning of the saying "The stage disappears and becomes one of the actors"; space-time as a "container" over which physics takes place has no objective physical meaning and instead the gravitational interaction is represented as just one of the fields forming the world. This is known as the relationalist interpretation of space-time. The realization by Einstein that general relativity should be interpreted this way is the origin of his remark "Beyond my wildest expectations".

In LQG this aspect of general relativity is taken seriously and this symmetry is preserved by requiring that the physical states remain invariant under the generators of diffeomorphisms. The interpretation of this condition is well understood for purely spatial diffeomorphisms. However, the understanding of diffeomorphisms involving time (the Hamiltonian constraint) is more subtle because it is related to dynamics and the so-called "problem of time" in general relativity.[6] A generally accepted calculational framework to account for this constraint has yet to be found.[7][8] A plausible candidate for the quantum hamiltonian constraint is the operator introduced by Thiemann.[9]

LQG is formally background independent. The equations of LQG are not embedded in, or dependent on, space and time (except for its invariant topology). Instead, they are expected to give rise to space and time at distances which are large compared to the Planck length. The issue of background independence in LQG still has some unresolved subtleties. For example, some derivations require a fixed choice of the topology, while any consistent quantum theory of gravity should include topology change as a dynamical process.

32.3 Constraints and their Poisson bracket algebra

Main articles: Poisson bracket and Hamiltonian constraint

32.3.1 The constraints of classical canonical general relativity

Main article: Lie derivative

In the Hamiltonian formulation of ordinary classical mechanics the Poisson bracket is an important concept. A "canonical coordinate system" consists of canonical position and momentum variables that satisfy canonical Poisson-bracket relations,

$$\{q_i, p_j\} = \delta_{ij}$$

where the Poisson bracket is given by

$$\{f, g\} = \sum_{i=1}^{N} \left(\frac{\partial f}{\partial q_i} \frac{\partial g}{\partial p_i} - \frac{\partial f}{\partial p_i} \frac{\partial g}{\partial q_i} \right).$$

for arbitrary phase space functions $f(q_i, p_j)$ and $g(q_i, p_j)$. With the use of Poisson brackets, the Hamilton's equations can be rewritten as,

$$\dot{q}_i = \{q_i, H\}.$$

$$\dot{p}_i = \{p_i, H\}.$$

These equations describe a "flow" or orbit in phase space generated by the Hamiltonian H. Given any phase space function $F(q, p)$, we have

$$\frac{d}{dt} F(q_i, p_i) = \{F, H\}.$$

Let us consider constrained systems, of which General relativity is an example. In a similar way the Poisson bracket between a constraint and the phase space variables generates a flow along an orbit in (the unconstrained) phase space generated by the constraint. There are three types of constraints in Ashtekar's reformulation of classical general relativity:

$SU(2)$ Gauss gauge constraints

The Gauss constraints

$$G_j(x) = 0.$$

This represents an infinite number of constraints one for each value of x. These come about from re-expressing General relativity as an $SU(2)$ Yang–Mills type gauge theory (Yang–Mills is a generalization of Maxwell's theory where the gauge field transforms as a vector under Gauss transformations, that is, the Gauge field is of the form $A_a^i(x)$ where i is an internal index. See Ashtekar variables). These infinite number of Gauss gauge constraints can be smeared with test fields with internal indices, $\lambda^j(x)$,

$$G(\lambda) = \int d^3 x G_j(x) \lambda^j(x).$$

which we demand vanish for any such function. These smeared constraints defined with respect to a suitable space of smearing functions give an equivalent description to the original constraints.

In fact Ashtekar's formulation may be thought of as ordinary $SU(2)$ Yang–Mills theory together with the following special constraints, resulting from diffeomorphism invariance, and a Hamiltonian that vanishes. The dynamics of such a theory are thus very different from that of ordinary Yang–Mills theory.

Spatial diffeomorphisms constraints

The spatial diffeomorphism constraints

$$C_a(x) = 0$$

can be smeared by the so-called shift functions $\vec{N}(x)$ to give an equivalent set of smeared spatial diffeomorphism constraints,

$$C(\vec{N}) = \int d^3x C_a(x) N^a(x) \,.$$

These generate spatial diffeomorphisms along orbits defined by the shift function $N^a(x)$.

Hamiltonian constraints

The Hamiltonian

$$H(x) = 0$$

can be smeared by the so-called lapse functions $N(x)$ to give an equivalent set of smeared Hamiltonian constraints,

$$H(N) = \int d^3x H(x) N(x) \,.$$

These generate time diffeomorphisms along orbits defined by the lapse function $N(x)$.

In Ashtekar formulation the gauge field $A_a^i(x)$ is the configuration variable (the configuration variable being analogous to q in ordinary mechanics) and its conjugate momentum is the (densitized) triad (electrical field) $\tilde{E}_i^a(x)$. The constraints are certain functions of these phase space variables.

We consider the action of the constraints on arbitrary phase space functions. An important notion here is the Lie derivative, $\mathcal{L}_V$, which is basically a derivative operation that infinitesimally "shifts" functions along some orbit with tangent vector V.

32.3.2 The Poisson bracket algebra

Of particular importance is the Poisson bracket algebra formed between the (smeared) constraints themselves as it completely determines the theory. In terms of the above smeared constraints the constraint algebra amongst the Gauss' law reads,

$$\{G(\lambda), G(\mu)\} = G([\lambda, \mu])$$

where $[\lambda, \mu]^k = \lambda_i \mu_j \epsilon^{ijk}$. And so we see that the Poisson bracket of two Gauss' law is equivalent to a single Gauss' law evaluated on the commutator of the smearings. The Poisson bracket amongst spatial diffeomorphisms constraints reads

$$\{C(\vec{N}), C(\vec{M})\} = C(\mathcal{L}_{\vec{N}} \vec{M})$$

and we see that its effect is to "shift the smearing". The reason for this is that the smearing functions are not functions of the canonical variables and so the spatial diffeomorphism does not generate diffeomorphims on them. They do however generate diffeomorphims on everything else. This is equivalent to leaving everything else fixed while shifting the smearing .The action of the spatial diffeomorphism on the Gauss law is

$$\{C(\vec{N}), G(\lambda)\} = G(\mathcal{L}_{\vec{N}} \lambda) \,.$$

again, it shifts the test field λ. The Gauss law has vanishing Poisson bracket with the Hamiltonian constraint. The spatial diffeomorphism constraint with a Hamiltonian gives a Hamiltonian with its smearing shifted,

$$\{C(\vec{N}), H(M)\} = H(\mathcal{L}_{\vec{N}} M) \,.$$

Finally, the poisson bracket of two Hamiltonians is a spatial diffeomorphism,

$$\{H(N), H(M)\} = C(K)$$

where K is some phase space function. That is, it is a sum over infinitesimal spatial diffeomorphisms constraints where the coefficients of proportionality are not constants but have non-trivial phase space dependence.

A (Poisson bracket) Lie algebra, with constraints C_I , is of the form

$$\{C_I, C_J\} = f_{IJ}^{K} C_K$$

where f_{IJ}^{K} are constants (the so-called structure constants). The above Poisson bracket algebra for General relativity does not form a true Lie algebra as we have structure functions rather than structure constants for the Poisson bracket between two Hamiltonians. This leads to difficulties.

32.3.3   Dirac observables

The constraints define a constraint surface in the original phase space. The gauge motions of the constraints apply to all phase space but have the feature that they leave the constraint surface where it is, and thus the orbit of a point in the hypersurface under gauge transformations will be an orbit entirely within it. Dirac observables are defined as phase space functions, O , that Poisson commute with all the constraints when the constraint equations are imposed,

$$\{G_j, O\}_{G_j = C_a = H = 0} = \{C_a, O\}_{G_j = C_a = H = 0} = \{H, O\}_{G_j = C_a = H = 0} = 0 ,$$

that is, they are quantities defined on the constraint surface that are invariant under the gauge transformations of the theory.

Then, solving only the constraint $G_j = 0$ and determining the Dirac observables with respect to it leads us back to the ADM phase space with constraints H, C_a . The dynamics of general relativity is generated by the constraints, it can be shown that six Einstein equations describing time evolution (really a gauge transformation) can be obtained by calculating the Poisson brackets of the three-metric and its conjugate momentum with a linear combination of the spatial diffeomorphism and Hamiltonian constraint. The vanishing of the constraints, giving the physical phase space, are the four other Einstein equations.[10]

32.4   Quantization of the constraints – the equations of quantum general relativity

32.4.1   Pre-history and Ashtekar new variables

Main articles: Frame fields in general relativity, Ashtekar variables and Self-dual Palatini action

Many of the technical problems in canonical quantum gravity revolve around the constraints. Canonical general relativity was originally formulated in terms of metric variables, but there seemed to be insurmountable mathematical difficulties in promoting the constraints to quantum operators because of their highly non-linear dependence on the canonical variables. The equations were much simplified with the introduction of Ashtekars new variables. Ashtekar variables describe canonical general relativity in terms of a new pair canonical variables closer to that of gauge theories. The first step consists of using densitized triads $\tilde{E}_i^a$ (a triad E_i^a is simply three orthogonal vector fields labeled by $i = 1, 2, 3$ and the densitized triad is defined by $\tilde{E}_i^a = \sqrt{\det(q)} E_i^a$) to encode information about the spatial metric,

$$\det(q) q^{ab} = \tilde{E}_i^a \tilde{E}_j^b \delta^{ij} .$$

(where δ^{ij} is the flat space metric, and the above equation expresses that q^{ab} , when written in terms of the basis E_i^a , is locally flat). (Formulating general relativity with triads instead of metrics was not new.) The densitized triads are not unique, and in fact one can perform a local in space rotation with respect to the internal indices i . The canonically conjugate variable is related to the extrinsic curvature by $K_a^i = K_{ab} \tilde{E}^{ai}/\sqrt{\det(q)}$. But problems similar to using the metric formulation arise when one tries to quantize the theory. Ashtekar's new insight was to introduce a new configuration variable,

$$A_a^i = \Gamma_a^i - i K_a^i$$

that behaves as a complex SU(2) connection where Γ_a^i is related to the so-called spin connection via $\Gamma_a^i = \Gamma_{ajk} \epsilon^{jki}$. Here A_a^i is called the chiral spin connection. It defines a covariant derivative $\mathcal{D}_a$. It turns out that $\tilde{E}_i^a$ is the conjugate momentum of A_a^i , and together these form Ashtekar's new variables.

The expressions for the constraints in Ashtekar variables: the Gauss's law, the spatial diffeomorphism constraint and the (densitized) Hamiltonian constraint then read:

$$G^i = \mathcal{D}_a \tilde{E}^a_i = 0$$

$$C_a = \tilde{E}^b_i F^i_{ab} - A^i_a (\mathcal{D}_b \tilde{E}^b_i) = V_a - A^i_a G^i = 0 ,$$

$$\tilde{H} = \epsilon_{ijk} \tilde{E}^a_i \tilde{E}^b_j F^i_{ab} = 0$$

respectively, where F^i_{ab} is the field strength tensor of the connection A^i_a and where V_a is referred to as the vector constraint. The above-mentioned local in space rotational invariance is the original of the SU(2) gauge invariance here expressed by the Gauss law. Note that these constraints are polynomial in the fundamental variables, unlike as with the constraints in the metric formulation. This dramatic simplification seemed to open up the way to quantizing the constraints. (See the article Self-dual Palatini action for a derivation of Ashtekar's formulism).

With Ashtekar's new variables, given the configuration variable A^i_a , it is natural to consider wavefunctions $\Psi(A^i_a)$. This is the connection representation. It is analogous to ordinary quantum mechanics with configuration variable q and wavefunctions $\psi(q)$. The configuration variable gets promoted to a quantum operator via:

$$\hat{A}^i_a \Psi(A) = A^i_a \Psi(A) ,$$

(analogous to $\hat{q}\psi(q) = q\psi(q)$) and the triads are (functional) derivatives,

$$\hat{\tilde{E}}^a_i \Psi(A) = -i \frac{\delta \Psi(A)}{\delta A^i_a} .$$

(analogous to $\hat{p}\psi(q) = -i\hbar d\psi(q)/dq$). In passing over to the quantum theory the constraints become operators on a kinematic Hilbert space (the unconstrained SU(2) Yang–Mills Hilbert space). Note that different ordering of the A 's and $\tilde{E}$'s when replacing the $\tilde{E}$'s with derivatives give rise to different operators - the choice made is called the factor ordering and should be chosen via physical reasoning. Formally they read

$$\hat{G}_j |\psi\rangle = 0$$

$$\hat{C}_a |\psi\rangle = 0$$

$$\hat{\tilde{H}} |\psi\rangle = 0 .$$

There are still problems in properly defining all these equations and solving them. For example the Hamiltonian constraint Ashtekar worked with was the densitized version instead of the original Hamiltonian, that is, he worked with $\tilde{H} = \sqrt{\det(q)}H$. There were serious difficulties in promoting this quantity to a quantum operator. Moreover, although Ashtekar variables had the virtue of simplifying the Hamiltonian, they are complex. When one quantizes the theory, it is difficult to ensure that one recovers real general relativity as opposed to complex general relativity.

32.4.2 Quantum constraints as the equations of quantum general relativity

We now move on to demonstrate an important aspect of the quantum constraints. We consider Gauss' law only. First we state the classical result that the Poisson bracket of the smeared Gauss' law $G(\lambda) = \int d^3x \lambda^j (D_a E^a)^j$ with the connections is

$$\{G(\lambda), A^i_a\} = \partial_a \lambda^i + g \epsilon^{ijk} A^j_a \lambda^k = (D_a \lambda)^i .$$

The quantum Gauss' law reads

$$\hat{G}_j \Psi(A) = -iD_a \frac{\delta \lambda \Psi[A]}{\delta A^j_a} = 0.$$

If one smears the quantum Gauss' law and study its action on the quantum state one finds that the action of the constraint on the quantum state is equivalent to shifting the argument of Ψ by an infinitesimal (in the sense of the parameter λ small) gauge transformation,

$$\left[1 + \int d^3x \lambda^j(x) \hat{G}_j \right] \Psi(A) = \Psi[A + D\lambda] = \Psi[A],$$

and the last identity comes from the fact that the constraint annihilates the state. So the constraint, as a quantum operator, is imposing the same symmetry that its vanishing imposed classically: it is telling us that the functions $\Psi[A]$ have to be gauge invariant functions of the connection. The same idea is true for the other constraints.

Therefore the two step process in the classical theory of solving the constraints $C_I = 0$ (equivalent to solving the admissibility conditions for the initial data) and looking for the gauge orbits (solving the 'evolution' equations) is replaced by a one step process in the quantum theory, namely looking for solutions Ψ of the quantum equations $\hat{C}_I \Psi = 0$. This is because it obviously solves the constraint at the quantum level and it simultaneously looks for states that are gauge invariant because $\hat{C}_I$ is the quantum generator of gauge transformations (gauge invariant functions are constant along the gauge orbits and thus characterize them).[111] Recall that, at the classical level, solving the admissibility conditions and evolution equations was equivalent to solving all of Einstein's field equations, this underlines the central role of the quantum constraint equations in canonical quantum gravity.

32.4.3   Introduction of the loop representation

Main articles: Holonomy, Wilson loop and Knot invariant

It was in particular the inability to have good control over the space of solutions to the Gauss' law and spacial diffeomorphism constraints that led Rovelli and Smolin to consider a new representation - the loop representation in gauge theories and quantum gravity.[112]

We need the notion of a holonomy. A holonomy is a measure of how much the initial and final values of a spinor or vector differ after parallel transport around a closed loop; it is denoted

$h_\gamma[A]$.

Knowledge of the holonomies is equivalent to knowledge of the connection, up to gauge equivalence. Holonomies can also be associated with an edge; under a Gauss Law these transform as

$(h'_e)_{\alpha\beta} = U^{-1}_{\alpha\gamma}(x)(h_e)_{\gamma\sigma}U_{\sigma\beta}(y)$.

For a closed loop $x = y$ if we take the trace of this, that is, putting $\alpha = \beta$ and summing we obtain

$(h'_e)_{\alpha\alpha} = U^{-1}_{\alpha\gamma}(x)(h_e)_{\gamma\sigma}U_{\sigma\alpha}(x) = [U_{\sigma\alpha}(x)U^{-1}_{\alpha\gamma}(x)](h_e)_{\gamma\sigma} = \delta_{\sigma\gamma}(h_e)_{\gamma\sigma} = (h_e)_{\gamma\gamma}$

or

$\operatorname{Tr} h'_\gamma = \operatorname{Tr} h_\gamma$.

The trace of an holonomy around a closed loop is written

$W_\gamma[A]$

and is called a Wilson loop. Thus Wilson loops are gauge invariant. The explicit form of the Holonomy is

$h_\gamma[A] = \mathcal{P} \exp\left\{ -\int_{\gamma_0}^{\gamma_1} ds \dot{\gamma}^a A^i_a(\gamma(s)) T_i \right\}$

where γ is the curve along which the holonomy is evaluated, and s is a parameter along the curve, $\mathcal{P}$ denotes path ordering meaning factors for smaller values of s appear to the left, and T_i are matrices that satisfy the SU(2) algebra

$[T^i, T^j] = 2i\epsilon^{ijk}T^k$.

The Pauli matrices satisfy the above relation. It turns out that there are infinitely many more examples of sets of matrices that satisfy these relations, where each set comprises $(N + 1) \times (N + 1)$ matrices with $N = 1, 2, 3, \ldots$, and where none of these can be thought to 'decompose' into two or more examples of lower dimension. They are called different irreducible representations of the SU(2) algebra. The most fundamental representation being the Pauli matrices. The holonomy is labelled by a half integer $N/2$ according to the irreducible representation used.

The use of Wilson loops explicitly solves the Gauss gauge constraint. To handle the spatial diffeomorphism constraint we need to go over to the loop representation. As Wilson loops form a basis we can formally expand any Gauss gauge invariant function as,

$\Psi[A] = \sum_\gamma \Psi[\gamma]W_\gamma[A]$.

This is called the loop transform. We can see the analogy with going to the momentum representation in quantum mechanics(see Position and momentum space). There one has a basis of states $\exp(ikx)$ labelled by a number k and one expands

$$\psi[x] = \int dk \psi(k) \exp(ikx) .$$

and works with the coefficients of the expansion $\psi(k)$.

The inverse loop transform is defined by

$$\Psi[\gamma] = \int [dA] \Psi[A] W_\gamma[A] .$$

This defines the loop representation. Given an operator $\hat{O}$ in the connection representation,

$$\Phi[A] = \hat{O} \Psi[A] \qquad Eq\ 1 .$$

one should define the corresponding operator $\hat{O}'$ on $\Psi[\gamma]$ in the loop representation via,

$$\Phi[\gamma] = \hat{O}' \Psi[\gamma] \qquad Eq\ 2 .$$

where $\Phi[\gamma]$ is defined by the usual inverse loop transform,

$$\Phi[\gamma] = \int [dA] \Phi[A] W_\gamma[A] \qquad Eq\ 3. .$$

A transformation formula giving the action of the operator $\hat{O}'$ on $\Psi[\gamma]$ in terms of the action of the operator $\hat{O}$ on $\Psi[A]$ is then obtained by equating the R.H.S. of Eq 2 with the R.H.S. of Eq 3 with Eq 1 substituted into Eq 3 , namely

$$\hat{O}' \Psi[\gamma] = \int [dA] W_\gamma[A] \hat{O} \Psi[A] ,$$

or

$$\hat{O}' \Psi[\gamma] = \int [dA] (\hat{O}^\dagger W_\gamma[A]) \Psi[A] .$$

where by $\hat{O}^\dagger$ we mean the operator $\hat{O}$ but with the reverse factor ordering (remember from simple quantum mechanics where the product of operators is reversed under conjugation). We evaluate the action of this operator on the Wilson loop as a calculation in the connection representation and rearranging the result as a manipulation purely in terms of loops (one should remember that when considering the action on the Wilson loop one should choose the operator one wishes to transform with the opposite factor ordering to the one chosen for its action on wavefunctions $\Psi[A]$). This gives the physical meaning of the operator $\hat{O}'$. For example if $\hat{O}^\dagger$ corresponded to a spatial diffeomorphism, then this can be thought of as keeping the connection field A of $W_\gamma[A]$ where it is while performing a spatial diffeomorphism on γ instead. Therefore the meaning of $\hat{O}'$ is a spatial diffeomorphism on γ , the argument of $\Psi[\gamma]$.

In the loop representation we can then solve the spatial diffeomorphism constraint by considering functions of loops $\Psi[\gamma]$ that are invariant under spatial diffeomorphisms of the loop γ . That is, we construct what mathematicians call knot invariants. This opened up an unexpected connection between knot theory and quantum gravity.

What about the Hamiltonian constraint? Let us go back to the connection representation. Any collection of non-intersecting Wilson loops satisfy Ashtekar's quantum Hamiltonian constraint. This can be seen from the following. With a particular ordering of terms and replacing $\hat{E}_i^a$ by a derivative, the action of the quantum Hamiltonian constraint on a Wilson loop is

$$\hat{H}^\dagger W_\gamma[A] = -\epsilon_{ijk} \hat{F}_{ab}^k \frac{\delta}{\delta A_a^i} \frac{\delta}{\delta A_b^j} W_\gamma[A] .$$

When a derivative is taken it brings down the tangent vector, $\dot{\gamma}^a$, of the loop, γ . So we have something like

$$\hat{F}_{ab}^i \dot{\gamma}^a \dot{\gamma}^b .$$

However, as F_{ab}^i is anti-symmetric in the indices a and b this vanishes (this assumes that γ is not discontinuous anywhere and so the tangent vector is unique). Now let us go back to the loop representation.

We consider wavefunctions $\Psi[\gamma]$ that vanish if the loop has discontinuities and that are knot invariants. Such functions solve the Gauss law, the spatial diffeomorphism constraint and (formally) the Hamiltonian constraint. Thus we have identified an infinite set of exact (if only formal) solutions to all the equations of quantum general relativity![12] This generated a lot of interest in the approach and eventually led to LQG.

32.4.4   Geometric operators, the need for intersecting Wilson loops and spin network states

The easiest geometric quantity is the area. Let us choose coordinates so that the surface Σ is characterized by $x^3 = 0$. The area of small parallelogram of the surface Σ is the product of length of each side times $\sin\theta$ where θ is the angle between the sides. Say one edge is given by the vector $\vec{u}$ and the other by $\vec{v}$ then,

$$A = \|\vec{u}\|\|\vec{v}\|\sin\theta = \sqrt{\|\vec{u}\|^2\|\vec{v}\|^2(1 - \cos^2\theta)}   = \sqrt{\|\vec{u}\|^2\|\vec{v}\|^2 - (\vec{u}\cdot\vec{v})^2}$$

From this we get the area of the surface Σ to be given by

$$A_\Sigma = \int_\Sigma dx^1 dx^2 \sqrt{\det(q^{(2)})}$$

where $\det(q^{(2)}) = q_{11}q_{22} - q_{12}^2$ and is the determinant of the metric induced on Σ. This can be rewritten as

$$\det(q^{(2)}) = \frac{\epsilon^{3ab}\epsilon^{3cd}q_{ac}q_{bd}}{2} .$$

The standard formula for an inverse matrix is

$$q^{ab} = \frac{\epsilon^{acd}\epsilon^{bef}q_{ce}q_{df}}{3!\det(q)}$$

Note the similarity between this and the expression for $\det(q^{(2)})$. But in Ashtekar variables we have $\tilde{E}_i^a \tilde{E}^{bi} = \det(q)q^{ab}$. Therefore

$$A_\Sigma = \int_\Sigma dx^1 dx^2 \sqrt{\tilde{E}_i^3 \tilde{E}^{3i}} .$$

According to the rules of canonical quantization we should promote the triads $\tilde{E}_i^3$ to quantum operators.

$$\hat{\tilde{E}}_i^3 \sim \frac{\delta}{\delta A_3^i} .$$

It turns out that the area A_Σ can be promoted to a well defined quantum operator despite the fact that we are dealing with product of two functional derivatives and worse we have a square-root to contend with as well.[13] Putting $N = 2J$, we talk of being in the J-th representation. We note that $\sum_i T^i T^i = J(J+1)\mathbf{1}$. This quantity is important in the final formula for the area spectrum. We simply state the result below.

$$\hat{A}_\Sigma W_\gamma[A] = 8\pi\ell_{\text{Planck}}^2\beta\sum_I \sqrt{j_I(j_I + 1)}W_\gamma[A]$$

where the sum is over all edges I of the Wilson loop that pierce the surface Σ.

The formula for the volume of a region R is given by

$$V = \int_R d^3x \sqrt{\det(q)} = \frac{1}{6}\int_R dx^3 \sqrt{\epsilon_{abc}\epsilon^{ijk}\tilde{E}_i^a \tilde{E}_j^b \tilde{E}_k^c} .$$

The quantization of the volume proceeds the same way as with the area. As we take the derivative, and each time we do so we bring down the tangent vector $\dot{\gamma}^a$, when the volume operator acts on non-intersecting Wilson loops the result vanishes. Quantum states with non-zero volume must therefore involve intersections. Given that the anti-symmetric summation is taken over in the formula for the volume we would need at least intersections with three non-coplanar lines. Actually it turns out that one needs at least four-valent vertices for the volume operator to be non-vanishing.

We now consider Wilson loops with intersections. We assume the real representation where the gauge group is $SU(2)$. Wilson loops are an over complete basis as there are identities relating different Wilson loops. These come about from the fact that Wilson loops are based on matrices (the holonomy) and these matrices satisfy identities. Given any two $SU(2)$ matrices $\mathbb{A}$ and $\mathbb{B}$ it is easy to check that,

$$\text{Tr}(\mathbb{A})\text{Tr}(\mathbb{B}) = \text{Tr}(\mathbb{A}\mathbb{B}) + \text{Tr}(\mathbb{A}\mathbb{B}^{-1}) .$$

This implies that given two loops γ and η that intersect, we will have,

$$W_\gamma[A]W_\eta[A] = W_{\gamma\circ\eta}[A] + W_{\gamma\circ\eta^{-1}}[A]$$

where by η^{-1} we mean the loop η traversed in the opposite direction and $\gamma\circ\eta$ means the loop obtained by going around the loop γ and then along η. See figure below. Given that the matrices are unitary one has that $W_\gamma[A] = W_{\gamma^{-1}}[A]$. Also given the cyclic property of the matrix traces (i.e. $Tr(\mathbb{A}\mathbb{B}) = Tr(\mathbb{B}\mathbb{A})$) one has that $W_{\gamma\circ\eta}[A] = W_{\eta\circ\gamma}[A]$. These identities can be combined with each other into further identities of increasing complexity adding more loops. These identities are the so-called Mandelstam identities. Spin networks certain are linear combinations of intersecting Wilson loops designed to address the over completeness introduced by the Mandelstam identities (for trivalent intersections they eliminate the over-completeness entirely) and actually constitute a basis for all gauge invariant functions.

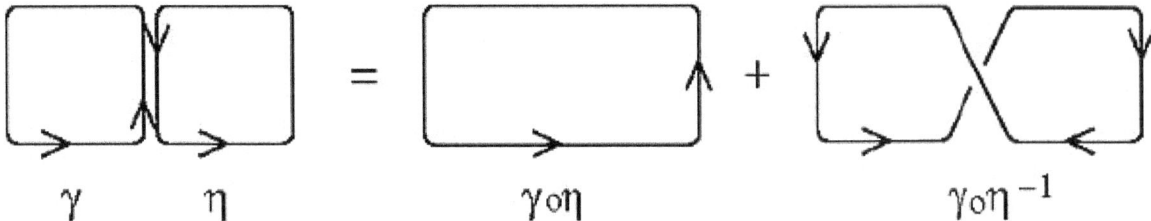

γ η $\gamma o \eta$ $\gamma o \eta^{-1}$

Graphical representation of the simplest non-trivial Mandestam identity relating different Wilson loops.

As mentioned above the holonomy tells you how to propagate test spin half particles. A spin network state assigns an amplitude to a set of spin half particles tracing out a path in space, merging and splitting. These are described by spin networks γ : the edges are labelled by spins together with `intertwiners' at the vertices which are prescription for how to sum over different ways the spins are rerouted. The sum over rerouting are chosen as such to make the form of the intertwiner invariant under Gauss gauge transformations.

32.4.5   Real variables, modern analysis and LQG

Main article: Hamiltonian constraint of LQG

Let us go into more detail about the technical difficulties associated with using Ashtekar's variables:

With Ashtekar's variables one uses a complex connection and so the relevant gauge group as actually $SL(2,\mathbb{C})$ and not $SU(2)$. As $SL(2,\mathbb{C})$ is non-compact it creates serious problems for the rigorous construction of the necessary mathematical machinery. The group $SU(2)$ is on the other hand is compact and the relevant constructions needed have been developed.

As mentioned above, because Ashtekar's variables are complex it results in complex general relativity. To recover the real theory one has to impose what are known as the reality conditions. These require that the densitized triad be real and that the real part of the Ashtekar connection equals the compatible spin connection (the compatibility condition being $\nabla_a e_b^I = 0$) determined by the desitized triad. The expression for compatible connection Γ_a^i is rather complicated and as such non-polynomial formula enters through the back door.

Before we state the next difficulty we should give a definition; a tensor density of weight W transforms like an ordinary tensor, except that in additional the W th power of the Jacobian,

$$J = \left| \frac{\partial x^a}{\partial x'^b} \right|$$

appears as a factor, i.e.

$$T'^{a\ldots}_{b\ldots} = J^W \frac{\partial x'^a}{\partial x^c} \cdots \frac{\partial x^d}{\partial x'^b} T^{c\ldots}_{d\ldots} .$$

It turns out that it is impossible, on general grounds, to construct a UV-finite, diffeomorphism non-violating operator corresponding to $\sqrt{\det(q)}H$. The reason is that the rescaled Hamiltonian constraint is a scalar density of weight two while it can be shown that only scalar densities of weight one have a chance to result in a well defined operator. Thus, one is forced to work with the original unrescaled, density one-valued, Hamiltonian constraint. However, this is non-polynomial and the whole virtue of the complex variables is questioned. In fact, all the solutions constructed for Ashtekar's Hamiltonian constraint only vanished for finite regularization (physics), however, this violates spatial diffeomorphism invariance.

Without the implementation and solution of the Hamiltonian constraint no progress can be made and no reliable predictions are possible!

To overcome the first problem one works with the configuration variable

$$A_a^i = \Gamma_a^i + \beta K_a^i$$

where β is real (as pointed out by Barbero, who introduced real variables some time after Ashtekar's variables[14][15]).

The Guass law and the spatial diffeomorphism constraints are the same. In real Ashtekar variables the Hamiltonian is

$$H = \frac{\epsilon_{ijk}F^k_{ab}\tilde{E}^a_i\tilde{E}^b_j}{\sqrt{\det(q)}} + 2\frac{\beta^2+1}{\beta^2}\frac{(\tilde{E}^a_i\tilde{E}^b_j-\tilde{E}^a_j\tilde{E}^b_i)}{\sqrt{\det(q)}}(A^i_a - \Gamma^i_a)(A^j_b - \Gamma^j_b) = H_E + H'.$$

The complicated relationship between Γ^i_a and the desitized triads causes serious problems upon quantization. It is with the choice $\beta = \pm i$ that the second more complicated term is made to vanish. However, as mentioned above Γ^i_a reappears in the reality conditions. Also we still have the problem of the $1/\sqrt{\det(q)}$ factor.

Thiemann was able to make it work for real β. First he could simplify the troublesome $1/\sqrt{\det(q)}$ by using the identity

$$\{A^k_c, V\} = \frac{\epsilon_{abc}\epsilon^{ijk}\tilde{E}^a_i\tilde{E}^b_j}{\sqrt{\det(q)}}$$

where V is the volume. The A^k_c and V can be promoted to well defined operators in the loop representation and the Poisson bracket is replaced by a commutator upon quantization; this takes care of the first term. It turns out that a similar trick can be used to treat the second term. One introduces the quantity

$$K = \int d^3x K^i_a \tilde{E}^a_i$$

and notes that

$$K^i_a = \{A^i_a, K\}.$$

We are then able to write

$$A^i_a - \Gamma^i_a = \beta K^i_a = \beta\{A^i_a, K\}.$$

The reason the quantity K is easier to work with at the time of quantization is that it can be written as

$$K = -\{V, \int d^3x H_E\}$$

where we have used that the integrated densitized trace of the extrinsic curvature, K, is the "time derivative of the volume".

In the long history of canonical quantum gravity formulating the Hamiltonian constraint as a quantum operator (Wheeler–DeWitt equation) in a mathematically rigorous manner has been a formidable problem. It was in the loop representation that a mathematically well defined Hamiltonian constraint was finally formulated in 1996.[9] We leave more details of its construction to the article Hamiltonian constraint of LQG. This together with the quantum versions of the Gauss law and spatial diffeomorphism constrains written in the loop representation are the central equations of LQG (modern canonical quantum General relativity).

Finding the states that are annihilated by these constraints (the physical states), and finding the corresponding physical inner product, and observables is the main goal of the technical side of LQG.

A very important aspect of the Hamiltonian operator is that it only acts at vertices (a consequence of this is that Thiemann's Hamiltonian operator, like Ashtekar's operator, annihilates non-intersecting loops except now it is not just formal and has rigorous mathematical meaning). More precisely, its action is non-zero on at least vertices of valence three and greater and results in a linear combination of new spin networks where the original graph has been modified by the addition of lines at each vertex together and a change in the labels of the adjacent links of the vertex.

32.4.6   Solving the quantum constraints

Main articles: spectrum, dual space and Rigged Hilbert space

We solve, at least approximately, all the quantum constraint equations and for the physical inner product to make physical predictions.

Before we move on to the constraints of LQG, lets us consider certain cases. We start with a kinematic Hilbert space $\mathcal{H}_{Kin}$ as so is equipped with an inner product—the kinematic inner product $\langle \phi, \psi \rangle_{Kin}$.

i) Say we have constraints $\hat{C}_I$ whose zero eigenvalues lie in their discrete spectrum. Solutions of the first constraint, $\hat{C}_1$, correspond to a subspace of the kinematic Hilbert space, $\mathcal{H}_1 \subset \mathcal{H}_{Kin}$. There will be a projection operator P_1 mapping

$\mathcal{H}_{\text{Kin}}$ onto $\mathcal{H}_1$. The kinematic inner product structure is easily employed to provide the inner product structure after solving this first constraint: the new inner product $\langle \phi, \psi \rangle_1$ is simply

$$\langle \phi, \psi \rangle_1 = \langle P\phi, P\psi \rangle_{\text{Kin}}$$

They are based on the same inner product and are states normalizable with respect to it.

ii) The zero point is not contained in the point spectrum of all the $\hat{C}_I$, there is then no non-trivial solution $\Psi \in \mathcal{H}_{\text{Kin}}$ to the system of quantum constraint equations $\hat{C}_I \Psi = 0$ for all I .

For example the zero eigenvalue of the operator

$$\hat{C} = \left(i\frac{d}{dx} - k \right)$$

on $L_2(\mathbb{R}, dx)$ lies in the continuous spectrum $\mathbb{R}$ but the formal "eigenstate" $\exp(-ikx)$ is not normalizable in the kinematic inner product.

$$\int_{-\infty}^{\infty} dx \psi^*(x)\psi(x) = \int_{-\infty}^{\infty} dx e^{ikx} e^{-ikx} = \int_{-\infty}^{\infty} dx = \infty$$

and so does not belong to the kinematic Hilbert space $\mathcal{H}_{\text{Kin}}$. In these cases we take a dense subset S of $\mathcal{H}_{\text{Kin}}$ (intuitively this means either any point in S is either in $\mathcal{H}_{\text{Kin}}$ or arbitrarily close to a point in $\mathcal{H}_{\text{Kin}}$) with very good convergence properties and consider its dual space S' (intuitively these map elements of S onto finite complex numbers in a linear manner), then $S \subset \mathcal{H}_{\text{Kin}} \subset S'$ (as S' contains distributional functions). The constraint operator is then implemented on this larger dual space, which contains distributional functions, under the adjoint action on the operator. One looks for solutions on this larger space. This comes at the price that the solutions must be given a new Hilbert space inner product with respect to which they are normalizable (see article on rigged Hilbert space). In this case we have a generalized projection operator on the new space of states. We cannot use the above formula for the new inner product as it diverges, instead the new inner product is given by the simply modification of the above.

$$\langle \phi, \psi \rangle_1 = \langle P\phi, \psi \rangle_{\text{Kin}}.$$

The generalized projector P is known as a rigging map.

Let us move to LQG, additional complications will arise from the fact the constraint algebra is not a Lie algebra due to the bracket between two Hamiltonian constraints.

The Gauss law is solved by the use of spin network states. They provide a basis for the Kinematic Hilbert space $\mathcal{H}_{\text{Kin}}$. The spatial diffeomorphism constraint has been solved. The induced inner product on $\mathcal{H}_{\text{Diff}}$ (we do not pursue the details) has a very simple description in terms of spin network states; given two spin networks s and s' , with associated spin network states ψ_s and $\psi_{s'}$, the inner product is 1 if s and s' are related to each other by a spatial diffeomorphism and zero otherwise.

The Hamiltonian constraint maps diffeomorphism invariant states onto non-diffeomorphism invaiant states as so does not preserve the diffeomorphism Hilbert space $\mathcal{H}_{\text{Diff}}$. This is an unavoidable consequence of the operator algebra, in particular the commutator:

$$[\hat{C}(\vec{N}), \hat{H}(M)] \propto \hat{H}(\mathcal{L}_{\vec{N}} M)$$

as can be seen by applying this to $\psi_s \in \mathcal{H}_{Diff}$.

$$(\vec{C}(\vec{N})\hat{H}(M) - \hat{H}(M)\vec{C}(\vec{N}))\psi_s \propto \hat{H}(\mathcal{L}_{\vec{N}} M)\psi_s$$

and using $\vec{C}(\vec{N})\psi_s = 0$ to obtain

$$\vec{C}(\vec{N})[\hat{H}(M)\psi_s] \propto \hat{H}(\mathcal{L}_{\vec{N}} M)\psi_s \neq 0$$

and so $\hat{H}(M)\psi_s$ is not in $\mathcal{H}_{Diff}$.

This means that you can't just solve the diffeomorphism constraint and then the Hamiltonian constraint. This problem can be circumvented by the introduction of the master constraint, with its trivial operator algebra, one is then able in principle to construct the physical inner product from $\mathcal{H}_{\text{Diff}}$.

32.5   Spin foams

Main articles: spin network, spin foam, BF model and Barrett–Crane model

In loop quantum gravity (LQG), a spin network represents a "quantum state" of the gravitational field on a 3-dimensional hypersurface. The set of all possible spin networks (or, more accurately, "s-knots" - that is, equivalence classes of spin networks under diffeomorphisms) is countable; it constitutes a basis of LQG Hilbert space.

In physics, a spin foam is a topological structure made out of two-dimensional faces that represents one of the configurations that must be summed to obtain a Feynman's path integral (functional integration) description of quantum gravity. It is closely related to loop quantum gravity.

32.5.1   Spin foam derived from the Hamiltonian constraint operator

The Hamiltonian constraint generates 'time' evolution. Solving the Hamiltonian constraint should tell us how quantum states evolve in 'time' from an initial spin network state to a final spin network state. One approach to solving the Hamiltonian constraint starts with what is called the Dirac delta function. This is a rather singular function of the real line, denoted $\delta(x)$, that is zero everywhere except at $x = 0$ but whose integral is finite and nonzero. It can be represented as a Fourier integral,

$$\delta(x) = \int e^{ikx} dk \ .$$

One can employ the idea of the delta function to impose the condition that the Hamiltonian constraint should vanish. It is obvious that

$$\prod_{x \in \Sigma} \delta(\hat{H}(x))$$

is non-zero only when $\hat{H}(x) = 0$ for all x in Σ. Using this we can 'project' out solutions to the Hamiltonian constraint. With analogy to the Fourier integral given above, this (generalized) projector can formally be written as

$$\int [dN] e^{i \int d^3 x N(x) \hat{H}(x)} \ .$$

Interestingly, this is formally spatially diffeomorphism-invariant. As such it can be applied at the spatially diffeomorphism-invariant level. Using this the physical inner product is formally given by

$$\left\langle \int [dN] e^{i \int d^3 x N(x) \hat{H}(x)} s_{\mathrm{int}} s_{\mathrm{fin}} \right\rangle_{\mathrm{Diff}}$$

where s_{int} are the initial spin network and s_{fin} is the final spin network.

The exponential can be expanded

$$\left\langle \int [dN] (1 + i \int d^3 x N(x) \hat{H}(x) + \tfrac{i^2}{2!} [\int d^3 x N(x) \hat{H}(x)][\int d^3 x' N(x') \hat{H}(x')] + \dots) s_{\mathrm{int}} \cdot s_{\mathrm{fin}} \right\rangle_{\mathrm{Diff}}$$

and each time a Hamiltonian operator acts it does so by adding a new edge at the vertex. The summation over different sequences of actions of $\hat{H}$ can be visualized as a summation over different histories of 'interaction vertices' in the 'time' evolution sending the initial spin network to the final spin network. This then naturally gives rise to the two-complex (a combinatorial set of faces that join along edges, which in turn join on vertices) underlying the spin foam description; we evolve forward an initial spin network sweeping out a surface, the action of the Hamiltonian constraint operator is to produce a new planar surface starting at the vertex. We are able to use the action of the Hamiltonian constraint on the vertex of a spin network state to associate an amplitude to each "interaction" (in analogy to Feynman diagrams). See figure below. This opens up a way of trying to directly link canonical LQG to a path integral description. Now just as a spin networks describe quantum space, each configuration contributing to these path integrals, or sums over history, describe 'quantum space-time'. Because of their resemblance to soap foams and the way they are labeled John Baez gave these 'quantum space-times' the name 'spin foams'.

There are however severe difficulties with this particular approach, for example the Hamiltonian operator is not self-adjoint, in fact it is not even a normal operator (i.e. the operator does not commute with its adjoint) and so the spectral theorem cannot be used to define the exponential in general. The most serious problem is that the $\hat{H}(x)$'s are not mutually

The action of the Hamiltonian constraint translated to the path integral or so-called spin foam description. A single node splits into three nodes, creating a spin foam vertex. $N(x_n)$ is the value of N at the vertex and H_{nop} are the matrix elements of the Hamiltonian constraint $\hat{H}$.

commuting, it can then be shown the formal quantity $\int [dN]e^{i\int d^3x N(x)\hat{H}(x)}$ cannot even define a (generalized) projector. The master constraint (see below) does not suffer from these problems and as such offers a way of connecting the canonical theory to the path integral formulation.

32.5.2 Spin foams from BF theory

It turns out there are alternative routes to formulating the path integral, however their connection to the Hamiltonian formalism is less clear. One way is to start with the BF theory. This is a simpler theory to general relativity. It has no local degrees of freedom and as such depends only on topological aspects of the fields. BF theory is what is known as a topological field theory. Surprisingly, it turns out that general relativity can be obtained from BF theory by imposing a constraint,[16] BF theory involves a field B_{ab}^{IJ} and if one chooses the field B to be the (anti-symmetric) product of two tetrads

$$B_{ab}^{IJ} = \tfrac{1}{2}(E_a^I E_b^J - E_b^I E_a^J)$$

(tetrads are like triads but in four spacetime dimensions), one recovers general relativity. The condition that the B field be given by the product of two tetrads is called the simplicity constraint. The spin foam dynamics of the topological field theory is well understood. Given the spin foam `interaction' amplitudes for this simple theory, one then tries to implement the simplicity conditions to obtain a path integral for general relativity. The non-trivial task of constructing a spin foam model is then reduced to the question of how this simplicity constraint should be imposed in the quantum theory. The first attempt at this was the famous Barrett–Crane model.[17] However this model was shown to be problematic, for example there did not seem to be enough degrees of freedom to ensure the correct classical limit.[18] It has been argued that the simplicity constraint was imposed too strongly at the quantum level and should only be imposed in the sense of expectation values just as with the Lorenz gauge condition $\partial_\mu \hat{A}^\mu$ in the Gupta–Bleuler formalism of quantum electrodynamics. New models have now been put forward, sometimes motivated by imposing the simplicity conditions in a weaker sense.

Another difficulty here is that spin foams are defined on a discretization of spacetime. While this presents no problems for a topological field theory as it has no local degrees of freedom, it presents problems for GR. This is known as the problem triangularization dependence.

32.5.3 Modern formulation of spin foams

Just as imposing the classical simplicity constraint recovers general relativity from BF theory, one expects an appropriate quantum simplicity constraint will recover quantum gravity from quantum BF theory.

Much progress has been made with regard to this issue by Engle, Pereira, and Rovelli[19] and Freidal and Krasnov[20] in defining spin foam interaction amplitudes with much better behaviour.

An attempt to make contact between EPRL-FK spin foam and the canonical formulation of LQG has been made.[21]

32.5.4 Spin foam derived from the master constraint operator

See below.

32.6 The semi-classical limit

32.6.1 What is the semiclassical limit?

Main articles: Correspondence principle and classical limit

The **classical limit** or **correspondence limit** is the ability of a physical theory to approximate or "recover" classical mechanics when considered over special values of its parameters.[22] The classical limit is used with physical theories that predict non-classical behavior.

In physics, the **correspondence principle** states that the behavior of systems described by the theory of quantum mechanics (or by the old quantum theory) reproduces classical physics in the limit of large quantum numbers. In other words, it says that for large orbits and for large energies, quantum calculations must agree with classical calculations.[23]

The principle was formulated by Niels Bohr in 1920,[24] though he had previously made use of it as early as 1913 in developing his model of the atom.[25]

There are two basic requirements in establishing the semi-classical limit of any quantum theory:

i) reproduction of the Poisson brackets (of the diffeomorphism constraints in the case of general relativity). This is extremely important because, as noted above, the Poisson bracket algebra formed between the (smeared) constraints themselves completely determines the classical theory. This is analogous to establishing Ehrenfest's theorem;

ii) the specification of a complete set of classical observables whose corresponding operators (see complete set of commuting observables for the quantum mechanical definition of a complete set of observables) when acted on by appropriate semi-classical states reproduce the same classical variables with small quantum corrections (a subtle point is that states that are semi-classical for one class of observables may not be semi-classical for a different class of observables[26]).

This may be easily done, for example, in ordinary quantum mechanics for a particle but in general relativity this becomes a highly non-trivial problem as we will see below.

32.6.2 Why might LQG not have general relativity as its semiclassical limit?

Any candidate theory of quantum gravity must be able to reproduce Einstein's theory of general relativity as a classical limit of a quantum theory. This is not guaranteed because of a feature of quantum field theories which is that they have different sectors, these are analogous to the different phases that come about in the thermodynamical limit of statistical systems. Just as different phases are physically different, so are different sectors of a quantum field theory. It may turn out that LQG belongs to an unphysical sector - one in which you do not recover general relativity in the semi classical limit (in fact there might not be any physical sector at all).

Theorems establishing the uniqueness of the loop representation as defined by Ashtekar et al. (i.e. a certain concrete realization of a Hilbert space and associated operators reproducing the correct loop algebra - the realization that everybody was using) have been given by two groups (Lewandowski, Okolow, Sahlmann and Thiemann;[27] and Christian Fleischhack[28]). Before this result was established it was not known whether there could be other examples of Hilbert spaces with operators invoking the same loop algebra, other realizations, not equivalent to the one that had been used so far. These uniqueness theorems imply no others exist and so if LQG does not have the correct semiclassical limit then this would mean the end of the loop representation of quantum gravity altogether.

32.6.3 Difficulties checking the semiclassical limit of LQG

There are difficulties in trying to establish LQG gives Einstein's theory of general relativity in the semi classical limit. There are a number of particular difficulties in establishing the semi-classical limit

1. There is no operator corresponding to infinitesimal spacial diffeomorphisms (it is not surprising that the theory has no generator of infinitesimal spatial `translations' as it predicts spatial geometry has a discrete nature, compare to the situation in condensed matter). Instead it must be approximated by finite spatial diffeomorphisms and so the Poisson bracket structure of the classical theory is not exactly reproduced. This problem can be circumvented with the introduction of the so-called master constraint (see below)[29]

2. There is the problem of reconciling the discrete combinatorial nature of the quantum states with the continuous nature of the fields of the classical theory.

3. There are serious difficulties arising from the structure of the Poisson brackets involving the spatial diffeomorphism and Hamiltonian constraints. In particular, the algebra of (smeared) Hamiltonian constraints does not close, it is proportional to a sum over infinitesimal spatial diffeomorphisms (which, as we have just noted, does not exist in the quantum theory) where the coefficients of proportionality are not constants but have non-trivial phase space dependence - as such it does not form a Lie algebra. However, the situation is much improved by the introduction of the master constraint.[29]

4. The semi-classical machinery developed so far is only appropriate to non-graph-changing operators, however, Thiemann's Hamiltonian constraint is a graph-changing operator - the new graph it generates has degrees of freedom upon which the coherent state does not depend and so their quantum fluctuations are not suppressed. There is also the restriction, so far, that these coherent states are only defined at the Kinematic level, and now one has to lift them to the level of $\mathcal{H}_{Diff}$ and $\mathcal{H}_{Phys}$. It can be shown that Thiemann's Hamiltonian constraint is required to be graph changing in order to resolve problem 3 in some sense. The master constraint algebra however is trivial and so the requirement that it be graph changing can be lifted and indeed non-graph changing master constraint operators have been defined.

5. Formulating observables for classical general relativity is a formidable problem by itself because of its non-linear nature and space-time diffeomorphism invariance. In fact a systematic approximation scheme to calculate observables has only been recently developed.[30][31]

Difficulties in trying to examine the semi classical limit of the theory should not be confused with it having the wrong semi classical limit.

32.6.4 Progress in demonstrating LQG has the correct semiclassical limit

Much details here to be written up...

Concerning issue number 2 above one can consider so-called weave states. Ordinary measurements of geometric quantities are macroscopic, and planckian discreteness is smoothed out. The fabric of a T-shirt is analogous. At a distance it is a smooth curved two-dimensional surface. But a closer inspection we see that it is actually composed of thousands of one-dimensional linked threads. The image of space given in LQG is similar, consider a very large spin network formed by a very large number of nodes and links, each of Planck scale. But probed at a macroscopic scale, it appears as a three-dimensional continuous metric geometry.

As far as the editor knows problem 4 of having semi-classical machinery for non-graph changing operators is as the moment still out of reach.

To make contact with familiar low energy physics it is mandatory to have to develop approximation schemes both for the physical inner product and for Dirac observables.

The spin foam models have been intensively studied can be viewed as avenues toward approximation schemes for the physical inner product.

Markopoulou et al. adopted the idea of noiseless subsystems in an attempt to solve the problem of the low energy limit in background independent quantum gravity theories[32][33][34] The idea has even led to the intriguing possibility of matter of the standard model being identified with emergent degrees of freedom from some versions of LQG (see section below: *LQG and related research programs*).

As Wightman emphasized in the 1950s, in Minkowski QFTs the $n-$ point functions

$$W(x_1, \ldots, x_n) = \langle 0|\phi(x_n) \ldots \phi(x_1)|0\rangle ,$$

completely determine the theory. In particular, one can calculate the scattering amplitudes from these quantities. As explained below in the section on the *Background independent scattering amplitudes*, in the background-independent context, the $n-$ point functions refer to a state and in gravity that state can naturally encode information about a specific geometry which can then appear in the expressions of these quantities. To leading order LQG calculations have been shown to agree in an appropriate sense with the $n-$ point functions calculated in the effective low energy quantum general relativity.

32.7   Improved dynamics and the master constraint

Main articles: Hamiltonian (quantum mechanics), Hamiltonian constraint of LQG and Friedrichs extension

32.7.1   The master constraint

Thiemann's master constraint should not be confused with the master equation which has to do with random processes. The Master Constraint Programme for Loop Quantum Gravity (LQG) was proposed as a classically equivalent way to impose the infinite number of Hamiltonian constraint equations

$$H(x) = 0$$

(x being a continuous index) in terms of a single master constraint,

$$M = \int d^3x \frac{|H(x)|^2}{\sqrt{\det(q(x))}} .$$

which involves the square of the constraints in question. Note that $H(x)$ were infinitely many whereas the master constraint is only one. It is clear that if M vanishes then so do the infinitely many $H(x)$'s. Conversely, if all the $H(x)$'s vanish then so does M , therefore they are equivalent. The master constraint M involves an appropriate averaging over all space and so is invariant under spatial diffeomorphisms (it is invariant under spatial "shifts" as it is a summation over all such spatial "shifts" of a quantity that transforms as a scalar). Hence its Poisson bracket with the (smeared) spacial diffeomorphism constraint, $C(\vec{N})$, is simple:

$$\{M, C(\vec{N})\} = 0 .$$

(it is $su(2)$ invariant as well). Also, obviously as any quantity Poisson commutes with itself, and the master constraint being a single constraint, it satisfies

$$\{M, M\} = 0 .$$

We also have the usual algebra between spatial diffeomorphisms. This represents a dramatic simplification of the Poisson bracket structure, and raises new hope in understanding the dynamics and establishing the semi-classical limit.[35]

An initial objection to the use of the master constraint was that on first sight it did not seem to encode information about the observables; because the Mater constraint is quadratic in the constraint, when you compute its Poisson bracket with any quantity, the result is proportional to the constraint, therefore it always vanishes when the constraints are imposed and as such does not select out particular phase space functions. However, it was realized that the condition

$$\{\{M, O\}, O\}_{M=0} = 0$$

is equivalent to O being a Dirac observable. So the master constraint does capture information about the observables. Because of its significance this is known as the Master equation.[35]

That the master constraint Poisson algebra is an honest Lie algebra opens up the possibility of using a certain method, known as group averaging, in order to construct solutions of the infinite number of Hamiltonian constraints, a physical inner product thereon and Dirac observables via what is known as refined algebraic quantization RAQ[36]

32.7.2 The quantum master constraint

Define the quantum master constraint (regularisation issues aside) as

$$\hat{M} := \int d^3x \left(\widetilde{\frac{H}{\det(q(x))^{1/4}}} \right)^\dagger (x) \left(\widetilde{\frac{H}{\det(q(x))^{1/4}}} \right) (x) \, .$$

Obviously,

$$\left(\widetilde{\frac{H}{\det(q(x))^{1/4}}} \right) (x)\Psi = 0$$

for all x implies $\hat{M}\Psi = 0$. Conversely, if $\hat{M}\Psi = 0$ then

$$0 = < \Psi, \hat{M}\Psi > = \int d^3x \left\| \left(\widetilde{\frac{H}{\det(q(x))^{1/4}}} \right)(x)\Psi \right\|^2 \qquad Eq \ 4$$

implies

$$\left(\widetilde{\frac{H}{\det(q(x))^{1/4}}} \right) (x)\Psi = 0 \, .$$

What is done first is, we are able to compute the matrix elements of the would-be operator $\hat{M}$, that is, we compute the quadratic form Q_M. It turns out that as Q_M is a graph changing, diffeomorphism invariant quadratic form it cannot exist on the kinematic Hilbert space H_{Kin}, and must be defined on H_{Diff}. The fact that the master constraint operator $\hat{M}$ is densely defined on H_{Diff}, it is obvious that $\hat{M}$ is a positive and symmetric operator in H_{Diff}. Therefore, the quadratic form Q_M associated with $\hat{M}$ is closable. The closure of Q_M is the quadratic form of a unique self-adjoint operator $\overline{M}$, called the Friedrichs extension of $\hat{M}$. We relabel $\overline{M}$ as $\hat{M}$ for simplicity. (Note that the presence of an inner product, viz Eq 4, means there are no superfluous solutions i.e. there are no Ψ such that $\left(\widetilde{\frac{H}{\det(q(x))^{1/4}}} \right)(x)\Psi \neq 0$ but for which $\hat{M}\Psi = 0$).

It is also possible to construct a quadratic form Q_{M_E} for what is called the extended master constraint (discussed below) on H_{Kin} which also involves the weighted integral of the square of the spatial diffeomorphism constraint (this is possible because Q_{M_E} is not graph changing).

The spectrum of the master constraint may not contain zero due to normal or factor ordering effects which are finite but similar in nature to the infinite vacuum energies of background-dependent quantum field theories. In this case it turns out to be physically correct to replace $\hat{M}$ with $\hat{M}' := \hat{M} - min(spec(\hat{M}))\hat{1}$ provided that the "normal ordering constant" vanishes in the classical limit, that is, $\lim_{h \to 0} min(spec(\hat{M})) = 0$, so that $\hat{M}'$ is a valid quantisation of M.

32.7.3 Testing the master constraint

The constraints in their primitive form are rather singular, this was the reason for integrating them over test functions to obtain smeared constraints. However, it would appear that the equation for the master constraint, given above, is even more singular involving the product of two primitive constraints (although integrated over space). Squaring the constraint is dangerous as it could lead to worsened ultraviolent behaviour of the corresponding operator and hence the master constraint programme must be approached with due care.

In doing so the master constraint programme has been satisfactorily tested in a number of model systems with non-trivial constraint algebras, free and interacting field theories.[37][38][39][40][41] The master constraint for LQG was established as a genuine positive self-adjoint operator and the physical Hilbert space of LQG was shown to be non-empty,[42] an obvious consistency test LQG must pass to be a viable theory of quantum General relativity.

32.7.4   Applications of the master constraint

The master constraint has been employed in attempts to approximate the physical inner product and define more rigorous path integrals.[43][44][45][46]

The Consistent Discretizations approach to LQG,[47][48] is an application of the master constraint program to construct the physical Hilbert space of the canonical theory.

32.7.5   Spin foam from the master constraint

It turns out that the master constraint is easily generalized to incorporate the other constraints. It is then referred to as the extended master constraint, denoted M_E. We can define the extended master constraint which imposes both the Hamiltonian constraint and spatial diffeomorphism constraint as a single operator,

$$M_E = \int_{\Sigma} d^3x \frac{H(x)^2 - q^{ab} V_a(x) V_b(x)}{\sqrt{det(q)}}.$$

Setting this single constraint to zero is equivalent to $H(x) = 0$ and $V_a(x) = 0$ for all x in Σ. This constraint implements the spatial diffeomorphism and Hamiltonian constraint at the same time on the Kinematic Hilbert space. The physical inner product is then defined as

$$\langle \phi, \psi \rangle_{\text{Phys}} = \lim_{T \to \infty} \left\langle \phi, \int_{-T}^{T} dt e^{it\hat{M}_E} \psi \right\rangle$$

(as $\delta(\hat{M}_E) = \lim_{T \to \infty} \int_{-T}^{T} dt e^{it\hat{M}_E}$). A spin foam representation of this expression is obtained by splitting the t -parameter in discrete steps and writing

$$e^{it\hat{M}_E} = \lim_{n \to \infty} [e^{it\hat{M}_E/n}]^n = \lim_{n \to \infty} [1 + it\hat{M}_E/n]^n.$$

The spin foam description then follows from the application of $[1 + it\hat{M}_E/n]$ on a spin network resulting in a linear combination of new spin networks whose graph and labels have been modified. Obviously an approximation is made by truncating the value of n to some finite integer. An advantage of the extended master constraint is that we are working at the kinematic level and so far it is only here we have access semi-classical coherent states. Moreover, one can find none graph changing versions of this master constraint operator, which are the only type of operators appropriate for these coherent states.

32.7.6   Algebraic quantum gravity

The master constraint programme has evolved into a fully combinatorial treatment of gravity known as Algebraic Quantum Gravity (AQG).[49] The non-graph changing master constraint operator is adapted in the framework of algebraic quantum gravity. While AQG is inspired by LQG, it differs drastically from it because in AQG there is fundamentally no topology or differential structure - it is background independent in a more generalized sense and could possibly have something to say about topology change. In this new formulation of quantum gravity AQG semiclassical states always control the fluctuations of all present degrees of freedom. This makes the AQG semiclassical analysis superior over that of LQG, and progress has been made in establishing it has the correct semiclassical limit and providing contact with familiar low energy physics.[50][51] See Thiemann's book for details.

32.8   Physical applications of LQG

32.8.1   Black hole entropy

Main articles: Black hole thermodynamics, Isolated horizon and Immirzi parameter

The Immirzi parameter (also known as the Barbero-Immirzi parameter) is a numerical coefficient appearing in loop quantum gravity. It may take real or imaginary values.

An artist depiction of two black holes merging, a process in which the laws of thermodynamics are upheld.

Black hole thermodynamics is the area of study that seeks to reconcile the laws of thermodynamics with the existence of black hole event horizons. The no hair conjecture of general relativity states that a black hole is characterized only by its mass, its charge, and its angular momentum; hence, it has no entropy. It appears, then, that one can violate the second law of thermodynamics by dropping an object with nonzero entropy into a black hole.[52] Work by Stephen Hawking and Jacob Bekenstein showed that one can preserve the second law of thermodynamics by assigning to each black hole a *black-hole entropy*

$$S_{\mathrm{BH}} = \frac{k_{\mathrm{B}} A}{4 \ell_{\mathrm{P}}^2},$$

where A is the area of the hole's event horizon, k_{B} is the Boltzmann constant, and $\ell_{\mathrm{P}} = \sqrt{G\hbar/c^3}$ is the Planck length.[53] The fact that the black hole entropy is also the maximal entropy that can be obtained by the Bekenstein bound (wherein the Bekenstein bound becomes an equality) was the main observation that led to the holographic principle.[52]

An oversight in the application of the no-hair theorem is the assumption that the relevant degrees of freedom accounting for the entropy of the black hole must be classical in nature; what if they were purely quantum mechanical instead and had non-zero entropy? Actually, this is what is realized in the LQG derivation of black hole entropy, and can be seen as a consequence of its background-independence – the classical black hole spacetime comes about from the semi-classical limit of the quantum state of the gravitational field, but there are many quantum states that have the same semiclassical limit. Specifically, in LQG[54] it is possible to associate a quantum geometrical interpretation to the microstates: These are the quantum geometries of the horizon which are consistent with the area, A, of the black hole and the topology of the horizon (i.e. spherical). LQG offers a geometric explanation of the finiteness of the entropy and of the proportionality of the area of the horizon.[55][56] These calculations have been generalized to rotating black holes.[57]

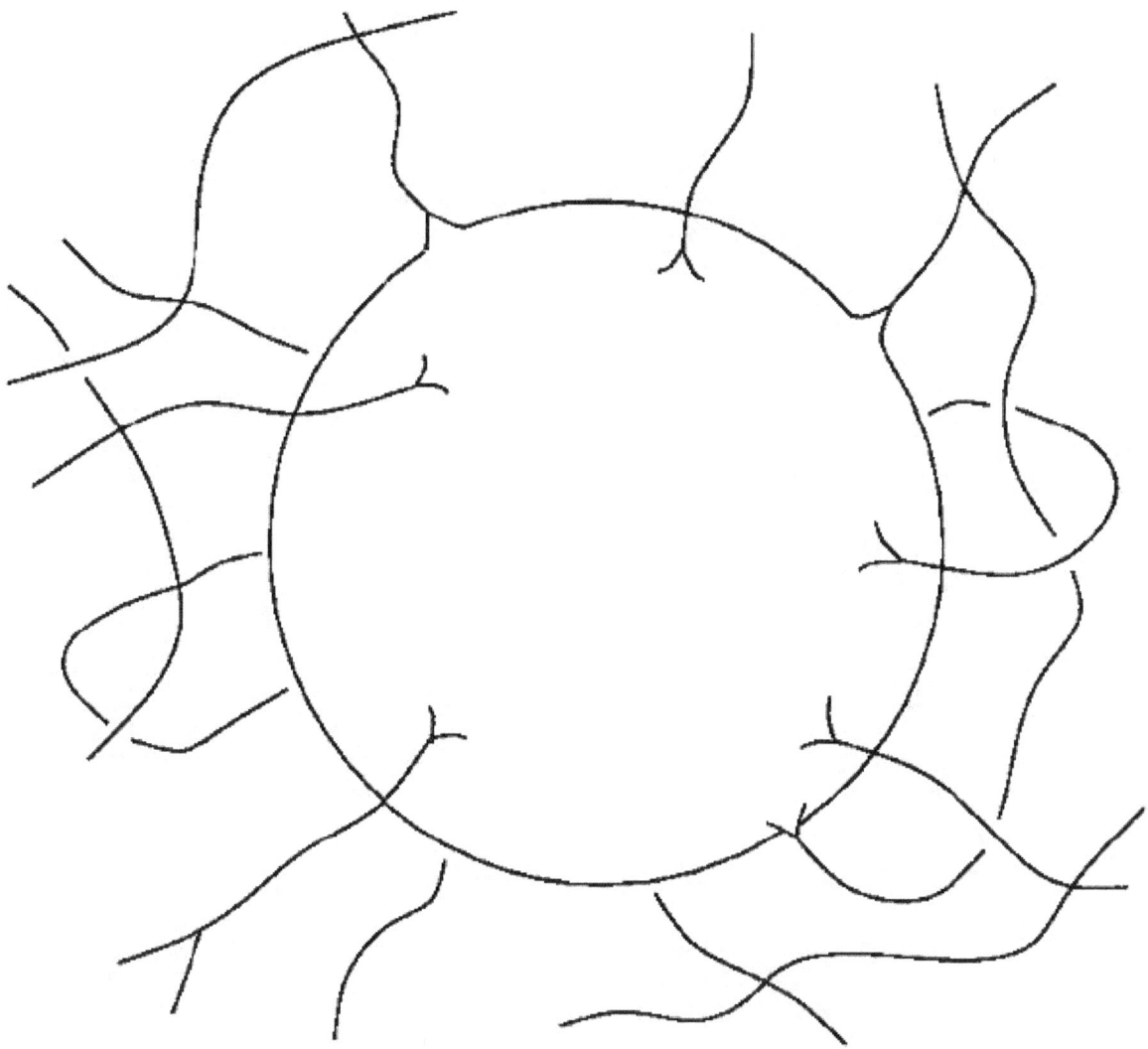

Representation of quantum geometries of the horizon. Polymer excitations in the bulk puncture the horizon, endowing it with quantized area. Intrinsically the horizon is flat except at punctures where it acquires a quantized deficit angle or quantized amount of curvature. These deficit angles add up to 4π .

It is possible to derive, from the covariant formulation of full quantum theory (Spinfoam) the correct relation between energy and area (1st law), the Unruh temperature and the distribution that yields Hawking entropy.[58] The calculation makes use of the notion of dynamical horizon and is done for non-extremal black holes.

A recent success of the theory in this direction is the computation of the entropy of all non singular black holes directly from theory and independent of Immirzi parameter.[59] The result is the expected formula $S = A/4$, where S is the entropy and A the area of the black hole, derived by Bekenstein and Hawking on heuristic grounds. This is the only known derivation of this formula from a fundamental theory, for the case of generic non singular black holes. Older attempts at this calculation had difficulties. The problem was that although Loop quantum gravity predicted that the entropy of a black hole is proportional to the area of the event horizon, the result depended on a crucial free parameter in the theory, the above-mentioned Immirzi parameter. However, there is no known computation of the Immirzi parameter, so it had to be fixed by demanding agreement with Bekenstein and Hawking's calculation of the black hole entropy.

32.8.2   Loop quantum cosmology

Main articles: loop quantum cosmology, Big bounce and inflation (cosmology)

The popular and technical literature makes extensive references to LQG-related topic of loop quantum cosmology. LQC was mainly developed by Martin Bojowald, it was popularized Loop quantum cosmology in *Scientific American* for predicting a Big Bounce prior to the Big Bang. Loop quantum cosmology (LQC) is a symmetry-reduced model of classical general relativity quantized using methods that mimic those of loop quantum gravity (LQG) that predicts a "quantum bridge" between contracting and expanding cosmological branches.

Achievements of LQC have been the resolution of the big bang singularity, the prediction of a Big Bounce, and a natural mechanism for inflation (cosmology).

LQC models share features of LQG and so is a useful toy model. However, the results obtained are subject to the usual restriction that a truncated classical theory, then quantized, might not display the true behaviour of the full theory due to artificial suppression of degrees of freedom that might have large quantum fluctuations in the full theory. It has been argued that singularity avoidance in LQC are by mechanisms only available in these restrictive models and that singularity avoidance in the full theory can still be obtained but by a more subtle feature of LQG.[60][61]

32.8.3   Loop quantum gravity phenomenology

Quantum gravity effects are notoriously difficult to measure because the Planck length is so incredibly small. However recently physicists have started to consider the possibility of measuring quantum gravity effects, mostly from astrophysical observations and gravitational wave detectors.The energy of those fluctuations at scales this small cause space-perturbations which are visible at higher scales.

32.8.4   Background independent scattering amplitudes

Loop quantum gravity is formulated in a background-independent language. No spacetime is assumed a priori, but rather it is built up by the states of theory themselves - however scattering amplitudes are derived from n -point functions (Correlation function (quantum field theory)) and these, formulated in conventional quantum field theory, are functions of points of a background space-time. The relation between the background-independent formalism and the conventional formalism of quantum field theory on a given spacetime is far from obvious, and it is far from obvious how to recover low-energy quantities from the full background-independent theory. One would like to derive the n -point functions of the theory from the background-independent formalism, in order to compare them with the standard perturbative expansion of quantum general relativity and therefore check that loop quantum gravity yields the correct low-energy limit.

A strategy for addressing this problem has been suggested:[62] the idea is to study the boundary amplitude, namely a path integral over a finite space-time region, seen as a function of the boundary value of the field.[63] In conventional quantum field theory, this boundary amplitude is well–defined[64][65] and codes the physical information of the theory; it does so in quantum gravity as well, but in a fully background–independent manner.[66] A generally covariant definition of n -point functions can then be based on the idea that the distance between physical points –arguments of the n -point function is determined by the state of the gravitational field on the boundary of the spacetime region considered.

Progress has been made in calculating background independent scattering amplitudes this way with the use of spin foams. This is a way to extract physical information from the theory. Claims to have reproduced the correct behaviour for graviton scattering amplitudes and to have recovered classical gravity have been made. "We have calculated Newton's law starting from a world with no space and no time." - Carlo Rovelli.

32.9   Gravitons, string theory, supersymmetry, extra dimensions in LQG

Main articles: graviton, string theory, supersymmetry, Kaluza–Klein theory and supergravity

Some quantum theories of gravity posit a spin-2 quantum field that is quantized, giving rise to gravitons. In string theory one generally starts with quantized excitations on top of a classically fixed background. This theory is thus described as background dependent. Particles like photons as well as changes in the spacetime geometry (gravitons) are both described as excitations on the string worldsheet. The background dependence of string theory can have important physical consequences, such as determining the number of quark generations. In contrast, loop quantum gravity, like general relativity, is manifestly background independent, eliminating the background required in string theory. Loop quantum gravity, like string theory, also aims to overcome the nonrenormalizable divergences of quantum field theories.

LQG never introduces a background and excitations living on this background, so LQG does not use gravitons as building blocks. Instead one expects that one may recover a kind of semiclassical limit or weak field limit where something like "gravitons" will show up again. In contrast, gravitons play a key role in string theory where they are among the first (massless) level of excitations of a superstring.

LQG differs from string theory in that it is formulated in 3 and 4 dimensions and without supersymmetry or Kaluza-Klein extra dimensions, while the latter requires both to be true. There is no experimental evidence to date that confirms string theory's predictions of supersymmetry and Kaluza–Klein extra dimensions. In a 2003 paper A dialog on quantum gravity,[67] Carlo Rovelli regards the fact LQG is formulated in 4 dimensions and without supersymmetry as a strength of the theory as it represents the most parsimonious explanation, consistent with current experimental results, over its rival string/M-theory. Proponents of string theory will often point to the fact that, among other things, it demonstrably reproduces the established theories of general relativity and quantum field theory in the appropriate limits, which Loop Quantum Gravity has struggled to do. In that sense string theory's connection to established physics may be considered more reliable and less speculative, at the mathematical level. Peter Woit in Not Even Wrong and Lee Smolin in The Trouble with Physics regard string/M-theory to be in conflict with current known experimental results.

Since LQG has been formulated in 4 dimensions (with and without supersymmetry), and M-theory requires supersymmetry and 11 dimensions, a direct comparison between the two has not been possible. It is possible to extend mainstream LQG formalism to higher-dimensional supergravity, general relativity with supersymmetry and Kaluza–Klein extra dimensions should experimental evidence establish their existence. It would therefore be desirable to have higher-dimensional Supergravity loop quantizations at one's disposal in order to compare these approaches. In fact a series of recent papers have been published attempting just this.[68][69][70][71][72][73][74][75] Most recently, Thiemann (and alumni) have made progress toward calculating black hole entropy for supergravity in higher dimensions. It will be interesting to compare these results to the corresponding super string calculations.[76][77]

As of April 2013 LHC has failed to find evidence of supersymmetry or Kaluza–Klein extra dimensions, which has encouraged LQG researchers. Shaposhnikov in his paper "Is there a new physics between electroweak and Planck scales?" has proposed the neutrino minimal standard model,[78] which claims the most parsimonious theory is a standard model extended with neutrinos, plus gravity, and that extra dimensions, GUT physics, and supersymmetry, string/M-theory physics are unrealized in nature, and that any theory of quantum gravity must be four dimensional, like loop quantum gravity.

32.10   LQG and related research programs

Main articles: noncommutative geometry, twistor theory, entropic gravity, Sundance Bilson-Thompson, Asymptotic safety in quantum gravity, Causal dynamical triangulation, group field theory and consistent discretizations

Several research groups have attempted to combine LQG with other research programs: Johannes Aastrup, Jesper M. Grimstrup et al. research combines noncommutative geometry with loop quantum gravity,[79] Laurent Freidel, Simone Speziale, et al., spinors and twistor theory with loop quantum gravity,[80] and Lee Smolin et al. with Verlinde entropic gravity and loop gravity.[81] Stephon Alexander, Antonino Marciano and Lee Smolin have attempted to explain the origins of weak force chirality in terms of Ashketar's variables, which describe gravity as chiral,[82] and LQG with Yang–Mills theory fields[83] in four dimensions. Sundance Bilson-Thompson, Hackett et al.,[84][85] has attempted to introduce standard model via LQG"s degrees of freedom as an emergent property (by employing the idea noiseless subsystems a useful notion introduced in more general situation for constrained systems by Fotini Markopoulou-Kalamara et al.[86]) LQG has also drawn philosophical comparisons with causal dynamical triangulation[87] and asymptotically safe gravity,[88] and the spinfoam with group field theory and AdS/CFT correspondence.[89] Smolin and Wen have suggested combining LQG with String-net liquid, tensors, and Smolin and Fotini Markopoulou-Kalamara Quantum Graphity. There is the consistent discretizations approach. In addition to what has already mentioned above, Pullin and Gambini provide a framework to connect the path integral and canonical approaches to quantum gravity. They may help reconcile the spin foam and canonical loop representation approaches. Recent research by Chris Duston and Matilde Marcolli introduces topology change via topspin networks.[90]

32.11   Problems and comparisons with alternative approaches

Main article: List of unsolved problems in physics

Some of the major unsolved problems in physics are theoretical, meaning that existing theories seem incapable of explaining a certain observed phenomenon or experimental result. The others are experimental, meaning that there is a difficulty in creating an experiment to test a proposed theory or investigate a phenomenon in greater detail.

Can quantum mechanics and general relativity be realized as a fully consistent theory (perhaps as a quantum field theory)? Is spacetime fundamentally continuous or discrete? Would a consistent theory involve a force mediated by a hypothetical graviton, or be a product of a discrete structure of spacetime itself (as in loop quantum gravity)? Are there deviations from the predictions of general relativity at very small or very large scales or in other extreme circumstances that flow from a quantum gravity theory?

The theory of LQG is one possible solution to the problem of quantum gravity, as is string theory. There are substantial differences however. For example, string theory also addresses unification, the understanding of all known forces and particles as manifestations of a single entity, by postulating extra dimensions and so-far unobserved additional particles and symmetries. Contrary to this, LQG is based only on quantum theory and general relativity and its scope is limited to understanding the quantum aspects of the gravitational interaction. On the other hand, the consequences of LQG are radical, because they fundamentally change the nature of space and time and provide a tentative but detailed physical and mathematical picture of quantum spacetime.

Presently, no semiclassical limit recovering general relativity has been shown to exist. This means it remains unproven that LQG's description of spacetime at the Planck scale has the right continuum limit (described by general relativity with possible quantum corrections). Specifically, the dynamics of the theory is encoded in the Hamiltonian constraint, but there is no candidate Hamiltonian.[91] Other technical problems include finding off-shell closure of the constraint algebra and physical inner product vector space, coupling to matter fields of Quantum field theory, fate of the renormalization of the graviton in perturbation theory that lead to ultraviolet divergence beyond 2-loops (see One-loop Feynman diagram in Feynman diagram).[91]

While there has been a recent proposal relating to observation of naked singularities,[92] and doubly special relativity as a part of a program called loop quantum cosmology, there is no experimental observation for which loop quantum gravity makes a prediction not made by the Standard Model or general relativity (a problem that plagues all current theories of quantum gravity). Because of the above-mentioned lack of a semiclassical limit, LQG has not yet even reproduced the predictions made by general relativity.

An alternative criticism is that general relativity may be an effective field theory, and therefore quantization ignores the fundamental degrees of freedom.

32.12   See also

32.13   Notes

[1] Rovelli, Carlo (August 2008). "Loop Quantum Gravity" (PDF). *CERN*. Retrieved 14 September 2014.

[2] Rovelli, C. (2011). "Zakopane lectures on loop gravity". arXiv:1102.3660 [gr-qc].

[3] Muxin, H. (2011). "Cosmological constant in loop quantum gravity vertex amplitude". *Physical Review D* **84** (6): 064010. arXiv:1105.2212. Bibcode:2011PhRvD..84f4010H. doi:10.1103/PhysRevD.84.064010.

[4] Fairbairn, W. J.; Meusburger, C. (2011). "q-Deformation of Lorentzian spin foam models". arXiv:1112.2511 [gr-qc].

[5] Rovelli, C. (2004). *Quantum Gravity*. Cambridge Monographs on Mathematical Physics. p. 71. ISBN 978-0-521-83733-0.

[6] Kauffman, S.; Smolin, L. (7 April 1997). "A Possible Solution For The Problem Of Time In Quantum Cosmology". *Edge.org*. Retrieved 2014-08-20.

[7] Smolin, L. (2006). "The Case for Background Independence". In Rickles, D.; French, S.; Saatsi, J. T. *The Structural Foundations of Quantum Gravity*. Clarendon Press. pp. 196*ff*. arXiv:hep-th/0507235. ISBN 978-0-19-926969-3.

[8] Rovelli, C. (2004). *Quantum Gravity*. Cambridge Monographs on Mathematical Physics. p. 13ff. ISBN 978-0-521-83733-0.

[9] Thiemann, T. (1996). "Anomaly-free formulation of non-perturbative, four-dimensional Lorentzian quantum gravity". *Physics Letters B* **380**: 257–264. arXiv:gr-qc/9606088. Bibcode:1996PhLB..380..257T. doi:10.1016/0370-2693(96)00532-1.

[10] Baez, J.; de Muniain, J. P. (1994). *Gauge Fields, Knots and Quantum Gravity*. Series on Knots and Everything. Vol. 4. World Scientific. Part III, chapter 4. ISBN 978-981-02-1729-7.

[11] Thiemann, T. (2003). "Lectures on Loop Quantum Gravity". *Lecture Notes in Physics* **631**: 41–135. arXiv:gr-qc/0210094. Bibcode:2003LNP...631...41T. doi:10.1007/978-3-540-45230-0_3.

[12] Rovelli, C.;Smolin, L.(1988). "Knot Theory and Quantum Gravity".*Physical Review Letters***61**(10): 1155–1958.Bibcode doi:10.1103/PhysRevLett.61.1155.

[13] Gambini, R.; Pullin, J. (2011). *A First Course in Loop Quantum Gravity*. Oxford University Press. Section 8.2. ISBN 978-0-19-959075-9.

[14] Fernando, J.; Barbero, G. (1995). "Reality Conditions and Ashtekar Variables: A Different Perspective". *Physical Review D* **51**: 5498–5506. arXiv:gr-qc/9410013. Bibcode:1995PhRvD..51.5498B. doi:10.1103/PhysRevD.51.5498.

[15] Fernando, J.; Barbero, G. (1995). "Real Ashtekar Variables for Lorentzian Signature Space-times". *Physical Review D* **51**: 5507–5520. arXiv:gr-qc/9410014. Bibcode:1995PhRvD..51.5507B. doi:10.1103/PhysRevD.51.5507.

[16] Bojowald, M.; Alejandro, P. "Spin Foam Quantization and Anomalies". arXiv:gr-qc/0303026 [gr-qc].

[17] Barrett, J.; Crane, L. (2000). "A Lorentzian signature model for quantum general relativity". *Classical and Quantum Gravity* **17**: 3101–3118. arXiv:gr-qc/9904025. Bibcode:2000CQGra..17.3101B. doi:10.1088/0264-9381/17/16/302..

[18] Rovelli, C.; Alesci, E. (2007). "The complete LQG propagator I. Difficulties with the Barrett–Crane vertex". *Physical Review D* **76**: 104012. arXiv:hep-th/0703074. Bibcode:2007PhRvD..76b4012B. doi:10.1103/PhysRevD.76.024012.

[19] Engle, J.; Pereira, R.; Rovelli, C. (2009). "Loop-Quantum-Gravity Vertex Amplitude". *Physical Review Letters* **99**: 161301. arXiv:0705.2388. Bibcode:2007PhRvL..99p1301E. doi:10.1103/physrevlett.99.161301.

[20] Freidal, L.; Krasnov, K. (2008). "A new spin foam model for 4D gravity". *Classical and Quantum Gravity* **25**: 125018. arXiv:0708.1595. Bibcode:2008CQGra..25l5018F. doi:10.1088/0264-9381/25/12/125018.

[21] Alesci, E.; Thiemann, T.; Zipfel, A. (2011). "Linking covariant and canonical LQG: new solutions to the Euclidean Scalar Constraint". arXiv:1109.1290.

[22] Bohm, D. (1989). *Quantum Theory*. Dover Publications. ISBN 978-0-486-65969-5.

[23] Tipler, P.; Llewellyn, R. (2008). *Modern Physics* (5th ed.). W. H. Freeman and Co. pp. 160–161. ISBN 978-0-7167-7550-8.

[24] Bohr, N. (1920). "Über die Serienspektra der Element". *Zeitschrift für Physik* **2** (5): 423–478. Bibcode:1920ZPhy....2..423B. doi:10.1007/BF01329978. (English translation in Bohr 1976, pp. 241–282)

[25] Jammer, M. (1989). *The Conceptual Development of Quantum Mechanics* (2nd ed.). Tomash Publishers. Section 3.2. ISBN 978-0-88318-617-6.

[26] Ashtekar, A.; Bombelli, L.; Corichi, A. (2005). "Semiclassical States for Constrained Systems". *Physical Review D* **72**: 025008. arXiv:hep-ph/0504114. Bibcode:2005PhRvD..72a5008C. doi:10.1103/PhysRevD.72.015008.

[27] Lewandowski, J.; Okołów, A.; Sahlmann, H.; Thiemann, T. (2005). "Uniqueness of Diffeomorphism Invariant States on Holonomy-Flux Algebras". *Communications in Mathematical Physics* **267**: 703–733. arXiv:gr-qc/0504147. Bibcode:2006CMaPh.267..703L.doi:10.1007/s00220-006-0100-7.

[28] Fleischhack, C. (2006). "Irreducibility of the Weyl algebra in loop quantum gravity". *Physical Review Letters* **97**: 061302. Bibcode:2006PhRvL..97f1302F. doi:10.1103/physrevlett.97.061302.

[29] Thiemann, T. (2008). *Modern Canonical General Relativity*. Cambridge Monographs on Mathematical Physics. Cambridge University Press. Section 10.6. ISBN 978-0-521-74187-3.

[30] "Partial and Complete Observables for Hamiltonian Constrained Systems". *General Relativity and Gravitation* **39**: 1891–1927. 2007. arXiv:gr-qc/0411013. Bibcode:2007GReGr..39.1891D. doi:10.1007/s10714-007-0495-2.

[31] "Partial and Complete Observables for Canonical General Relativity". *Classical and Quantum Gravity* **23**: 6155–6184. arXiv:gr-qc/0507106. Bibcode:2006CQGra..23.6155D. doi:10.1088/0264-9381/23/22/006.

[32] Dreyer, O.; Markopoulou, f.; Smolin, L. (2006). "Symmetry and entropy of black hole horizons". *Nuclear Physics B* **774**: 1–13. arXiv:hep-th/0409056. Bibcode:2006NuPhB.744....1D. doi:10.1016/j.nuclphysb.2006.02.045.

[33] Kribs, D. W.; Markopoulou, F. "Geometry from quantum particles". arXiv:gr-qc/0510052.

[34] Markopoulou, F.; Poulin, D. "Noiseless subsystems and the low energy limit of spin foam models" (unpublished).

[35] *The Phoenix Project: Master Constraint Programme for Loop Quantum Gravity*, Class.Quant.Grav.23:2211-2248,2006 or http://fr.arxiv.org/pdf/gr-qc/0305080

[36] *Modern Canonical Quantum General Relativity* by Thomas Thiemann

[37] *Testing the Master Constraint Programme for Loop Quantum Gravity I. General Framework*, Bianca Dittrich, Thomas Thiemann, Class.Quant.Grav. 23 (2006) 1025-1066.

[38] *Testing the Master Constraint Programme for Loop Quantum Gravity II. Finite Dimensional Systems*, Bianca Dittrich, Thomas Thiemann, Class.Quant.Grav. 23 (2006) 1067-1088.

[39] *Testing the Master Constraint Programme for Loop Quantum Gravity III. SL(2,R) Models*, Bianca Dittrich, Thomas Thiemann, Class.Quant.Grav. 23 (2006) 1089-1120.

[40] *Testing the Master Constraint Programme for Loop Quantum Gravity IV. Free Field Theories*, Bianca Dittrich, Thomas Thiemann, Class.Quant.Grav. 23 (2006) 1121-1142.

[41] *Testing the Master Constraint Programme for Loop Quantum Gravity V. Interacting Field Theories*, Bianca Dittrich, Thomas Thiemann, Class.Quant.Grav. 23 (2006) 1143-1162.

[42] *Quantum Spin Dynamics VIII. The Master Constraint*, Thomas Thiemann. Class.Quant.Grav. 23 (2006) 2249-2266.

[43] *Approximating the physical inner product of Loop Quantum Cosmology*, Benjamin Bahr, Thomas Thiemann, Class.Quant.Grav.24: 2138,2007.

[44] *On the Relation between Operator Constraint --, Master Constraint --, Reduced Phase Space --, and Path Integral Quantisation*, Muxin Han, Thomas Thiemann, Class.Quant.Grav.27:225019,2010.

[45] *On the Relation between Rigging Inner Product and Master Constraint Direct Integral Decomposition*, Muxin Han, Thomas Thiemann, J.Math.Phys.51:092501,2010.

[46] *A Path-integral for the Master Constraint of Loop Quantum Gravity*, Muxin Han, Class.Quant.Grav.27:215009,2010

[47] *Emergent diffeomorphism invariance in a discrete loop quantum gravity model*, Rodolfo Gambini, Jorge Pullin, Class.Quant.Grav.

[48] Section 10.2.2 *A First Course in Loop quantum Gravity*, Rodolfo Gambinni, Jorge Pullin, Oxford University Press, first published 2011.

[49] *Algebraic Quantum Gravity (AQG) I. Conceptual Setup*, K. Giesel, T. Thiemann, Class.Quant.Grav.24:2465-2498,2007.

[50] *Algebraic Quantum Gravity (AQG) II. Semiclassical Analysis*, K. Giesel, T. Thiemann, Class.Quant.Grav.24:2499-2564,2007.

[51] *Algebraic Quantum Gravity (AQG) III. Semiclassical Perturbation Theory*, K. Giesel, T. Thiemann, Class.Quant.Grav.24:2565-2588,2007.

[52] Bousso, Raphael (2002). "The Holographic Principle". *Reviews of Modern Physics* **74** (3): 825–874. arXiv:hep-th/0203101. Bibcode:2002RvMP...74..825B. doi:10.1103/RevModPhys.74.825.

[53] Majumdar, Parthasarathi (1998). "Black Hole Entropy and Quantum Gravity"**73**. p. 147.arXiv:gr-qc/9807045.Bibcode:199

[54] See List of loop quantum gravity researchers

[55] Rovelli, Carlo (1996). "Black Hole Entropy from Loop Quantum Gravity". *Physical Review Letters* **77** (16): 3288–3291. arXiv:gr-qc/9603063. Bibcode:1996PhRvL..77.3288R. doi:10.1103/PhysRevLett.77.3288.

[56] Ashtekar, Abhay; Baez, John; Corichi, Alejandro; Krasnov, Kirill (1998). "Quantum Geometry and Black Hole Entropy". *Physical Review Letters* **80** (5): 904–907. arXiv:gr-qc/9710007. Bibcode:1998PhRvL..80..904A. doi:10.1103/PhysRevLett.80.904.

[57] *Quantum horizons and black hole entropy: Inclusion of distortion and rotation*, Abhay Ashtekar, Jonathan Engle, Chris Van Den Broeck, Class.Quant.Grav.22:L27-L34, 2005.

[58] Bianchi, Eugenio (2012). "Entropy of Non-Extremal Black Holes from Loop Gravity". arXiv:1204.5122.

[59] http://inspirehep.net/record/940357?ln=en. http://inspirehep.net/record/1111991.

[60] *On (Cosmological) Singularity Avoidance in Loop Quantum Gravity*, Johannes Brunnemann, Thomas Thiemann, Class.Quant.Grav. 23 (2006) 1395-1428.

[61] *Unboundedness of Triad -- Like Operators in Loop Quantum Gravity*, Johannes Brunnemann, Thomas Thiemann, Class.Quant.Grav. 23 (2006) 1429-1484.

[62] L. Modesto, C. Rovelli:*Particle scattering in loop quantum gravity*, Phys Rev Lett 95 (2005) 191301

[63] R Oeckl, *A 'general boundary' formulation for quantum mechanics and quantum gravity*, Phys Lett B575 (2003) 318-324 ; *Schrodinger's cat and the clock: lessons for quantum gravity*, Class Quant Grav 20 (2003) 5371-5380l

[64] F. Conrady, C. Rovelli *Generalized Schrodinger equation in Euclidean field theory*", Int J Mod Phys A 19, (2004) 1-32.

[65] L. Doplicher, *Generalized Tomonaga-Schwinger equation from the Hadamard formula*, Phys Rev D70 (2004) 064037

[66] F. Conrady, L. Doplicher, R. Oeckl, C. Rovelli, M. Testa, *Minkowski vacuum in background independent quantum gravity*, Phys Rev D69 (2004) 064019.

[67] http://arxiv.org/abs/arXiv:hep-th/0310077

[68] *New Variables for Classical and Quantum Gravity in all Dimensions I. Hamiltonian Analysis*, Norbert Bodendorfer, Thomas Thiemann, Andreas Thurn, Class. Quantum Grav. 30 (2013) 045001

[69] *New Variables for Classical and Quantum Gravity in all Dimensions II. Lagrangian Analysis*, Norbert Bodendorfer, Thomas Thiemann, Andreas Thurn, Quantum Grav. 30 (2013) 045002

[70] *New Variables for Classical and Quantum Gravity in all Dimensions III. Quantum Theory*, Norbert Bodendorfer, Thomas Thiemann, Andreas Thurn, Class. Quantum Grav. 30 (2013) 045003

[71] *New Variables for Classical and Quantum Gravity in all Dimensions IV. Matter Coupling*, Norbert Bodendorfer, Thomas Thiemann, Andreas Thurn, Class. Quantum Grav. 30 (2013) 045004

[72] *On the Implementation of the Canonical Quantum Simplicity Constraint*, Norbert Bodendorfer, Thomas Thiemann, Andreas Thurn, Class. Quantum Grav. 30 (2013) 045005

[73] *Towards Loop Quantum Supergravity (LQSG) I. Rarita-Schwinger Sector*, Norbert Bodendorfer, Thomas Thiemann, Andreas Thurn, Class. Quantum Grav. 30 (2013) 045006

[74] *Towards Loop Quantum Supergravity (LQSG) II. p-Form Sector*, Norbert Bodendorfer, Thomas Thiemann, Andreas Thurn, Class. Quantum Grav. 30 (2013) 045007

[75] *Towards Loop Quantum Supergravity (LQSG)*, Norbert Bodendorfer, Thomas Thiemann, Andreas Thurn, Phys. Lett. B 711: 205-211 (2012)

[76] *New Variables for Classical and Quantum Gravity in all Dimensions V. Isolated Horizon Boundary Degrees of Freedom*, Norbert Bodendorfer, Thomas Thiemann, Andreas Thurn, http://uk.arxiv.org/pdf/1304.2679.

[77] *Black hole entropy from loop quantum gravity in higher dimensions*, Norbert Bodendorfer http://uk.arxiv.org/pdf/1307.5029

[78] http://arxiv.org/abs/0708.3550

[79] http://arxiv.org/abs/1203.6164

[80] http://arxiv.org/abs/1006.0199

[81] http://arxiv.org/abs/1001.3668

[82] http://arxiv.org/abs/1212.5246

[83] http://arxiv.org/abs/1105.3480

[84] *Quantum gravity and the standard model*, Sundance O. Bilson-Thompson, Fotini Markopoulou, Lee Smolin, Class.Quant.Grav.24: 3994,2007.

[85] For a precise review and outlook of this research see: *Emergent Braided Matter of Quantum Geometry*, Sundance Bilson-Thompson, Jonathan Hackett, Louis Kauffman, Yidun Wan, SIGMA 8 (2012), 014, 43 pages.

[86] *Constrained Mechanics and Noiseless Subsystems*, Tomasz Konopka, Fotini Markopoulou, arXiv:gr-qc/0601028.

[87] http://www.perimeterinstitute.ca/people/renate-loll

[88] wwnpqft.inln.cnrs.fr/pdf/Bianchi.pdf

[89] http://arxiv.org/abs/0804.0632

[90] http://arxiv.org/abs/1308.2934

[91] Nicolai, Hermann; Peeters, Kasper; Zamaklar, Marija (2005). "Loop quantum gravity: an outside view". *Classical and Quantum Gravity* **22** (19): R193–R247. arXiv:hep-th/0501114. Bibcode:2005CQGra..22R.193N. doi:10.1088/0264-9381/22/19/R01.

[92] Goswami; Joshi, Pankaj S.; Singh, Parampreet; et al. (2006). "Quantum evaporation of a naked singularity". *Physical Review Letters* **96** (3): 31302. arXiv:gr-qc/0506129. Bibcode:2006PhRvL..96c1302G. doi:10.1103/PhysRevLett.96.031302.

32.14 References

- Topical Reviews

    - Rovelli, Carlo (2011). "Zakopane lectures on loop gravity". arXiv:1102.3660.

    - Rovelli, Carlo (1998). "Loop Quantum Gravity". *Living Reviews in Relativity* **1**. Retrieved 2008-03-13.

    - Thiemann, Thomas (2003). "Lectures on Loop Quantum Gravity". *Lectures Notes in Physics*. Lecture Notes in Physics **631**: 41–135. arXiv:gr-qc/0210094. Bibcode:2003LNP...631...41T. doi:10.1007/978-3-540-45230-0_3. ISBN 978-3-540-40810-9.

    - Ashtekar, Abhay; Lewandowski, Jerzy (2004). "Background Independent Quantum Gravity: A Status Report". *Classical and Quantum Gravity* **21** (15): R53–R152. arXiv:gr-qc/0404018. Bibcode:2004CQGra..21R ..53A.doi:10.1088/0264-9381/21/15/R01.

    - Carlo Rovelli and Marcus Gaul, *Loop Quantum Gravity and the Meaning of Diffeomorphism Invariance*, e-print available as gr-qc/9910079.

    - Lee Smolin, *The case for background independence*, e-print available as hep-th/0507235.

    - Alejandro Corichi, *Loop Quantum Geometry: A primer*, e-print available as .

    - Alejandro Perez, *Introduction to loop quantum gravity and spin foams*, e-print available as .

    - Hermann Nicolai and Kasper Peeters *Loop and spin foam quantum gravity: A Brief guide for beginners.*, e-print available as .

- Popular books:

  - Lee Smolin, *Three Roads to Quantum Gravity*

  - Carlo Rovelli, *Che cos'è il tempo? Che cos'è lo spazio?*, Di Renzo Editore, Roma, 2004. French translation: *Qu'est ce que le temps? Qu'est ce que l'espace?*, Bernard Gilson ed, Brussel, 2006. English translation: *What is Time? What is space?*, Di Renzo Editore, Roma, 2006.

  - Julian Barbour, *The End of Time: The Next Revolution in Our Understanding of the Universe*

  - Musser, George (2008). "The Complete Idiot's Guide to String Theory". *The Physics Teacher* (Indianapolis: Alpha) **47** (2): 368. Bibcode:2009PhTea..47Q.128H. doi:10.1119/1.3072469. ISBN 978-1-59257-702-6. – Focuses on string theory but has an extended discussion of loop gravity as well.

- Magazine articles:

  - Lee Smolin, "Atoms of Space and Time", *Scientific American*, January 2004

  - Martin Bojowald, "Following the Bouncing Universe", *Scientific American*, October 2008

- Easier introductory, expository or critical works:

  - Abhay Ashtekar, *Gravity and the quantum*, e-print available as gr-qc/0410054 (2004)

  - John C. Baez and Javier Perez de Muniain, *Gauge Fields, Knots and Quantum Gravity*, World Scientific (1994)

  - Carlo Rovelli, *A Dialog on Quantum Gravity*, e-print available as hep-th/0310077 (2003)

  - Rodolfo Gambini and Jorge Pullin, *A First Course in Loop Quantum Gravity*, Oxford (2011)

  - Carlo Rovelli and Francesca Vidotto, *Covariant Loop Quantum Gravity*, Cambridge (2014); draft available online

- More advanced introductory/expository works:

  - Carlo Rovelli, *Quantum Gravity*, Cambridge University Press (2004); draft available online

  - Thomas Thiemann, *Introduction to modern canonical quantum general relativity*, e-print available as gr-qc/0110034

  - Thomas Thiemann, *Introduction to Modern Canonical Quantum General Relativity*, Cambridge University Press (2007)

  - Abhay Ashtekar, *New Perspectives in Canonical Gravity*, Bibliopolis (1988).

  - Abhay Ashtekar, *Lectures on Non-Perturbative Canonical Gravity*, World Scientific (1991)

  - Rodolfo Gambini and Jorge Pullin, *Loops, Knots, Gauge Theories and Quantum Gravity*, Cambridge University Press (1996)

  - Hermann Nicolai, Kasper Peeters, Marija Zamaklar, *Loop quantum gravity: an outside view*, e-print available as hep-th/0501114

  - H. Nicolai and K. Peeters, *Loop and Spin Foam Quantum Gravity: A Brief Guide for Beginners*, e-print available as hep-th/0601129

  - T. Thiemann The LQG – String: Loop Quantum Gravity Quantization of String Theory (2004)

- Conference proceedings:

  - John C. Baez (ed.), *Knots and Quantum Gravity*

- Fundamental research papers:

  - Ashtekar, Abhay (1986). "New variables for classical and quantum gravity". *Physical Review Letters* **57** (18): 2244–2247. Bibcode:1986PhRvL..57.2244A. doi:10.1103/PhysRevLett.57.2244. PMID 10033673

  - Ashtekar, Abhay (1987). "New Hamiltonian formulation of general relativity". *Physical Review D* **36** (6): 1587–1602. Bibcode:1987PhRvD..36.1587A. doi:10.1103/PhysRevD.36.1587

- Roger Penrose, *Angular momentum: an approach to combinatorial space-time* in *Quantum Theory and Beyond*, ed. Ted Bastin, Cambridge University Press, 1971

- Rovelli, Carlo; Smolin, Lee (1988). "Knot theory and quantum gravity". *Physical Review Letters* **61** (10): 1155–1158. Bibcode:1988PhRvL..61.1155R. doi:10.1103/PhysRevLett.61.1155.

- Rovelli, Carlo; Smolin, Lee (1990). "Loop space representation of quantum general relativity". *Nuclear Physics* **B331**: 80–152.

- Carlo Rovelli and Lee Smolin, *Discreteness of area and volume in quantum gravity*, Nucl. Phys., **B442** (1995) 593-622, e-print available as gr-qc/9411005

- Kuchař, Karel (1973). "Canonical Quantization of Gravity". In Israel, Werner. *Relativity, Astrophysics and Cosmology*. D. Reidel. pp. 237–288. ISBN 90-277-0369-8.

- Thiemann, Thomas (2006). "Loop Quantum Gravity: An Inside View". *Approaches to Fundamental Physics*. Lecture Notes in Physics **721**: 185–263. arXiv:hep-th/0608210. Bibcode:2007LNP...721..185T. doi:10.1007/978-3-540-71117-9_10.ISBN978-3-540-71115-5.

32.15 External links

- "Loop Quantum Gravity" by Carlo Rovelli Physics World, November 2003

- Quantum Foam and Loop Quantum Gravity

- Abhay Ashtekar: Semi-Popular Articles . Some excellent popular articles suitable for beginners about space, time, GR, and LQG.

- Loop Quantum Gravity: Lee Smolin.

- Loop Quantum Gravity on arxiv.org

- A list of LQG references catered to fresh graduates

- Loop Quantum Gravity Lectures Online by Lee Smolin

- Spin networks, spin foams and loop quantum gravity

- Wired magazine, News: *Moving Beyond String Theory*

- April 2006 Scientific American Special Issue, *A Matter of Time*, has Lee Smolin LQG Article *Atoms of Space and Time*

- September 2006, The Economist, article *Looping the loop*

- Gamma-ray Large Area Space Telescope: http://glast.gsfc.nasa.gov/

- Zeno meets modern science. Article from Acta Physica Polonica B by Z.K. Silagadze.

- Did pre-big bang universe leave its mark on the sky? - According to a model based on "loop quantum gravity" theory, a parent universe that existed before ours may have left an imprint (*New Scientist*, 10 April 2008)

Chapter 33

Lovelock theory of gravity

In theoretical physics, **Lovelock's theory of gravity** (often referred to as **Lovelock gravity**) is a generalization of Einstein's theory of general relativity introduced by David Lovelock in 1971.[1] It is the most general metric theory of gravity yielding conserved second order equations of motion in arbitrary number of spacetime dimensions D. In this sense, Lovelock's theory is the natural generalization of Einstein's General Relativity to higher dimensions. In three and four dimensions ($D = 3, 4$), Lovelock's theory coincides with Einstein's theory, but in higher dimensions the theories are different. In fact, for $D > 4$ Einstein gravity can be thought of as a particular case of Lovelock gravity since the Einstein–Hilbert action is one of several terms that constitute the Lovelock action.

33.1 Lagrangian density

The Lagrangian of the theory is given by a sum of dimensionally extended Euler densities, and it can be written as follows

$$\mathcal{L} = \sqrt{-g} \sum_{n=0}^{t} \alpha_n \, \mathcal{R}^n, \qquad \mathcal{R}^n = \frac{1}{2^n} \delta^{\mu_1 \nu_1 \ldots \mu_n \nu_n}_{\alpha_1 \beta_1 \ldots \alpha_n \beta_n} \prod_{r=1}^{n} R^{\alpha_r \beta_r}_{\mu_r \nu_r}$$

where $R_{\mu\nu}{}^{ij}$ represents the Riemann tensor, and where the generalized Kronecker delta δ is defined as the antisymmetric product

$$\delta^{\mu_1 \nu_1 \ldots \mu_n \nu_n}_{\alpha_1 \beta_1 \ldots \alpha_n \beta_n} = \frac{1}{n!} \delta^{\mu_1}_{[\alpha_1} \delta^{\nu_1}_{\beta_1} \cdots \delta^{\mu_n}_{\alpha_n} \delta^{\nu_n}_{\beta_n]}.$$

Each term $\mathcal{R}^n$ in $\mathcal{L}$ corresponds to the dimensional extension of the Euler density in $2n$ dimensions, so that these only contribute to the equations of motion for $n < D/2$. Consequently, without lack of generality, t in the equation above can be taken to be $D = 2t + 2$ for even dimensions and $D = 2t + 1$ for odd dimensions.

33.2 Coupling constants

The coupling constants αn in the Lagrangian $\mathcal{L}$ have dimensions of [length]$^{2n - D}$, although it is usual to normalize the Lagrangian density in units of the Planck scale

$$\alpha_1 = (16\pi G)^{-1} = l_P^{2-D}.$$

Expanding the product in $\mathcal{L}$, the Lovelock Lagrangian takes the form

$$\mathcal{L} = \sqrt{-g}\left(\alpha_0 + \alpha_1 R + \alpha_2 \left(R^2 + R_{\alpha\beta\mu\nu}R^{\alpha\beta\mu\nu} - 4R_{\mu\nu}R^{\mu\nu}\right) + \alpha_3\mathcal{O}(R^3)\right).$$

where one sees that coupling α_0 corresponds to the cosmological constant Λ, while an with $n \neq 2$ are coupling constants of additional terms that represent ultraviolet corrections to Einstein theory, involving higher order contractions of the Riemann tensor $R\mu\nu^{\alpha\beta}$. In particular, the second order term

$$\mathcal{R}^2 = R^2 + R_{\alpha\beta\mu\nu}R^{\alpha\beta\mu\nu} - 4R_{\mu\nu}R^{\mu\nu}$$

is precisely the quadratic Gauss–Bonnet term, which is the dimensionally extended version of the four-dimensional Euler density.

33.3 Other contexts

Because Lovelock action contains, among others, the quadratic Gauss–Bonnet term (i.e. the four-dimensional Euler characteristic extended to D dimensions), it is usually said that Lovelock theory resembles string theory inspired models of gravity. This is because a quadratic term is present in the low energy effective action of heterotic string theory, and it also appears in six-dimensional Calabi–Yau compactifications of M-theory. In the mid 1980s, a decade after Lovelock proposed his generalization of the Einstein tensor, physicists began to discuss the quadratic Gauss–Bonnet term within the context of string theory, with particular attention to its property of being ghost-free in Minkowski space. The theory is known to be free of ghosts about other exact backgrounds as well, e.g. about one of the branches of the spherically symmetric solution found by Boulware and Deser in 1985. In general, Lovelock's theory represents a very interesting scenario to study how the physics of gravity is corrected at short distance due to the presence of higher order curvature terms in the action, and in the mid-2000s the theory was considered as a testing ground to investigate the effects of introducing higher-curvature terms in the context of AdS/CFT correspondence.

33.4 See also

- f(R) gravity

- Gauss–Bonnet gravity

- Curtright field

33.5 Notes

[1] D. Lovelock, The Einstein tensor and its generalizations, J. Math. Phys. 12 (1971) 498.

33.6 References

- Lovelock, D. (1971). "The Einstein tensor and its generalizations". *Journal of Mathematical Physics* **12** (3): 498–502. Bibcode:1971JMP....12..498L. doi:10.1063/1.1665613.

- Lovelock, D. (1969). "The uniqueness of the Einstein field equations in a four-dimensional space". *Archive for Rational Mechanics and Analysis* **33** (1): 54–70. Bibcode:1969ArRMA..33...54L. doi:10.1007/BF00248156.

- Lovelock, D. (1972). "The four-dimensionality of space and the Einstein tensor". *Journal of Mathematical Physics* **13**: 874–876.

- Lovelock, David; Rund, Hanno (1989), *Tensors, Differential Forms, and Variational Principles*, Dover, ISBN 0-486-65840-6

- Navarro, A.; Navarro, J. (2011). "Lovelock's theorem revisited". *Journal of Mathematical Physics* **61**: 1950–1956.

- B. Zwiebach, Curvature Squared Terms and String Theories, Phys. Lett. B156 (1985) 315.

- D. Boulware and S. Deser, String Generated Gravity Models, Phys. Rev. Lett. 55 (1985) 2656.

Chapter 34

Massive gravity

In theoretical physics, **massive gravity** is a theory of gravity that modifies general relativity by endowing the graviton with a nonzero mass. In the classical theory, this means that gravitational waves obey a massive wave equation and hence travel at speeds below the speed of light.

Massive gravity has a long and winding history, dating back to the 1930s when Wolfgang Pauli and Markus Fierz first developed a theory of a massive spin-2 field propagating on a flat spacetime background. It was later realized in the 1970s that theories of a massive graviton suffered from dangerous pathologies, including a ghost mode and a discontinuity with general relativity in the limit where the graviton mass goes to zero. While solutions to these problems had existed for some time in three spacetime dimensions,[1][2] they were not solved in four dimensions and higher until the work of Claudia de Rham, Gregory Gabadadze, and Andrew Tolley in 2010.

The fact that general relativity is modified at large distances in massive gravity provides a possible explanation for the accelerated expansion of the Universe that does not require any dark energy. Massive gravity and its extensions, such as bimetric gravity,[3] can yield cosmological solutions which do in fact display late-time acceleration in agreement with observations.[4][5][6]

34.1   Linearized massive gravity

At the linear level, one can construct a theory of a massive spin-2 field $h_{\mu\nu}$ propagating on Minkowski space. This can be seen as an extension of linearized gravity in the following way. Linearized gravity is obtained by linearizing general relativity around flat space, $g_{\mu\nu} = \eta_{\mu\nu} + M_{\text{Pl}}^{-1} h_{\mu\nu}$, where $M_{\text{Pl}} = (8\pi G)^{-1/2}$ is the Planck mass with G the gravitational constant. This leads to a kinetic term in the Lagrangian for $h_{\mu\nu}$, which is consistent with diffeomorphism invariance, as well as a coupling to matter of the form

$$h^{\mu\nu} T_{\mu\nu}$$

where $T_{\mu\nu}$ is the stress–energy tensor.

Massive gravity is obtained by adding nonderivative interaction terms for $h_{\mu\nu}$. At the linear level (i.e., second order in $h_{\mu\nu}$), there are only two possible mass terms:

$$\mathcal{L}_{\text{int}} = a h^{\mu\nu} h_{\mu\nu} + b \left(\eta^{\mu\nu} h_{\mu\nu} \right)^2 .$$

Fierz and Pauli[7] showed in 1939 that this only propagates the expected five polarizations of a massive graviton (as compared to two for the massless case) if the coefficients are chosen so that $a = -b$. Any other choice will unlock a sixth, ghostly degree of freedom. A ghost is a mode with a negative kinetic energy. Its Hamiltonian is unbounded from below and it is therefore unstable to decay into particles of arbitrarily large positive and negative energies. The *Fierz-Pauli mass term,*

$$\mathcal{L}_{\text{FP}} = m^2 \left(h^{\mu\nu} h_{\mu\nu} - (\eta^{\mu\nu} h_{\mu\nu})^2 \right)$$

is therefore the unique consistent linear theory of a massive spin-2 field.

34.2   The vDVZ discontinuity and Vainshtein screening

In the 1970s Hendrik van Dam and Martinus J. G. Veltman[8] and, independently, Vladimir E. Zakharov[9] discovered a peculiar property of Fierz-Pauli massive gravity: its predictions do not uniformly reduce to those of general relativity in the limit $m \to 0$. In particular, while at small scales (shorter than the Compton wavelength of the graviton mass), Newton's gravitational law is recovered, the bending of light is only three quarters of the result Albert Einstein obtained in general relativity. This is known as the *vDVZ discontinuity.*

We may understand the smaller light bending as follows. The Fierz-Pauli massive graviton,due to broken diffeomorphism invariance,propagates three extra degrees of freedom compared to the massless graviton of linearized general relativity. These three degrees of freedom package themselves into a vectorfield,which is irrelevant for our purposes,and a scalar field.This scalar mode exerts an extra attraction in the massive case compared to the massless case.Hence,if one wants measurements of the force exerted between nonrelativistic masses to agree,the coupling constant of the massive theory should be smaller than that of the massless theory.But light bending is blind to the scalar sector,because the stress-energy tensor of light is traceless.Hence,provided the two theories agree on the force between nonrelativistic probes,the massive theory would predict a smaller light bending than the massless one.

It was argued by Vainshtein[10] two years later that the vDVZ discontinuity is an artifact of the linear theory, and that the predictions of general relativity are in fact recovered at small scales when one takes into account nonlinear effects, i.e., higher than quadratic terms in $h_{\mu\nu}$. Heuristically speaking, within a region known as the *Vainshtein radius*, fluctuations of the scalar mode become nonlinear, and its higher-order derivative terms become larger than the canonical kinetic term. Canonically normalizing the scalar around this background therefore leads to a heavily-suppressed kinetic term, which damps fluctuations of the scalar within the Vainshtein radius. Because the extra force mediated by the scalar is proportional to (minus) its gradient, this leads to a much smaller extra force than we would have calculated just using the linear Fierz-Pauli theory.

This phenomenon, known as *Vainshtein screening*, is at play not just in massive gravity, but also in related theories of modified gravity such as DGP and certain scalar-tensor theories, where it is crucial for hiding the effects of modified gravity in the solar system. This allows these theories to match terrestrial and solar-system tests of gravity as well as general relativity does, while maintaining large deviations at larger distances. In this way these theories can lead to cosmic acceleration and have observable imprints on the large-scale structure of the Universe without running afoul of other, much more stringent constraints from observations closer to home.

34.3   The Boulware-Deser ghost

Around the same time as the vDVZ discontinuity and Vainshtein mechanism were discovered, David Boulware and Stanley Deser found in 1972 that generic nonlinear extensions of the Fierz-Pauli theory reintroduced the dangerous ghost mode;[11] the tuning $a = -b$ which ensured this mode's absence at quadratic order was, they found, generally broken at cubic and higher orders, reintroducing the ghost at those orders. As a result, this *Boulware-Deser ghost* would be present around, for example, highly inhomogeneous backgrounds.

This is problematic because a linearized theory of gravity, like Fierz-Pauli, is well-defined on its own but cannot interact with matter, as the coupling $h^{\mu\nu} T_{\mu\nu}$ breaks diffeomorphism invariance. This must be remedied by adding new terms at higher and higher orders, *ad infinitum*. For a massless graviton, this process converges and the end result is well-known: one simply arrives at general relativity. This is the meaning of the statement that general relativity is the unique theory (up to conditions on dimensionality, locality, etc.) of a massless spin-2 field.

In order for massive gravity to actually describe gravity, i.e., a massive spin-2 field coupling to matter and thereby mediating the gravitational force, a nonlinear completion must similarly be obtained. The Boulware-Deser ghost presents a serious obstacle to such an endeavor. The vast majority of theories of massive and interacting spin-2 fields will suffer from this ghost and therefore not be viable. In fact, until 2010 it was widely believed that *all* Lorentz-invariant massive gravity theories possessed the Boulware-Deser ghost.[12]

34.4   Ghost-free massive gravity

In 2010 a breakthrough was achieved when de Rham, Gabadadze, and Tolley constructed, order by order, a theory of massive gravity with coefficients tuned to avoid the Boulware-Deser ghost by packaging all ghostly (i.e., higher-derivative) operators into total derivatives which do not contribute to the equations of motion.[13][14] The complete absence of the Boulware-Deser ghost, to all orders and beyond the decoupling limit, was subsequently proven by Fawad Hassan and Rachel Rosen.[15][16]

The action for the ghost-free *de Rham-Gabadadze-Tolley (dRGT) massive gravity* is given by[17]

$$S = \int d^4x \sqrt{-g} \left(-\frac{M_{\mathrm{Pl}}^2}{2} R + m^2 M_{\mathrm{Pl}}^2 \sum_{n=0}^{4} \alpha_n e_n(\mathbb{K}) + \mathcal{L}_{\mathrm{m}}(g, \Phi_i) \right) ,$$

or, equivalently,

$$S = \int d^4x \sqrt{-g} \left(-\frac{M_{\mathrm{Pl}}^2}{2} R + m^2 M_{\mathrm{Pl}}^2 \sum_{n=0}^{4} \beta_n e_n(\mathbb{X}) + \mathcal{L}_{\mathrm{m}}(g, \Phi_i) \right) .$$

The ingredients require some explanation. As in standard general relativity, there is an Einstein-Hilbert kinetic term proportional to the Ricci scalar R and a minimal coupling to the matter Lagrangian $\mathcal{L}_{\mathrm{m}}$, with Φ_i representing all of the matter fields, such as those of the Standard Model. The new piece is a mass term, or interaction potential, constructed carefully to avoid the Boulware-Deser ghost, with an interaction strength m which is (if the nonzero β_i are $\mathcal{O}(1)$) closely related to the mass of the graviton.

The interaction potential is built out of the elementary symmetric polynomials e_n of the eigenvalues of the matrices $\mathbb{K} = \mathbb{I} - \sqrt{g^{-1}f}$ or $\mathbb{X} = \sqrt{g^{-1}f}$, parametrized by dimensionless coupling constants α_i or β_i, respectively. Here $\sqrt{g^{-1}f}$ is the matrix square root of the matrix $g^{-1}f$. Written in index notation, $\mathbb{X}$ is defined by the relation

$$X^\mu{}_\alpha X^\alpha{}_\nu = g^{\mu\alpha} f_{\nu\alpha} .$$

We have introduced a *reference metric* $f_{\mu\nu}$ in order to construct the interaction term. There is a simple reason for this: it is impossible to construct a nontrivial interaction (i.e., nonderivative) term from $g_{\mu\nu}$ alone. The only possibilities are $g^{\mu\alpha} g_{\alpha\nu} = \delta^\mu_\nu$ and $\det g$, both of which lead to a cosmological constant term rather than a bona fide interaction. Physically, $f_{\mu\nu}$ corresponds to the *background metric* around which fluctuations take the Fierz-Pauli form. This means that, for instance, nonlinearly completing the Fierz-Pauli theory around Minkowski space given above will lead to dRGT massive gravity with $f_{\mu\nu} = \eta_{\mu\nu}$, although the proof of absence of the Boulware-Deser ghost holds for general $f_{\mu\nu}$.[18]

In principle, the reference metric must be specified by hand, and therefore there is no single dRGT massive gravity theory, as the theory with a flat reference metric is different from one with a de Sitter reference metric, etc. Alternatively, one can think of $f_{\mu\nu}$ as a constant of the theory, much like m or M_{Pl}. Instead of specifying a reference metric from the start, one can allow it to have its own dynamics. If the kinetic term for $f_{\mu\nu}$ is also Einstein-Hilbert, then the theory remains ghost-free and we are left with a theory of *massive bigravity*,[3] propagating the two degrees of freedom of a massless graviton in addition to the five of a massive one.

In practice it is unnecessary to compute the eigenvalues of $\mathbb{X}$ (or $\mathbb{K}$) in order to obtain the e_n. They can be written directly in terms of $\mathbb{X}$ as

$$e_0(\mathbb{X}) = 1,$$

$$e_1(\mathbb{X}) = [\mathbb{X}].$$

$$e_2(\mathbb{X}) = \frac{1}{2}\left([\mathbb{X}]^2 - [\mathbb{X}^2]\right).$$

$$e_3(\mathbb{X}) = \frac{1}{6}\left([\mathbb{X}]^3 - 3[\mathbb{X}][\mathbb{X}^2] + 2[\mathbb{X}^3]\right).$$

$$e_4(\mathbb{X}) = \det \mathbb{X}.$$

where brackets indicate a trace, $[\mathbb{X}] \equiv X^\mu{}_\mu$. It is the particular antisymmetric combination of terms in each of the e_n which is responsible for rendering the Boulware-Deser ghost nondynamical.

The choice to use $\mathbb{X}$ or $\mathbb{K} = \mathbb{I} - \mathbb{X}$, with $\mathbb{I}$ the identity matrix, is a convention, as in both cases the ghost-free mass term is a linear combination of the elementary symmetric polynomials of the chosen matrix. One can transform from one basis to the other, in which case the coefficients satisfy the relationship[17]

$$\beta_n = (4 - n)! \sum_{i=n}^{4} \frac{(-1)^{i+n}}{(4-i)!(i-n)!}\alpha_i.$$

34.5   Massive gravity in the vielbein language

The presence of a square-root matrix is somewhat awkward and points to an alternative, simpler formulation in terms of vielbeins. Splitting the metrics into vielbeins as

$$g_{\mu\nu} = \eta_{ab}e^a{}_\mu e^b{}_\nu.$$

$$f_{\mu\nu} = \eta_{ab}f^a{}_\mu e^b{}_\nu.$$

and then defining one-forms

$$\mathbf{e}^a = e^a{}_\mu dx^\mu.$$

$$\mathbf{f}^a = f^a{}_\mu dx^\mu.$$

the ghost-free interaction terms above can be written simply as (up to numerical factors)[19]

$$e_0(\mathbb{X}) \propto \epsilon_{abcd}\mathbf{e}^a \wedge \mathbf{e}^b \wedge \mathbf{e}^c \wedge \mathbf{e}^d$$

$$e_1(\mathbb{X}) \propto \epsilon_{abcd}\mathbf{e}^a \wedge \mathbf{e}^b \wedge \mathbf{e}^c \wedge \mathbf{f}^d$$

$$e_2(\mathbb{X}) \propto \epsilon_{abcd}\mathbf{e}^a \wedge \mathbf{e}^b \wedge \mathbf{f}^c \wedge \mathbf{f}^d$$

$$e_3(\mathbb{X}) \propto \epsilon_{abcd}\mathbf{e}^a \wedge \mathbf{f}^b \wedge \mathbf{f}^c \wedge \mathbf{f}^d$$

$$e_4(\mathbb{X}) \propto \epsilon_{abcd}\mathbf{f}^a \wedge \mathbf{f}^b \wedge \mathbf{f}^c \wedge \mathbf{f}^d$$

In terms of vielbeins, rather than metrics, we can therefore see the physical significance of the ghost-free dRGT potential terms quite clearly: they are simply all the different possible combinations of wedge products of the vielbeins of the two metrics.

Note that massive gravity in the metric and vielbein formulations are only equivalent if the symmetry condition

$$(e^{-1})_a{}^\mu f_{b\nu} = (e^{-1})_b{}^\mu f_{a\nu}$$

is satisfied. While this is true for most physical situations, there may be cases, such as when matter couples to both metrics or in multimetric theories with interaction cycles, in which it is not. In these cases the metric and vielbein formulations are distinct physical theories, although each propagates a healthy massive graviton.

34.6 Cosmology

If the graviton mass m is comparable to the Hubble rate H_0, then at cosmological distances the mass term can produce a repulsive gravitational effect that leads to cosmic acceleration. Because, roughly speaking, the enhanced diffeomorphism symmetry in the limit $m = 0$ protects a small graviton mass from large quantum corrections, the choice $m \sim H_0$ is in fact *technically natural*.[20] Massive gravity thus may provide a solution to the cosmological constant problem: why do quantum corrections not cause the Universe to accelerate at extremely early times?

However, it turns out that flat and closed Friedmann–Lemaître–Robertson–Walker cosmological solutions do not exist in dRGT massive gravity with a flat reference metric.[4] Open solutions and solutions with general reference metrics suffer from instabilities.[21] Therefore viable cosmologies can only be found in massive gravity if one abandons the cosmological principle that the Universe is uniform on large scales, or otherwise generalizes dRGT. For instance, cosmological solutions are better behaved in bigravity,[5] the theory which extends dRGT by giving $f_{\mu\nu}$ dynamics. While these tend to possess instabilities as well,[22][23] those instabilities might find a resolution in the nonlinear dynamics (through a Vainshtein-like mechanism) or by pushing the era of instability to the very early Universe.[6]

34.7 3D massive gravity

A special case exists in three dimensions, where a massless graviton does not propagate any degrees of freedom. Here several ghost-free theories of a massive graviton, propagating two degrees of freedom, can be defined. In the case of *topologically massive gravity*[1] one has the action

$$ S = \frac{M_3}{2} \int d^3x \sqrt{-g}(R - 2\Lambda) + \frac{1}{4\mu} \epsilon^{\lambda\mu\nu} \Gamma^\rho_{\lambda\sigma} \left(\partial_\mu \Gamma^\sigma_{\rho\nu} + \frac{2}{3} \Gamma^\sigma_{\mu\alpha} \Gamma^\alpha_{\nu\rho} \right). $$

with M_3 the three-dimensional Planck mass. This is three-dimensional general relativity supplemented by a Chern-Simons-like term built out of the Christoffel symbols.

More recently, a theory referred to as *new massive gravity* has been developed,[2] which is described by the action

$$ S = M_3 \int d^3x \sqrt{-g} \left[\pm R + \frac{1}{m^2} \left(R_{\mu\nu} R^{\mu\nu} - \frac{3}{8} R^2 \right) \right]. $$

34.8 See also

- Accelerating universe
- Alternatives to general relativity
- Bimetric gravity
- DGP model
- Scalar–tensor theory

34.9   Further reading

Review articles

- •de Rham, Claudia(2014),"Massive Gravity",*Living Reviews in Relativity***17**: 7,arXiv:1401.4173,Bibcode:2014LRR doi:10.12942/lrr-2014-7

- • Hinterbichler, Kurt (2012). "Theoretical Aspects of Massive Gravity", *Reviews of Modern Physics* **84**: 671–710, arXiv:1105.3735, Bibcode:2012RvMP...84..671H, doi:10.1103/RevModPhys.84.671

34.10   References

[1] Deser, Stanley; Jackiw, R.; Templeton, S. (1982). "Topologically Massive Gauge Theories". *Annals Phys.* **140**: 372–411. Bibcode:1982AnPhy.140..372D. doi:10.1016/0003-4916(82)90164-6.

[2] Bergshoeff, Eric A.; Hohm, Olaf; Townsend, Paul K.. (2009). "Massive Gravity in Three Dimensions". *Phys.Rev.Lett.* **102**: 201301. arXiv:0901.1766. Bibcode:2009PhRvL.102t1301B. doi:10.1103/PhysRevLett.102.201301.

[3]Hassan, S.F.; Rosen, Rachel A. (2012). "Bimetric Gravity from Ghost-free Massive Gravity".*JHEP***1202**: 126.arXiv:1109.3 Bibcode:2012JHEP...02..126H. doi:10.1007/JHEP02(2012)126.

[4] D'Amico, G.; de Rham, C.; Dubovsky, S.; Gabadadze, G.; Pirtskhalava, D.; Tolley, A.J. (2011). "Massive Cosmologies". *Phys.Rev.* **D84**: 124046. arXiv:1108.5231. Bibcode:2011PhRvD..84l4046D. doi:10.1103/PhysRevD.84.124046.

[5] Akrami, Yashar; Koivisto, Tomi S.; Sandstad, Marit (2013). "Accelerated expansion from ghost-free bigravity: a statistical analysis with improved generality".*JHEP***1303**: 099.arXiv:1209.0457.Bibcode:2013JHEP...03..099A.doi:10.1007/JHEP03

[6] Akrami, Yashar; Hassan, S.F.; Könnig, Frank; Schmidt-May, Angnis; Solomon, Adam R. (2015). "Bimetric gravity is cosmologically viable". arXiv:1503.07521. Bibcode:2015arXiv150307521A.

[7] Fierz, Markus; Pauli, Wolfgang (1939). "On relativistic wave equations for particles of arbitrary spin in an electromagnetic field". *Proc.Roy.Soc.Lond.* **A173**: 211–232. Bibcode:1939RSPSA.173..211F. doi:10.1098/rspa.1939.0140.

[8] van Dam, Hendrik; Veltman, Martinus J. G. (1970). "Massive and massless Yang-Mills and gravitational fields". *Nucl.Phys.* **B22**: 397–411. Bibcode:1970NuPhB..22..397V. doi:10.1016/0550-3213(70)90416-5.

[9]Zakharov, Vladimir I. (1970). "Linearized gravitation theory and the graviton mass".*JETP Lett.***12**: 312.Bibcode:1970JETPL...

[10]Vainshtein, A.I. (1972). "To the problem of nonvanishing gravitation mass".*Phys.Lett.***B39**: 393–394.Bibcode:1972PhLB... doi:10.1016/0370-2693(72)90147-5.

[11]Boulware, David G.; Deser, Stanley (1972). "Can gravitation have a finite range?".*Phys.Rev.* **D6**: 3368–3382.Bibcode:1972 doi:10.1103/PhysRevD.6.3368.

[12] Creminelli, Paolo; Nicolis, Alberto; Papucci, Michele; Trincherini, Enrico (2005). "Ghosts in massive gravity". *JHEP* **0509**: 003. arXiv:hep-th/0505147. Bibcode:2005JHEP...09..003C. doi:10.1088/1126-6708/2005/09/003.

[13]de Rham, Claudia; Gabadadze, Gregory (2010). "Generalization of the Fierz-Pauli Action".*Phys.Rev.***D82**: 044020.arXiv: Bibcode:2010PhRvD..82d4020D. doi:10.1103/PhysRevD.82.044020.

[14] de Rham, Claudia; Gabadadze, Gregory; Tolley, Andrew J. (2010). "Resummation of Massive Gravity". *Phys.Rev.Lett.* **106**: 231101. arXiv:1011.1232. Bibcode:2011PhRvL.106w1101D. doi:10.1103/PhysRevLett.106.231101.

[15] Hassan, S.F.; Rosen, Rachel A. (2012). "Resolving the Ghost Problem in non-Linear Massive Gravity". *Phys.Rev.Lett.* **108**: 041101. arXiv:1106.3344. Bibcode:2012PhRvL.108d1101H. doi:10.1103/PhysRevLett.108.041101.

[16] Hassan, S.F.; Rosen, Rachel A. (2012). "Confirmation of the Secondary Constraint and Absence of Ghost in Massive Gravity and Bimetric Gravity". *JHEP* **1204**: 123. arXiv:1111.2070. Bibcode:2012JHEP...04..123H. doi:10.1007/JHEP04(2012)123.

[17] Hassan, S.F.; Rosen, Rachel A. (2011). "On Non-Linear Actions for Massive Gravity". *JHEP* **1107**: 009. arXiv:1103.6055. Bibcode:2011JHEP...07..009H. doi:10.1007/JHEP07(2011)009.

[18] Hassan, S.F.; Rosen, Rachel A.; Schmidt-May, Angnis (2012). "Ghost-free Massive Gravity with a General Reference Metric". *JHEP* **1202**: 026. arXiv:1109.3230. Bibcode:2012JHEP...02..026H. doi:10.1007/JHEP02(2012)026.

[19]Hinterbichler, Kurt; Rosen, Rachel A. (2012). "Interacting Spin-2 Fields".*JHEP***1207**: 047.arXiv:1203.5783.Bibcode:2012 doi:10.1007/JHEP07(2012)047.

[20] de Rham, Claudia; Heisenberg, Lavinia; Ribeiro, Raquel H. (2013). "Quantum Corrections in Massive Gravity". *Phys.Rev.* **D88**: 084058. arXiv:1307.7169. Bibcode:2013PhRvD..88h4058D. doi:10.1103/PhysRevD.88.084058.

[21] de Felice, Antonio; Gümrükçüoğlu, A. Emir; Lin, Chunshan; Mukohyama, Shinji (2013). "On the cosmology of massive gravity".*Class.Quant.Grav.***30**: 184004.arXiv:1304.0484.Bibcode:2013CQGra..30r4004D.doi:10.1088/0264-9381/30/18/184

[22] Comelli, Denis; Crisostomi, Marco; Pilo, Luigi (2012). "Perturbations in Massive Gravity Cosmology". *JHEP* **1206**: 085. arXiv:1202.1986. Bibcode:2012JHEP...06..085C. doi:10.1007/JHEP06(2012)085.

[23] Könnig, Frank; Akrami, Yashar; Amendola, Luca; Motta, Mariele; Solomon, Adam R. (2014). "Stable and unstable cosmological models in bimetric massive gravity". *Phys.Rev.* **D90**: 124014. arXiv:1407.4331. Bibcode:2014PhRvD..90l4014K. doi:10.1103/PhysRevD.90.124014.

Chapter 35

Mechanical explanations of gravitation

Mechanical explanations of gravitation (or **kinetic theories of gravitation**) are attempts to explain the action of gravity by aid of basic mechanical processes, such as pressure forces caused by pushes, without the use of any action at a distance. These theories were developed from the 16th until the 19th century in connection with the Aether. However, such models are no longer regarded as viable theories within the mainstream scientific community and general relativity is now the standard model to describe gravitation without the use of actions at a distance. Modern "quantum gravity" hypotheses also attempt to describe gravity by more fundamental processes such as particle fields, but they are not based on classical mechanics.

35.1 Screening

Main article: Le Sage's theory of gravitation

This theory is probably[1] the best-known mechanical explanation, and was developed for the first time by Nicolas Fatio de Duillier in 1690, and re-invented, among others, by Georges-Louis Le Sage (1748), Lord Kelvin (1872), and Hendrik Lorentz (1900), and criticized by James Clerk Maxwell (1875), and Henri Poincaré (1908).

The theory posits that the force of gravity is the result of tiny particles or waves moving at high speed in all directions, throughout the universe. The intensity of the flux of particles is assumed to be the same in all directions, so an isolated object A is struck equally from all sides, resulting in only an inward-directed pressure but no net directional force. With a second object B present, however, a fraction of the particles that would otherwise have struck A from the direction of B is intercepted, so B works as a shield, so-to-speak—that is, from the direction of B, A will be struck by fewer particles than from the opposite direction. Likewise, B will be struck by fewer particles from the direction of A than from the opposite direction. One can say that A and B are "shadowing" each other, and the two bodies are pushed toward each other by the resulting imbalance of forces.

This shadow obeys the inverse square law, because the imbalance of momentum flow over an entire spherical surface enclosing the object is independent of the size of the enclosing sphere, whereas the surface area of the sphere increases in proportion to the square of the radius. To satisfy the need for mass proportionality, the theory posits that a) the basic elements of matter are very small so that gross matter consists mostly of empty space, and b) that the particles are so small, that only a small fraction of them would be intercepted by gross matter. The result is, that the "shadow" of each body is proportional to the surface of every single element of matter.

Criticism: This theory was declined primarily for thermodynamic reasons because a shadow only appears in this model if the particles or waves are at least partly absorbed, which should lead to an enormous heating of the bodies. Also drag, *i.e.* the resistance of the particle streams in the direction of motion, is a great problem too. This problem can be solved by assuming superluminal speeds, but this solution largely increases the thermal problems and contradicts special relativity.[2][3]

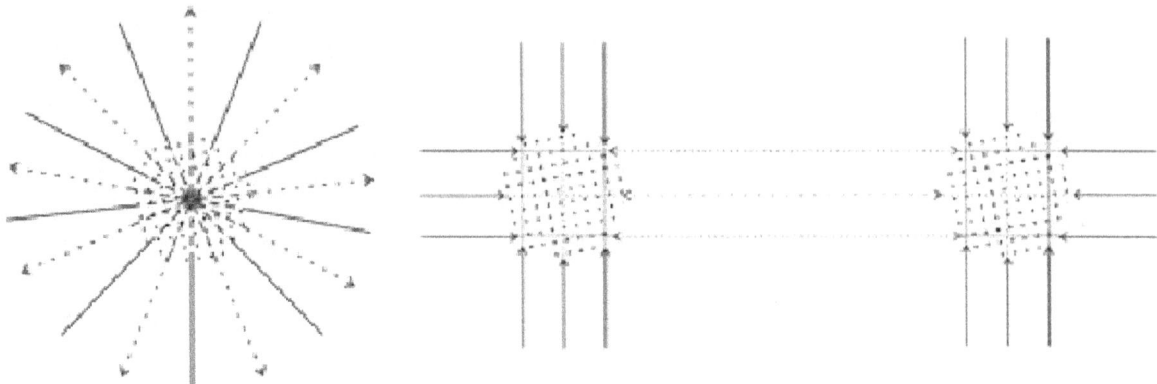

P5: Permeability, attenuation and mass proportionality

35.2 Vortex

Because of his philosophical beliefs, René Descartes proposed in 1644 that no empty space can exist and that space must consequently be filled with matter. The parts of this matter tend to move in straight paths, but because they lie close together, they can not move freely, which according to Descartes implies that every motion is circular, so the aether is filled with vortices. Descartes also distinguishes between different forms and sizes of matter in which rough matter resists the circular movement more strongly than fine matter. Due to centrifugal force, matter tends towards the outer edges of the vortex, which causes a condensation of this matter there. The rough matter cannot follow this movement due to its greater inertia—so due to the pressure of the condensed outer matter those parts will be pushed into the center of the vortex. According to Descartes, this inward pressure is nothing else than gravity. He compared this mechanism with the fact that if a rotating, liquid filled vessel is stopped, the liquid goes on to rotate. Now, if one drops small pieces of light matter (e.g. wood) into the vessel, the pieces move to the middle of the vessel.[4] [5] [6]

Following the basic premises of Descartes, Christiaan Huygens between 1669 and 1690 designed a much more exact vortex model. This model was the first theory of gravitation which was worked out mathematically. He assumed that the aether particles are moving in every direction, but were thrown back at the outer borders of the vortex and this causes (as in the case of Descartes) a greater concentration of fine matter at the outer borders. So also in his model the fine matter presses the rough matter into the center of the vortex. Huygens also found out that the centrifugal force is equal to the force, which acts in the direction of the center of the vortex (centripetal force). He also posited that bodies must consist mostly of empty space so that the aether can penetrate the bodies easily, which is necessary for mass proportionality. He further concluded that the aether moves much faster than the falling bodies. At this time, Newton developed his theory of gravitation which is based on attraction, and although Huygens agreed with the mathematical formalism, he said the model was insufficient due to the lack of a mechanical explanation of the force law. Newton's discovery that gravity obeys the inverse square law surprised Huygens and he tried to take this into account by assuming that the speed of the aether is smaller in greater distance.[6][7][8]

Criticism: Newton objected to the theory because drag must lead to noticeable deviations of the orbits which were not observed.[9] Another problem was that moons often move in different directions, against the direction of the vortex motion. Also, Huygens' explanation of the inverse square law is circular, because this means that the aether obeys Kepler's third law. But a theory of gravitation has to explain those laws and must not presuppose them.[6][9]

35.3 Streams

In a 1675 letter to Henry Oldenburg, and later to Robert Boyle, Newton wrote the following: [Gravity is the result of] "a condensation causing a flow of ether with a corresponding thinning of the ether density associated with the increased velocity of flow." He also asserted that such a process was consistent with all his other work and Kepler's Laws of Motion.[10] Newtons' idea of a pressure drop associated with increased velocity of flow was mathematically formalised as Bernoulli's principle published in Daniel Bernoulli's book *Hydrodynamica* in 1738.

However, although he later proposed a second explanation (see section below), Newton's comments to that question remained ambiguous. In the third letter to Bentley in 1692 he wrote:[11]

> It is inconceivable that inanimate brute matter should, without the mediation of something else which is not material, operate upon and affect other matter, without mutual contact, as it must do if gravitation in the sense of Epicurus be essential and inherent in it. And this is one reason why I desired you would not ascribe 'innate gravity' to me. That gravity should be innate, inherent, and essential to matter, so that one body may act upon another at a distance, through a vacuum, without the mediation of anything else, by and through which their action and force may be conveyed from one to another, is to me so great an absurdity, that I believe no man who has in philosophical matters a competent faculty of thinking can ever fall into it. Gravity must be caused by an agent acting constantly according to certain laws; but whether this agent be material or immaterial, I have left to the consideration of my readers.

On the other hand, Newton is also well known for the phrase Hypotheses non fingo, written in 1713:[12]

> I have not as yet been able to discover the reason for these properties of gravity from phenomena, and I do not feign hypotheses. For whatever is not deduced from the phenomena must be called a hypothesis; and hypotheses, whether metaphysical or physical, or based on occult qualities, or mechanical, have no place in experimental philosophy. In this philosophy particular propositions are inferred from the phenomena, and afterwards rendered general by induction.

And according to the testimony of some of his friends, such as Nicolas Fatio de Duillier or David Gregory, Newton thought that gravitation is based directly on divine influence.[8]

Similar to Newton, but mathematically in greater detail, Bernhard Riemann assumed in 1853 that the gravitational aether is an incompressible fluid and normal matter represents sinks in this aether. So if the aether is destroyed or absorbed proportionally to the masses within the bodies, a stream arises and carries all surrounding bodies into the direction of the central mass. Riemann speculated that the absorbed aether is transferred into another world or dimension.[13]

Another attempt to solve the energy problem was made by Ivan Osipovich Yarkovsky in 1888. Based on his aether stream model, which was similar to that of Riemann, he argued that the absorbed aether might be converted into new matter, leading to a mass increase of the celestial bodies.[14]

Criticism: As in the case of Le Sage's theory, the disappearance of energy without explanation violates the energy conservation law. Also some drag must arise, and no process which leads to a creation of matter is known.

35.4   Static pressure

Newton updated the second edition of *Optics* (1717) with another mechanical-ether theory of gravity. Unlike his first explanation (1675 - see Streams), he proposed a stationary aether which gets thinner and thinner nearby the celestial bodies. On the analogy of the lift (force), a force arises, which pushes all bodies to the central mass. He minimized drag by stating an extremely low density of the gravitational aether.

Like Newton, Leonhard Euler presupposed in 1760 that the gravitational aether loses density in accordance with the inverse square law. Similarly to others, Euler also assumed that to maintain mass proportionality, matter consists mostly of empty space.[15]

Criticism: Both Newton and Euler gave no reason why the density of that static aether should change. Furthermore, James Clerk Maxwell pointed out that in this "hydrostatic" model *"the state of stress... which we must suppose to exist in the invisible medium, is 3000 times greater than that which the strongest steel could support"*.[16]

35.5   Waves

Robert Hooke speculated in 1671 that gravitation is the result of all bodies emitting waves in all directions through the aether. Other bodies, which interchange with these waves, move in the direction of the source of the waves. Hooke saw an analogy to the fact that small objects on a disturbed surface of water move to the center of the disturbance.[17]

A similar theory was worked out mathematically by James Challis from 1859 to 1876. He calculated that the case of attraction occurs if the wavelength is large in comparison with the distance between the gravitating bodies. If the wavelength is small, the bodies repel each other. By a combination of these effects, he also tried to explain all other forces.[18]

Criticism: Maxwell objected that this theory requires a steady production of waves, which must be accompanied by an infinite consumption of energy.[16] Challis himself admitted, that he hadn't reached a definite result due to the complexity of the processes.[17]

35.6   Pulsation

Lord Kelvin (1871) and Carl Anton Bjerknes (1871) assumed that all bodies pulsate in the aether. This was in analogy to the fact that, if the pulsation of two spheres in a fluid is in phase, they will attract each other; and if the pulsation of two spheres is *not* in phase, they will repel each other. This mechanism was also used for explaining the nature of electric charges. Among others, this hypothesis has also been examined by George Gabriel Stokes and Woldemar Voigt.[19]

Criticism : To explain universal gravitation, one is forced to assume that all pulsations in the universe are in phase—which appears very implausible. In addition, the aether should be incompressible to ensure that attraction also arises at greater distances.[19] And Maxwell argued that this process must be accompanied by a permanent new production and destruction of aether.[16]

35.7   Other historical speculations

In 1690, Pierre Varignon assumed that all bodies are exposed to pushes by aether particles from all directions, and that there is some sort of limitation at a certain distance from the Earth's surface which cannot be passed by the particles. He assumed that if a body is closer to the Earth than to the limitation boundary, then the body would experience a greater push from above than from below, causing it to fall toward the Earth.[20]

In 1748, Mikhail Lomonosov assumed that the effect of the aether is proportional to the complete surface of the elementary components of which matter consists (similar to Huygens and Fatio before him). He also assumed an enormous penetrability of the bodies. However, no clear description was given by him as to how exactly the aether interchanges with matter so that the law of gravitation arises.[21]

In 1821, John Herapath tried to apply his co-developed model of the kinetic theory of gases on gravitation. He assumed that the aether is heated by the bodies and loses density so that other bodies are pushed to these regions of lower density.[22] However, it was shown by Taylor that the decreased density due to thermal expansion is compensated for by the increased speed of the heated particles; therefore, no attraction arises.[17]

35.8   Recent theorizing

These mechanical explanations for gravity never gained widespread acceptance, although such ideas continued to be studied occasionally by physicists until the beginning of the twentieth century, by which time it was generally considered to be conclusively discredited. However, some researchers outside the scientific mainstream still try to work out some consequences of those theories.

Le Sage's theory was studied by Radzievskii and Kagalnikova (1960),[23] Shneiderov (1961),[24] Buonomano and Engels (1976),[25] Adamut (1982),[26] Jaakkola (1996),[27] Tom Van Flandern (1999),[28] and Edwards (2007).[29] A variety of

Le Sage models and related topics are discussed in Edwards, et al.[30]

Gravity due to static pressure was recently studied by Arminjon.[31]

35.9   References

[1] Taylor (1876), Peck (1903), secondary sources

[2] Poincaré (1908), Secondary sources

[3] Maxwell (1875, Atom), Secondary sources

[4] Descartes, R. (1824–1826), Cousin, V., ed., "Les principes de la philosophie (1644)", *Oeuvres de Descartes* (Paris: F.-G. Levrault) **3**

[5] Descartes, 1644; Zehe, 1980, pp. 65–70; Van Lunteren, p. 47

[6] Zehe (1980), Secondary sources

[7] Huygens, C. (1944), Société Hollandaise des Sciences, ed., "Discours de la Cause de la Pesanteur (1690)", *Oeuvres complètes de Christiaan Huygens* (Den Haag) **21**: 443–488

[8] Van Lunteren (2002), Secondary sources

[9] Newton, I. (1846), *Newton's Principia : the mathematical principles of natural philosophy (1687)*, New York: Daniel Adee

[10] I. Newton, letters quoted in detail in The Metaphysical Foundations of Modern Physical Science by Edwin Arthur Burtt, Double day Anchor Books.

[11] Newton, 1692, 3rd letter to Bentley

[12] Isaac Newton (1726), *Philosophiae Naturalis Principia Mathematica*, General Scholium. Third edition, page 943 of I. Bernard Cohen and Anne Whitman's 1999 translation, University of California Press ISBN 0-520-08817-4, 974 pages.

[13] Riemann, B. (1876), Dedekind, R. & Weber, W., ed., "Neue mathematische Prinzipien der Naturphilosophie", *Bernhard Riemanns Werke und gesammelter Nachlass* (Leipzig): 528–538

[14] Yarkovsky, I. O. (1888), *Hypothese cinetique de la Gravitation universelle et connexion avec la formation des elements chimiques*, Moscow

[15] Euler, L. (1776), *Briefe an eine deutsche Prinzessin, Nr. 50, 30. August 1760*, Leipzig, pp. 173–176

[16] Maxwell (1875, Attraction), Secondary sources

[17] Taylor (1876), Secondary sources

[18] Challis, J. (1869), *Notes of the Principles of Pure and Applied Calculation*, Cambridge

[19] Zenneck (1903), Secondary sources

[20] Varignon, P. (1690), *Nouvelles conjectures sur la Pesanteur*, Paris

[21] Lomonosow, M. (1970), Henry M. Leicester, ed., "On the Relation of the Amount of Material and Weight (1758)", *Mikhail Vasil'evich Lomonosov on the Corpuscular Theory* (Cambridge: Harvard University Press): 224–233

[22] Herapath, J. (1821), "On the Causes, Laws and Phenomena of Heat, Gases, Gravitation", *Annals of Philosophy* (Paris) **9**: 273–293

[23] Radzievskii, V.V. & Kagalnikova, I.I. (1960), "The nature of gravitation", *Vsesoyuz. Astronom.-Geodezich. Obsch. Byull.* **26** (33): 3–14 A rough English translation appeared in a U.S. government technical report: FTD TT64 323; TT 64 11801 (1964), Foreign Tech. Div., Air Force Systems Command, Wright-Patterson AFB, Ohio (reprinted in *Pushing Gravity*)

[24] Shneiderov, A. J. (1961), "On the internal temperature of the earth", *Bollettino di Geofisica Teorica ed Applicata* **3**: 137–159

[25] Buonomano, V. & Engel, E. (1976), "Some speculations on a causal unification of relativity, gravitation, and quantum mechanics", *Int. J. Theor. Phys.* **15** (3): 231–246, Bibcode:1976IJTP...15..231B, doi:10.1007/BF01807095

[26] Adamut, I. A. (1982), "The screen effect of the earth in the TETG. Theory of a screening experiment of a sample body at the equator using the earth as a screen", *Nuovo Cimento C* **5**(2): 189–208, Bibcode:1982NCimC...5..189A, doi:10.1007/BF025090

[27] Jaakkola, T. (1996), "Action-at-a-distance and local action in gravitation: discussion and possible solution of the dilemma" (PDF), *Apeiron* **3** (3–4): 61–75

[28] Van Flandern, T. (1999), *Dark Matter, Missing Planets and New Comets* (2 ed.), Berkeley: North Atlantic Books, pp. Chapters 2–4

[29] Edwards, M .R. (2007), "Photon-Graviton Recycling as Cause of Gravitation" (PDF), *Apeiron* **14** (3): 214–233

[30] Edwards, M. R., ed. (2002), *Pushing Gravity: New Perspectives on Le Sage's Theory of Gravitation*, Montreal: C. Roy Keys Inc.

[31] Mayeul Arminjon (11 November 2004), "Gravity as Archimedes' Thrust and a Bifurcation in that Theory", *Foundations of Physics* **34** (11): 1703–1724, arXiv:physics/0404103, Bibcode:2004FoPh...34.1703A, doi:10.1007/s10701-004-1312-3

35.10 Sources

- Aiton, E.J. (1969), "Newton's Aether-Stream Hypothesis and the Inverse Square Law of Gravitation", *Annals of Science* **25** (3): 255–260, doi:10.1080/00033796900200151

- Carrington, Hereward (1913), Sugden, Sherwood J. B. ed., "Earlier Theories of Gravity", *The Monist* **23** (3): 445–458, doi:10.5840/monist19132332

- Drude, Paul (1897), "Ueber Fernewirkungen", *Annalen der Physik* **298**(12): I–XLIX, Bibcode:1897AnP...298D doi:10.1002/andp.18972981220

- Hall, Thomas Proctor (1895), "Physical Theories of Gravitation", *Proceedings of the Iowa Academy of Science* **3**: 47–52

- Helm, Georg (1881), "Ueber die Vermittelung der Fernewirkungen durch den Aether", *Annalen der Physik* **250** (9): 149–176, Bibcode:1881AnP...250..149H, doi:10.1002/andp.18812500912

- Isenkrahe, Caspar (1892), "Über die Rückführung der Schwere auf Absorption und die daraus abgeleiteten Gesetze", *Abhandlungen zur Geschichte der Mathematik* **6**, Leipzig, pp. 161–204

- Maxwell, James Clerk (1875), "Atom", *Encyclopædia Britannica Ninth Edition* **3**: 36–49

- Maxwell, James Clerk (1875), "Attraction", *Encyclopædia Britannica Ninth Edition* **3**: 63–65

- Peck, J. W. (1903), "The Corpuscular Theories of Gravitation", *Proceedings of the Royal Philosophical Society of Glasgow* **34**: 17–44

- Poincaré, Henri (1914) [1908], "Lesage's theory", *Science and Method*, London, New York: Nelson & Sons, pp. 246–253

- Preston, Samuel Tolver (1895), "Comparative Review of some Dynamical Theories of Gravitation", *Philosophical Magazine*, 5th series **39** (237): 145–159, doi:10.1080/14786449508620698

- Taylor, William Bower (1876), "Kinetic Theories of Gravitation", *Smithsonian report*: 205–282

- Van Lunteren, F. (2002), "Nicolas Fatio de Duillier on the mechanical cause of Gravitation", in Edwards, M.R., *Pushing Gravity: New Perspectives on Le Sage's Theory of Gravitation*, Montreal: C. Roy Keys Inc., pp. 41–59

- Zehe, Horst (1980), "Die Gravitationstheorie des Nicolas Fatio de Duillier", *Arch. Hist. Exact Sci* (Hildesheim: Gerstenberg) **28**: 1, Bibcode:1983AHES...28....1Z, doi:10.1007/BF00327787, ISBN 3-8067-0862-2

- Zenneck, Jonathan (1903), "Gravitation", *Encyklopädie der mathematischen Wissenschaften mit Einschluss ihrer Anwendungen* **5** (1): 25–67, doi:10.1007/978-3-663-16016-8_2

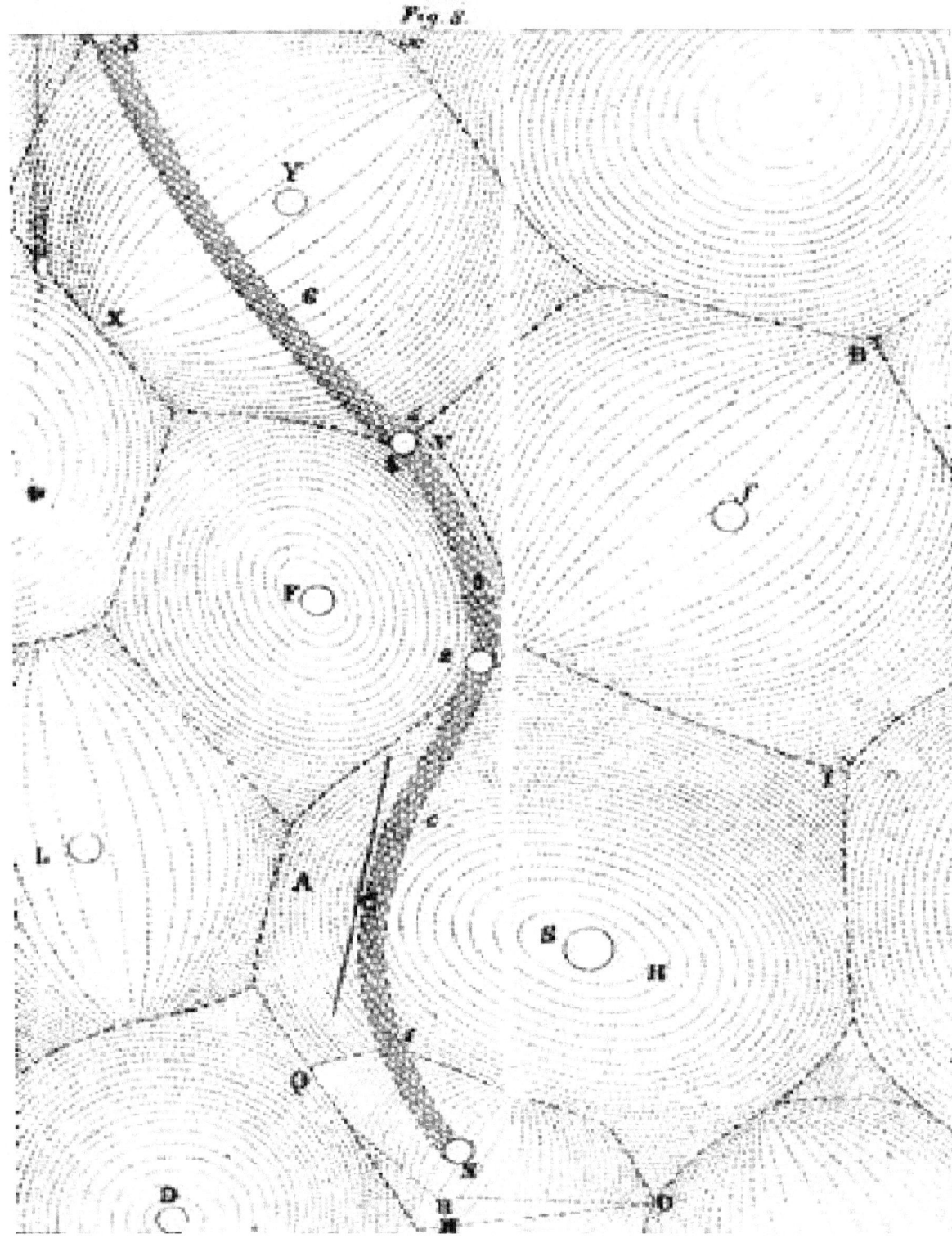

Aether vortices around celestial bodies

Chapter 36

Metric-affine gravitation theory

In comparison with General Relativity, dynamic variables of **metric-affine gravitation theory** are both a pseudo-Riemannian metric and a general linear connection on a world manifold X. Metric-affine gravitation theory has been suggested as a natural generalization of Einstein–Cartan theory of gravity with torsion where a linear connection obeys the condition that a covariant derivative of a metric equals zero.

Metric-affine gravitation theory straightforwardly comes from gauge gravitation theory where a general linear connection plays the role of a gauge field. Let TX be the tangent bundle over a manifold X provided with bundle coordinates $(x^\mu, \dot{x}^\mu)$. A general linear connection on TX is represented by a connection tangent-valued form

$$\Gamma = dx^\lambda \otimes (\partial_\lambda + \Gamma_\lambda{}^\mu{}_\nu \dot{x}^\nu \dot{\partial}_\mu).$$

It is associated to a principal connection on the principal frame bundle FX of frames in the tangent spaces to X whose structure group is a general linear group $GL(4, \mathbb{R})$. Consequently, it can be treated as a gauge field. A pseudo-Riemannian metric $g = g_{\mu\nu} dx^\mu \otimes dx^\nu$ on TX is defined as a global section of the quotient bundle $FX/SO(1,3) \to X$, where $SO(1,3)$ is the Lorentz group. Therefore, on can regard it as a classical Higgs field in gauge gravitation theory. Gauge symmetries of metric-affine gravitation theory are general covariant transformations.

It is essential that, given a pseudo-Riemannian metric g, any linear connection Γ on TX admits a splitting

$$\Gamma_{\mu\nu\alpha} = \{\mu\nu\alpha\} + S_{\mu\nu\alpha} + \frac{1}{2} C_{\mu\nu\alpha}$$

in the Christoffel symbols

$$\{\mu\nu\alpha\} = -\frac{1}{2}(\partial_\mu g_{\nu\alpha} + \partial_\alpha g_{\nu\mu} - \partial_\nu g_{\mu\alpha}),$$

a nonmetricity tensor

$$C_{\mu\nu\alpha} = C_{\mu\alpha\nu} = \nabla_\mu^\Gamma g_{\nu\alpha} = \partial_\mu g_{\nu\alpha} + \Gamma_{\mu\nu\alpha} + \Gamma_{\mu\alpha\nu}$$

and a contorsion tensor

$$S_{\mu\nu\alpha} = -S_{\mu\alpha\nu} = \frac{1}{2}(T_{\nu\mu\alpha} + T_{\nu\alpha\mu} + T_{\mu\nu\alpha} + C_{\alpha\nu\mu} - C_{\nu\alpha\mu}),$$

where

$$T_{\mu\nu\alpha} = \frac{1}{2}(\Gamma_{\mu\nu\alpha} - \Gamma_{\alpha\nu\mu})$$

is the torsion tensor of Γ.

Due to this splitting, metric-affine gravitation theory possesses a different collection of dynamic variables which are a pseudo-Riemannian metric, a non-metricity tensor and a torsion tensor. As a consequence, a Lagrangian of metric-affine gravitation theory can contain different terms expressed both in a curvature of a connection Γ and its torsion and non-metricity tensors. In particular, a metric-affine f(R) gravity, whose Lagrangian is an arbitrary function of a scalar curvature R of Γ, is considered.

A linear connection Γ is called the *metric connection* for a pseudo-Riemannian metric g if g is its integral section, i.e., the metricity condition

$$\nabla^{\Gamma}_{\mu} g_{\nu\alpha} = 0$$

holds. A metric connection reads

$$\Gamma_{\mu\nu\alpha} = \{_{\mu\nu\alpha}\} + \frac{1}{2}(T_{\nu\mu\alpha} + T_{\nu\alpha\mu} + T_{\mu\nu\alpha}).$$

For instance, the Levi-Civita connection in General Relativity is a torsion-free metric connection.

A metric connection is associated to a principal connection on a Lorentz reduced subbundle $F^g X$ of the frame bundle FX corresponding to a section g of the quotient bundle $FX/SO(1,3) \to X$. Restricted to metric connections, metric-affine gravitation theory comes to the above-mentioned Einstein – Cartan gravitation theory.

At the same time, any linear connection Γ defines a principal adapted connection Γ^g on a Lorentz reduced subbundle $F^g X$ by its restriction to a Lorentz subalgebra of a Lie algebra of a general linear group $GL(4, \mathbb{R})$. For instance, the Dirac operator in metric-affine gravitation theory in the presence of a general linear connection Γ is well defined, and it depends just of the adapted connection Γ^g. Therefore, Einstein – Cartan gravitation theory can be formulated as the metric-affine one, without appealing to the metricity constraint.

In metric-affine gravitation theory, in comparison with the Einstein - Cartan one, a question on a matter source of a non-metricity tensor arises. It is so called hypermomentum, e.g., a Noether current of a scaling symmetry.

36.1 References

- F.Hehl, J. McCrea, E. Mielke, Y. Ne'eman, Metric-affine gauge theory of gravity: field equations, Noether identities, world spinors, and breaking of dilaton invariance, *Physics Reports* **258** (1995) 1-171; arXiv: gr-qc/9402012

- V. Vitagliano, T. Sotiriou, S. Liberati, The dynamics of metric-affine gravity, *Annals of Physics* **326** (2011) 1259-1273; arXiv: 1008.0171

- G. Sardanashvily, Classical gauge gravitation theory, *Int. J. Geom. Methods Mod. Phys.* **8** (2011) 1869-1895; arXiv: 1110.1176

- C. Karahan, A. Altas, D. Demir, Scalars, vectors and tensors from metric-affine gravity, *General Relativity and Gravitation* **45** (2013) 319-343; arXiv: 1110.5168

36.2 See also

- Gauge gravitation theory

- Einstein-Cartan theory

- Affine gauge theory

Chapter 37

Modified models of gravity

The existence of anomalous astrophysical and cosmological phenomena like the cosmic acceleration and the dynamics of clusters of galaxies and gas in galaxies themselves recently boosted the growth of several long-range modifications of the usual laws of gravitation. Indeed, the aforementioned phenomena did not find satisfactory explanations in terms of the standard Newton–Einstein gravitational physics, unless exotic and still undetected forms of matter-energy are postulated. Examples of such entities are, e.g., dark matter and dark energy, identified with the Cosmological Constant. At galactic scales, the most famous modified model of gravity is, perhaps, the MOdified Newtonian Dynamics (MOND) proposed in 1983 by Mordehai Milgrom.[1] More recently, also the MOdified Gravity (MOG) by John Moffat [2] and Dynamic Newtonian Advanced gravity (DNAg)[3] took place.

37.1 MOND and MOG

While MOND modifies Newton's inverse square law at a characteristic acceleration scale $a_0 \approx 1 \times 10^{-10} \text{m s}^{-2}$, MOG's modifications occur at large distances, i.e. MOG is a length scale-dependent model.

37.2 DGP model

Another proposal to accommodate the cosmic acceleration without invoking the concept of dark energy is the multidimensional braneworld model by Gia Dvali, Gregory Gabadadze and Massimo Porrati (DGP model). In it our ordinary space-time is a 3+1 dimensional brane embedded in a five-dimensional bulk space.

37.3 References

[1] Milgrom, M. (July 1983). "A modification of the Newtonian dynamics as a possible alternative to the hidden mass hypothesis". *The Astrophysical Journal* **270**: 365–370. Bibcode:1983ApJ...270..365M. doi:10.1086/161130.

[2] Moffat, J.W. (2006). "Scalar–tensor–vector gravity theory". *Journal of Cosmology and Astroparticle Physics* **3**. arXiv:gr-qc/0506021. Bibcode:2006JCAP...03..004M. doi:10.1088/1475-7516/2006/03/004. 004.

[3] "The formulation of dynamic Newtonian advanced gravity, DNAg".

Chapter 38

Nonsymmetric gravitational theory

In theoretical physics, the **nonsymmetric gravitational theory**[1] (NGT) of John Moffat is a classical theory of gravitation that tries to explain the observation of the flat rotation curves of galaxies.

In general relativity, the gravitational field is characterized by a symmetric rank-2 tensor, the metric tensor. The possibility of generalizing the metric tensor has been considered by many, including Einstein and others. A general (nonsymmetric) tensor can always be decomposed into a symmetric and an antisymmetric part. As the electromagnetic field is characterized by an antisymmetric rank-2 tensor, there is an obvious possibility for a unified theory: a nonsymmetric tensor composed of a symmetric part representing gravity, and an antisymmetric part that represents electromagnetism. Research in this direction ultimately proved fruitless; the desired classical unified field theory was not found.

In 1979, Moffat made the observation[2] that the antisymmetric part of the generalized metric tensor need not necessarily represent electromagnetism; it may represent a new, hypothetical force. Later, in 1995, Moffat noted[1] that the field corresponding with the antisymmetric part need not be massless, like the electromagnetic (or gravitational) fields.

In its original form, the theory may be unstable, although this has only been shown in the case of the linearized version.[3][4]

In the weak field approximation where interaction between fields is not taken into account, NGT is characterized by a symmetric rank-2 tensor field (gravity), an antisymmetric tensor field, and a constant characterizing the mass of the antisymmetric tensor field. The antisymmetric tensor field is found to satisfy the equations of a Maxwell–Proca massive antisymmetric tensor field. This led Moffat to propose Metric Skew Tensor Gravity (MSTG),[5] in which a skew symmetric tensor field postulated as part of the gravitational action.

A newer version of MSTG, in which the skew symmetric tensor field was replaced by a vector field, is scalar–tensor–vector gravity (STVG). STVG, like Milgrom's Modified Newtonian Dynamics (MOND), can provide an explanation for flat rotation curves of galaxies.

38.1   See also

- Nonsymmetric gravitational theory on arxiv.org

38.2   References

[1] J. W. Moffat (1995), "Nonsymmetric Gravitational Theory", *Phys. Lett.* *B***355**(3-4): 447–452, arXiv:gr-qc/9411006, Bibcode doi:10.1016/0370-2693(95)00670-G

[2] J. W. Moffat (1979), "New theory of gravitation", *Phys. Rev.* *D* **19** (12): 3554–3558, Bibcode:1979PhRvD..19.3554M, doi:10.1103/PhysRevD.19.3554

[3] S. Ragusa (1997), "Nonsymmetric Theory of Gravitation", *Phys. Rev.* *D* **56** (2): 864–873, Bibcode:1997PhRvD..56..864R, doi:10.1103/PhysRevD.56.864

[4] Janssen, T.; Prokopec, T. (2007). "Problems and hopes in nonsymmetric gravity", *J. Phys. A* **40** (25): 7067–7074, arXiv:gr-qc/0611005, Bibcode:2007JPhA...40.7067J, doi:10.1088/1751-8113/40/25/S63

[5] J. W. Moffat (2005). "Gravitational Theory, Galaxy Rotation Curves and Cosmology without Dark Matter", *Journal of Cosmology and Astroparticle Physics* **2005** (05): 3, arXiv:astro-ph/0412195, Bibcode:2005JCAP...05..003M, doi:10.1088/1475-7516/2005/05/003

Chapter 39

Nordström's theory of gravitation

In theoretical physics, **Nordström's theory of gravitation** was a predecessor of general relativity. Strictly speaking, there were actually *two* distinct theories proposed by the Finnish theoretical physicist Gunnar Nordström, in 1912 and 1913 respectively. The first was quickly dismissed, but the second became the first known example of a metric theory of gravitation, in which the effects of gravitation are treated entirely in terms of the geometry of a curved spacetime.

Neither of Nordström's theories are in agreement with observation and experiment. Nonetheless, the first remains of interest insofar as it led to the second. The second remains of interest both as an important milestone on the road to the current theory of gravitation, general relativity, and as a simple example of a self-consistent relativistic theory of gravitation. As an example, this theory is particularly useful in the context of pedagogical discussions of how to derive and test the predictions of a metric theory of gravitation.

39.1   Development of the theories

Nordström's theories arose at a time when several leading physicists, including Nordström in Helsinki, Max Abraham in Milan, Gustav Mie in Greifswald, Germany, and Albert Einstein in Prague, were all trying to create competing relativistic theories of gravitation.

All of these researchers began by trying to suitably modify the existing theory, the field theory version of Newton's theory of gravitation. In this theory, the field equation is the Poisson equation $\Delta\phi = 4\pi\rho$, where ϕ is the gravitational potential and ρ is the density of matter, augmented by an equation of motion for a test particle in an ambient gravitational field, which we can derive from Newton's force law and which states that the acceleration of the test particle is given by the gradient of the potential

$$\frac{d\vec{u}}{dt} = -\nabla\phi$$

This theory is not relativistic because the equation of motion refers to coordinate time rather than proper time, and because, should the matter in some isolated object suddenly be redistributed by an explosion, the field equation requires that the potential everywhere in "space" must be "updated" *instantaneously*, which violates the principle that any "news" which has a physical effect (in this case, an effect on test particle motion far from the source of the field) cannot be transmitted faster than the speed of light. Einstein's former calculus professor, Hermann Minkowski had sketched a vector theory of gravitation as early as 1908, but in 1912, Abraham pointed out that no such theory would admit stable planetary orbits. This was one reason why Nordström turned to scalar theories of gravitation (while Einstein explored tensor theories).

Nordström's first attempt to propose a suitable relativistic scalar field equation of gravitation was the simplest and most natural choice imaginable: simply replace the Laplacian in the Newtonian field equation with the D'Alembertian or wave operator, which gives $\Box\phi = 4\pi\rho$. This has the result of changing the vacuum field equation from the Laplace equation to the wave equation, which means that any "news" concerning redistribution of matter in one location is transmitted at the speed of light to other locations. Correspondingly, the simplest guess for a suitable equation of motion for test particles might seem to be $\dot{u}_a = -\phi_{,a}$ where the dot signifies differentiation with respect to proper time, subscripts following the comma denote partial differentiation with respect to the indexed coordinate, and where u^a is the velocity four-vector of the test particle. This force law had earlier been proposed by Abraham, and Nordström knew that it wouldn't work. Instead he proposed $\dot{u}_a = -\phi_{,a} - \dot{\phi}\, u_a$.

However, this theory is unacceptable for a variety of reasons. Two objections are theoretical. First, this theory is not derivable from a Lagrangian, unlike the Newtonian field theory (or most metric theories of gravitation). Second, the proposed field equation is linear. But by analogy with electromagnetism, we should expect the gravitational field to carry energy, and on the basis of Einstein's work on relativity theory, we should expect this energy to be equivalent to mass and therefore, to gravitate. This implies that the field equation should be *nonlinear*. Another objection is more practical: this theory disagrees drastically with observation.

Einstein and von Laue proposed that the problem might lie with the field equation, which, they suggested, should have the linear form $FT_{\text{matter}} = \rho$, where F is some yet unknown function of ϕ , and where T_{matter} is the trace of the stress–energy tensor describing the density, momentum, and stress of any matter present.

In response to these criticisms, Nordström proposed his second theory in 1913. From the proportionality of inertial and gravitational mass, he deduced that the field equation should be $\phi \Box \phi = -4\pi\, T_{\text{matter}}$, which is nonlinear. Nordström now took the equation of motion to be

$$\frac{d\,(\phi\, u_a)}{ds} = -\phi_{,a}$$

or $\phi\, \dot{u}_a = -\phi_{,a} - \dot{\phi}\, u_a$.

Einstein took the first opportunity to proclaim his approval of the new theory. In a keynote address to the annual meeting of the Society of German Scientists and Physicians, given in Vienna on September 23, 1913, Einstein surveyed the state of the art, declaring that only his own work with Marcel Grossmann and the second theory of Nordström were worthy of consideration. (Mie, who was in the audience, rose to protest, but Einstein explained his criteria and Mie was forced to admit that his own theory did not meet them.) Einstein considered the special case when the only matter present is a cloud of *dust* (that is, a perfect fluid in which the pressure is assumed to be negligible). He argued that the contribution of this matter to the stress–energy tensor should be:

$$(T_{\text{matter}})_{ab} = \phi\, \rho\, u_a\, u_b$$

He then derived an expression for the stress–energy tensor of the gravitational field in Nordström's second theory,

$$4\pi\,(T_{\text{grav}})_{ab} = \phi_{,a}\, \phi_{,b} - 1/2\, \eta_{ab}\, \phi_{,m}\, \phi^{,m}$$

which he proposed should hold in general, and showed that the sum of the contributions to the stress–energy tensor from the gravitational field energy and from matter would be *conserved*, as should be the case. Furthermore, he showed, the field equation of Nordström's second theory follows from the Lagrangian

$$L = \frac{1}{8\pi}\, \eta^{ab}\, \phi_{,a}\, \phi_{,b} - \rho\, \phi$$

Since Nordström's equation of motion for test particles in an ambient gravitational field also follows from a Lagrangian, this shows that Nordström's second theory can be derived from an action principle and also shows that it obeys other properties we must demand from a self-consistent field theory.

Meanwhile, a gifted Dutch student, Adriaan Fokker had written a Ph.D. thesis under Hendrik Lorentz in which he derived what is now called the Fokker-Planck equation. Lorentz, delighted by his former student's success, arranged for Fokker to pursue post-doctoral study with Einstein in Prague. The result was a historic paper which appeared in 1914, in which Einstein and Fokker observed that the Lagrangian for Nordström's equation of motion for test particles, $L = \phi^2\, \eta_{ab}\, \dot{u}^a\, \dot{u}^b$, is the geodesic Lagrangian for a curved Lorentzian manifold with metric tensor $g_{ab} = \phi^2 \eta_{ab}$. If we adopt Cartesian coordinates with line element $d\sigma^2 = \eta_{ab} \, dx^a \, dx^b$ with corresponding wave operator $\Box$ on the flat background, or Minkowski spacetime, so that the line element of the curved spacetime is $ds^2 = \phi^2 \eta_{ab} \, dx^a \, dx^b$, then the Ricci scalar of this curved spacetime is just

$$R = -\frac{6\,\Box\phi}{\phi^3}$$

Therefore Nordström's field equation becomes simply

$$R = 24\pi T$$

where on the right hand side, we have taken the trace of the stress–energy tensor (with contributions from matter plus any non-gravitational fields) using the metric tensor g_{ab}. This is a historic result, because here for the first time we have a field equation in which on the left hand side stands a purely geometrical quantity (the Ricci scalar is the trace of the Ricci tensor, which is itself a kind of trace of the fourth rank Riemann curvature tensor), and on the right hand stands a purely physical quantity, the trace of the stress–energy tensor. Einstein gleefully pointed out that this equation now takes the form which he had earlier proposed with von Laue, and gives a concrete example of a class of theories which he had studied with Grossmann.

Some time later, Hermann Weyl introduced the Weyl curvature tensor C_{abcd}, which measures the deviation of a Lorentzian manifold from being *conformally flat*, i.e. with metric tensor having the form of the product of some scalar function with the metric tensor of flat spacetime. This is exactly the special form of the metric proposed in Nordström's second theory, so the entire content of this theory can be summarized in the following two equations:

$$R = 24\pi T, \quad C_{abcd} = 0$$

39.2   Features of Nordström's theory

Einstein was attracted to Nordström's second theory by its simplicity. The *vacuum* field equations in Nordström's theory are simply

$$R = 0, \quad C_{abcd} = 0$$

We can immediately write down the *general* vacuum solution in Nordström's theory:

$$ds^2 = \exp(2\psi)\, \eta_{ab} \, dx^a \, dx^b, \quad \Box\varphi = 0$$

where $\phi = \exp(\psi)$ and $d\sigma^2 = \eta_{ab} \, dx^a \, dx^b$ is the line element for flat spacetime in any convenient coordinate chart (such as cylindrical, polar spherical, or double null coordinates), and where $\Box$ is the ordinary wave operator on flat spacetime (expressed in cylindrical, polar spherical, or double null coordinates, respectively). But the general solution of the ordinary three-dimensional wave equation is well known, and can be given rather explicit form. Specifically, for certain charts such as cylindrical or polar spherical charts on flat spacetime (which induce corresponding charts on our curved Lorentzian manifold), we can write the general solution in terms of a power series, and we can write the general solution of certain Cauchy problems in the manner familiar from the Lienard-Wiechert potentials in electromagnetism.

In any solution to Nordström's field equations (vacuum or otherwise), if we consider ψ as controlling a *conformal perturbation from flat spacetime*, then to first order in ψ we have

$$ds^2 = \exp(2\,\psi)\,\eta_{ab}\,dx^a\,dx^b \approx (1+2\psi)\,\eta_{ab}\,dx^a\,dx^b$$

Thus, in the weak field approximation, we can identify ψ with the Newtonian gravitational potential, and we can regard it as controlling a *small* conformal perturbation from a *flat spacetime background*.

In any metric theory of gravitation, all gravitational effects arise from the curvature of the metric. In a spacetime model in Nordström's theory (but not in general relativity), this depends only on the *trace* of the stress–energy tensor. But the field energy of an electromagnetic field contributes a term to the stress–energy tensor which is *traceless*, so *in Nordström's theory, electromagnetic field energy does not gravitate!* Indeed, since every solution to the field equations of this theory is a spacetime which is among other things conformally equivalent to flat spacetime, null geodesics must agree with the null geodesics of the flat background, so *this theory can exhibit no light bending*.

Incidentally, the fact that the trace of the stress–energy tensor for an electrovacuum solution (a solution in which there is no matter present, nor any non-gravitational fields except for an electromagnetic field) vanishes shows that in the general *electrovacuum solution* in Nordström's theory, the metric tensor has the same form as in a vacuum solution, so we need only write down and solve the curved spacetime Maxwell field equations. But these are *conformally invariant*, so we can also write down the *general electrovacuum solution*, say in terms of a power series.

In any Lorentzian manifold (with appropriate tensor fields describing any matter and physical fields) which stands as a solution to Nordström's field equations, the conformal part of the Riemann tensor (i.e. the Weyl tensor) always vanishes. The Ricci scalar also vanishes identically in any vacuum region (or even, any region free of matter but containing an electromagnetic field). Are there any further restrictions on the Riemann tensor in Nordström's theory?

To find out, note that an important identity from the theory of manifolds, the Ricci decomposition, splits the Riemann tensor into three pieces, which are each fourth-rank tensors, built out of, respectively, the Ricci scalar, the trace-free Ricci tensor

$$S_{ab} = R_{ab} - \frac{1}{4}\,R\,g_{ab}$$

and the Weyl tensor. It immediately follows that Nordström's theory *leaves the trace-free Ricci tensor entirely unconstrained by algebraic relations* (other than the symmetric property, which this second rank tensor always enjoys). But taking account of the twice-contracted and detraced Bianchi identity, a differential identity which holds for the Riemann tensor in any (semi)-Riemannian manifold, we see that in Nordström's theory, as a consequence of the field equations, we have the *first-order covariant differential equation*

$$S_a{}^b{}_{;b} = 6\,\pi\,T_{;a}$$

which constrains the semi-traceless part of the Riemann tensor (the one built out of the trace-free Ricci tensor).

Thus, according to Nordström's theory, in a vacuum region only the semi-traceless part of the Riemann tensor can be nonvanishing. Then our covariant differential constraint on S_{ab} shows how variations in the trace of the stress–energy tensor in our spacetime model can generate a nonzero trace-free Ricci tensor, and thus nonzero semi-traceless curvature, which can propagate into a vacuum region. This is critically important, because *otherwise gravitation would not, according to this theory, be a long-range force capable of propagating through a vacuum*.

In general relativity, something somewhat analogous happens, but there it is the *Ricci tensor* which vanishes in any vacuum region (but *not* in a region which is matter-free but contains an electromagnetic field), and it is the *Weyl curvature* which is generated (via another first order covariant differential equation) by variations in the stress–energy tensor and which then propagates into vacuum regions, rendering gravitation a long-range force capable of propagating through a vacuum.

We can tabulate the most basic differences between Nordström's theory and general relativity, as follows:

Another feature of Nordström's theory is that it while it can be written as the theory of a certain scalar field in Minkowski spacetime, and in this form enjoys the expected conservation law for nongravitational mass-energy *together with* gravitational field energy, but suffers from a not very memorable force law, in the curved spacetime formulation the motion of test particles is described (the world line of a free test particle is a timelike geodesic, and by an obvious limit, the world line of a laser pulse is a null geodesic), but we lose the conservation law. So which interpretation is correct? In other words, which metric is the one which according to Nordström can be measured locally by physical experiments? The answer is: the curved spacetime is the physically observable one in this theory (as in all metric theories of gravitation); the flat background is a mere mathematical fiction which is however of inestimable value for such purposes as writing down the general vacuum solution, or studying the weak field limit.

At this point, we could show that in the limit of slowly moving test particles and slowly evolving weak gravitational fields, Nordström's theory of gravitation reduces to the Newtonian theory of gravitation. Rather than showing this in detail, we will proceed to a detailed study of the two most important solutions in this theory:

- the spherically symmetric static asymptotically flat vacuum solutions

- the general vacuum gravitational plane wave solution in this theory.

We will use the first to obtain the predictions of Nordström's theory for the four classic solar system tests of relativistic gravitation theories (in the ambient field of an isolated spherically symmetric object), and we will use the second to compare gravitational radiation in Nordström's theory and in Einstein's general theory of relativity.

39.3   The static spherically symmetric asymptotically flat vacuum solution

The static vacuum solutions in Nordström's theory are the Lorentzian manifolds with metrics of the form

$$ds^2 = \exp(2\psi)\, \eta_{ab}\, dx^a\, dx^b, \quad \Delta\psi = 0$$

where we can take the flat spacetime Laplace operator on the right. To first order in ψ, the metric becomes

$$ds^2 = (1 + 2\psi)\, \eta_{ab}\, dx^a\, dx^b$$

where $\eta_{ab}\, dx^a\, dx^b$ is the metric of Minkowski spacetime (the flat background).

39.3.1   The metric

Adopting polar spherical coordinates, and using the known spherically symmetric asymptotically vanishing solutions of the Laplace equation, we can write the desired *exact solution* as

$$ds^2 = (1 - m/\rho)\left(-dt^2 + d\rho^2 + \rho^2\left(d\theta^2 + \sin(\theta)^2\, d\phi^2\right)\right)$$

where we justify our choice of integration constants by the fact that this is the unique choice giving the correct Newtonian limit. This gives the solution in terms of coordinates which directly exhibit the fact that this spacetime is conformally equivalent to Minkowski spacetime, but the radial coordinate in this chart does not readily admit a direct geometric interpretation. Therefore, we adopt instead Schwarzschild coordinates, using the transformation $r = \rho\,(1 - m/\rho)$, which brings the metric into the form

$$ds^2 = (1 + m/r)^{-2}\left(-dt^2 + dr^2\right) + r^2\left(d\theta^2 + \sin(\theta)^2\, d\phi^2\right)$$

$$-\infty < t < \infty, \; 0 < r < \infty, \; 0 < \theta < \pi, \; -\pi < \phi < \pi$$

Here, r now has the simple geometric interpretation that the surface area of the coordinate sphere $r = r_0$ is just $4\pi r_0^2$.

Just as happens in the corresponding static spherically symmetric asymptotically flat solution of general relativity, this solution admits a four-dimensional Lie group of isometries, or equivalently, a four-dimensional (real) Lie algebra of Killing vector fields. These are readily determined to be

$$\partial_t$$

$$\partial_\phi$$

$$-\cos(\phi)\,\partial_\theta + \cot(\theta)\,\sin(\phi)\,\partial_\phi$$

$$\sin(\phi)\,\partial_\theta + \cot(\theta)\,\cos(\phi)\,\partial_\phi$$

These are exactly the same vector fields which arise in the Schwarzschild coordinate chart for the Schwarzschild vacuum solution of general relativity, and they simply express the fact that this spacetime is static and spherically symmetric.

39.3.2 Geodesics

The geodesic equations are readily obtained from the geodesic Lagrangian. As always, these are second order nonlinear ordinary differential equations.

If we set $\theta = \pi/2$ we find that test particle motion confined to the equatorial plane is possible, and in this case first integrals (first order ordinary differential equations) are readily obtained. First, we have

$$\dot{t} = E\,(1 + m/r)^2 \approx E\,(1 + 2m/r)$$

where to first order in m we have the same result as for the Schwarzschild vacuum. This also shows that Nordström's theory agrees with the result of the Pound–Rebka experiment. Second, we have

$$\dot{\phi} = L/r^2$$

which is the same result as for the Schwarzschild vacuum. This expresses conservation of orbital anglar momentum of test particles moving in the equatorial plane, and shows that the period of a nearly circular orbit (as observed by a distant observer) will be same as for the Schwarzschild vacuum. Third, with $\epsilon = -1, 0, 1$ for timelike, null, spacelike geodesics, we find

$$\frac{\dot{r}^2}{(1 + m/r)^4} = E^2 - V$$

where

$$V = \frac{L^2/r^2 - \epsilon}{(1 + m/r)^2}$$

is a kind of *effective potential*. In the timelike case, we see from this that there exist *stable circular orbits* at $r_c = L^2/m$, which agrees perfectly with Newtonian theory (if we ignore the fact that now the *angular* but not the *radial* distance interpretation of r agrees with flat space notions). In contrast, in the Schwarzschild vacuum we have to first order in m the expression $r_c \approx L^2/m - 3m$. In a sense, the extra term here results from the nonlinearity of the vacuum Einstein field equation.

39.3.3   Static observers

It makes sense to ask how much force is required to hold a test particle with a given mass over the massive object which we assume is the source of this static spherically symmetric gravitational field. To find out, we need only adopt the simple frame field

$$\vec{e}_0 = (1 + m/r)\, \partial_t$$
$$\vec{e}_1 = (1 + m/r)\, \partial_r$$
$$\vec{e}_2 = \frac{1}{r}\, \partial_\theta$$
$$\vec{e}_3 = \frac{1}{r\,\sin(\theta)}\, \partial_\phi$$

Then, the acceleration of the world line of our test particle is simply

$$\nabla_{\vec{e}_0} \vec{e}_0 = \frac{m}{r^2}\, \vec{e}_2$$

Thus, the particle must maintain radially outward to maintain its position, with a magnitude given by the familiar Newtonian expression (but again we must bear in mind that the radial coordinate here cannot quite be identified with a flat space radial coordinate). Put in other words, this is the "gravitational acceleration" measured by a static observer who uses a rocket engine to maintain his position. In contrast, to *second* order in m, in the Schwarzschild vacuum the magnitude of the radially outward acceleration of a static observer is m r^{-2} + m^2 r^{-3}; here too, the second term expresses the fact that Einstein gravity is slightly stronger "at corresponding points" than Nordström gravity.

The tidal tensor measured by a static observer is

$$E[\vec{X}]_{ab} = \frac{m}{r^3}\, \mathrm{diag}(-2, 1, 1) + \frac{m^2}{r^4}\, \mathrm{diag}(-1, 1, 1)$$

where we take $\vec{X} = \vec{e}_0$. The first term agrees with the corresponding solution in the Newtonian theory of gravitation and the one in general relativity. The second term shows that the tidal forces are a bit *stronger* in Nordström gravity than in Einstein gravity.

39.3.4   Extra-Newtonian precession of periastria

In our discussion of the geodesic equations, we showed that in the equatorial coordinate plane $\theta = \pi/2$ we have

$$\dot{r}^2 = (E^2 - V)\,(1 + m/r)^4$$

where $V = (1 + L^2/r^2)/(1 + m/r)^2$ for a timelike geodesic. Differentiating with respect to proper time s, we obtain

$$2\dot{r}\ddot{r} = \frac{d}{dr}\left((E^2 - V)\,(1 + m/r)^4\right)\, \dot{r}$$

Dividing both sides by $\dot{r}$ gives

$$\ddot{r} = \frac{1}{2} \frac{d}{dr}\left((E^2 - V)\,(1 + m/r)^4\right)$$

We found earlier that the minimum of V occurs at $r_c = L^2/m$ where $E_c = L^2/(L^2 + m^2)$. Evaluating the derivative, using our earlier results, and setting $\varepsilon = r - L^2/m^2$, we find

$$\ddot{\varepsilon} = -\frac{m^4}{L^8}(m^2 + L^2)\varepsilon + O(\varepsilon^2)$$

which is (to first order) the equation of simple harmonic motion.

In other words, nearly circular orbits will exhibit a radial oscillation. However, unlike what happens in Newtonian gravitation, the period of this oscillation will not quite match the orbital period. This will result in slow precession of the periastria (points of closest approach) of our nearly circular orbit, or more vividly, in a slow rotation of the long axis of a quasi-Keplerian nearly elliptical orbit. Specifically,

$$\omega_{shm} \approx \frac{m^2}{L^4}\sqrt{m^2 + L^2} = \frac{1}{r^2}\sqrt{m^2 + mr}$$

(where we used $L = \sqrt{mr}$ and removed the subscript from r_c), whereas

$$\omega_{orb} = \frac{L}{r^2} = \sqrt{m/r^3}$$

The discrepancy is

$$\Delta\omega = \omega_{orb} - \omega_{shm} = \sqrt{\frac{m}{r^3}} - \sqrt{\frac{m^2}{r^4} + \frac{m}{r^3}} \approx -\frac{1}{2}\sqrt{\frac{m^3}{r^5}}$$

so the periastrion lag per orbit is

$$\Delta\phi = 2\pi\,\Delta\omega \approx -\pi\sqrt{\frac{m^3}{r^5}}$$

and to first order in m, the long axis of the nearly elliptical orbit rotates with the rate

$$\frac{\Delta\phi}{\omega_{orb}} \approx -\frac{\pi m}{r}$$

This can be compared with the corresponding expression for the Schwarzschild vacuum solution in general relativity, which is (to first order in m)

$$\frac{\Delta\phi}{\omega_{orb}} \approx \frac{6\pi m}{r}$$

Thus, in Nordström's theory, if the nearly elliptical orbit is tranversed counterclockwise, the long axis slowly rotates *clockwise*, whereas in general relativity, it rotates *counterclockwise* six times faster. In the first case we may speak of a periastrion *lag* and in the second case, a periastrion *advance*. In either theory, with more work, we can derive more general expressions, but we shall be satisfied here with treating the special case of nearly circular orbits.

For example, according to Nordström's theory, the perihelia of Mercury should *lag* at a rate of about 7 seconds of arc per century, whereas according to general relativity, the perihelia should *advance* at a rate of about 43 seconds of arc per century.

39.3.5   Light delay

Null geodesics in the equatorial plane of our solution satisfy

$$0 = \frac{-dt^2 + dr^2}{(1 + m/r)^2} + r^2\, d\phi^2$$

Consider two events on a null geodesic, before and after its point of closest approach to the origin. Let these distances be R_1, R, R_2 with R_1, $R_2 \gg R$. We wish to eliminate ϕ, so put $R = r\cos\phi$ (the equation of a straight line in polar coordinates) and differentiate to obtain

$$0 = -r\sin\phi\, d\phi + \cos\phi\, dr$$

Thus

$$r^2\, d\phi^2 = \cot(\phi)^2\, dr^2 = \frac{R^2}{r^2 - R^2}\, dr^2$$

Plugging this into the line element and solving for dt, we obtain

$$dt \approx \frac{1}{\sqrt{r^2 - R^2}}\ \left(r + m\,\frac{R^2}{r^2}\right)\ dr$$

Thus the coordinate time from the first event to the event of closest approach is

$$(\Delta t)_1 = \int_R^{R_1} dt \approx \frac{m + R_1}{R_1}\sqrt{R_1^2 - R^2} = \sqrt{R_1^2 - R^2} + m\sqrt{1 - (R/R_1)^2}$$

and likewise

$$(\Delta t)_2 = \int_R^{R_2} dt \approx \frac{m + R_2}{R_2}\sqrt{R_2^2 - R^2} = \sqrt{R_2^2 - R^2} + m\sqrt{1 - (R/R_2)^2}$$

Here the elapsed coordinate time expected from Newtonian theory is of course

$$\sqrt{R_1^2 - R^2} + \sqrt{R_2^2 - R^2}$$

so the relativistic time delay, according to Nordström's theory, is

$$\Delta t = m\left(\sqrt{1 - (R/R_1)^2} + \sqrt{1 - (R/R_2)^2}\right)$$

To first order in the small ratios R/R_1, R/R_2 this is just $\Delta t = 2m$.

The corresponding result in general relativity is

$$\Delta t = 2m + 2m\,\log\left(\frac{4\,R_1\,R_2}{R^2}\right)$$

which depends logarithmically on the small ratios R/R_1, R/R_2. For example, in the classic experiment in which, at a time when, as viewed from Earth, Venus is just about to pass *behind* the Sun, a radar signal emitted from Earth which grazes the limb of the Sun, bounces off Venus, and returns to Earth (once again grazing the limb of the Sun), the relativistic time delay is about 20 microseconds according to Nordström's theory and about 240 microseconds according to general relativity.

39.3.6   Summary of results

We can summarize the results we found above in the following table, in which the given expressions represent appropriate approximations:

The last four lines in this table list the so-called *four classic solar system tests* of relativistic theories of gravitation. Of the three theories appearing in the table, only general relativity is in agreement with the results of experiments and observations in the solar system. Nordström's theory gives the correct result only for the Pound–Rebka experiment; not surprisingly, Newton's theory flunks all four relativistic tests.

39.4   Vacuum gravitational plane wave

In the double null chart for Minkowski spacetime,

$$ds^2 = 2\,du\,dv + dx^2 + dy^2, \quad -\infty < u,\, v,\, x,\, y < \infty$$

a simple solution of the wave equation

$$2\,\psi_{uv} + \psi_{xx} + \psi_{yy} = 0$$

is $\psi = f(u)$, where f is an *arbitrary* smooth function. This represents a plane wave traveling in the z direction. Therefore, Nordström's theory admits the *exact vacuum solution*

$$ds^2 = \exp(2f(u))\,\left(2\,du\,dv + dx^2 + dy^2\right), \quad -\infty < u,\, v,\, x,\, y < \infty$$

which we can interpret in terms of the propagation of a gravitational plane wave.

This Lorentzian manifold admits a *six-dimensional Lie group of isometries*, or equivalently, a six-dimensional Lie algebra of Killing vector fields:

∂_v (a null translation, *"opposing"* the wave vector field ∂_u)

∂_x, ∂_y (spatial translation orthogonal to the wavefronts)

$-y\,\partial_x + x\,\partial_y$ (rotation about axis parallel to direction of propagation)

$x\,\partial_v + u\,\partial_x$, $y\,\partial_v + u\,\partial_y$

For example, the Killing vector field $x\,\partial_v + u\,\partial_x$ integrates to give the one parameter family of isometries

$$(u, v, x, y) \longrightarrow (u,\ v + x\,\lambda + \frac{u}{2}\,\lambda^2,\ x + u\,\lambda,\ y)$$

Just as in special relativity (and general relativity), it is always possible to change coordinates, without disturbing the form of the solution, so that the wave propagates in any direction transverse to ∂_z. Note that our isometry group is transitive on the hypersurfaces $u = u_0$.

In contract, the generic gravitational plane wave in general relativity has only a *five-dimensional Lie group of isometries*. (In both theories, special plane waves may have extra symmetries.) We'll say a bit more about why this is so in a moment.

Adopting the frame field

$$\vec{e}_0 = \frac{1}{\sqrt{2}}\,\left(\partial_v + \exp(-2f)\,\partial_u\right)$$

$$\vec{e}_1 = \frac{1}{\sqrt{2}} \left(\partial_v - \exp(-2f) \, \partial_u \right)$$

$$\vec{e}_2 = \partial_x$$

$$\vec{e}_3 = \partial_y$$

we find that the corresponding family of test particles are *inertial* (freely falling), since the acceleration vector vanishes

$$\nabla_{\vec{e}_0} \vec{e}_0 = 0$$

Notice that if f vanishes, this family becomes a family of mutually stationary test particles in flat (Minkowski) spacetime. With respect to the timelike geodesic congruence of world lines obtained by integrating the timelike unit vector field $\vec{X} = \vec{e}_0$, the *expansion tensor*

$$\theta[\vec{X}]_{\hat{p}\hat{q}} = \frac{1}{\sqrt{2}} f'(u) \, \exp(-2f(u)) \, \mathrm{diag}(0,1,1)$$

shows that our test particles are expanding or contracting *isotropically* and *transversely to the direction of propagation*. This is exactly what we would expect for a transverse *spin-0 wave*; the behavior of analogous families of test particles which encounter a gravitational plane wave in general relativity is quite different, because these are *spin-2 waves*. This is due to the fact that Nordström's theory of gravitation is a *scalar theory*, whereas Einstein's theory of gravitation (general relativity) is a *tensor theory*. On the other hand, gravitational waves in both theories are *transverse* waves. Electromagnetic plane waves are of course also *transverse*. The *tidal tensor*

$$E[\vec{X}]_{\hat{p}\hat{q}} = \frac{1}{2} \exp(-4f(u)) \left(f'(u)^2 - f''(u) \right) \mathrm{diag}(0,1,1)$$

further exhibits the spin-0 character of the gravitational plane wave in Nordström's theory. (The tidal tensor and expansion tensor are three-dimensional tensors which "live" in the hyperplane elements orthogonal to $\vec{e}_0$, which in this case happens to be irrotational, so we can regard these tensors as defined on orthogonal hyperslices.)

The exact solution we are discussing here, which we interpret as a propagating gravitational plane wave, gives some basic insight into the *propagation* of gravitational radiation in Nordström's theory, but it does not yield any insight into the *generation* of gravitational radiation in this theory. At this point, it would be natural to discuss the analog for Nordström's theory of gravitation of the standard linearized gravitational wave theory in general relativity, but we shall not pursue this.

39.5   See also

- Classical theories of gravitation
- Congruence (general relativity)
- Gunnar Nordström
- Obsolete physical theories
- General Theory of Relativity

39.6   References

- Ravndal, Finn (2004). Scalar Gravitation and Extra Dimensions

- Pais, Abraham (1982). *Subtle is the Lord: The Science and the Life of Albert Einstein*. Oxford: Oxford University Press. ISBN 0-19-280672-6. See *Chapter 13*.

- Lightman, Alan P.; Press, William H.; Price, Richard H. & Teukolsky, Saul A. (1975). *Problem Book in Relativity and Gravitation*. Princeton: Princeton University Press. ISBN 0-691-08162-X. See *problem 13.2*.

Chapter 40

Rainbow Gravity theory

For other uses of "Gravity's Rainbow", see Gravity's Rainbow (disambiguation).

Rainbow Gravity theory (sometimes called "Gravity's Rainbow"[1]) is a proposed theory which suggests that the

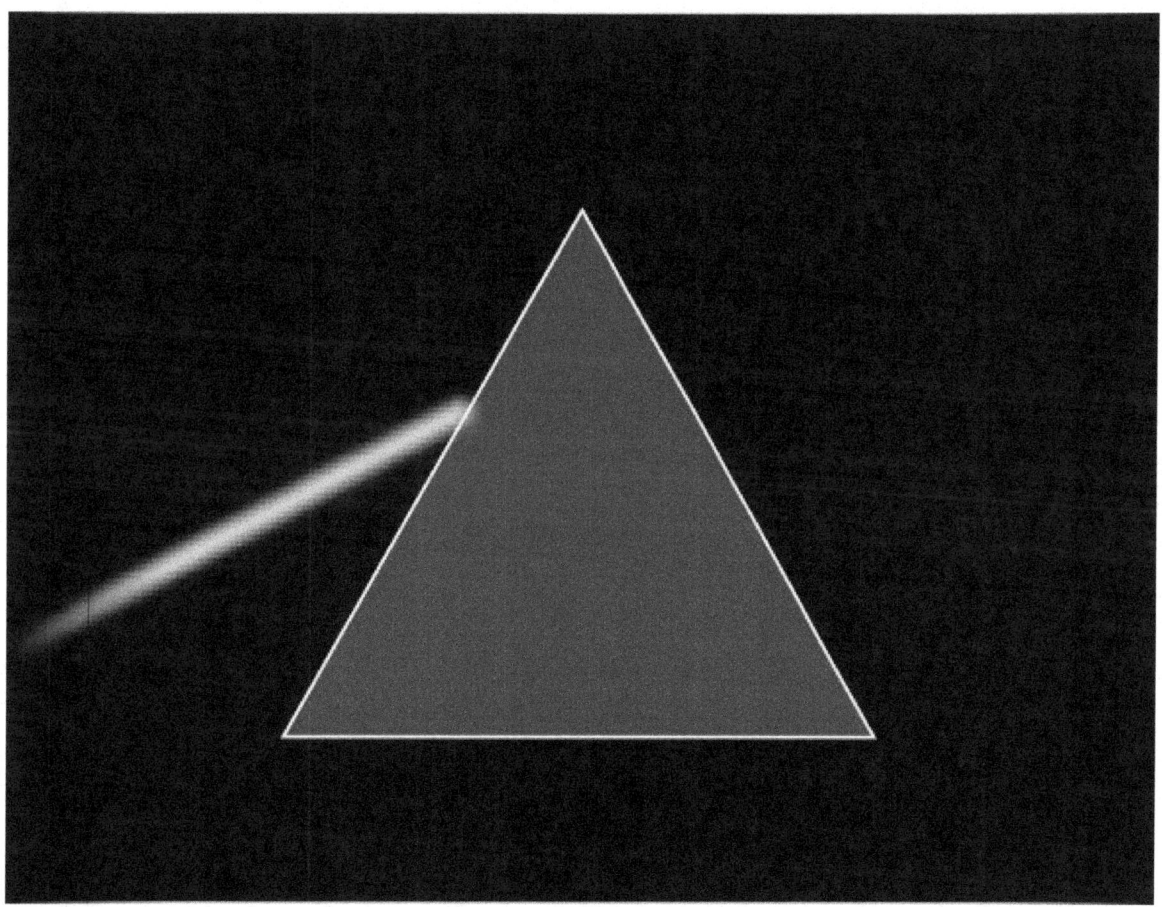

A prism which according to the Rainbow Gravity theory affects different light wavelengths in a similar way as the gravity

universe has no beginning and stretches back in time infinitely with no Big Bang. The theory was first put forth at the beginning of the 21st century.[2] It is said to fix the gaps between the general relativity theory and quantum mechanics. The theory of Rainbow Gravity states that gravity affects different wavelengths of light differently similarly to a prism creating a rainbow.[3] This effect is very small for Earth-sized objects but might be detectable for objects with extremely big energy such as black holes.[4] Scientists are trying to detect the Rainbow Gravity using the Large Hadron Collider.[4]

40.1 See also

- Steady State theory
- Eternal inflation
- Cyclic model

40.2 References

[1] see, e.g., Zyga, Lisa (January 15, 2015). "Black holes do not exist where space and time do not exist, says new theory". *phys.org*. Retrieved March 28, 2015.

[2] Clara Moskowitz (9 Dec 2013). "In a "Rainbow" Universe Time May Have No Beginning". *Scientific American*.

[3] Greg Kestin (13 Dec 2013). "Universe May Have Been Around Since Forever, According to Rainbow Gravity Theory". *NovaNext*.

[4] Sarah Knapton (23 Mar 2015). "Big Bang theory could be debunked by Large Hadron Collider". *The Telegraph*.

Chapter 41

Scalar theories of gravitation

Scalar theories of gravitation are field theories of gravitation in which the gravitational field is described using a scalar field, which is required to satisfy some field equation.

Note: This article focuses on relativistic classical field theories of gravitation. The best known relativistic classical field theory of gravitation, general relativity, is a tensor theory, in which the gravitational interaction is described using a tensor field.

41.1   Newtonian gravity

The prototypical scalar theory of gravitation is Newtonian gravitation. In this theory, the gravitational interaction is completely described by the potential Φ, which is required to satisfy the Poisson equation (with the mass density acting as the source of the field). To wit:

$\Delta \Phi = 4\pi G \rho$, where

- G is the gravitational constant and

- ρ is the mass density.

This field theory formulation leads directly to the familiar law of universal gravitation, $F = m_1 m_2 G / r^2$.

41.2   Nordström's theories of gravitation

The first attempts to present a relativistic (classical) field theory of gravitation were also scalar theories. Gunnar Nordström created two such theories.[1]

Nordström's first idea (1912) was to simply replace the divergence operator in the field equation of Newtonian gravity with the d'Alembertian operator $\Box = \partial_t^2 - \nabla^2$. This gives the field equation

$\Box \Phi = 4\pi G \rho$

However, several theoretical difficulties with this theory quickly arose, and Nordström dropped it.

A year later, Nordström tried again, presenting the field equation

$\Phi \Box \Phi = -4\pi G T$

where T is the trace of the stress–energy tensor.

Solutions of Nordström's second theory are conformally flat Lorentzian spacetimes. That is, the metric tensor can be written as $g_{\mu\nu} = A\eta_{\mu\nu}$, where

- η$\mu\nu$ is the Minkowski metric, and

- A is a scalar which is a function of position.

This suggestion signifies that the inertial mass should depend on the scalar field.

Nordström's second theory satisfies the weak equivalence principle. However:

- The theory fails to predict any deflection of light passing near a massive body (contrary to observation)

- The theory predicts an anomalous perihelion precession of Mercury, but this disagrees in both sign and magnitude with the observed anomalous precession (the part which cannot be explained using Newtonian gravitation).

Despite these disappointing results, Einstein's critiques of Nordström's second theory played an important role in his development of general relativity.

41.3 Einstein's scalar theory

In 1913, Einstein (erroneously) concluded from his hole argument that general covariance was not viable.[2] Inspired by Nordström's work, he proposed his own scalar theory.[3] This theory employs a massless scalar field coupled to the stress–energy tensor, which is the sum of two terms. The first,

$$T_g^{\mu\nu} = \frac{1}{4\pi G}\left[\partial^\mu\phi\,\partial^\nu\phi - \frac{1}{2}\eta^{\mu\nu}\partial_\lambda\phi\,\partial^\lambda\phi\right]$$

represents the stress–momentum–energy of the scalar field itself. The second represents the stress-momentum-energy of any matter which may be present:

$$T_m^{\mu\nu} = \rho\phi u^\mu u^\nu$$

where u^μ is the velocity vector of an observer, or tangent vector to the world line of the observer. (Einstein made no attempt, in this theory, to take account of possible gravitational effects of the field energy of the electromagnetic field.)

Unfortunately, this theory is not diffeomorphism covariant. This is an important consistency condition, so Einstein dropped this theory in late 1914.[4] Associating the scalar field with the metric leads to Einstein's later conclusions that the theory of gravitation he sought could not be a scalar theory. Indeed, the theory he finally arrived at in 1915, general relativity, is a tensor theory, not a scalar theory, with a 2-tensor, the metric, as the potential. Unlike his 1913 scalar theory, it is generally covariant, and it does take into account the field energy–momentum–stress of the electromagnetic field (or any other nongravitational field).

41.4 Additional variations

- Kaluza–Klein theory involves the use of a scalar gravitational field in addition to the electromagnetic field potential A^μ in an attempt to create a five-dimensional unification of gravity and electromagnetism. Its generalization with a 5th variable component of the metric that leads to a variable gravitational constant was first given by Pascual Jordan.[5][6]

- Brans–Dicke theory is a scalar-tensor theory, not a scalar theory, meaning that it represents the gravitational interaction using both a scalar field and a tensor field. We mention it here because one of the field equations of this theory involves only the scalar field and the trace of the stress–energy tensor, as in Nordström's theory. Moreover, the Brans–Dicke theory is equal to the independently derived theory of Jordan (hence it is often referred to as the Jordan-Brans–Dicke or JBD theory). The Brans–Dicke theory couples a scalar field with the curvature of space-time and is self-consistent and, assuming appropriate values for a tunable constant, this theory has not been ruled out by observation. The Brans–Dicke theory is generally regarded as a leading competitor of general relativity, which is a pure tensor theory. However, the Brans–Dicke theory seems to need too high a parameter, which favours general relativity).[5]

- Zee combined the idea of the BD theory with the Higgs-Mechanism of Symmetry Breakdown for mass generation, which led to a scalar-tensor theory with Higgs field as scalar field, in which the scalar field is massive (short-ranged). An example of this theory was proposed by H. Dehnen and H. Frommert 1991, parting from the nature of Higgs field interacting gravitational- and Yukawa (long-ranged)-like with the particles that get mass through it.[7][8][9]

- The Watt–Misner theory (1999) is a recent example of a scalar theory of gravitation. It is not intended as a viable theory of gravitation (since, as Watt and Misner point out, it is not consistent with observation), but as a toy theory which can be useful in testing numerical relativity schemes. It also has pedagogical value.[10]

41.5   See also

- Nordström's theory of gravitation

41.6   References

[1] Norton, John D. (1992). "Einstein, Nordström and the early demise of scalar, Lorentz-covariant theories of gravitation" (PDF). *Archive for History of Exact Sciences* **45** (1): 17–94. Retrieved 20 April 2015.

[2] Stachel, John (2014). "The Hole Argument and Some Physical and Philosophical Implications". *Living Reviews in Relativity* **17** (1). Retrieved 20 April 2015.

[3] Janssen, Michel (2007). "What did Einstein know and when did He know it? A Besso Memo Dated August 1913". *Boston Studies in the Philosophy of Science* **250**: 787–837.

[4] Norton, John (1984). "How Einstein found his field equations: 1912-1915" (PDF). *Historical studies in the physical sciences*: 253–316.

[5] Brans, Carl H. (2005). "The roots of scalar-tensor theory: an approximate history". arXiv:gr-qc/0506063.

[6] Goenner, Hubert (2012). "Some remarks on the genesis of scalar-tensor theories". arXiv:1204.3455v1.

[7] Dehnen, H.; Frommert, H. (1990). "Scalar gravity and Higgs potential". *International Journal of Theoretical Physics* **29** (4): 361–370. Bibcode:1990IJTP...29..361D. doi:10.1007/BF00674437.

[8] Dehnen, H.; Frommert, H. (1991). "Higgs-Field Gravity within the Standard Model". *International Journal of Theoretical Physics* **30** (7): 995–998.

[9] Dehnen, H.; Frommert, H.; Ghaboussi, F. (1992). "Higgs field and a new scalar-tensor theory of gravity". *International Journal of Theoretical Physics* **31** (1): 109–114. Bibcode:1992IJTP...31..109D. doi:10.1007/BF00674344.

[10] Watt, Keith & Misner, Charles W. (1999). "Relativistic Scalar Gravity: A Laboratory for Numerical Relativity". arXiv:gr-qc/9910032.

41.7 External links

- Goenner, Hubert F. M., "On the History of Unified Field Theories"; *Living Rev. Relativity* *7(2)*, 2004, lrr-2004-2. Retrieved August 10, 2005.

- Ravndal, Finn (2004). "Scalar Gravitation and Extra Dimensions". arXiv:gr-qc/0405030 [gr-qc].

- P. Jordan, *Schwerkraft und Weltall*, Vieweg (Braunschweig) 1955.

Chapter 42

Scalar–tensor–vector gravity

Not to be confused with Tensor–vector–scalar gravity or Bi-scalar tensor vector gravity.

Scalar–tensor–vector gravity (STVG)[1] is a modified theory of gravity developed by John Moffat, a researcher at the Perimeter Institute for Theoretical Physics in Waterloo, Ontario. The theory is also often referred to by the acronym **MOG** (*MO*dified *G*ravity).

42.1 Overview

Scalar–tensor–vector gravity theory,[2] also known as MOdified Gravity (MOG), is based on an action principle and postulates the existence of a vector field, while elevating the three constants of the theory to scalar fields. In the weak-field approximation, STVG produces a Yukawa-like modification of the gravitational force due to a point source. Intuitively, this result can be described as follows: far from a source gravity is stronger than the Newtonian prediction, but at shorter distances, it is counteracted by a repulsive fifth force due to the vector field.

STVG has been used successfully to explain galaxy rotation curves,[3] the mass profiles of galaxy clusters,[4] gravitational lensing in the Bullet Cluster,[5] and cosmological observations[6] without the need for dark matter. On a smaller scale, in the Solar System, STVG predicts no observable deviation from general relativity.[7] The theory may also offer an explanation for the origin of inertia.[8]

42.2 Mathematical details

STVG is formulated using the action principle. In the following discussion, a metric signature of $[+, -, -, -]$ will be used; the speed of light is set to $c = 1$, and we are using the following definition for the Ricci tensor: $R_{\mu\nu} = \partial_\alpha \Gamma^\alpha_{\mu\nu} - \partial_\nu \Gamma^\alpha_{\mu\alpha} + \Gamma^\alpha_{\mu\nu}\Gamma^\beta_{\alpha\beta} - \Gamma^\alpha_{\mu\beta}\Gamma^\beta_{\alpha\nu}$.

We begin with the Einstein-Hilbert Lagrangian:

$$\mathcal{L}_G = -\frac{1}{16\pi G}\left(R + 2\Lambda\right)\sqrt{-g}.$$

where R is the trace of the Ricci tensor, G is the gravitational constant, g is the determinant of the metric tensor $g_{\mu\nu}$, while Λ is the cosmological constant.

We introduce the Maxwell-Proca Lagrangian for the STVG vector field ϕ_μ:

$$\mathcal{L}_\phi = -\frac{1}{4\pi}\omega\left[\frac{1}{4}B^{\mu\nu}B_{\mu\nu} - \frac{1}{2}\mu^2\phi_\mu\phi^\mu + V_\phi(\phi)\right]\sqrt{-g}.$$

where $B_{\mu\nu} = \partial_\mu\phi_\nu - \partial_\nu\phi_\mu$, μ is the mass of the vector field, ω characterizes the strength of the coupling between the fifth force and matter, and V_ϕ is a self-interaction potential.

The three constants of the theory, G, μ and ω, are promoted to scalar fields by introducing associated kinetic and potential terms in the Lagrangian density:

$$\mathcal{L}_S = -\frac{1}{G} \left[\frac{1}{2} g^{\mu\nu} \left(\frac{\nabla_\mu G \nabla_\nu G}{G^2} + \frac{\nabla_\mu \mu \nabla_\nu \mu}{\mu^2} - \nabla_\mu \omega \nabla_\nu \omega \right) + \frac{V_G(G)}{G^2} + \frac{V_\mu(\mu)}{\mu^2} + V_\omega(\omega) \right] \sqrt{-g},$$

where ∇_μ denotes covariant differentiation with respect to the metric $g_{\mu\nu}$, while V_G, V_μ, and V_ω are the self-interaction potentials associated with the scalar fields.

The STVG action integral takes the form

$$S = \int \left(\mathcal{L}_G + \mathcal{L}_\phi + \mathcal{L}_S + \mathcal{L}_M \right) d^4x,$$

where $\mathcal{L}_M$ is the ordinary matter Lagrangian density.

42.3 Spherically symmetric, static vacuum solution

The field equations of STVG can be developed from the action integral using the variational principle. First a test particle Lagrangian is postulated in the form

$$\mathcal{L}_{TP} = -m + \alpha \omega q_5 \phi_\mu u^\mu,$$

where m is the test particle mass, α is a factor representing the nonlinearity of the theory, q_5 is the test particle's fifth-force charge, and $u^\mu = dx^\mu/ds$ is its four-velocity. Assuming that the fifth-force charge is proportional to mass, i.e., $q_5 = \kappa m$, the value of $\kappa = \sqrt{G_N/\omega}$ is determined and the following equation of motion is obtained in the spherically symmetric, static gravitational field of a point mass of mass M:

$$\ddot{r} = -\frac{G_N M}{r^2} \left[1 + \alpha - \alpha(1 + \mu r) e^{-\mu r} \right],$$

where G_N is Newton's constant of gravitation. Further study of the field equations allows a determination of α and μ for a point gravitational source of mass M in the form[9]

$$\mu = \frac{D}{\sqrt{M}},$$

$$\alpha = \frac{G_\infty - G_N}{G_N} \frac{M}{(\sqrt{M} + E)^2}.$$

where $G_\infty \simeq 20 G_N$ is determined from cosmological observations, while for the constants D and E galaxy rotation curves yield the following values:

$$D \simeq 6250 M_\odot^{1/2} \text{kpc}^{-1},$$

$$E \simeq 25000 M_\odot^{1/2},$$

where $M_\odot$ is the mass of the Sun. These results form the basis of a series of calculations that are used to confront the theory with observation.

42.4 Observations

STVG/MOG has been applied successfully to a range of astronomical, astrophysical, and cosmological phenomena.

On the scale of the Solar System, the theory predicts no deviation[7] from the results of Newton and Einstein. This is also true for star clusters containing no more than a maximum of a few million solar masses.

The theory accounts for the rotation curves of spiral galaxies,[3] correctly reproducing the Tully-Fisher law.[9]

STVG is in good agreement with the mass profiles of galaxy clusters.[4]

STVG can also account for key cosmological observations, including:[6]

- The acoustic peaks in the cosmic microwave background radiation;

- The accelerating expansion of the universe that is apparent from type Ia supernova observations;

- The matter power spectrum of the universe that is observed in the form of galaxy-galaxy correlations.

42.5   See also

- Modified Newtonian Dynamics

- Tensor–vector–scalar gravity

- Nonsymmetric Gravitational Theory

42.6   References

[1] McKee, M. (25 January 2006). "Gravity theory dispenses with dark matter". *New Scientist*. Retrieved 2008-07-26.

[2] Moffat, J. W. (2006). "Scalar-Tensor-Vector Gravity Theory". *Journal of Cosmology and Astroparticle Physics* **3**: 4. arXiv:gr-qc/0506021. Bibcode:2006JCAP...03..004M. doi:10.1088/1475-7516/2006/03/004.

[3] Brownstein, J. R.; Moffat, J. W. (2006). "Galaxy Rotation Curves Without Non-Baryonic Dark Matter". *Astrophysical Journal* **636**: 721–741. arXiv:astro-ph/0506370. Bibcode:2006ApJ...636..721B. doi:10.1086/498208.

[4] Brownstein, J. R.; Moffat, J. W. (2006). "Galaxy Cluster Masses Without Non-Baryonic Dark Matter". *Monthly Notices of the Royal Astronomical Society* **367**: 527–540. arXiv:astro-ph/0507222. Bibcode:2006MNRAS.367..527B. doi:10.1111/j.1365-2966.2006.09996.x.

[5] Brownstein, J. R.; Moffat, J. W. (2007). "The Bullet Cluster 1E0657-558 evidence shows Modified Gravity in the absence of Dark Matter". *Monthly Notices of the Royal Astronomical Society* **382**: 29–47. arXiv:astro-ph/0702146. Bibcode:2007MNRAS.382...29B.doi:10.1111/j.1365-2966.2007.12275.x.

[6] Moffat, J. W.; Toth, V. T. (2007). "Modified Gravity: Cosmology without dark matter or Einstein's cosmological constant". arXiv:0710.0364 [astro-ph].

[7] Moffat, J. W.; Toth, V. T. (2008). "Testing modified gravity with globular cluster velocity dispersions". *Astrophysical Journal* **680**: 1158–1161. arXiv:0708.1935. Bibcode:2008ApJ...680.1158M. doi:10.1086/587926.

[8] Moffat, J. W.; Toth, V. T. (2009). "Modified gravity and the origin of inertia". *Monthly Notices of the Royal Astronomical Society Letters* **395**: L25. arXiv:0710.3415. Bibcode:2009MNRAS.395L..25M. doi:10.1111/j.1745-3933.2009.00633.x.

[9] Moffat, J. W.; Toth, V. T. (2009). "Fundamental parameter-free solutions in Modified Gravity". *Classical and Quantum Gravity* **26**: 085002. arXiv:0712.1796. Bibcode:2009CQGra..26h5002M. doi:10.1088/0264-9381/26/8/085002.

Chapter 43

Supergravity

In theoretical physics, **supergravity** (**supergravity theory**; **SUGRA** for short) is a field theory that combines the principles of supersymmetry and general relativity. Together, these imply that, in supergravity, the supersymmetry is a local symmetry (in contrast to non-gravitational supersymmetric theories, such as the Minimal Supersymmetric Standard Model). Since the generators of supersymmetry (SUSY) are convoluted with the Poincaré group to form a super-Poincaré algebra, it can be seen that supergravity follows naturally from supersymmetry.[1]

43.1 Gravitons

Like any field theory of gravity, a supergravity theory contains a spin-2 field whose quantum is the graviton. Supersymmetry requires the graviton field to have a superpartner. This field has spin 3/2 and its quantum is the gravitino. The number of gravitino fields is equal to the number of supersymmetries.

43.2 History

43.2.1 Gauge supersymmetry

The first theory[1] of local supersymmetry was proposed in 1975 by Dick Arnowitt and Pran Nath and was called **gauge supersymmetry**.

43.2.2 SUGRA

SUGRA, or supergravity, was discovered in 1976 by Dan Freedman, Sergio Ferrara and Peter van Nieuwenhuizen,[1] but was quickly generalized to many different theories in various numbers of dimensions and additional (N) supersymmetry charges. Supergravity theories with N>1 are usually referred to as extended supergravity (SUEGRA). Some supergravity theories were shown to be equivalent to certain higher-dimensional supergravity theories via dimensional reduction (e.g. $N = 1$ **11-dimensional** supergravity is dimensionally reduced on S^7 to $N = 8$, $d = 4$ SUGRA). The resulting theories were sometimes referred to as Kaluza–Klein theories as Kaluza and Klein constructed in 1919 a 5-dimensional gravitational theory, that when dimensionally reduced on circle, its 4-dimensional non-massive modes describe electromagnetism coupled to gravity.

43.2.3   mSUGRA

mSUGRA means minimal SUper GRAvity. The construction of a realistic model of particle interactions within the $N = 1$ supergravity framework where supersymmetry (SUSY) is broken by a super Higgs mechanism was carried out by Ali Chamseddine, Richard Arnowitt and Pran Nath in 1982. In these classes of models collectively now known as minimal supergravity Grand Unification Theories (mSUGRA GUT), gravity mediates the breaking of SUSY through the existence of a hidden sector. mSUGRA naturally generates the Soft SUSY breaking terms which are a consequence of the Super Higgs effect. Radiative breaking of electroweak symmetry through Renormalization Group Equations (RGEs) follows as an immediate consequence. mSUGRA is one of the most widely investigated models of particle physics due to its predictive power—requiring only four input parameters and a sign to determine the low energy phenomenology from the scale of Grand Unification.

See also: Gravity-Mediated Supersymmetry Breaking in the MSSM

43.2.4   11d: the maximal SUGRA

One of these supergravities, the 11-dimensional theory, generated considerable excitement as the first potential candidate for the theory of everything. This excitement was built on four pillars, two of which have now been largely discredited:

- Werner Nahm showed[2] that 11 dimensions was the largest number of dimensions consistent with a single graviton, and that a theory with more dimensions would also have particles with spins greater than 2. These problems are avoided in 12 dimensions if two of these dimensions are timelike, as has been often emphasized by Itzhak Bars.

- In 1981, Ed Witten showed[3] that 11 was the smallest number of dimensions that was big enough to contain the gauge groups of the Standard Model, namely SU(3) for the strong interactions and SU(2) times U(1) for the electroweak interactions. Today many techniques exist to embed the standard model gauge group in supergravity in any number of dimensions. For example, in the mid and late 1980s, the obligatory gauge symmetry in type I and heterotic string theories was often used. In type II string theory they could also be obtained by compactifying on certain Calabi–Yau manifolds. Today one may also use D-branes to engineer gauge symmetries.

- In 1978, Eugène Cremmer, Bernard Julia and Joël Scherk (CJS) found[4] the classical action for an 11-dimensional supergravity theory. This remains today the only known classical 11-dimensional theory with local supersymmetry and no fields of spin higher than two. Other 11-dimensional theories are known that are quantum-mechanically inequivalent to the CJS theory, but classically equivalent (that is, they reduce to the CJS theory when one imposes the classical equations of motion). For example, in the mid 1980s Bernard de Wit and Hermann Nicolai found an alternate theory in D=11 Supergravity with Local SU(8) Invariance. This theory, while not manifestly Lorentz-invariant, is in many ways superior to the CJS theory in that, for example, it dimensionally-reduces to the 4-dimensional theory without recourse to the classical equations of motion.

- In 1980, Peter Freund and M. A. Rubin showed that compactification from 11 dimensions preserving all the SUSY generators could occur in two ways, leaving only 4 or 7 macroscopic dimensions (the other 7 or 4 being compact).[5] Unfortunately, the noncompact dimensions have to form an anti-de Sitter space. Today it is understood that there are many possible compactifications, but that the Freund-Rubin compactifications are invariant under all of the supersymmetry transformations that preserve the action.

Thus, the first two results appeared to establish 11 dimensions uniquely, the third result appeared to specify the theory, and the last result explained why the observed universe appears to be four-dimensional.

Many of the details of the theory were fleshed out by Peter van Nieuwenhuizen, Sergio Ferrara and Daniel Z. Freedman.

43.2.5   The end of the SUGRA era

The initial excitement over 11-dimensional supergravity soon waned, as various failings were discovered, and attempts to repair the model failed as well. Problems included:

- The compact manifolds which were known at the time and which contained the standard model were not compatible with supersymmetry, and could not hold quarks or leptons. One suggestion was to replace the compact dimensions with the 7-sphere, with the symmetry group SO(8), or the squashed 7-sphere, with symmetry group SO(5) times SU(2).

- Until recently, the physical neutrinos seen in experiments were believed to be massless, and appeared to be left-handed, a phenomenon referred to as the chirality of the Standard Model. It was very difficult to construct a chiral fermion from a compactification — the compactified manifold needed to have singularities, but physics near singularities did not begin to be understood until the advent of orbifold conformal field theories in the late 1980s.

- Supergravity models generically result in an unrealistically large cosmological constant in four dimensions, and that constant is difficult to remove, and so require fine-tuning. This is still a problem today.

- Quantization of the theory led to quantum field theory gauge anomalies rendering the theory inconsistent. In the intervening years physicists have learned how to cancel these anomalies.

Some of these difficulties could be avoided by moving to a 10-dimensional theory involving superstrings. However, by moving to 10 dimensions one loses the sense of uniqueness of the 11-dimensional theory.

The core breakthrough for the 10-dimensional theory, known as the first superstring revolution, was a demonstration by Michael B. Green, John H. Schwarz and David Gross that there are only three supergravity models in 10 dimensions which have gauge symmetries and in which all of the gauge and gravitational anomalies cancel. These were theories built on the groups SO(32) and $E_8 \times E_8$, the direct product of two copies of E_8. Today we know that, using D-branes for example, gauge symmetries can be introduced in other 10-dimensional theories as well.[6]

43.2.6   The second superstring revolution

Initial excitement about the 10-dimensional theories, and the string theories that provide their quantum completion, died by the end of the 1980s. There were too many Calabi–Yaus to compactify on, many more than Yau had estimated, as he admitted in December 2005 at the 23rd International Solvay Conference in Physics. None quite gave the standard model, but it seemed as though one could get close with enough effort in many distinct ways. Plus no one understood the theory beyond the regime of applicability of string perturbation theory.

There was a comparatively quiet period at the beginning of the 1990s; however, several important tools were developed. For example, it became apparent that the various superstring theories were related by "string dualities", some of which relate weak string-coupling (i.e. perturbative) physics in one model with strong string-coupling (i.e. non-perturbative) in another.

Then it all changed, in what is known as the second superstring revolution. Joseph Polchinski realized that obscure string theory objects, called D-branes, which he had discovered six years earlier, are stringy versions of the p-branes that were known in supergravity theories. The treatment of these p-branes was not restricted by string perturbation theory; in fact, thanks to supersymmetry, p-branes in supergravity were understood well beyond the limits in which string theory was understood.

Armed with this new nonperturbative tool, Edward Witten and many others were able to show that all of the perturbative string theories were descriptions of different states in a single theory which Witten named M-theory. Furthermore he argued that M-theory's long wavelength limit (i.e. when the quantum wavelength associated to objects in the theory are much larger than the size of the 11th dimension) should be described by the 11-dimensional supergravity that had fallen out of favor with the first superstring revolution 10 years earlier, accompanied by the 2- and 5-branes.

Historically, then, supergravity has come "full circle". It is a commonly used framework in understanding features of string theories, M-theory and their compactifications to lower spacetime dimensions.

43.3   Relation to superstrings

Particular 10-dimensional supergravity theories are considered "low energy limits" of the 10-dimensional superstring theories; more precisely, these arise as the massless, tree-level approximation of string theories. True effective field theories of string theories, rather than truncations, are rarely available. Due to string dualities, the conjectured 11-dimensional M-theory is required to have 11-dimensional supergravity as a "low energy limit". However, this doesn't necessarily mean that string theory/M-theory is the only possible UV completion of supergravity; supergravity research is useful independent of those relations.

43.4   4D $N = 1$ SUGRA

Before we move on to SUGRA proper, let's recapitulate some important details about general relativity. We have a 4D differentiable manifold M with a Spin(3,1) principal bundle over it. This principal bundle represents the local Lorentz symmetry. In addition, we have a vector bundle T over the manifold with the fiber having four real dimensions and transforming as a vector under Spin(3,1). We have an invertible linear map from the tangent bundle TM to T. This map is the vierbein. The local Lorentz symmetry has a gauge connection associated with it, the spin connection.

The following discussion will be in superspace notation, as opposed to the component notation, which isn't manifestly covariant under SUSY. There are actually *many* different versions of SUGRA out there which are inequivalent in the sense that their actions and constraints upon the torsion tensor are different, but ultimately equivalent in that we can always perform a field redefinition of the supervierbeins and spin connection to get from one version to another.

In 4D N=1 SUGRA, we have a 4|4 real differentiable supermanifold M, i.e. we have 4 real bosonic dimensions and 4 real fermionic dimensions. As in the nonsupersymmetric case, we have a Spin(3,1) principal bundle over M. We have an $\mathbf{R}^{4|4}$ vector bundle T over M. The fiber of T transforms under the local Lorentz group as follows: the four real bosonic dimensions transform as a vector and the four real fermionic dimensions transform as a Majorana spinor. This Majorana spinor can be reexpressed as a complex left-handed Weyl spinor and its complex conjugate right-handed Weyl spinor (they're not independent of each other). We also have a spin connection as before.

We will use the following conventions: the spatial (both bosonic and fermionic) indices will be indicated by M, N, ... The bosonic spatial indices will be indicated by μ, ν, ..., the left-handed Weyl spatial indices by α, β,..., and the right-handed Weyl spatial indices by $\dot\alpha$, $\dot\beta$, ... The indices for the fiber of T will follow a similar notation, except that they will be hatted like this: $\hat{M}, \hat\alpha$. See van der Waerden notation for more details. $M = (\mu, \alpha, \dot\alpha)$. The supervierbein is denoted by $e_N^{\hat M}$, and the spin connection by $\omega_{\hat M \hat N \hat P}$. The *inverse* supervierbein is denoted by $E_{\hat M}^N$.

The supervierbein and spin connection are real in the sense that they satisfy the reality conditions

$$e_N^{\hat M}(x, \overline\theta, \theta)^* = e_N^{\hat M'}(x, \theta, \overline\theta) \text{ where } \mu^* = \mu , \alpha^* = \dot\alpha , \text{ and } \dot\alpha^* = \alpha \text{ and } \omega(x, \overline\theta, \theta)^* = \omega(x, \theta, \overline\theta) .$$

The covariant derivative is defined as

$$D_M f = E_{\hat M}^N (\partial_N f + \omega_N [f])$$

The covariant exterior derivative as defined over supermanifolds needs to be super graded. This means that every time we interchange two fermionic indices, we pick up a +1 sign factor, instead of −1.

The presence or absence of R symmetries is optional, but if R-symmetry exists, the integrand over the full superspace has to have an R-charge of 0 and the integrand over chiral superspace has to have an R-charge of 2.

A chiral superfield X is a superfield which satisfies $\overline D_{\dot\alpha} X = 0$. In order for this constraint to be consistent, we require the integrability conditions that $\left\{ \overline D_{\dot\alpha}, \overline D_{\dot\beta} \right\} = c^{\dot\gamma}_{\dot\alpha\dot\beta} \overline D_{\dot\gamma}$ for some coefficients c.

Unlike nonSUSY GR, the torsion has to be nonzero, at least with respect to the fermionic directions. Already, even in flat superspace, $D_{\dot\alpha} e_{\hat\alpha} + \overline D_{\dot\alpha} e_{\hat\alpha} \neq 0$. In one version of SUGRA (but certainly not the only one), we have the following constraints upon the torsion tensor:

$$T^{\dot{\gamma}}_{\underline{\dot{\alpha}}\dot{\beta}} = 0$$

$$T^{\dot{\mu}}_{\dot{\alpha}\dot{\beta}} = 0$$

$$T^{\dot{\mu}}_{\dot{\alpha}\dot{\beta}} = 0$$

$$T^{\dot{\mu}}_{\dot{\alpha}\dot{\beta}} = 2i\sigma^{\dot{\mu}}_{\dot{\alpha}\dot{\beta}}$$

$$T^{\dot{\nu}}_{\mu\underline{\dot{\alpha}}} = 0$$

$$T^{\dot{\rho}}_{\mu\dot{\nu}} = 0$$

Here, $\underline{\alpha}$ is a shorthand notation to mean the index runs over either the left or right Weyl spinors.

The superdeterminant of the supervierbein, $|e|$, gives us the volume factor for M. Equivalently, we have the volume 4|4-superform $e^{\dot{\mu}=0} \wedge \cdots \wedge e^{\dot{\mu}=3} \wedge e^{\dot{\alpha}=1} \wedge e^{\dot{\alpha}=2} \wedge e^{\dot{\alpha}=1} \wedge e^{\dot{\alpha}=2}$.

If we complexify the superdiffeomorphisms, there is a gauge where $E^{\dot{\mu}}_{\dot{\alpha}} = 0$, $E^{\beta}_{\dot{\alpha}} = 0$ and $E^{\dot{\beta}}_{\dot{\alpha}} = \delta^{\dot{\beta}}_{\dot{\alpha}}$. The resulting chiral superspace has the coordinates x and Θ.

R is a scalar valued chiral superfield derivable from the supervielbeins and spin connection. If f is any superfield, $(\bar{D}^2 - 8R) f$ is always a chiral superfield.

The action for a SUGRA theory with chiral superfields X, is given by

$$S = \int d^4x d^2\Theta 2\mathcal{E} \left[\frac{3}{8} \left(\bar{D}^2 - 8R \right) e^{-K(\bar{X},X)/3} + W(X) \right] + c.c.$$

where K is the Kähler potential and W is the superpotential, and $\mathcal{E}$ is the chiral volume factor.

Unlike the case for flat superspace, adding a constant to either the Kähler or superpotential is now physical. A constant shift to the Kähler potential changes the effective Planck constant, while a constant shift to the superpotential changes the effective cosmological constant. As the effective Planck constant now depends upon the value of the chiral superfield X, we need to rescale the supervierbeins (a field redefinition) to get a constant Planck constant. This is called the **Einstein frame**.

43.5 N = 8 supergravity in 4 dimensions

N=8 Supergravity is the most symmetric quantum field theory which involves gravity and a finite number of fields. It can be found from a dimensional reduction of 11D supergravity by making the size of 7 of the dimensions go to zero. It has 8 supersymmetries which is the most any gravitational theory can have since there are 8 half-steps between spin 2 and spin -2. (A graviton has the highest spin in this theory which is a spin 2 particle). More supersymmetries would mean the particles would have superpartners with spins higher than 2. The only theories with spins higher than 2 which are consistent involve an infinite number of particles (such as String Theory). Stephen Hawking in his *A Brief History of Time* speculated that this theory could be the Theory of Everything. However in later years this was abandoned in favour of String Theory. There has been renewed interest in the 21st century with the possibility that this theory may be finite.

43.6 Higher-dimensional SUGRA

Main article: Higher-dimensional supergravity

Higher-dimensional SUGRA is the higher-dimensional, supersymmetric generalization of general relativity. Supergravity can be formulated in any number of dimensions up to eleven. Higher-dimensional SUGRA focuses upon supergravity in greater than four dimensions.

The number of supercharges in a spinor depends on the dimension and the signature of spacetime. The supercharges occur in spinors. Thus the limit on the number of supercharges cannot be satisfied in a spacetime of arbitrary dimension. Some theoretical examples in which this is satisfied are:

- 12-dimensional two-time theory

- 11-dimensional maximal SUGRA

- 10-dimensional SUGRA theories

    - Type IIA SUGRA: N = (1, 1)

    - IIA SUGRA from 11d SUGRA

    - Type IIB SUGRA: N = (2, 0)

    - Type I gauged SUGRA: N = (1, 0)

- 9d SUGRA theories

    - Maximal 9d SUGRA from 10d

    - T-duality

    - N = 1 Gauged SUGRA

The supergravity theories that have attracted the most interest contain no spins higher than two. This means, in particular, that they do not contain any fields that transform as symmetric tensors of rank higher than two under Lorentz transformations. The consistency of interacting higher spin field theories is, however, presently a field of very active interest.

43.7   See also

43.8   Notes

[1] P. van Nieuwenhuizen, Phys. Rep. 68, 189 (1981)

[2] Werner Nahm, "Supersymmetries and their representations". *Nuclear Physics B* **135** no 1 (1978) pp 149-166, doi:10.1016/0550-3213(78)90218-3

[3] Ed Witten, "Search for a realistic Kaluza-Klein theory". *Nuclear Physics B* **186** no 3 (1981) pp 412-428, doi:10.1016/0550-3213(81)90021-3

[4] E. Cremmer, B. Julia and J. Scherk, "Supergravity theory in eleven dimensions", *Physics Letters* **B76** (1978) pp 409-412,

[5] Peter G.O. Freund; Mark A. Rubin (1980). "Dynamics of dimensional reduction". *Physics Letters B* **97** (2): 233–235. Bibcode:1980PhLB...97..233F. doi:10.1016/0370-2693(80)90590-0.

[6] Blumenhagen, R.; Cvetic, M.; Langacker, P.; Shiu, G. (2005). "Toward Realistic Intersecting D-Brane Models". arXiv:hep-th/0502005 [hep-th].

43.9 References

43.9.1 Historical

- P. Nath and R. Arnowitt, "Generalized Super-Gauge Symmetry as a New Framework for Unified Gauge Theories", *Physics Letters B* '56 (1975) 177.

- D.Z. Freedman, P. van Nieuwenhuizen and S. Ferrara, "Progress Toward A Theory Of Supergravity", *Physical Review* **D13** (1976) pp 3214–3218.

- E. Cremmer, B. Julia and J. Scherk, "Supergravity theory in eleven dimensions", *Physics Letters* **B76** (1978) pp 409–412. scanned version

- P. Freund and M. Rubin, "Dynamics of dimensional reduction", *Physics Letters* **B97** (1980) pp 233–235.

- Ali H. Chamseddine, R. Arnowitt, Pran Nath, "Locally Supersymmetric Grand Unification", " Phys. Rev.Lett.49:9

- Michael B. Green, John H. Schwarz, "Anomaly Cancellation in Supersymmetric D=10 Gauge Theory and Superstring Theory", *Physics Letters* **B149** (1984) pp117–122.

43.9.2 General

- Bernard de Wit(2002) Supergravity

- A Supersymmetry Primer (1998); updated in (2006).

- Adel Bilal, Introduction to supersymmetry (2001) ArXiv hep-th/0101055, (*a comprehensive introduction to supersymmetry*).

- Friedemann Brandt, Lectures on supergravity (2002) ArXiv hep-th/0204035, (*an introduction to 4-dimensional N = 1 supergravity*).

- Wess, Julius; Bagger, Jonathan (1992). *Supersymmetry and Supergravity*. Princeton University Press. p. 260. ISBN 0-691-02530-4.

Chapter 44

Tensor–vector–scalar gravity

Not to be confused with Scalar–tensor–vector gravity or Bi-scalar tensor vector gravity.

Tensor–vector–scalar gravity (**TeVeS**),[1] developed by Jacob Bekenstein in 2004, is a relativistic generalization of Mordehai Milgrom's Modified Newtonian dynamics (MOND) paradigm.[2] [3]

The main features of TeVeS can be summarized as follows:

- As it is derived from the action principle, TeVeS respects conservation laws;

- In the weak-field approximation of the spherically symmetric, static solution, TeVeS reproduces the MOND acceleration formula;

- TeVeS avoids the problems of earlier attempts to generalize MOND, such as superluminal propagation;

- As it is a relativistic theory it can accommodate gravitational lensing.

The theory is based on the following ingredients:

- A unit vector field;

- A dynamical scalar field;

- A nondynamical scalar field;

- A matter Lagrangian constructed using an alternate metric;

- An arbitrary dimensionless function.

These components are combined into a relativistic Lagrangian density, which forms the basis of TeVeS theory.

44.1 Details

MOND[2] is a phenomenological modification of the Newtonian acceleration law. In Newtonian gravity theory, the gravitational acceleration in the spherically symmetric, static field of a point mass M at distance r from the source can be written as

$$a = -\frac{GM}{r^2},$$

where G is Newton's constant of gravitation. The corresponding force acting on a test mass m is

$F = ma$.

To account for the anomalous rotation curves of spiral galaxies, Milgrom proposed a modification of this force law in the form

$F = \mu(a/a_0)ma$,

where $\mu(x)$ is an arbitrary function subject to the following conditions:

$\mu(x) = 1$ if $|x| \gg 1$.

$\mu(x) = x$ if $|x| \ll 1$.

In this form, MOND is not a complete theory: for instance, it violates the law of momentum conservation.

However, such conservation laws are automatically satisfied for physical theories that are derived using an action principle. This led Bekenstein[1] to a first, nonrelativistic generalization of MOND. This theory, called AQUAL (for A QUAdratic Lagrangian) is based on the Lagrangian

$\mathcal{L} = -\frac{a_0^2}{8\pi G} f\left(\frac{|\nabla \Phi|^2}{a_0^2}\right) - \rho\Phi$,

where Φ is the Newtonian gravitational potential, ρ is the mass density, and $f(y)$ is a dimensionless function.

In the case of a spherically symmetric, static gravitational field, this Lagrangian reproduces the MOND acceleration law after the substitutions $a = -\nabla\Phi$ and $\mu(\sqrt{y}) = df(y)/dy$ are made.

Bekenstein further found that AQUAL can be obtained as the nonrelativistic limit of a relativistic field theory. This theory is written in terms of a Lagrangian that contains, in addition to the Einstein–Hilbert action for the metric field $g_{\mu\nu}$, terms pertaining to a unit vector field u^α and two scalar fields σ and ϕ, of which only ϕ is dynamical. The TeVeS action, therefore, can be written as

$S_{\text{TeVeS}} = \int (\mathcal{L}_g + \mathcal{L}_s + \mathcal{L}_v)\, d^4x$.

The terms in this action include the Einstein–Hilbert Lagrangian (using a metric signature $[+, -, -, -]$ and setting the speed of light, $c = 1$):

$\mathcal{L}_g = -\frac{1}{16\pi G} R\sqrt{-g}$,

where R is the Ricci scalar and g is the determinant of the metric tensor.

The scalar field Lagrangian is

$\mathcal{L}_s = -\frac{1}{2}\left[\sigma^2 h^{\alpha\beta}\partial_\alpha\phi\partial_\beta\phi + \frac{1}{2}\frac{G}{l^2}\sigma^4 F(kG\sigma^2)\right]\sqrt{-g}$,

with $h^{\alpha\beta} = g^{\alpha\beta} - u^\alpha u^\beta$, l is a constant length, k is the dimensionless parameter and F an unspecified dimensionless function; while the vector field Lagrangian is

$\mathcal{L}_v = -\frac{K}{32\pi G}\left[g^{\alpha\beta}g^{\mu\nu}(B_{\alpha\mu}B_{\beta\nu}) + 2\frac{\lambda}{K}(g^{\mu\nu}u_\mu u_\nu - 1)\right]\sqrt{-g}$

where $B_{\alpha\beta} = \partial_\alpha u_\beta - \partial_\beta u_\alpha$, while K is a dimensionless parameter. k and K are respectively called the scalar and vector coupling constants of the theory. The consistency between the Gravitoelectromagnetism of the TeVeS theory and that predicted and measured by the general relativity leads to $K = \frac{k}{2\pi}$.[4]

In particular, $\mathcal{L}_v$ incorporates a Lagrange multiplier term that guarantees that the vector field remains a unit vector field.

The function F in TeVeS is unspecified.

TeVeS also introduces a "physical metric" in the form

$\hat{g}^{\mu\nu} = e^{2\phi}g^{\mu\nu} - 2u^\alpha u^\beta \sinh(2\phi)$.

The action of ordinary matter is defined using the physical metric:

$S_m = \int \mathcal{L}(\hat{g}_{\mu\nu}, f^\alpha, f^\alpha_{|\mu}, ...)\sqrt{-g}d^4x$,

where covariant derivatives with respect to $\hat{g}_{\mu\nu}$ are denoted by $|$.

TeVeS solves problems associated with earlier attempts to generalize MOND, such as superluminal propagation. In his paper, Bekenstein also investigated the consequences of TeVeS in relation to gravitational lensing and cosmology.

44.2   Problems and criticisms

In addition to its ability to account for the flat rotation curves of galaxies (which is what MOND was originally designed to address), TeVeS is claimed to be consistent with a range of other phenomena, such as gravitational lensing and cosmological observations. However, Seifert[5] shows that with Bekenstein's proposed parameters, a TeVeS star is highly unstable, on the scale of approximately 10^6 seconds (two weeks). The ability of the theory to simultaneously account for galactic dynamics and lensing is also challenged.[6] A possible resolution may be in the form of massive (around 2eV) neutrinos.[7]

A study in August 2006 reported an observation of a pair of colliding galaxy clusters, the Bullet Cluster, whose behavior, it was reported, was not compatible with any current modified gravity theories.[8]

A quantity E_G [9] probing General Relativity (GR) on large scales (a hundred billion times the size of the solar system) for the first time has been measured with data from the Sloan Digital Sky Survey to be[10] $E_G = 0.392 \pm 0.065$ (~16%) consistent with GR, GR plus Lambda CDM and the extended form of GR known as $f(R)$ theory, but ruling out a particular TeVeS model predicting $E_G = 0.22$. This estimate should improve to ~1% with the next generation of sky surveys and may put tighter constraints on the parameter space of all modified gravity theories.

44.3   See also

- Modified Newtonian Dynamics
- Gauge Vector-Tensor gravity[11]
- Scalar–tensor–vector gravity
- General theory of relativity
- Law of universal gravitation
- Pioneer anomaly
- Nonsymmetric Gravitational Theory
- Dark matter
- Dark energy
- Dark fluid
- Tensor
- Vector
- Scalar - scalar field

44.4   References

[1] Bekenstein, J. D. (2004), "Relativistic gravitation theory for the modified Newtonian dynamics paradigm", *Physical Review D* **70** (8): 083509, arXiv:astro-ph/0403694, Bibcode:2004PhRvD..70h3509B, doi:10.1103/PhysRevD.70.083509

[2] Milgrom, M. (1983), "A modification of the Newtonian dynamics as a possible alternative to the hidden mass hypothesis", *The Astrophysical Journal* **270**: 365–370, Bibcode:1983ApJ...270..365M, doi:10.1086/161130

[3] Famaey, B.; McGaugh, S. S. (2012), "Modified Newtonian Dynamics (MOND): Observational Phenomenology and Relativistic Extensions", *Living Rev. Relativity* **15** (10), arXiv:1112.3960, Bibcode:2012LRR....15...10F, doi:10.12942/lrr-2012-10, ISSN 1433-8351

[4] Exirifard, Q. (2013), "GravitoMagnetic Field in Tensor-Vector-Scalar Theory", *Journal of Cosmology and Astroparticle Physics*, JCAP04: 034, arXiv:1111.5210, Bibcode:2013JCAP...04..034E, doi:10.1088/1475-7516/2013/04/034

[5] Seifert, M. D. (2007), "Stability of spherically symmetric solutions in modified theories of gravity", *Physical Review D* **76** (6): 064002, arXiv:gr-qc/0703060, Bibcode:2007PhRvD..76f4002S, doi:10.1103/PhysRevD.76.064002

[6] Mavromatos, Nick E.; Sakellariadou, Mairi; Yusaf, Muhammad Furqaan (2009), "Can TeVeS avoid Dark Matter on galactic scales?", *Physical Review D* **79**(8): 081301, arXiv:0901.3932, Bibcode:2009PhRvD..79h1301M, doi:10.1103/PhysRevD.79.08

[7] Angus, G. W.; Shan, H. Y.; Zhao, H. S.; Famaey, B. (2007), "On the Proof of Dark Matter, the Law of Gravity, and the Mass of Neutrinos", *The Astrophysical Journal Letters* **654** (1): L13–L16, arXiv:astro-ph/0609125, Bibcode:2007ApJ...654L..13A, doi:10.1086/510738

[8] Clowe, D.; Bradač, M.; Gonzalez, A. H.; Markevitch, M.; Randall, S. W.; Jones, C.; Zaritsky, D. (2006), "A Direct Empirical Proof of the Existence of Dark Matter", *The Astrophysical Journal Letters* **648** (2): L109, arXiv:astro-ph/0608407, Bibcode:2006ApJ...648L.109C, doi:10.1086/508162

[9] Zhang, P.; Liguori, M.; Bean, R.; Dodelson, S. (2007), "Probing Gravity at Cosmological Scales by Measurements which Test the Relationship between Gravitational Lensing and Matter Overdensity", *Physical Review Letters* **99** (14): 141302, arXiv:0704.1932, Bibcode:2007PhRvL..99n1302Z, doi:10.1103/PhysRevLett.99.141302

[10] Reyes, R.; Mandelbaum, R.; Seljak, U.; Baldauf, T.; Gunn, J. E.; Lombriser, L.; Smith, R. E. (2010), "Confirmation of general relativity on large scales from weak lensing and galaxy velocities", *Nature* **464** (7286): 256–258, arXiv:1003.2185, Bibcode:2010Natur.464..256R, doi:10.1038/nature08857, PMID 20220843

[11] Exirifard, Q. (2013), "GravitoMagnetic force in modified Newtonian dynamics", *Journal of Cosmology and Astroparticle Physics*, JCAP08: 046–046, arXiv:1107.2109, Bibcode:2013JCAP...08..046E, doi:10.1088/1475-7516/2013/08/046

44.5 Further reading

- Bekenstein, J. D.; Sanders, R. H. (2006), "A Primer to Relativistic MOND Theory", *EAS Publications Series* **20**: 225–230, arXiv:astro-ph/0509519, Bibcode:2006EAS....20..225B, doi:10.1051/eas:2006075

- Zhao, H. S.; Famaey, B. (2006), "Refining the MOND Interpolating Function and TeVeS Lagrangian", *The Astrophysical Journal* **638** (1): L9–L12, arXiv:astro-ph/0512425, Bibcode:2006ApJ...638L...9Z, doi:10.1086/500805

- Dark Matter Observed (SLAC Today)

- Einstein's Theory 'Improved'? (PPARC)

- Einstein Was Right: General Relativity Confirmed ' TeVeS, however, made predictions that fell outside the observational error limits', (Space.com)

Chapter 45

Whitehead's theory of gravitation

In theoretical physics, **Whitehead's theory of gravitation** was introduced by the distinguished mathematician and philosopher Alfred North Whitehead in 1922.

45.1   Principal features of the theory

Clifford M. Will argued that Whitehead's theory features a *prior geometry*,[1] but this is disputed by Dean R. Fowler, since it contradicts Whitehead's philosophy of nature. For Whitehead, the geometric structure of nature grows out of the relations among actual occasions. Fowler's interpretation of Whitehead's theory makes it an alternate, mathematically equivalent, presentation of general relativity.[2]

Under Will's presentation (which was inspired by John Lighton Synge's interpretation of the theory[3][4]), Whitehead's theory has the curious feature that electromagnetic waves propagate along null geodesics of the physical spacetime (as defined by the metric determined from geometrical measurements and timing experiments), while gravitational waves propagate along null geodesics of a *flat background* represented by the metric tensor of Minkowski spacetime. The gravitational potential can be expressed entirely in terms of waves retarded along the background metric, like the Liénard–Wiechert potential in electromagnetic theory.

A cosmological constant can be introduced by changing the background metric to a de Sitter or anti-de Sitter metric. This was first suggested by G. Temple in 1923.[5] Temple's suggestions on how to do this were criticized by C. B. Rayner in 1955.[6][7]

45.2   Tests of Whitehead's theory

Whitehead's theory is equivalent with the Schwarzschild metric[8] and makes the same predictions as general relativity regarding the four classical solar system tests (gravitational red shift, light bending, perihelion shift, Shapiro time delay), and was regarded as a viable competitor of general relativity for several decades. In 1971, young Clifford M. Will thanks Ni Wei-Tou to comprehend Whitehead's theory[9] and claims that the theory makes predictions concerning ordinary ocean tides on Earth (suggested to him by Jim Peebles) which are in violent disagreement with observation (specifically, the theory predicts a "sidereal tide", induced by the gravitational field of the Milky Way, which is hundreds of times stronger than the solar and lunar tides) which immediately nullified this theory.[10] As mentioned previously, the interpretation of the theory used by Will has been criticized by Fowler, who has also argued that different tidal predictions can be obtained by a more realistic model of the galaxy.[2][11]

In 1989, a new interpretation of Whitehead's theory was proposed that eliminated the unobserved sidereal tide effects.[12] However, the new interpretation predicted a new, unobserved, effect, called the "Nordtvedt Effect."

45.3 Footnotes

[1]Will, Clifford (1972). "Einstein on the Firing Line".*Physics Today***25**: 23–29.Bibcode:1972PhT....25j..23W.doi:10.1063/1.30

[2] Fowler, Dean (Winter 1974). "Disconfirmation of Whitehead's Relativity Theory -- A Critical Reply". *Process Studies* **4** (4): 288–290. doi:10.5840/process19744432.

[3] Synge, John (1951). *Relativity Theory of A. N. Whitehead*. Baltimore: University of Maryland.

[4] Tanaka, Yutaka (1987). "Einstein and Whitehead-The Comparison between Einstein's and Whitehead's Theories of Relativity". *Historia Scientiarum* **32**.

[5] Temple, G. (1924). "Central Orbit in Relativistic Dynamics Treated by the Hamilton-Jacobi Method". *Philosophical Magazine*. 6 **48**: 277–292. doi:10.1080/14786442408634491.

[6] Rayner, C. (1954). "The Application of the Whitehead Theory of Relativity to Non-static Spherically Symmetrical Systems". *Proceedings of the Royal Society of London* **222**: 509–526. Bibcode:1954RSPSA.222..509R. doi:10.1098/rspa.1954.0092.

[7] Rayner, C. (1955). "The Effects of Rotation in the Central Body on its Planetary Orbits after the Whitehead Theory of Gravitation". *Proceedings of the Royal Society of London* **232**: 135–148. Bibcode:1955RSPSA.232..135R. doi:10.1098/rspa.1955.0206.

[8] Eddington, A.S. (1924). "A comparison of Whitehead's and Einstein's formulas". *Nature* **113**: 192.

[9] http://articles.adsabs.harvard.edu//full/1971ApJ...169..141W/0000152.000.html - Relativistic Gravity in the Solar System. II. Anisotropy in the Newtonian Gravitational Constant

[10] Will, Clifford & Gibbons, Gary. "On the Multiple Deaths of Whitehead's Theory of Gravity", to be submitted to *Studies In History And Philosophy Of Modern Physics* (2006).

[11] Bain, Jonathan (1998). "Whitehead's Theory of Gravity". *Stud. Hist. Phil. Mod. Phys.* **29** (4): 547–574.

[12] Hyman, Andrew (1989). "A New Interpretation of Whitehead's Theory", 104B". *Il Nuovo Cimento* **387**.

45.4 See also

- Alfred North Whitehead
- Classical theories of gravitation
- Eddington–Finkelstein coordinates

45.5 References

- Will, Clifford M. (1993). *Was Einstein Right?: Putting General Relativity to the Test* (2nd ed.). Basic Books. ISBN 0-465-09086-9.

- Misner, Charles; Thorne, Kip S. & Wheeler, John Archibald (1973). *Gravitation*. San Francisco: W. H. Freeman. ISBN 0-7167-0344-0. discusses Whitehead's theory in various places.

Chapter 46

Yilmaz theory of gravitation

The **Yilmaz theory of gravitation** is an attempt by **Huseyin Yilmaz** (1924-2013) (Turkish: *Hüseyin Yılmaz*) and his coworkers to formulate a classical field theory of gravitation which is similar to general relativity in weak-field conditions, but in which event horizons cannot appear.

Yilmaz's work has been criticized on the grounds that

- his proposed field equation is ill-defined,

- event horizons can occur in weak field situations according to the general theory of relativity, in the case of a supermassive black hole.

- the theory is consistent only with either a completely empty universe or a negative energy vacuum[1]

It is well known that attempts to quantize general relativity along the same lines which lead from Maxwell's classical field theory of electromagnetism to quantum electrodynamics fail, and that it has proven very difficult to construct a theory of quantum gravity which goes over to general relativity in an appropriate limit. However Yilmaz has claimed that his theory is 'compatible with quantum mechanics'. He suggests that it might be an alternative to superstring theory.

In his theory, Yilmaz wishes to retain the left hand side of the Einstein field equation (namely the Einstein tensor, which is well-defined for any Lorentzian manifold, independent of general relativity) but to modify the right hand side, the stress–energy tensor, by adding a kind of gravitational contribution. According to Yilmaz's critics, this additional term is not well-defined, and cannot be made well defined.

No astronomers have tested his ideas, although some have tested competitors of general relativity; see Category:Tests of general relativity.

46.1   External links

- One page in the website Relativity on the World Wide Web (archived link) lists some apparent misstatements by Yilmaz concerning the general theory of relativity, similar to those discussed by Fackerell.

46.2   References

[1] Ibison, M. (2006). "Cosmological test of the Yilmz theory of gravity ".*Classical Quantum Gravity* **23**(3): 577–589.arXiv:0705 Bibcode:2006CQGra..23..577I. doi:10.1088/0264-9381/23/3/001.

- Yilmaz, H. (1992). "Toward a field theory of gravitation".*Nuovo Cimento B* **107**(8): 941–960.Bibcode:1992N doi:10.1007/BF02899296.

- Misner, C. W. (1999). "Yilmaz Cancels Newton". *Nuovo Cimento B* **114**: 1079–1085. arXiv:gr-qc/9504050. Bibcode:1999NCimB.114.1079M. In this paper, Charles Misner argues that Yilmaz's field equation is ill-defined.

- Alley, C.O.; Aschan, P. K.; Yilmaz, H. (1995). "Refutation of C. W. Misner's claims in his article "Yilmaz Cancels Newton"". arXiv:gr-qc/9506082 [gr-qc].

- Fackerell, E. D. (2006). "Remarks on the Yilmaz and Alley papers". School of Mathematics and Statistics F07, University of Sydney. In this preprint, Edward Fackerell criticizes several claims by Yilmaz concerning gtr

- Alley, C. O.; Yilmaz, H. (2000). "Response to Fackerell's Article". arXiv:gr-qc/0008040 [gr-qc].

- Misner, C.; Thorne, K. S.; Wheeler, J. A. (1973). *Gravitation*. W. H. Freeman. ISBN 0-7167-0344-0. See *section 20.4* for nonlocal nature of gravitational field energy, and all of chapter 20 for relation between integration, Bianchi identities, and 'conservation laws' in curved spacetimes.

Chapter 47

Theory of everything

This article is about the physical concept.

A **theory of everything** (**ToE**) or **final theory**, **ultimate theory**, or**master theory**is a hypothetical ,all-encompassing, coherent theoretical framework of physics that fully explains and links together all physical aspects of the universe. Finding a ToE is one of the major unsolved problems in physics. Over the past few centuries, two theoretical frameworks have been developed that, as a whole,most closely resemble a ToE. The two theories upon which all modern physics rests are general relativity(GR) and quantum field theory(QFT). GR is a theoretical framework that only focuses on the force of gravity for understanding the universe in regions of both large-scale and high-mass: stars, galaxies, clusters of galaxies,etc. On the other hand, QFT is a theoretical framework that only focuses on three non-gravitational forces for understanding the universe in regions of both small scale and low mass:sub-atomic particles, atoms, molecules, etc. QFT successfully implemented the Standard Model and unified the interactions (so-calledGrand Unified Theory) between the three non-gravitational forces: weak, strong, and electromagnetic force.[2]:122

Through years of research, physicists have experimentally confirmed with tremendous accuracy virtually every prediction made by these two theories when in their appropriate domains of applicability. In accordance with their findings, scientists also learned that GR and QFT, as they are currently formulated, are mutually incompatible - they cannot both be right. Since the usual domains of applicability of GR and QFT are so different, most situations require that only one of the two theories be used.[3][4]:842–844 As it turns out, this incompatibility between GR and QFT is only an apparent issue in regions of extremely small-scale and high-mass, such as those that exist within a black hole or during the beginning stages of the universe (i.e., the moment immediately following the Big Bang). To resolve this conflict, a theoretical framework revealing a deeper underlying reality, unifying gravity with the other three interactions, must be discovered to harmoniously integrate the realms of GR and QFT into a seamless whole: a single theory that, in principle, is capable of describing all phenomena. In pursuit of this goal, quantum gravity has recently become an area of active research.

Over the past few decades, a single explanatory framework, called "string theory", has emerged that may turn out to be the ultimate theory of the universe. Many physicists believe that, at the beginning of the universe (up to 10^{-43} seconds after the Big Bang), the four fundamental forces were once a single fundamental force. Unlike most (if not all) other theories, string theory may be on its way to successfully incorporating each of the four fundamental forces into a unified whole. According to string theory, every particle in the universe, at its most microscopic level (Planck length), consists of varying combinations of vibrating strings (or strands) with preferred patterns of vibration. String theory claims that it is through these specific oscillatory patterns of strings that a particle of unique mass and force charge is created (that is to say, the electron is a type of string that vibrates one way, while the up-quark is a type of string vibrating another way, and so forth).

Initially, the term *theory of everything* was used with an ironic connotation to refer to various overgeneralized theories. For example, a grandfather of Ijon Tichy — a character from a cycle of Stanisław Lem's science fiction stories of the 1960s — was known to work on the "General Theory of Everything". Physicist John Ellis[5] claims to have introduced the term into the technical literature in an article in *Nature* in 1986.[6] Over time, the term stuck in popularizations of theoretical physics research.

47.1 Historical antecedents

47.1.1 From ancient Greece to Einstein

Archimedes was possibly the first scientist known to have described nature with axioms (or principles) and then deduce new results from them.[7] He thus tried to describe "everything" starting from a few axioms. Any "theory of everything" is similarly expected to be based on axioms and to deduce all observable phenomena from them.[8]:340

The concept of 'atom', introduced by Democritus, unified all phenomena observed in nature as the motion of atoms. In ancient Greek times philosophers speculated that the apparent diversity of observed phenomena was due to a single type of interaction, namely the collisions of atoms. Following atomism, the mechanical philosophy of the 17th century posited that all forces could be ultimately reduced to contact forces between the atoms, then imagined as tiny solid particles.[9]:184[10]

In the late 17th century, Isaac Newton's description of the long-distance force of gravity implied that not all forces in nature result from things coming into contact. Newton's work in his *Principia* dealt with this in a further example of unification, in this case unifying Galileo's work on terrestrial gravity, Kepler's laws of planetary motion and the phenomenon of tides by explaining these apparent actions at a distance under one single law: the law of universal gravitation.[11]

In 1814, building on these results, Laplace famously suggested that a sufficiently powerful intellect could, if it knew the position and velocity of every particle at a given time, along with the laws of nature, calculate the position of any particle at any other time:[12]:ch 7

> An intellect which at a certain moment would know all forces that set nature in motion, and all positions of all items of which nature is composed, if this intellect were also vast enough to submit these data to analysis, it would embrace in a single formula the movements of the greatest bodies of the universe and those of the tiniest atom; for such an intellect nothing would be uncertain and the future just like the past would be present before its eyes.
> — *Essai philosophique sur les probabilités*, Introduction. 1814

Laplace thus envisaged a combination of gravitation and mechanics as a theory of everything. Modern quantum mechanics implies that uncertainty is inescapable, and thus that Laplace's vision has to be amended: a theory of everything must include gravitation and quantum mechanics.

In 1820, Hans Christian Ørsted discovered a connection between electricity and magnetism, triggering decades of work that culminated in 1865, in James Clerk Maxwell's theory of electromagnetism. During the 19th and early 20th centuries, it gradually became apparent that many common examples of forces – contact forces, elasticity, viscosity, friction, and pressure – result from electrical interactions between the smallest particles of matter.

In his experiments of 1849–50, Michael Faraday was the first to search for a unification of gravity with electricity and magnetism.[13] However, he found no connection.

In 1900, David Hilbert published a famous list of mathematical problems. In Hilbert's sixth problem, he challenged researchers to find an axiomatic basis to all of physics. In this problem he thus asked for what today would be called a theory of everything.[14]

In the late 1920s, the new quantum mechanics showed that the chemical bonds between atoms were examples of (quantum) electrical forces, justifying Dirac's boast that "the underlying physical laws necessary for the mathematical theory of a large part of physics and the whole of chemistry are thus completely known".[15]

After 1915, when Albert Einstein published the theory of gravity (general relativity), the search for a unified field theory combining gravity with electromagnetism began with a renewed interest. In Einstein's day, the strong and the weak forces had not yet been discovered, yet, he found the potential existence of two other distinct forces -gravity and electromagnetism- far more alluring. This launched his thirty-year voyage in search of the so-called "unified field theory" that he hoped would show that these two forces are really manifestations of one grand underlying principle. During these last few decades of his life, this quixotic quest isolated Einstein from the mainstream of physics. Understandably, the mainstream was instead far more excited about the newly emerging framework of quantum mechanics. Einstein wrote to a friend in the early 1940s, "I have become a lonely old chap who is mainly known because he doesn't wear socks and who is exhibited as a curiosity on special occasions." Prominent contributors were Gunnar Nordström, Hermann Weyl, Arthur Eddington, Theodor Kaluza, Oskar Klein, and most notably, Albert Einstein and his collaborators. Einstein intensely searched for, but ultimately failed to find, a unifying theory.[16]:ch 17 (But see:Einstein–Maxwell–Dirac equations.) More than a half a century later, Einstein's dream of discovering a unified theory has become the Holy Grail of modern physics.

47.1.2 Twentieth century and the nuclear interactions

In the twentieth century, the search for a unifying theory was interrupted by the discovery of the strong and weak nuclear forces (or interactions), which differ both from gravity and from electromagnetism. A further hurdle was the acceptance that in a ToE, quantum mechanics had to be incorporated from the start, rather than emerging as a consequence of a deterministic unified theory, as Einstein had hoped.

Gravity and electromagnetism could always peacefully coexist as entries in a list of classical forces, but for many years it seemed that gravity could not even be incorporated into the quantum framework, let alone unified with the other fundamental forces. For this reason, work on unification, for much of the twentieth century, focused on understanding the three "quantum" forces: electromagnetism and the weak and strong forces. The first two were combined in 1967–68 by Sheldon Glashow, Steven Weinberg, and Abdus Salam into the "electroweak" force.[17] Electroweak unification is a broken symmetry: the electromagnetic and weak forces appear distinct at low energies because the particles carrying the weak force, the W and Z bosons, have non-zero masses of 80.4 GeV/c^2 and 91.2 GeV/c^2, whereas the photon, which carries the electromagnetic force, is massless. At higher energies Ws and Zs can be created easily and the unified nature of the force becomes apparent.

While the strong and electroweak forces peacefully coexist in the Standard Model of particle physics, they remain distinct. So far, the quest for a theory of everything is thus unsuccessful on two points: neither a unification of the strong and electroweak forces – which Laplace would have called 'contact forces' – has been achieved, nor has a unification of these forces with gravitation been achieved.

47.2 Modern physics

47.2.1 Conventional sequence of theories

A Theory of Everything would unify all the fundamental interactions of nature: gravitation, strong interaction, weak interaction, and electromagnetism. Because the weak interaction can transform elementary particles from one kind into another, the ToE should also yield a deep understanding of the various different kinds of possible particles. The usual assumed path of theories is given in the following graph, where each unification step leads one level up:

In this graph, electroweak unification occurs at around 100 GeV, grand unification is predicted to occur at 10^{16} GeV, and unification of the GUT force with gravity is expected at the Planck energy, roughly 10^{19} GeV.

Several Grand Unified Theories (GUTs) have been proposed to unify electromagnetism and the weak and strong forces. Grand unification would imply the existence of an electronuclear force; it is expected to set in at energies of the order of 10^{16} GeV, far greater than could be reached by any possible Earth-based particle accelerator. Although the simplest GUTs have been experimentally ruled out, the general idea, especially when linked with supersymmetry, remains a favorite candidate in the theoretical physics community. Supersymmetric GUTs seem plausible not only for their theoretical "beauty", but because they naturally produce large quantities of dark matter, and because the inflationary force may be related to GUT physics (although it does not seem to form an inevitable part of the theory). Yet GUTs are clearly not the final answer; both the current standard model and all proposed GUTs are quantum field theories which require the problematic technique of renormalization to yield sensible answers. This is usually regarded as a sign that these are only effective field theories, omitting crucial phenomena relevant only at very high energies.[3]

The final step in the graph requires resolving the separation between quantum mechanics and gravitation, often equated with general relativity. Numerous researchers concentrate their efforts on this specific step; nevertheless, no accepted theory of quantum gravity – and thus no accepted theory of everything – has emerged yet. It is usually assumed that the ToE will also solve the remaining problems of GUTs.

In addition to explaining the forces listed in the graph, a ToE may also explain the status of at least two candidate forces suggested by modern cosmology: an inflationary force and dark energy. Furthermore, cosmological experiments also suggest the existence of dark matter, supposedly composed of fundamental particles outside the scheme of the standard model. However, the existence of these forces and particles has not been proven yet.

47.2.2 String theory and M-theory

Since the 1990s, many physicists believe that 11-dimensional M-theory, which is described in some limits by one of the five perturbative superstring theories, and in another by the maximally-supersymmetric 11-dimensional supergravity, is the theory of everything. However, there is no widespread consensus on this issue.

A surprising property of string/M-theory is that extra dimensions are required for the theory's consistency. In this regard, string theory can be seen as building on the insights of the Kaluza–Klein theory, in which it was realized that applying general relativity to a five-dimensional universe (with one of them small and curled up) looks from the four-dimensional perspective like the usual general relativity together with Maxwell's electrodynamics. This lent credence to the idea of unifying gauge and gravity interactions, and to extra dimensions, but didn't address the detailed experimental requirements. Another important property of string theory is its supersymmetry, which together with extra dimensions are the two main proposals for resolving the hierarchy problem of the standard model, which is (roughly) the question of why gravity is so much weaker than any other force. The extra-dimensional solution involves allowing gravity to propagate into the other dimensions while keeping other forces confined to a four-dimensional spacetime, an idea that has been realized with explicit stringy mechanisms.[18]

Research into string theory has been encouraged by a variety of theoretical and experimental factors. On the experimental side, the particle content of the standard model supplemented with neutrino masses fits into a spinor representation of $SO(10)$, a subgroup of E8 that routinely emerges in string theory, such as in heterotic string theory[19] or (sometimes equivalently) in F-theory.[20][21] String theory has mechanisms that may explain why fermions come in three hierarchical generations, and explain the mixing rates between quark generations.[22] On the theoretical side, it has begun to address some of the key questions in quantum gravity, such as resolving the black hole information paradox, counting the correct entropy of black holes[23][24] and allowing for topology-changing processes.[25][26][27] It has also led to many insights in pure mathematics and in ordinary, strongly-coupled gauge theory due to the Gauge/String duality.

In the late 1990s, it was noted that one major hurdle in this endeavor is that the number of possible four-dimensional universes is incredibly large. The small, "curled up" extra dimensions can be compactified in an enormous number of different ways (one estimate is 10^{500}) each of which leads to different properties for the low-energy particles and forces. This array of models is known as the string theory landscape.[8]:347

One proposed solution is that many or all of these possibilities are realised in one or another of a huge number of universes, but that only a small number of them are habitable, and hence the fundamental constants of the universe are ultimately the result of the anthropic principle rather than dictated by theory. This has led to criticism of string theory,[28] arguing that it cannot make useful (i.e., original, falsifiable, and verifiable) predictions and regarding it as a pseudoscience. Others disagree,[29] and string theory remains an extremely active topic of investigation in theoretical physics.

47.2.3 Loop quantum gravity

Current research on loop quantum gravity may eventually play a fundamental role in a ToE, but that is not its primary aim.[30] Also loop quantum gravity introduces a lower bound on the possible length scales.

There have been recent claims that loop quantum gravity may be able to reproduce features resembling the Standard Model. So far only the first generation of fermions (leptons and quarks) with correct parity properties have been modelled by Sundance Bilson-Thompson using preons constituted of braids of spacetime as the building blocks.[31] However, there is no derivation of the Lagrangian that would describe the interactions of such particles, nor is it possible to show that such particles are fermions, nor that the gauge groups or interactions of the Standard Model are realised. Utilization of quantum computing concepts made it possible to demonstrate that the particles are able to survive quantum fluctuations.[32]

This model leads to an interpretation of electric and colour charge as topological quantities (electric as number and chirality of twists carried on the individual ribbons and colour as variants of such twisting for fixed electric charge).

Bilson-Thompson's original paper suggested that the higher-generation fermions could be represented by more complicated braidings, although explicit constructions of these structures were not given. The electric charge, colour, and parity properties of such fermions would arise in the same way as for the first generation. The model was expressly generalized for an infinite number of generations and for the weak force bosons (but not for photons or gluons) in a 2008 paper by Bilson-Thompson, Hackett, Kauffman and Smolin.[33]

47.2.4   Other attempts

A recent development is the theory of causal fermion systems,[34] giving all three current physical theories (quantum mechanics, general relativity and quantum field theory) as limiting cases.

A recent and very prolific attempt is called Causal Sets. As some of the approaches mentioned above, its direct goal isn't necessarily to achieve a ToE but primarily a working theory of quantum gravity, which might eventually include the standard model and become a candidate for a ToE. Its founding principle is that spacetime is fundamentally discrete and that the spacetime events are related by a partial order. This partial order has the physical meaning of the causality relations between relative past and future distinguishing spacetime events.

Outside the previously mentioned attempts there is Garrett Lisi's E8 proposal. This theory provides an attempt of identifying general relativity and the standard model within the Lie group E8. The theory doesn't provide a novel quantization procedure and the author suggests its quantization might follow the Loop Quantum Gravity approach above mentioned.[35]

Christoph Schiller's Strand Model attempts to account for the gauge symmetry of the Standard Model of particle physics, $U(1) \times SU(2) \times SU(3)$, with the three Reidemeister moves of knot theory by equating each elementary particle to a different tangle of one, two, or three strands (selectively a long prime knot or unknotted curve, a rational tangle, or a braided tangle respectively).

47.2.5   Present status

At present, there is no candidate theory of everything that includes the standard model of particle physics and general relativity. For example, no candidate theory is able to calculate the fine structure constant or the mass of the electron. Most particle physicists expect that the outcome of the ongoing experiments – the search for new particles at the large particle accelerators and for dark matter – are needed in order to provide further input for a ToE.

47.3   Theory of everything and philosophy

Main article: Theory of everything (philosophy)

The philosophical implications of a physical ToE are frequently debated. For example, if philosophical physicalism is true, a physical ToE will coincide with a philosophical theory of everything.

The "system building" style of metaphysics attempts to answer *all* the important questions in a coherent way, providing a complete picture of the world. Plato and Aristotle could be said to have created early examples of comprehensive systems. In the early modern period (17th and 18th centuries), the system-building *scope* of philosophy is often linked to the rationalist *method* of philosophy, which is the technique of deducing the nature of the world by pure *a priori* reason. Examples from the early modern period include the Leibniz's Monadology, Descarte's Dualism, and Spinoza's Monism. Hegel's Absolute idealism and Whitehead's Process philosophy were later systems.

47.4   Arguments against a theory of everything

In parallel to the intense search for a ToE, various scholars have seriously debated the possibility of its discovery.

47.4.1 Gödel's incompleteness theorem

A number of scholars claim that Gödel's incompleteness theorem suggests that any attempt to construct a ToE is bound to fail. Gödel's theorem, informally stated, asserts that any formal theory expressive enough for elementary arithmetical facts to be expressed and strong enough for them to be proved is either inconsistent (both a statement and its denial can be derived from its axioms) or incomplete, in the sense that there is a true statement that can't be derived in the formal theory.

Stanley Jaki, in his 1966 book *The Relevance of Physics*, pointed out that, because any "theory of everything" will certainly be a consistent non-trivial mathematical theory, it must be incomplete. He claims that this dooms searches for a deterministic theory of everything.[36] In a later reflection, Jaki states that it is wrong to say that a final theory is impossible, but rather that "when it is on hand one cannot know rigorously that it is a final theory."[37]

Freeman Dyson has stated that "Gödel's theorem implies that pure mathematics is inexhaustible. No matter how many problems we solve, there will always be other problems that cannot be solved within the existing rules. [...] Because of Gödel's theorem, physics is inexhaustible too. The laws of physics are a finite set of rules, and include the rules for doing mathematics, so that Gödel's theorem applies to them."[38]

Stephen Hawking was originally a believer in the Theory of Everything but, after considering Gödel's Theorem, concluded that one was not obtainable: "Some people will be very disappointed if there is not an ultimate theory, that can be formulated as a finite number of principles. I used to belong to that camp, but I have changed my mind."[39]

Jürgen Schmidhuber (1997) has argued against this view; he points out that Gödel's theorems are irrelevant for computable physics.[40] In 2000, Schmidhuber explicitly constructed limit-computable, deterministic universes whose pseudo-randomness based on undecidable, Gödel-like halting problems
is extremely hard to detect but does not at all prevent formal ToEs describable by very few bits of information.[41]

Related critique was offered by Solomon Feferman,[42] among others. Douglas S. Robertson offers Conway's game of life as an example:[43] The underlying rules are simple and complete, but there are formally undecidable questions about the game's behaviors. Analogously, it may (or may not) be possible to completely state the underlying rules of physics with a finite number of well-defined laws, but there is little doubt that there are questions about the behavior of physical systems which are formally undecidable on the basis of those underlying laws.

Since most physicists would consider the statement of the underlying rules to suffice as the definition of a "theory of everything", most physicists argue that Gödel's Theorem does *not* mean that a ToE cannot exist. On the other hand, the scholars invoking Gödel's Theorem appear, at least in some cases, to be referring not to the underlying rules, but to the understandability of the behavior of all physical systems, as when Hawking mentions arranging blocks into rectangles, turning the computation of prime numbers into a physical question.[44] This definitional discrepancy may explain some of the disagreement among researchers.

47.4.2 Fundamental limits in accuracy

No physical theory to date is believed to be precisely accurate. Instead, physics has proceeded by a series of "successive approximations" allowing more and more accurate predictions over a wider and wider range of phenomena. Some physicists believe that it is therefore a mistake to confuse theoretical models with the true nature of reality, and hold that the series of approximations will never terminate in the "truth". Einstein himself expressed this view on occasions.[45] Following this view, we may reasonably hope for *a* theory of everything which self-consistently incorporates all currently known forces, but we should not expect it to be the final answer.

On the other hand it is often claimed that, despite the apparently ever-increasing complexity of the mathematics of each new theory, in a deep sense associated with their underlying gauge symmetry and the number of fundamental physical constants, the theories are becoming simpler. If this is the case, the process of simplification cannot continue indefinitely.

47.4.3 Lack of fundamental laws

There is a philosophical debate within the physics community as to whether a theory of everything deserves to be called *the* fundamental law of the universe.[46] One view is the hard reductionist position that the ToE is the fundamental law and that all other theories that apply within the universe are a consequence of the ToE. Another view is that emergent laws, which govern the behavior of complex systems, should be seen as equally fundamental. Examples of emergent laws are the second law of thermodynamics and the theory of natural selection. The advocates of emergence argue that emergent laws, especially those describing complex or living systems are independent of the low-level, microscopic laws. In this view, emergent laws are as fundamental as a ToE.

The debates do not make the point at issue clear. Possibly the only issue at stake is the right to apply the high-status term "fundamental" to the respective subjects of research. A well-known one took place between Steven Weinberg and Philip Anderson

47.4.4   Impossibility of being "of everything"

Although the name "theory of everything" suggests the determinism of Laplace's quotation, this gives a very misleading impression. Determinism is frustrated by the probabilistic nature of quantum mechanical predictions, by the extreme sensitivity to initial conditions that leads to mathematical chaos, by the limitations due to event horizons, and by the extreme mathematical difficulty of applying the theory. Thus, although the current standard model of particle physics "in principle" predicts almost all known non-gravitational phenomena, in practice only a few quantitative results have been derived from the full theory (e.g., the masses of some of the simplest hadrons), and these results (especially the particle masses which are most relevant for low-energy physics) are less accurate than existing experimental measurements. The ToE would almost certainly be even harder to apply for the prediction of experimental results, and thus might be of limited use.

A motive for seeking a ToE, apart from the pure intellectual satisfaction of completing a centuries-long quest, is that prior examples of unification have predicted new phenomena, some of which (e.g., electrical generators) have proved of great practical importance. And like in these prior examples of unification, the ToE would probably allow us to confidently define the domain of validity and residual error of low-energy approximations to the full theory.

47.4.5   Infinite number of onion layers

Lee Smolin regularly argues that the layers of nature may be like the layers of an onion, and that the number of layers might be infinite. This would imply an infinite sequence of physical theories.

The argument is not universally accepted, because it is not obvious that infinity is a concept that applies to the foundations of nature.

47.4.6   Impossibility of calculation

Weinberg[47] points out that calculating the precise motion of an actual projectile in the Earth's atmosphere is impossible. So how can we know we have an adequate theory for describing the motion of projectiles? Weinberg suggests that we know *principles* (Newton's laws of motion and gravitation) that work "well enough" for simple examples, like the motion of planets in empty space. These principles have worked so well on simple examples that we can be reasonably confident they will work for more complex examples. For example, although general relativity includes equations that do not have exact solutions, it is widely accepted as a valid theory because all of its equations with exact solutions have been experimentally verified. Likewise, a ToE must work for a wide range of simple examples in such a way that we can be reasonably confident it will work for every situation in physics.

47.5   See also

- Absolute (philosophy)

- An Exceptionally Simple Theory of Everything

- Argument from beauty

- Attractor

- Beyond black holes

- Beyond the standard model

- Big Bang

- Brownian motion

- Chaos theory

- Chronology of the universe

- Electroweak interaction

- Holographic principle

- Mathematical beauty

- Mathematical universe hypothesis

- Multiverse

- Standard Model (mathematical formulation)

- *The Theory of Everything (2014 film)* - a feature film about Prof.Stephen Hawking and his first wife Jane Hawking.

- Timeline of the Big Bang

- Zero-energy universe

47.6 References

47.6.1 Footnotes

[1] Steven Weinberg. *Dreams of a Final Theory: The Scientist's Search for the Ultimate Laws of Nature*. Knopf Doubleday Publishing Group. ISBN 978-0-307-78786-6.

[2] Stephen W. Hawking (28 February 2006). *The Theory of Everything: The Origin and Fate of the Universe*. Phoenix Books: Special Anniv. ISBN 978-1-59777-508-3.

[3] Carlip, Steven (2001). "Quantum Gravity: a Progress Report". *Reports on Progress in Physics* **64** (8). arXiv:gr-qc/0108040. Bibcode:2001RPPh...64..885C. doi:10.1088/0034-4885/64/8/301.

[4] Susanna Hornig Priest (14 July 2010). *Encyclopedia of Science and Technology Communication*. SAGE Publications. ISBN 978-1-4522-6578-0.

[5]Ellis, John (2002). "Physics gets physical (correspondence)".*Nature***415**(6875): 957.Bibcode:2002Natur.415..957E.doi:10.

[6]Ellis, John (1986). "The Superstring: Theory of Everything, or of Nothing?".*Nature***323**(6089): 595–598.Bibcode:1986 doi:10.1038/323595a0.

[7] Rorres, Chris (2009). "ARCHIMEDES AND THE QUEST FOR THE THEORY OF EVERYTHING".

[8] Chris Impey (26 March 2012). *How It Began: A Time-Traveler's Guide to the Universe*. W. W. Norton. ISBN 978-0-393-08002-5.

[9] William E. Burns (1 January 2001). *The Scientific Revolution: An Encyclopedia*. ABC-CLIO. ISBN 978-0-87436-875-8.

[10] Shapin, Steven (1996). *The Scientific Revolution*. University of Chicago Press. ISBN 0-226-75021-3.

[11] Newton, Sir Isaac (1729). *The Mathematical Principles of Natural Philosophy* **II**. p. 255.

[12] Sean Carroll (7 January 2010). *From Eternity to Here: The Quest for the Ultimate Theory of Time*. Penguin Group US. ISBN 978-1-101-15215-7.

[13] Faraday, M. (1850). "Experimental Researches in Electricity. Twenty-Fourth Series. On the Possible Relation of Gravity to Electricity". *Abstracts of the Papers Communicated to the Royal Society of London* 5: 994–995. doi:10.1098/rspl.1843.0267.

[14] A.N. Gorban, I. Karlin, Hilbert's 6th Problem: exact and approximate hydrodynamic manifolds for kinetic equations, Bull. Amer. Math. Soc., 51 (2014), no. 2, 186-246, doi:10.1090/S0273-0979-2013-01439-3

[15] Dirac, P.A.M. (1929). "Quantum mechanics of many-electron systems". *Proceedings of the Royal Society of London A* **123** (792): 714. Bibcode:1929RSPSA.123..714D. doi:10.1098/rspa.1929.0094.

[16] Abraham Pais (23 September 1982). *Subtle is the Lord : The Science and the Life of Albert Einstein: The Science and the Life of Albert Einstein*. Oxford University Press. ISBN 978-0-19-152402-8.

[17] Weinberg (1993). Ch. 5

[18] Holloway, M (2005)."The Beauty of Branes"(PDF).*Scientific American*(Scientific American)**293**(4): 38.Bibcode:2005Sci doi:10.1038/scientificamerican1005-38. PMID 16196251. Retrieved August 13, 2012.

[19] Nilles, Hans Peter; Ramos-Sánchez, Saúl; Ratz, Michael; Vaudrevange, Patrick K. S. (2008). "From strings to the MSSM". *The European Physical Journal C* **59** (2): 249. arXiv:0806.3905. Bibcode:2009EPJC...59..249N. doi:10.1140/epjc/s10052-008-0740-1.

[20] Beasley, Chris; Heckman, Jonathan J; Vafa, Cumrun (2009). "GUTs and exceptional branes in F-theory — I". *Journal of High Energy Physics* **2009**: 058. arXiv:0802.3391. Bibcode:2009JHEP...01..058B. doi:10.1088/1126-6708/2009/01/058.

[21] Donagi, Ron and Wijnholt, Martijn (2008) Model Building with F-Theory

[22] Heckman, Jonathan J. and Vafa, Cumrun (2008) Flavor Hierarchy From F-theory

[23] Strominger, Andrew; Vafa, Cumrun (1996). "Microscopic origin of the Bekenstein-Hawking entropy". *Physics Letters B* **379**: 99. arXiv:hep-th/9601029. Bibcode:1996PhLB..379...99S. doi:10.1016/0370-2693(96)00345-0.

[24] Horowitz, Gary (1996) The Origin of Black Hole Entropy in String Theory

[25] Greene, Brian R.; Morrison, David R.; Strominger, Andrew (1995). "Black hole condensation and the unification of string vacua". *Nuclear Physics B* **451**: 109. arXiv:hep-th/9504145. Bibcode:1995NuPhB.451..109G. doi:10.1016/0550-3213(95)00371-X.

[26] Aspinwall, Paul S.; Greene, Brian R.; Morrison, David R. (1994). "Calabi-Yau moduli space, mirror manifolds and spacetime topology change in string theory". *Nuclear Physics B* **416** (2): 414. arXiv:hep-th/9309097. Bibcode:1994NuPhB.416..414A. doi:10.1016/0550-3213(94)90321-2.

[27] Adams, Allan; Liu, Xiao; McGreevy, John; Saltman, Alex; Silverstein, Eva (2005). "Things fall apart: Topology change from winding tachyons". *Journal of High Energy Physics* **2005** (10): 033. arXiv:hep-th/0502021. Bibcode:2005JHEP...10..033A. doi:10.1088/1126-6708/2005/10/033.

[28] Smolin, Lee (2006). *The Trouble With Physics: The Rise of String Theory, the Fall of a Science, and What Comes Next*. Houghton Mifflin. ISBN 978-0-618-55105-7.

[29] Duff, M. J. (2011). "String and M-Theory: Answering the Critics". *Foundations of Physics* **43**: 182. arXiv:1112.0788. Bibcode:2013FoPh...43..182D. doi:10.1007/s10701-011-9618-4.

[30] Potter, Franklin (15 February 2005). "Leptons And Quarks In A Discrete Spacetime" (PDF). *Frank Potter's Science Gems*. Retrieved 2009-12-01.

[31] Bilson-Thompson, Sundance O.; Markopoulou, Fotini; Smolin, Lee (2007). "Quantum gravity and the standard model". *Classical and Quantum Gravity* **24** (16): 3975–3994. arXiv:hep-th/0603022. Bibcode:2007CQGra..24.3975B. doi:10.1088/0264-9381/24/16/002.

[32] Castelvecchi, Davide; Valerie Jamieson (August 12, 2006). "You are made of space-time". *New Scientist* (2564).

[33] Sundance Bilson-Thompson; Jonathan Hackett; Lou Kauffman; Lee Smolin (2008). "Particle Identifications from Symmetries of Braided Ribbon Network Invariants". arXiv:0804.0037 [hep-th].

[34] F. Finster; J. Kleiner (2015). "Causal fermion systems as a candidate for a unified physical theory". arXiv:1502.03587 [math-ph].

[35] A. G. Lisi (2007). "An Exceptionally Simple Theory of Everything". arXiv:0711.0770 [hep-th].

[36] Jaki, S.L. (1966). *The Relevance of Physics*. Chicago Press. pp. 127–130.

[37] Stanley L. Jaki (2004) "A Late Awakening to Gödel in Physics", pp. 8–9.

[38] Freeman Dyson, NYRB, May 13, 2004

[39] Stephen Hawking, Gödel and the end of physics, July 20, 2002

[40] Schmidhuber, Jürgen (1997). *A Computer Scientist's View of Life, the Universe, and Everything. Lecture Notes in Computer Science*. Springer. pp. 201–208. doi:10.1007/BFb0052071. ISBN 978-3-540-63746-2.

[41] Schmidhuber, Jürgen (2002). "Hierarchies of generalized Kolmogorov complexities and nonenumerable universal measures computable in the limit". arXiv:quant-ph/0011122.

[42] Feferman, Solomon (17 November 2006). "The nature and significance of Gödel's incompleteness theorems" (PDF). Institute for Advanced Study. Retrieved 2009-01-12.

[43] Robertson, Douglas S. (2007). "Goedel's Theorem, the Theory of Everything, and the Future of Science and Mathematics". *Complexity* 5 (5): 22–27. doi:10.1002/1099-0526(200005/06)5:5<22::AID-CPLX4>3.0.CO;2-0.

[44] Hawking, Stephen (20 July 2002). "Gödel and the end of physics". Retrieved 2009-12-01.

[45] Einstein, letter to Felix Klein, 1917. (On determinism and approximations.) Quoted in Pais (1982), Ch. 17.

[46] Weinberg (1993), Ch 2.

[47] Weinberg (1993) p. 5

47.6.2 Bibliography

- Pais, Abraham (1982) *Subtle is the Lord...: The Science and the Life of Albert Einstein* (Oxford University Press, Oxford, . Ch. 17, ISBN 0-19-853907-X

- Weinberg, Steven (1993) *Dreams of a Final Theory: The Search for the Fundamental Laws of Nature*, Hutchinson Radius, London, ISBN 0-09-177395-4

47.7 External links

- The Elegant Universe, *Nova* episode about the search for the theory of everything and string theory.

- Theory of Everything, freeview video by the Vega Science Trust, BBC and Open University.

- The Theory of Everything: Are we getting closer, or is a final theory of matter and the universe impossible? Debate between John Ellis (physicist), Frank Close and Nicholas Maxwell.

- Why The World Exists, a discussion between physicist Laura Mersini-Houghton, cosmologist George Francis Rayner Ellis and philosopher David Wallace about dark matter, parallel universes and explaining why these and the present Universe exist.

47.8   Text and image sources, contributors, and licenses

47.8.1   Text

- **Gravity** *Source:* https://en.wikipedia.org/wiki/Gravity?oldid=683336844 *Contributors:* Bryan Derksen, Danny, Rmhermen, Caltrop, Heron, Montrealais, Stevertigo, Patrick, D, Michael Hardy, Ixfd64, TakuyaMurata, GTBacchus, Wintran, Snarfies, Ahoerstemeier, Mac, Dgaubin, William M. Connolley, Jeff Relf, TonyClarke, Smack, Ec5618, Tpbradbury, Phoebe, Fairandbalanced, BenRG, Pollinator, Lumos3, Jni, Northgrove, Donarreiskoffer, Robbot, Sander123, Jakohn, TimothyPilgrim, Academic Challenger, Caknuck, Bkell, Nerval, Aetheling, Ruakh, Dina, Alan Liefting, Cedars, Ancheta Wis, Giftlite, Mikez, Haeleth, Wolfkeeper, Inkling, Fropuff, Everyking, Frencheigh, Aechols, Bobblewik, Utcursch, Keith Edkins, Antandrus, Phe, Quarl, Elembis, Kiteinthewind, Jossi, Karol Langner, Lindberg G Williams Jr, Demiurge, Trevor MacInnis, Grstain, Mike Rosoft, &Delta, DanielCD, Shipmaster, JimJast, Discospinster, Zaheen, Supercoop, Vsmith, Smyth, Notinasnaid, Dbachmann, Bender235, Rubicon, Loren36, El C, Huntster, Kwamikagami, Aude, RoyBoy, Nrbelex, Gershwinrb, Bobo192, Smalljim, Elipongo, I9Q79oL78KiL0QTFHgyc, Larry V, MPerel, Danski14, Alansohn, JYolkowski, Anthony Appleyard, Davetcoleman, Atlant, Maya Levy, Paleorthid, Andrew Gray, Cjthellama, Riana, Goldom, Mac Davis, Wdfarmer, Hdeasy, Snowolf, Mononoke-enwiki, BRW, Yuckfoo, RainbowOfLight, Jesvane, Bookandcoffee, Kazvorpal, Falcorian, Stephen, Feezo, Richard Arthur Norton (1958-), OwenX, Linas, Camw, LOL, Benhocking, JFG, MONGO, Mpatel, Tabletop, GregorB, Eaolson, Isnow, Scm83x, SDC, Philbarker, TheAlphaWolf, Brownsteve, Radiant!, Dysepsion, Sin-man, Ashmoo, Chrispasquale, Chun-hian, Rjwilmsi, Tangotango, Nichiran, Fel64, Kazrak, Ligulem, Ems57fcva, Boccobrock, Bhadani, MarnetteD, Matt Deres, Sango123, Syced, Yamamoto Ichiro, Fish and karate, JanSuchy, Magmafox, Titoxd, RobertG, Old Moonraker, Nihiltres, Nivix, RexNL, Gurch, Fresheneesz, Alphachimp, Tardis, Srleffler, King of Hearts, DVdm, Guliolopez, Hall Monitor, Bomb319, Gwernol, EamonnPKeane, Raelx, The Rambling Man, Wavelength, Drdisque, Sceptre, Hairy Dude, Jimp, Hillman, Brandmeister (old), Ohwilleke, Red Slash, Musicpvm, Anonymous editor, Loom91, Bhny, Pigman, Epolk, Philip Hazelden, DanMS, Scott5834, CambridgeBayWeather, Wimt, Cemcem, NawlinWiki, Nowa, Wiki alf, Ytcracker, SigPig, SCZenz, Nick, Brandon, Katrina Graziano, Semperf, Aaron Schulz, Roy Brumback, Addps4cat, Jessemerriman, Rayc, Mgnbar, Dna-webmaster, Eurosong, Hecroyuy135, Enormousdude, 2over0, SFGiants, Chase me ladies, I'm the Cavalry, Lappado, Endomion, E Wing, Fmyhr, Smurrayinchester, Kevin, Geoffrey.landis, Anclation-enwiki, Eme2, Willtron, Kungfuadam, RG2, JDspeeder1, Bo Jacoby, Draicone, FyzixFighter, Mejor Los Indios, Sbyrnes321, DVD R W, Hide&Reason, That Guy, From That Show!, Shenhemu, Luk, TravisTX, Sardanaphalus, SmackBot, Ulterior19802005, Incnis Mrsi, Reedy, KnowledgeOfSelf, Hydrogen Iodide, NaiPiak, David Shear, Kilo-Lima, Jagged 85, Thunderboltz, Delldot, Hardyplants, Cessator, Syckls, BiT, Timotheus Canens, GraemeMcRae, HalfShadow, Typhoonchaser, Yamaguchi先生, Algont, Hmains, Ppntori, ERcheck, Andy M. Wang, Kmarinas86, The monkeyhate, Saros136, Bluebot, Cush, Keegan, Raymond arritt, Fplay, Silly rabbit, Lehkost, Complexica, Bbq332, Jeff5102, Sbharris, Hallenrm, CharonM72, Sewlong, Can't sleep, clown will eat me, Scott3, UNHchabo, MJCdetroit, Apostolos Margaritis, Lesnail, Cryocide, Rsm99833, Amazins490, Jmlk17, Cybercobra, Nakon, Steve Pucci, TedE, Red1-enwiki, Jiddisch-enwiki, Dreadstar, Dave-ros, Weregerbil, Cockneyite, Crd721, Bryanmcdonald, Jklin, DMacks, Wizardman, Where, LeoNomis, Risker, Sadi Carnot, Carlosp420, TTE, Yevgeny Kats, Will Beback, Jonpalmer, Nathanael Bar-Aur L., Beasterline, Geoffrey Wickham, Rklawton, Djeneba, Sophia, Dsantesteban, Kuru, Thefro552, Titus III, Richard L. Peterson, John, Scientizzle, Stephane Yelle, DocRocks1, Jaganath, Thegathering, Skoobieschnax, JorisvS, LestatdeLioncourt, Coredesat, Accurizer, Minna Sora no Shita, Mgiganteus1, A5y, Nonsuch, Ridershydelta, Mr. Lefty, AtD, Jess Mars, Ben Moore, JHunterJ, MarkSutton, Slakr, Special-T, Momolee, LuYiSi, Mr Stephen, Samaster1991, Spiel496, Buttle, Novangelis, PSUMark2006, Inquisitus, DI2000, ShakingSpirit, Hgrobe, Ginkgo100, Vanished user, JMK, Craigboy, Lakers, Newone, MOBle, J Di, StephenBuxton, Matt Bernius, Igoldste, Taucetiman, Tofoo, Tawkerbot2, Lsskys, George100, Kurtan-enwiki, Lahiru k, CalebNoble, SkyWalker, JForget, CmdrObot, Tanthalas39, Ghobson, Porterjoh, Ale jrb, Scohoust, Aherunar, Galo1969X, Picaroon, Shakespeare87, User92361, Zureks, Basawala, Ruslik0, GHe, Dgw, OMGsplosion, ShelfSkewed, WHATaintNOcountryIeverHEARDofDOtheySPEAKenglishINwhat, MarsRover, Hi There, Groosh, Myasuda, Anthony Bradbury, Gregbard, Logicus, Cydebot, Steel, Travelbird, Red Director, Jon Stockton, A Softer Answer, Adolphus79, Nicesai, Rracecarr, Codingmasters, Ch0rx, Tawkerbot4, Alexnye, Christian75, M a s, DumbBOT, DarkLink, Interwiki gl, FastLizard4, Optimist on the run, Jimip, Waxigloo, SteveMcCluskey, Omicronpersei8, Stoked, Gimmetrow, Sevenaces, Raoul NK, FrancoGG, Thijs!bot, Epbr123, DarlingFriend, Opabinia regalis, Pajz, Ishdarian, Jamesluster, Andyjsmith, 24fan24, Gamer007, ClosedEyesSeeing, Headbomb, John254, Bobblehead, Neil916, Pogogunner, Grayshi, EdJohnston, HistoryMaster 1, Zachary, The Hybrid, Nick Number, Lithpiperpilot, CarbonX, MichaelMaggs, Sam42, J.S.B.Anderson, Escarbot, Eleuther, Mentifisto, Hmrox, AntiVandalBot, Yonatan, Luna Santin, CodeWeasel, Themaxeditor, Prolog, Yay unto the Chicken, Dylan Lake, Gdo01, Myanw, Ioeth, JAnDbot, Leuko, Husond, Vorpal blade, ThomasO1989, Roman à clef, MER-C, Nevadacall, Andonic, Hut 8.5, 100110100, N shaji, Pablothegreat85, Magioladitis, Foobird107, Murgh, Bennybp, Bongwarrior, VoABot II, AuburnPilot, Xn4, Wikidudeman, Hendrixjoseph, Careless hx, Aerographer1981, Crazytonyi, 9holdss, ThomasThePolishMan, Bubba hotep, Mi6agent00g off, BatteryIncluded, Beetfarm Louie, Adrian J. Hunter, Alexei Kojenov, Kane1047, LorenzoB, Mollwollfumble, Scot.parker, Andykass, Talon Artaine, Chris G, DerHexer, MeEricYay, WLU, Seph Vellus, TheRanger, Patstuart, Seba5618, Oroso, NatureA16, FisherQueen, Hdt83, E.vondarkmoor, MartinBot, Sinfear, Shimwell, Flamingpanda, The Ubik, Unfriendly-Fire, APT, Rettetast, Juansidious, Anaxial, Sm8900, David J Wilson, Mschel, R'n'B, GarrisonGreen, AlexiusHoratius, Pekaje, LittleOldMe old, PrestonH, J.delanoy, Filll, Trusilver, Tonmoy Chowdhury, Bloomingiris, 72Dino, MoogleEXE, Lhynard, Ginsengbomb, WarthogDemon, Willow123-enwiki, SubwayEater, Yeti Hunter, Hisagi, James Mead, M C Y 1008, Wandering Ghost, Redmotherfive, Vertigo900, Mr Rookles, Samtheboy, Gurchzilla, Supuhstar, Pyrospirit, AntiSpamBot, GhostPirate, Belovedfreak, Raichu Trainer, Ohms law, Mitchell is hollywood, SJP, Policron, Touch Of Light, Pwnasaurusrex38, MKoltnow, Mufka, FJPB, Blekavnger, Cmichael, Mohrflies, Stoned Proffesor, Kenneth M Burke, Cosmictinker, RB972, Tiggerjay, U.S.A.U.S.A.U.S.A., Treisijs, Mike V, Redrocket, Gtg204y, MtyQuinn, Darkfrog24, Jxzj, Annax3, Smartman10, Ronbo76, Micmic28, Yecril, Missphysics, GoldenGolem, Xiahou, Lorax835, Steel1943, Washboard6, Sheliak, Funandtrvl, Gravityc, Teesup, Black Kite, Chinneeb, Deor, VolkovBot, TreasuryTag, TJ Elliot Scott, Meaningful Username, Danwills, DSRH, Mtesm, Indubitably, Gseletko, Veddan, Boaex, Dominics Fire, Heerojuy, Philip Trueman, Childhoodsend, TXiKiBoT, GVIlleneuve27, Davecas0, Adamwang, BuickCenturyDriver, 99DBSIMLR, MeStevo, Lolerballer-enwiki, Andrius.v, Z.E.R.O., Anonymous Dissident, Jonnymagic, Ask123, Trente, MattCarterSurrealist, IPSOS, Sarthella, Seraphim, Fizzackerly, Wvogeler, The One Cause, Tallcreek, Markp93, Drappel, PDFbot, Freak104, Manticore55, Cremepuff222, Blackdragon 1002, Wikiisawesome, Vrsixfire, Liberal Classic, Witchzilla, Noel rebeira, Inductiveload, Gladiator2155, Sydweighz, Spiral5800, Larklight, RobertFritzius, Dirkbb, SQL, SallyBoseman, Frag1983, Synthebot, ChillDeity, Speria, Heroesrule17, Enviroboy, Hollop09, Sylent, Kaseman519, Ballsucker22, Sesshomaru, Brianga, Skins88, Chickyfuzz14, AlleborgoBot, GavinTing, Shaidar cuebiyar, Happyhacker101, NHRHS2010, Steven Weston, D. Recorder, Al.Glitch, Xarr, The Random Editor, Mr. dick

008, SieBot, Tosun, Crunchedfor6, High2lowo, K. Annioyomous, Thong123456789, Work permit, Scarian, Invmog, Gerakibot, Mazza uk04, Mickeyd24, LealandA, Caltas, BreakfastTom, Beethoven314, Pieman123456789, RJaguar3, Triwbe, Rangutan, Chs^3, Tubular bells83, The way, the truth, and the light, Garrett gagne, Keilana, Elliott Fontain, CaptainIraq, RadicalOne, Tiptoety, CombatCraig, Bhahn0125, Oda Mari, Sunayanaa, Cnormansen, Thin joe, M keshe, Jirachipokemon, Kein213, Chives4life, Luskj, Oxymoron83, Faradayplank, Smilesfozwood, AngelOfSadness, Nuttycoconut, Edwardwittenstein, Zharradan.angelfire, John fromer, Lightmouse, AMCKen, Iain99, Techman224, Skateboards fly, Mr mimises, Blobbucket, BenoniBot~enwiki, Dillard421, Panicum, Fedosin, C'est moi, Hamiltondaniel, ShexRox, Dolphin51, M2Ys4U, Into The Fray, Canglesea, C0nanPayne, Albin&dani, Quinling, Muhends, Moony1000, Atif.t2, Tomasz Prochownik, ClueBot, Thejman123456789, Artichoker, PipepBot, The Thing That Should Not Be, JavaJesus, Mr.Pinklesworth, Meekywiki, Artist7337, Drmies, Firth m, Mild Bill Hiccup, CounterVandalismBot, Kitty9992, VandalCruncher, Harland1, Otolemur crassicaudatus, Jackey0105, Andwor9, Aik-Bkj, Another Matt, RogerEllman, Puchiko, Shocky95, Gakusha, DragonBot, Joshisamazing, Chris earl 89, Alexbot, Mccann tom, Icreaser, Robbie098, Sct5333, Shinkolobwe, Fsunka, Itel94, Ola Hansen, Ice Cold Beer, Jotterbot, PhySusie, Glacialvortex, Razorflame, Handcannonbeast, Applejacks47, Chaosdruid, Thingg, Venera 7, Wnt, Myagooshki666, Deproduction, MasterOfHisOwnDomain, Jaykay0424, Ioannes93, TimothyRias, Misterbeal, InternetMeme, Mhamhamha, XLinkBot, Royboturso, Gwandoya, Crustillicus, BodhisattvaBot, Rror, Ance.cdas, Baconlover13, Ost316, Srossman07, Noctibus, JinJian, Truthnlove, Ttimespan, Airplaneman, Infonation101, EEng, Freestyle-69, Kbdankbot, Xerbaycom, Addbot, Nyw195, Willking1979, NateDres23, Bobocheese, Betterusername, Olli Niemitalo, Cre84u, TutterMouse, Presentabsent, Eedlee, Veraptor, WFPM, Ashanda, MrOllie, Lost on Belmont, FerrousTigrus, Ld100, Delaszk, Glass Sword, Maddox1, Jasper Deng, Harvardstudent, Kicka, Tide rolls, EugeneKantarovich, Cesiumfrog, Krano, WikiDreamer Bot, Hartz, Narutolovehinata5, Legobot, Drpickem, Luckas-bot, Dov Henis, Senator Palpatine, Il MusLiM HyBRiD II, JustWong, Becky Sayles, THEN WHO WAS PHONE?, AnakngAraw, Solo Zone, Azcolvin429, Atqueamemus, MacTire02, AnomieBOT, Jdiyef, Ahmediq152, Jt16733, Gatoradeparade, Kanwaraj, Piano non troppo, Chaosmaker39, Gsd65, LlywelynII, Rhlowe, Unicornlad, VCleemput, Kanat Abiklinov, Typesships, Are you ready for IPv6?, Dje 8, Citation bot, Oddball.bfi, Nyrox395, Maxis ftw, Persistent76, Chase4813, ArthurBot, Snorlax Monster, Gravityforce, Xqbot, TinucherianBot II, TechBot, Millahnna, Markell West, GrouchoBot, Celebration1981, Gsard, Amaury, Charvest, Doulos Christos, A. di M., Acannas, CES1596, Ninjainventor, LucienBOT, Paine Ellsworth, Tobby72, TedlyW, Lookang, 陳宏賢, Slawomir Biały, ArkianNWM, Allen Jesus, Parvons, Cannolis, Orion 8, Citation bot 1, Pinethicket, Tom.Reding, RedBot, IceBlade710, SpaceFlight89, Rohitphy, Jauhienij, RockSolidCosmo, FoxBot, TobeBot, Yunshui, Gafferuk, Comet Tuttle, Le Docteur, Schwede66, MitchLay, Oms22, Earthandmoon, Tbhotch, RobertMfromLI, Brambleclawx, RjwilmsiBot, Androstachys, DASHBot, EmausBot, John of Reading, Mnkyman, Atwarwiththem, Qurq, Gfoley4, Kueller1, Dewritech, Baguettes, Syncategoremata, Jmencisom, Sheeana, The Blade of the Northern Lights, Solomonfromfinland, JSquish, Ryan.vilbig, ZéroBot, Crua9, GoldRenet, Dffgd, Everard Proudfoot, Quondum, Wikfr, Confession0791, Oeaasi, Wtsbeynon, BrokenAnchorBot, Brandmeister, I. Kensington, Zayzya, Ally1604, Checkmark56, Fluctuating metric, RockMagnetist, Teapeat, Laned130, Rememberway, ClueBot NG, Nebulosus, CocuBot, PoqVaUSA, Jj1236, Lilptrsn, Megalobingosaurus, MerllwBot, Helpful Pixie Bot, BG19bot, Negativecharge, Furkhaocean, GKFX, Cadiomals, Mr.viktor.stepanov, RGloucester, BattyBot, Tutelary, GravityForce, Padenton, Khazar2, JYBot, DavidIwinkler, LightandDark2000, Mogism, Rudrene, Reatlas, CsDix, Everymorning, Yuan Jullian Morales, Gacman67, DavidLeighEllis, Prokaryotes, Jwratner1, My name is not dave, Mfb, Konveyor Belt, Mahusha, Monkbot, Filedelinkerbot, Gronk Oz, Oiyarbepsy, Stefania.deluca, Loraof, Ps20231131, Pishcal, Freedom2003, Tetra quark, James 123234, Inorout, Supdiop, KasparBot, The oracle 2015, Sweepy, Jeffryan123, ImperatorRomanorvm, Addycrisp, The golden colten, J.A.Witt (Tony), Hiimemilylol, Dan6233 and Anonymous: 1221

- **History of gravitational theory** *Source:* https://en.wikipedia.org/wiki/History_of_gravitational_theory?oldid=683169622 *Contributors:* Ewen, Rorro, Ancheta Wis, Karol Langner, Discospinster, Vsmith, Dbachmann, Neko-chan, Theant2000, Rjwilmsi, Ems57fcva, DVdm, Dialectric, Schlafly, Doldrums, SmackBot, Jagged 85, Gilliam, Hmains, Colonies Chris, Wizardman, Sadi Carnot, Ohconfucius, Csladic, Special-T, Chris55, Tareq.khatib, Sakurambo, CmdrObot, TheTito, Alaibot, SteveMcCluskey, D.H, MarshBot, Email4mobile, WatchingYouLikeA-Hawk, David J Wilson, Hans Dunkelberg, KylieTastic, Rising*From*Ashes, Celebrei, Philip Trueman, Kgun5, Noveltyghost, Madhavacharya, Tomasz Prochownik, J8079s, Auntof6, Djr32, SchreiberBike, Addbot, Yoyoyow, MrOllie, Alanscottwalker, Luckas-bot, Pigetrational, Jim1138, Piano non troppo, Citation bot, LilHelpa, Almuhammedi, Citation bot 1, Pinethicket, Tom.Reding, Diiscool, Syncategoremata, Bollyjeff, Knight1993, JoeSperrazza, Donner60, Weirdly, ClueBot NG, Widr, Helpful Pixie Bot, Jayadevivenugopal, Kawaii-Soft, Monkbot and Anonymous: 63

- **Newton's law of universal gravitation** *Source:* https://en.wikipedia.org/wiki/Newton'[}s_law_of_universal_gravitation?oldid=683302022 *Contributors:* XJaM, William Avery, Caltrop, Patrick, Michael Hardy, SebastianHelm, Pizza Puzzle, Charles Matthews, Geraki, Robbot, Piels, Ancheta Wis, Giftlite, Sj, Wolfkeeper, Everyking, Antandrus, Beland, MisfitToys, EricJamesStone, Shotwell, Mike Rosoft, Rfl, JimJast, Discospinster, Rich Farmbrough, FT2, Vsmith, Dbachmann, WegianWarrior, Bender235, Lycurgus, Spoon!, Bobo192, Stesmo, Smalljim, I9Q79oL78KiL0QTFHgyc, Kjkolb, Jeodesic, Sam Korn, Nsaa, Danski14, Alansohn, Arthena, MarkGallagher, Snowolf, Wtmitchell, Wtshymanski, Allen McC.~enwiki, Gene Nygaard, Vadim Makarov, Brookie, WilliamKF, Linas, Camw, StradivariusTV, Drostie, Mpatel, JRHorse, Zzyzx11, Paxsimius, Ashmoo, Thierry Dugnolle~enwiki, Kbdank71, Sjö, Coemgenus, Sdornan, Salix alba, Ligulem, Ems57fcva, Matt Deres, Anskas, Latka, GnuDoyng, Rbonvall, Gurch, Chris D'Amato, Wgfcrafty, Jared Preston, DVdm, Random user 39849958, Sanpaz, Gwernol, YurikBot, TexasAndroid, Sceptre, RussBot, Bhny, Madkayaker, Howcheng, Gadget850, Dna-webmaster, Wknight94, Enormousdude, Ketsuekigata, Josh3580, GraemeL, CWenger, MagneticFlux, RG2, AssistantX, FyzixFighter, Sbyrnes321, SmackBot, MattieTK, Yamaguchi先生, Gilliam, Iskander32, Thumperward, Complexica, Tianxiaozhang~enwiki, Baa, Sbharris, Firetrap9254, NYKevin, Krich, Fuhghettaboutit, Kingdon, T-borg, Pwjb, Greg.collver, Xiutwel, Filpaul, Sigma 7, Stefano85, ElizabethFong, Sadi Carnot, Yevgeny Kats, SashatoBot, Lambiam, Nishkid64, Gobonobo, WhiteHatLurker, Mr Stephen, Onionmon, Mthsmith, Paul venter, Newone, Pathosbot, Chetvorno, Chris55, Mellery, McVities, Karenjc, Cydebot, Rracecarr, BishopOcelot, SteveMcCluskey, Thijs!bot, Epbr123, Daa89563, Andyjsmith, Headbomb, Marek69, Codee240, John254, Kathovo, Invitatious, D.H, Stannered, Luna Santin, Deeplogic, Mogzig, Storkk, Myanw, Shambolic Entity, JAnDbot, Seddon, Acroterion, Magioladitis, VoABot II, JNW, Appraiser, Twsx, Aa35te, Cardamon, CodeCat, Causesobad, Duckysmokton, Ptrpro, FisherQueen, MartinBot, Nthitz, Jay Litman, J.delanoy, Trusilver, Ali, P.wormer, NewEnglandYankee, Wesino, Sarregouset, Baseball-bob, Timboyk12, Jarry1250, Yecril, Quiet Silent Bob, T prev, VolkovBot, Larryisgood, ABF, Jeff G., Philip Trueman, TXiKiBoT, Asuffield, Red Act, Hqb, Gian-2, Canaima, Jackfork, PDFbot, Windrixx, Maxw41, Gunnar Berlin, Falcon8765, Newsaholic, Brianga, SieBot, Paradoctor, Gerakibot, JerrySteal, Darkwarlord95, Zbvhs, Quest for Truth, Masgatotkaca, Captain Yankee, James.Denholm, Paolo.dL, Prestonmag, Pac72, WWStone, Kenkenko, OKBot, Duae Quartunciae, Anchor Link Bot, Rotovia, Denisarona, Escape Orbit, 678right, ClueBot, ICAPTCHA, Ice-Unshattered, Spoladore, Donteras, Drmies, CounterVandalismBot, Make91, Maymay, GrapeSmuckers, Abdullah Köroğlu~enwiki, Ktr101, Excirial, Jusdafax, Erebus Morgaine, SpikeToronto, Iohannes Animosus, Muro Bot, Thehelpfulone, CarlosPatiño, Thingg, Redrocketboy, SoxBot

III, DumZiBoT, MadameBouvier, AgnosticPreachersKid, Terry0051, Little Mountain 5, Jakezing, Kbdankbot, Addbot, Proofreader77, Willking1979, D c weber, Guoguo12, Kung foo masta, Ronhjones, TutterMouse, MrOllie, Glass Sword, Tron78, Favonian, Jasper Deng, Barak Sh, Tide rolls, Lightbot, Krano, من إب, Alfie66, Angrysockhop, Johncolton, Luckas-bot, Yobot, Stamcose, ArchonMagnus, KamikazeBot, Dr-Trigon, عربى جزء, Tempodivalse, AnomieBOT, JackieBot, Rudolf.hellmuth, Csigabi, Materialscientist, E2eamon, Alex Yuwen, Arthur-Bot, Snorlax Monster, Xqbot, პოვიანი, Tranxodox, The sock that should not be, Capricorn42, Grim23, Ched, AbigailAbernathy, NOrbeck, GrouchoBot, AVBOT, RibotBOT, SassoBot, Mathonius, Amaury, Rickproser, 78.26, Kongkokhaw, A. di M., A.amitkumar, Ashik, GliderMaven, Paine Ellsworth, Lookang, HJ Mitchell, Steve Quinn, XXx xD LeGeNd MoJo xXx, HamburgerRadio, Citation bot 1, Pinethicket, I dream of horses, PrincessofLlyr, Elpiades, Tom.Reding, MastiBot, SpaceFlight89, Jauhienij, ActivExpression, TobeBot, Info4sina, EdFalzer, Vrenator, J0z777, Bluefist, Reaper Eternal, JLincoln, Diannaa, I change stuff ha, Sjsharksfan, Seanlacroosed, Onel5969, TjBot, Hajatvrc, Jb-whitmore, Deagle AP, EmausBot, Gfoley4, Razor2988, Vjga, Infringement153, Winner 42, Mmeijeri, Wikipelli, Zegod, Thecheesykid, Mz7, JSquish, John Cline, PRABHAT PINGREJA, Josve05a, Urgent01, Sakapraia, Druzhnik, Quondum, Christina Silverman, Dennis Kwaria, Jay-Sebastos, L Kensington, Donner60, Danlevy100, Zueignung, Sailsbystars, Carmichael, RockMagnetist, LGSlayer1127, DASHBotAV, Nathanlongan, Daniel55423, ClueBot NG, Iiii I I I, CocuBot, Ijrmoney1, Heimdallen, Millermk, Rokkyo13, EnglishTea4me, Braincricket, Thomask0, Widr, Moooo1234, Helpful Pixie Bot, BG19bot, Jwchong, Northamerica1000, Tiger42653, Mifter Public, Dan653, AwamerT, Piguy101, Jobin RV, Mark Arsten, Joydeep, Snow Blizzard, YVSREDDY, Glacialfox, BattyBot, Toumajk, Zhaofeng Li, BrightStarSky, Webclient101, SteenthIWbot, New York Resident, Sozopol, Sokotrof10, Reatlas, Cfarrar27, Vanamonde93, PhantomTech, Tentinator, Kogmaw, Gmantheladiesman, Alexbassist, DavidLeighEllis, Glaisher, Prokaryotes, Jwratner1, Ginsuloft, Macks2008, Technomite, Internetconservationist, KhaiYuen.TAN, Monkbot, Deshmukhswapnil97, Jr513, G6w000, Stefania.deluca, Tylerrigss01, Ducketduckduck, Poopnoop123, Sir IssacNewton2yy73y843yywy3, Arghyadeep Acharya, TheUniversalist, Superdupersmartdude, Inorout, KasparBot, Akhil1033 and Anonymous: 676

• **Semiclassical gravity** *Source:* https://en.wikipedia.org/wiki/Semiclassical_gravity?oldid=599458476 *Contributors:* Charles Matthews, Dmr2, Daniel Arteaga~enwiki, Mpatel, Joke137, BD2412, Bgwhite, Hillman, Teply, Colonies Chris, Vampus, Al Lemos, Pqnelson, Addbot, Yobot, Wireader, Carlog3, Raidr and Anonymous: 4

• **Introduction to general relativity** *Source:* https://en.wikipedia.org/wiki/Introduction_to_general_relativity?oldid=673756302 *Contributors:* Maury Markowitz, Michael Hardy, William M. Connolley, Schneelocke, Doradus, Tpbradbury, Gandalf61, Ancheta Wis, Giftlite, Wolfkeeper, Michael Devore, Piotrus, Thincat, JimJast, Pjacobi, Wadewitz, Bender235, Art LaPella, Circeus, I9Q79oL78KiL0QTFHgyc, Mpulier, Viridian, InShanee, Snowolf, Vashti, Mindmatrix, Ganeshk, Benhocking, Carcharoth, Robert K S, WadeSimMiser, Mpatel, Christopher Thomas, Graham87, Kbdank71, Rjwilmsi, Ems57fcva, Brighterorange, Remurmur, RobertG, Fresheneesz, Goudzovski, Spencerk, DVdm, Dresdnhope, Loom91, KSmrq, Eleassar, Madcoverboy, SCZenz, Beanyk, Vonfraginoff, Paul Magnussen, Masatran, Petri Krohn, Junglecat, Sardanaphalus, SmackBot, Mitchan, Onebravemonkey, Tony7444, Bluebot, Silly rabbit, CSWarren, Colonies Chris, Easwarno1, Grover cleveland, William Ackerman, T-borg, Speedplane, Richard001, Komojo, Sadi Carnot, Lester, Harryboyles, Caim, JorisvS, 16@r, SandyGeorgia, Novangelis, JoeBot, Tanthalas39, Olaf Davis, Linus M., Outriggr, WillowW, UncleBubba, Capedia, Juansempere, Crum375, Markus Pössel, Wdaylaporte, Headbomb, D.H, Proximo.xv~enwiki, AntiVandalBot, Bogolov, Billscottbob, Dodecahedron~enwiki, Darklilac, Easchiff, WolfmanSF, Bongwarrior, VoABot II, Fusionmix, Brianlanter, Stijn Vermeeren, SwiftBot, User A1, Charles Edward, CommonsDelinker, ZRV, Pbroks13, DrKiernan, Lantonov, It Is Me Here, Sarregouset, Sheliak, VolkovBot, Joopercoopers, Nxavar, Cody-7, Rjm at sleepers, Geometry guy, Lamro, Arcfrk, Gustav von Humpelschmumpel, Dnarby, SieBot, Timb66, ConfuciusOrnis, Spartan-James, Ward20, Randomblue, Anyeverybody, WingedPig, Dabomb87, ClueBot, The Thing That Should Not Be, Wwheaton, Piledhigheranddeeper, Ajoykt, NuclearWarfare, JDPhD, Brainman365, Ladsgroup, MystBot, Kbdankbot, Addbot, Mortense, DOI bot, EconoPhysicist, Albuseer, Bob K31416, Lightbot, Gail, Ret.Prof, Luckas-bot, Yobot, Materialscientist, The High Fin Sperm Whale, Citation bot, PerfWise, Xqbot, Mlpearc, Rainald62, 梁, Paine Ellsworth, Nojan, Steve Quinn, Citation bot 1, Citation bot 4, AstaBOTh15, Jordgette, Lotje, Chrysostome~enwiki, Cardinality, Earthandmoon, RjwilmsiBot, ScottyBerg, Azscottish, Graffacake2, H3llBot, Quondum, Gaarmyvet, Maschen, RockMagnetist, Sona11235, ClueBot NG, Frietjes, Mouse20080706, Helpful Pixie Bot, Gob Lofa, Bibcode Bot, Harizotoh9, Dexbot, Colinddelia, Fessi, Jamesx12345, Ram2003, Mark viking, Eyesnore, Rogerkoehler, Ginsuloft, Frinthruit, Filedelinkerbot, Anupkashyap7 and Anonymous: 114

• **History of general relativity** *Source:* https://en.wikipedia.org/wiki/History_of_general_relativity?oldid=657093238 *Contributors:* Malcolm Farmer, Michael Hardy, Bcrowell, Charles Matthews, Robbot, Fropuff, Alvestrand, Karol Langner, Rich Farmbrough, ThomasK, Kipton, Rgdboer, Plumbago, RJFJR, Capecodeph, Billhpike, Linas, PatGallacher, Carcharoth, Mpatel, Joke137, Christopher Thomas, Rjwilmsi, Mike Peel, Ligulem, Ems57fcva, Tedder, DVdm, Hillman, KSmrq, Sardanaphalus, SmackBot, Hbackman, Edgar181, Chris the speller, Colonies Chris, Georg-Johann, GeorgeMoney, Addshore, William Ackerman, E4mmacro, Fuhghettaboutit, Ligulembot, Ohconfucius, Lambiam, JRSpriggs, Pgr94, Tonyle, D.H, Rico402, Easchiff, DAGwyn, Mollwollfumble, CommonsDelinker, Vanwhistler, TheSeven, Lantonov, Teisco, JohnBlackburne, Hqb, Wmpearl, SockPuppetForTomruen, Addbot, AnomieBOT, Materialscientist, Citation bot, Xqbot, Spartan S58, Troglo, Steve Quinn, SkinnyPrude, Full-date unlinking bot, Lotje, Earthandmoon, EmausBot, Stibu, RockMagnetist, Will Beback Auto, Helpful Pixie Bot, Bibcode Bot, BG19bot, Mark Arsten, Harizotoh9, Mcn999, ChrisGualtieri, Filedelinkerbot and Anonymous: 35

• **Mathematics of general relativity** *Source:* https://en.wikipedia.org/wiki/Mathematics_of_general_relativity?oldid=679833636 *Contributors:* Stevertigo, Patrick, Charles Matthews, Phil Boswell, Giftlite, Fropuff, Karol Langner, Anythingyouwant, Deewiant, PhotoBox, Masudr, El C, Pharos, Falcorian, Linas, Carcharoth, Jeff3000, Mpatel, GregorB, Joke137, Ligulem, Ems57fcva, Mathbot, Itinerant1, DVdm, Dresdnhope, Hillman, Loom91, Bhny, Netrapt, Tevildo, Sardanaphalus, SmackBot, Silly rabbit, Complexica, Colonies Chris, Nat2, Spartanfox86, Elb2000, JRSpriggs, Vanisaac, Sakurambo, Cuneas, Harej bot, Jrgetsin, Cyclonenim, Thebard, Magioladitis, RogierBrussee, Atonita, DAGwyn, Arturj, HEL, Filll, Bobxii, GregWoodhouse, VolkovBot, The Duke of Waltham, Butwhatdoiknow, TXiKiBoT, Gobofro, YohanN7, VVVBot, Matthew Yeager, ClueBot, Ferred, Jpsfitz, Addbot, Jacopo Werther, Davy Cielen~enwiki, Albuseer, Yobot, Fraggle81, Citation bot, NOrbeck, Carlog3, Jamesmelody, FrescoBot, Tom.Reding, Francisco Quiumento, Tonyxty, ZéroBot, Eniagrom, Staszek Lem, Maschen, Ego White Tray, Wcherowi, Atomician, Joydeep, F=q(E+v^B), Enyokoyama, Dcollins3.1415, Frinthruit, Grifftron3000, Theoretical wormhole and Anonymous: 38

• **General relativity** *Source:* https://en.wikipedia.org/wiki/General_relativity?oldid=683739649 *Contributors:* AxelBoldt, Mav, Bryan Derksen, The Anome, AstroNomer~enwiki, Ap, RK, Andre Engels, XJaM, Chrislintott, JeLuF, Christian List, William Avery, Roadrunner, Kisquare, B4hand, Stevertigo, Frecklefoot, Patrick, Boud, Michael Hardy, Menchi, Ixfd64, Bcrowell, Nimrod~enwiki, TakuyaMurata, Mcarling, Minesweeper, Alfio, Looxix~enwiki, ArnoLagrange, Ellywa, Ahoerstemeier, Stevenj, William M. Connolley, Snoyes, Angela, Mark Foskey,

Julesd, Salsa Shark, AugPi, Andres, Evercat, Hectorthebat, Hick ninja, A.Tigges–enwiki, Gingekerr, Jitse Niesen, Gutza, Rednblu, Doradus, Wik, Dragons flight, Tero–enwiki, Phys, Shizhao, Elwoz, BenRG, Banno, Northgrove, Phil Boswell, Robbot, Craig Stuntz, Sdedeo, Bvc2000, Goethean, Altenmann, Romanm, Lowellian, Mayooranathan, Gandalf61, Blainster, Diderot, DHN, Hadal, Alba, Johnstone, Fuelbottle, Isopropyl, Xanzzibar, Carnildo, Tobias Bergemann, Enochlau, Ancheta Wis, Tosha, Giftlite, JamesMLane, Graeme Bartlett, Mikez, BenFrantzDale, Lethe, Tom harrison, Fropuff, Everyking, Physman, Curps, Michael Devore, Jason Quinn, Alvestrand, SWAdair, Glengarry, Bobblewik, Edcolins, DefLog–enwiki, Pgan002, Knutux, GeneralPatton, HorsePunchKid, Robert Brockway, Kaldari, Madlee, Karol Langner, Rjpetti, Rdsmith4, JimWae, Anythingyouwant, Martin Wisse, Thincat, Euphoria, Icairns, Zfr, AmarChandra, Zondor, Econrad, JimJast, Discospinster, Rich Farmbrough, Guanabot, Pak21, ThomasK, Masudr, Pjacobi, Vsmith, Cdyson37, Jowr, Paul August, SpookyMulder, Dmr2, Bender235, Dcabrilo, Ground, Ben Standeven, Nabla, Livajo, El C, Worldtraveller, Shanes, Etimbo, Causa sui, Bobo192, Robotje, Smalljim, Rbj, JW1805, ParticleMan, 19Q79oL78KiL0QTFHgyc, Mr2001, Matt McIrvin, PWilkinson, Haham hanuka, Schnolle, Varuna, Jumbuck, Jérôme, Alansohn, Hackwrench, Cctoide, Crebbin, Wikidea, SlimVirgin, Benefros, Alexwg, Wtmitchell, Orionix, CloudNine, Bsadowski1, DV8 2XL, LordLoki, HenryLi, Oleg Alexandrov, Kelly Martin, Linas, FeanorStar7, Sabejias, Moneky, Kzollman, Cleonis, Mpatel, Jok2000, Schzmo, Pdn–enwiki, GregorB, Plrk, Wayward, Joke137, Christopher Thomas, Mandarax, Colodia, Canderson7, Rjwilmsi, WCFrancis, MarSch, Eyu100, JoshuacUK, JHMM13, Mike Peel, SanitysEdge, R.e.b., Ems57fcva, Bubba73, Gringo300, Ian Pitchford, RobertG, Mishuletz, Arnero, Mathbot, Nihiltres, Vsion, Perfect Tommy–enwiki, Itinerant1, Alfred Centauri, Gparker, Slant, Carrionluggage, Srleffler, Chobot, DVdm, Bgwhite, Dresdnhope, Manscher, Roboto de Ajvol, YurikBot, Wavelength, Bcarm1185, Splintercellguy, Hillman, EDG, MattWright, RussBot, Loom91, AVM, KSmrq, DanMS, SpuriousQ, Shawn81, Eleassar, Shanel, Syth, Madcoverboy, Tailpig, Schlafly, Dputig07, Beanyk, Tony1, Dna-webmaster, Enormousdude, 2over0, KGasso, Petri Krohn, GraemeL, Rlove, Sambc, LeonardoRob0t, Geoffrey.landis, HereToHelp, Willtron, Meegs, Bsod2, Finell, Luk, Sardanaphalus, SmackBot, Kurochka, Hydrogen Iodide, Pavlović, Gnangarra, Unyoyega, Nickst, Delldot, Motorneuron, Cessator, Harald88, Edgar181, Shai-kun, Sectryan, Gilliam, Skizzik, Dauto, Saros136, Silly rabbit, Complexica, Colonies Chris, Zven, Abyssal, RProgrammer, Hve, RedHillian, BeatSm, Phaedriel, Khoikhoi, Cybercobra, Downwards, Coolbho3000, Nakon, Peterwhy, SkyWriter, DMacks, Nairebis, Henning Makholm, UncleFester, Bidabadi–enwiki, Byelf2007, SashatoBot, Lambiam, Lapaz, Cronholm144, Gizzakk, CPMcE, JorisvS, Goodnightmush, Ckatz, Frokor, Garthbarber, SirFozzie, SandyGeorgia, Midnightblueowl, RichardF, Novangelis, Peter Horn, MTSbot–enwiki, Kvng, JarahE, Licorne, Quaeler, Fan-1967, Editor.singapore, MFago, JoeBot, ShyK, MOBle, RekishiEJ, CapitalR, MD.astronomer, Courcelles, Tawkerbot2, JRSpriggs, Kurtan–enwiki, Harold f, JForget, Sakurambo, Thermochap, Avanu, NickW557, MarsRover, Harrigan, Ian Beynon, Cydebot, Jasperdoomen, WillowW, Fl, MC10, Mato, Pascal.Tesson, Michael C Price, Christian75, DumbBOT, Biblbroks, Omicronpersei8, Crum375, N. Macchiavelli, Epbr123, Fisherjs, Markus Pössel, Martin Hogbin, MrXow, Oliver202, Headbomb, Pjvpjv, Tom Barlow, Davidhorman, D.H, AntiVandalBot, Abu-Fool Danyal ibn Amir al-Makhiri, Tkirkman, Gnixon, VectorPosse, TimVickers, Scepia, Dawz, Billevans–enwiki, Tim Shuba, Rico402, Archmagusrm, Jaredroberts, JAnDbot, Vorpal blade, Hut 8.5, YK Times, Acroterion, Pervect, Magioladitis, Connormah, RogierBrussee, WolfmanSF, JamesBWatson, Swpb, Ling.Nut, Soulbot, Pixel ;-), KConWiki, WhatamIdoing, Eldumpo, Allstarecho, User A1, Mollwollfumble, Chris G, Archen–enwiki, Thompson.matthew, STBot, Mermaid from the Baltic Sea, Shentino, Mschel, CommonsDelinker, Pbroks13, J.delanoy, DrKiernan, R. Baley, Numbo3, Leafsfan85, Lantonov, M C Y 1008, Mathlabster, Zedmelon, Aboutmovies, C quest000, Tcisco, Marrilpet, Aatomic1, Potatoswatter, Kolja21, Lseixas, Rémih, Caracalocelot, DemonicInfluence, Sheliak, Deor, Part Deux, JohnBlackburne, Philip Trueman, TXiKiBoT, Coder Dan, GimmeBot, Gombo, Hqb, Rei-bot, IPSOS, Qxz, T doffing, Molinogi, Fizzackerly, JhsBot, Leafyplant, Geometry guy, Ilyushka88, Thebigbendizzle, SwordSmurf, Andy Dingley, Gabrielsleitao, Lamro, Antixt, Vector Potential, James-Chin, Arcfrk, Ccheese4, StevenJohnston, Katzmik, YohanN7, Dnarby, SieBot, Tiddly Tom, Work permit, Yintan, RadicalOne, Wizzard2k, SteakNShake, Arbor to SJ, Babareddeer, JSpung, Phil Bridger, Wmpearl, Oxymoron83, Henry Delforn (old), Csloomis, Thehotelambush, Lightmouse, BrightRoundCircle, OpTioNiGhT, The-G-Unit-Boss, Emgg, AWeishaupt, Divinestuff, Coldcreation, Adam Cuerden, Duae Quartunciae, Heptarchy of teh Anglo-Saxons, baby, Randomblue, TFCforever, Danthewhale, Martarius, Sfan00 IMG, ClueBot, The Thing That Should Not Be, Rjd0060, Metaprimer, Wwheaton, Der Golem, JTBX, TheAmigo42, CounterVandalismBot, Viran, Blanchardb, Rotational, Agge1000, Itzguru, Tanketz, CohesionBot, Eeekster, Stealth500, Brews ohare, NuclearWarfare, PhySusie, SockPuppetForTomruen, SchreiberBike, Another Believer, RubenGarciaHernandez, AC+79 3888, MasterOfHisOwnDomain, He6kd, TimothyRias, Lazyrussian, PseudoOne, Skarebo, NellieBly, JinJian, Truthnlove, Everydayidiot, Tayste, Balungifrancis, Addbot, Mortense, Some jerk on the Internet, Fizzycyst, DOI bot, Mistyocean3, Metagraph, Stariki, Fluffernutter, Schmoolik, MrOllie, Download, EconoPhysicist, Delaszk, Favonian, LinkFA-Bot, Tuition, Tassedethe, Nnedass, Tide rolls, Lightbot, Knutls, Luckas-bot, Ptbotgourou, Legobot II, Julia W, Trickyboarder93, Superamoeba, AnomieBOT, Kristen Eriksen, Giordano.ferdinandi, Jim1138, Jo3sampl, Materialscientist, Wandering Courier, The High Fin Sperm Whale, Citation bot, Xqbot, Stlwebs, Sionus, Amareto2, Unigfjkl, Nickkid5, Stsang, GrouchoBot, Collin21594, RibotBOT, Rucko123, GhalyBot, Acannas, LucienBOT, Paine Ellsworth, Lagelspeil, Steve Quinn, Knowandgive, Pokyrek, Citation bot 1, Citation bot 4, Electrozity8, Pinethicket, LittleWink, Jonesey95, A412, Tom.Reding, Yougeeaw, Barras, Jauhienij, Meier99, Citator, Comet Tuttle, Hughston, Defender of torch, Duoduoduo, Aribashka, libbmm, Diannaa, Earthandmoon, Tbhotch, Brambleclawx, Marie Poise, RjwilmsiBot, Aznhero3793, Ripchip Bot, EmausBot, WikitanvirBot, Immunize, Zhaskey, Fly by Night, DuKu, GoingBatty, Jmencisom, Slightsmile, Hhhippo, JSquish, ZéroBot, Cogiati, Stanford96, Empty Buffer, Sanford123456, H3llBot, Quondum, REkaxkjdsc, Monterey Bay, Mr little irish, TonyMath, Brandmeister, Maschen, Puffin, Carmichael, Newstv11, RockMagnetist, Sona11235, WizardofCalculus, Milk Coffee, Whoop whoop pull up, Mjbmrbot, Helpsome, ClueBot NG, Manubot, Hagenfeldt, This lousy T-shirt, SusikMkr, Ggonzalm, Jj1236, Mgvongoeden, Snotbot, Widr, Jamester234, Pluma, Ginger.spice14, Bibcode Bot, Jeraphine Gryphon, Lowercase sigmabot, Quarkgluonsoup, Bolatbek, Marsambe, Amp71, Mark Arsten, Lovepool1220, Marsambe1, Benzband, ENG.F.Younis, 123matt123, DeviantFrog, IrishDevil2, F=q(E+v^B), Egbertus2, Harizotoh9, Doctor Lipschitz, Snow Blizzard, Zoldyick, Roozitaa, BattyBot, Reed07, Vanobamo, JoshuSasori, Stigmatella aurantiaca, Cyberbot II, Abhay ravi, ChrisGualtieri, Maestro814, Deathlasersonline, Plokijnu, Billyshiverstick, Read Blooded, Theeditor6079, Flyer1997, Dexbot, Suffian Akhtar, Kryomaxim, Twhitguy14, CuriousMind01, J0437-4715, Jamesx12345, Among Men, WorldWideJuan, Devinray1991, 1888software, EvergreenFir, Enchantedscience, Mohamed F. El-Hewie, Vai ra'a toa Taina, NeapleBerlina, Jwratner1, Gigantmozg, Ginsuloft, SirKesuma, Anrrusna, JaconaFrere, Osamabin7, Juenni32, Filedelinkerbot, SantiLak, Aryabhatt 21, Willbh15, S11027158, Cris Cyborg, PeterShawhan, Evgeniy E., Sweeeeeeeed, Tetra quark, Praveece, JuanLT2045, Jf2839, GeneralizationsAreBad, KasparBot, Lemonberry622, Pizzaman62, Dgray101 and Anonymous: 700

- **Tests of general relativity** *Source:* https://en.wikipedia.org/wiki/Tests_of_general_relativity?oldid=679100991 *Contributors:* Bryan Derksen, FvdP, Rsabbatini, Michael Hardy, Tjfulopp, Stevenj, Msablic, Charles Matthews, Tpbradbury, Raul654, FlyByPC, Cluth, Shantavira, Phil Boswell, Rursus, Ancheta Wis, Giftlite, Fastfission, Herbee, FrYGuY, Wmahan, Kaldari, Csmiller, Icairns, Urhixidur, Rich Farmbrough, Tristan Schmelcher, Jamadagni, Bender235, Indio–enwiki, 19Q79oL78KiL0QTFHgyc, Splat, Stack, Bucephalus, Carcharoth, JFG, Mpatel, Pdn–enwiki, Joke137, Hyperzonk, MarcoTolo, Alcoved id, Rnt20, RxS, Rjwilmsi, Ems57fcva, Bubba73, STarry, Drrngrvy, Fivemack,

Old Moonraker, Itinerant1, Gparker, Dimator, GangofOne, DVdm, YurikBot, Hairy Dude, Hillman, Gaius Cornelius, David R. Ingham, Ragesoss, Tony1, Enormousdude, Citynoise, Modify, Geoffrey.landis, JLaTondre, Ilmari Karonen, Deuar, Sardanaphalus, SmackBot, Ququ, Spireguy, Edgar181, FredA, Njerseyguy, Silly rabbit, Colonies Chris, Dual Freq, Can't sleep, clown will eat me, HPSCHD, Tesseran, Tim bates, Kopeikins, Michaelbusch, Achoo5000, JRSpriggs, JForget, CmdrObot, Ruslik0, Delta407, Gregbard, Logicus, Cydebot, WillowW, BobQQ, TenthEagle, Headbomb, Tchannon, Mikeeg555, D.H. Komponisto, Ericoides, Botfob, Mollwollfumble, Mermaid from the Baltic Sea, R'n'B, Jarhed, J.delanoy, Tuanomsoc, Heyitspeter, Equazcion, Burzmali, Homo logos, JohnBlackburne, Red Act, Rei-bot, Michael H 34, Seb az86556, PDFbot, Geometry guy, Enviroboy, Codairem, PaddyLeahy, Phe-bot, Bigbearbooth, ClueBot, Trojancowboy, Deanlaw, Der Golem, Agge1000, Oxnard27, Alexbot, LarryMorseDCOhio, Crnorizec, Jane Bennet, Terry0051, Doc9871, Addbot, DOI bot, D c weber, Haskellguy, Favonian, Ozzie42, Lightbot, Luckas-bot, Yobot, 2D, Legobot II, Bhagwad, AnomieBOT, Jim1138, Materialscientist, Citation bot, TheAMmolluse, NOrbeck, Privalov, 图图, Steve Quinn, Citation bot 1, Momergil, Biblioscience, Jonesey95, Tom.Reding, RelatGene, Full-date unlinking bot, Tim1357, Earthandmoon, RjwilmsiBot, WildBot, John of Reading, WikitanvirBot, Racerx11, Jmencisom, H3llBot, AManWithNoPlan, Brandmeister, Ystalyfera, ClueBot NG, Euty, Bulbul99, Helpful Pixie Bot, Bibcode Bot, BG19bot, Quandom Angel, Mdy66, Snow Blizzard, Qwerty0800, BattyBot, Dexbot, Danguard00, Jweisber, Anti-McK, Joseaugustocale, Anrrusna, Monkbot and Anonymous: 96

- **Parameterized post-Newtonian formalism** *Source:* https://en.wikipedia.org/wiki/Parameterized_post-Newtonian_formalism?oldid=647221 *Contributors:* The Anome, SebastianHelm, Mdob, Oliver Jennrich, Ben Standeven, Cmdrjameson, Danski14, Keenan Pepper, Mpatel, Joke137, MarSch, RE, Hairy Dude, Hillman, Gaius Cornelius, Geoffrey.landis, Sardanaphalus, SmackBot, Colonies Chris, Ligulembot, MOBle, JR-Spriggs, Kurtan-enwiki, Gregbard, BobQQ, Thijs!bot, Mollwollfumble, Sheliak, Addbot, DWHalliday, LyricV, Shadowjams, Carlog3, Obankston, Mjbmrbot, ChrisGualtieri, Dexbot, Frinthruit and Anonymous:14

- **Linearized gravity** *Source:* https://en.wikipedia.org/wiki/Linearized_gravity?oldid=666746701 *Contributors:* Phys, Lumidek, Pearle, Pol098, Mpatel, Eyu100, Archelon, Geoffrey.landis, KasugaHuang, Sardanaphalus, SmackBot, Commander Keane bot, Radagast83, JRSpriggs, Spiffyzha, Nick Number, Liquid-aim-bot, Alphachimpbot, Spartaz, Sikory, R'n'B, Sheliak, Red Act, Barkeep, 1ForTheMoney, Addbot, Citation bot, Carlog3, Ribashka, Laurifer, Quondum, Vanished user lt94ma34le12, Frinthruit and Anonymous: 7

- **ADM formalism** *Source:* https://en.wikipedia.org/wiki/ADM_formalism?oldid=674750732 *Contributors:* SimonP, Chuunen Baka, Tobias Bergemann, Kareeser, Rich Farmbrough, Causa sui, Physicistjedi, Schaefer, GregorB, Joke137, Reedbeta, Bgwhite, Hillman, Salsb, Sardanaphalus, SmackBot, Bluebot, Jmax-, Leo C Stein, JorisvS, B3nz ek, P199, Gregbard, Xxanthippe, Markus Pössel, Headbomb, Smartcat, Wesino, Sheliak, Puzhok, Smenge32, Lisatwo, Djr32, Addbot, Luckas-bot, Yobot, AnomieBOT, Citation bot, Erik9bot, RelativelyCertain, Vhsatheeshkumar, Quondum, Λ, Maschen, Fluctuating metric, Doenermaster, Helpful Pixie Bot, Steve86au, Bibcode Bot, Melirius, Enyokoyama, Frinthruit and Anonymous: 23

- **Alternatives to general relativity** *Source:* https://en.wikipedia.org/wiki/Alternatives_to_general_relativity?oldid=679214551 *Contributors:* The Anome, Fxmastermind, Edward, Michael Hardy, SebastianHelm, William M. Connolley, Charles Matthews, Vsmith, Ben Standeven, Frankenschulz, Russ3Z, Gene Nygaard, Mpatel, Edison, Rjwilmsi, Ems57fcva, DVdm, Jpfagerback, Roboto de Ajvol, Doctorsundar, Welsh, Długosz, Teply, KasugaHuang, SmackBot, Frasor, Silly rabbit, Colonies Chris, Aldaron, Loodog, Ryulong, JRSpriggs, CRGreathouse, Olaf Davis, Vyznev Xnebara, Gregbard, Vttoth, Mckinlayr, Crum375, Headbomb, Davidhorman, D.H. MarshBot, AntiVandalBot, Gioto, Georgert, Spartaz, Igodard, Cardamon, Mollwollfumble, J.delanoy, BigrTex, Athaenara, Lantonov, Fences and windows, Red Act, Bass fishing physicist, Michael H 34, Meilenweit, Rep07, GirasoleDE, Noveltyghost, Rangutan, Sun Creator, XLinkBot, Addbot, DWHalliday, Yobot, AnomieBOT, InsufficientData, Materialscientist, Citation bot, Gravityforce, Omnipaedista, FrescoBot, Citation bot 1, Jonesey95, Haael, Loftpo, EmausBot, Maashatra11, Suslindisambiguator, Fluctuating metric, ClueBot NG, Bibcode Bot, ServiceAT, DarrenFuture, Peter Donald Rodgers, GravityForce, Garamond Lethe, Jamesx12345, Friedlicherkoenig, Anrrusna, Monkbot, NormDrez, Crakem, HFD90 and Anonymous: 35

- **Quantum gravity** *Source:* https://en.wikipedia.org/wiki/Quantum_gravity?oldid=684103921 *Contributors:* AstroNomer~enwiki, Matusz, Roadrunner, Stevertigo, Ubiquity, Bobby D. Bryant, Mcarling, NuclearWinner, Anders Feder, Susurrus, Coren, Charles Matthews, Timwi, Reddi, Tpbradbury, Phys, Bevo, Raul654, BenRG, Frazzydee, Jeffq, Sdedeo, Rholton, Wereon, Ilya (usurped), Seth Ilys, Ancheta Wis, Giftlite, Herbee, Fropuff, Endlessnameless, Malyctenar, Jason Quinn, Finn-Zoltan, YapaTi~enwiki, Lumidek, Marcus2, Joyous!, TJSwoboda, Vitaleyes, Davidclifford, JimJast, Guanabot, FT2, Masudr, Pjacobi, Pie4all88, David Schaich, Bender235, Clement Cherlin, El C, PhilHibbs, Army1987, Apyule, VBGFscJUn3, PWilkinson, Daniel Arteaga~enwiki, Keenan Pepper, Cjthellama, DonJStevens, Velella, Dabbler, Tycho, Cal 1234, RJFJR, Count Iblis, ThomasWinwood, Anarchimede, Scarykitty, Woohookitty, Igny, ToddFincannon, Mpatel, GregorB, Joke137, Christopher Thomas, Marudubshinki, Graham87, Yurik, Kroggz, Rjwilmsi, Eoghanacht, Jrasowsky, JHMM13, Smithfarm, Ems57fcva, FayssalF, Itinerant1, Lmatt, Chobot, Hmonroe, YurikBot, Hillman, ErkDemon, JocK, SCZenz, Roy Brumback, Bota47, Zunaid, JonathanD, 2over0, Arthur Rubin, Modify, LeonardoRob0t, Caco de vidro, RG2, KasugaHuang, Resolute, SmackBot, Samdutton, Vald, Eskimbot, Hbackman, Onebravemonkey, Chris the speller, Ben.c.roberts, Cthuljew, Silly rabbit, Complexica, Colonies Chris, QFT, Soosed, Theanphibian, Shushruth, Ck lostsword, Yevgeny Kats, DJIndica, Lambiam, Vampus, Vincenzo.romano, Jaganath, JorisvS, RoboDick~enwiki, IronGargoyle, Dicklyon, SirFozzie, Treyp, Twunchy, Piccor, Kurtan~enwiki, Harold f, CalebNoble, Duduong, Paulmlieberman, TVC 15, UncleBubba, TAz69x, Sam Staton, ST47, B, Patrick O'Leary, Epbr123, Koeplinger, Klasovsky, Markus Pössel, Keraunos, Headbomb, Marek69, MichaelMaggs, Tim Shuba, MER-C, ParadiZio, Clementvidal, Perlygatekeeper, VoABot II, Alvatros~enwiki, Bdalevin, SHCarter, Jpod2, DAGwyn, Nucleophilic, LorenzoB, Rickard Vogelberg, DancingPenguin, Rettetast, Victor Blacus, AstroHurricane001, Yonidebot, Acalamari, Mstuomel, Fullmetal2887, NewEnglandYankee, DorganBot, CardinalDan, Idioma-bot, Sheliak, VolkovBot, Pleasantville, Seattle Skier, AlnoktaBOT, TXiKiBoT, Dllahr, Rdekleer, Saibod, Cyberchip, Wikiwikimoore, Carlorovelli, LoreMiles, StevenJohnston, SieBot, LeadSongDog, Bentogoa, Coldcreation, ReluctantPhilosopher, StaticG, GarbagEcol, ClueBot, The Thing That Should Not Be, EoGuy, Polyamorph, Andwor9, Notburnt, Tms9, Alexbot, Resoru, Eeekster, Tamaratrouts, Brews ohare, SchreiberBike, Askahrc, BOTarate, Lambtron, DumZiBoT, XLinkBot, Rror, Facts707, SilvonenBot, Theonlydavewilliams, Mhsb, Truthnlove, Ttimespan, Trifonov~enwiki, Addbot, Mortense, Grayfell, Eric Drexler, Gravitophoton, DOI bot, AkhtaBot, CanadianLinuxUser, Frosty726, LaaknorBot, Delaszk, Tassedethe, Tide rolls, Taketa, Titan1129, Krano, Luckas-bot, Yobot, WikiDan61, Pigetrational, Wireader, Allowgolf~enwiki, Wiki Roxor, Jim1138, IRP, Sz-iwbot, Quantity, Materialscientist, Citation bot, ArthurBot, LilHelpa, Amareto2, Ekwos, KrisBogdanov, Rolfguthmann, StealthCopyEditor, 图图, Dan6hell66, Rabsmith, Hep thinker, Paine Ellsworth, DrArthurRubinPHD, Lagelspeil, Nunc aut numquam, Vacuunaut, Van Speijk, Knowandgive, Craig Pemberton, Udifuchs, Citation bot 2, Citation bot 1, Citation bot 4, Jonesey95, Hirvenkürpa, Tom Reding, Pmokeefe, Casimir9999, Dac04, Dude1818, Valeriy Pischenko, Follyland, TrueTeargem, N0814444, Earthandmoon, Korepin, DARTH SIDIOUS 2, Musictme4me, RjwilmsiBot, EmausBot, Francophile124, Octaazacubane, Fotoni, Slightsmile, Garfield Salazar, Hhhippo, JSquish, John Cline, Fæ, LostAlone, Brazmyth, Throwmeaway, Arbnos, Ebrambot, Kusername, DanielBurnstein, TonyMath, L Kensington, Maschen, Donner60, Parusaro, Apratim07, Terra Novus,

Isocliff, Googledin!, ClueBot NG, SpikeTorontoRCP, Science writer, Preon, Raidr, Jhmmok, 336, Widr, Helpful Pixie Bot, Bibcode Bot, Bardsley Rides a Segway, Apelikedawg, FiveColourMap, Trevayne08, Mr.viktor.stepanov, Brainssturm, BattyBot, Jimw338, Ryanr666, Kryomaxim, Garuda0001, Saehry, TwoTwoHello, Sanathdevalapurkar, Andyhowlett, GabeIglesia, Sanathlab, Roiwallace, Spencer.mccormick, Spencerfjase, MrShiongNo1, Marc D. Garrett, D00d00ballz, Gigantmozg, Susan.grayeff, Polytope24, Frinthruit, Anrnusna, Dfyytj, Monkbot, Umut Alihan Dikel, Amortias, Klj1234, Pfpguy, KasparBot and Anonymous: 294

- **Antimatter gravity measurement** *Source:* https://en.wikipedia.org/wiki/Antimatter_gravity_measurement?oldid=664075643 *Contributors:* Headbomb, Citation bot, Steve Quinn, Timetraveler3.14, Bibcode Bot, Debouch, OccultZone, Epigogue and Anonymous: 1

- **Bi-scalar tensor vector gravity** *Source:* https://en.wikipedia.org/wiki/Bi-scalar_tensor_vector_gravity?oldid=532049370 *Contributors:* Bbbl Rich Farmbrough, Malcolma, SmackBot, JorisvS, Headbomb, Yobot, Rammaum, Bibcode Bot, BG19bot and Anonymous: 1

- **Bimetric gravity** *Source:* https://en.wikipedia.org/wiki/Bimetric_gravity?oldid=677870681 *Contributors:* AdamSolomon, Red Act, C628, Trappist the monk, GoingBatty, Bibcode Bot, Monkbot and Ambulevy

- **Composite gravity** *Source:* https://en.wikipedia.org/wiki/Composite_gravity?oldid=650403276 *Contributors:* Phys, Lumidek, Mpatel, SmackBot, Colonies Chris, 100110100, PhilKnight, Yobot and Anonymous: 1

- **Conformal gravity** *Source:* https://en.wikipedia.org/wiki/Conformal_gravity?oldid=683226474 *Contributors:* The Anome, Michael Hardy, Jason Quinn, Gene s, Rich Farmbrough, Dmr2, Reuben, Physicistjedi, Siafu, Mpatel, Tone, Gadget850, SmackBot, QFT, JorisvS, Chhajjusandeep, Headbomb, Chris goulet, Igodard, Speaker to wolves, Addbot, DOI bot, Yobot, Baxxterr, Omnipaedista, FrescoBot, Citation bot 1, Tobi - Tobsen, Janiefar, ZéroBot, Cogiati, Maschen, ClueBot NG, Mgvongoeden, Bibcode Bot, Lee.boston, Rhlozier and Anonymous: 19

- **Entropic gravity** *Source:* https://en.wikipedia.org/wiki/Entropic_gravity?oldid=676840040 *Contributors:* Boud, AugPi, Dratman, Jason Quinn, Siroxo, Rich Farmbrough, Danski14, BRW, Christopher Thomas, Rjwilmsi, Koavf, Helvetius, JocK, SmackBot, Marklaramee, Apostolos Margaritis, Jmnbatista, Brienanni, Devourer09, Quibik, Raoul NK, Qwyrxian, Headbomb, Fetchcomms, DMOinL.A, Rickard Vogelberg, Mickwilson20, Maurice Carbonaro, TomS TDotO, Aqwis, Shawn in Montreal, Stan J Klimas, VanishedUserABC, Jportway, Cory Donnelly, Addbot, Basilicofresco, AndersBot, Scott MacDonald, AnomieBOT, Citation bot, Xqbot, Charvest, Finncarey, FrescoBot, Sanpitch, Machine Elf 1735, Nastasyuk v, Deniskrasnov, RjwilmsiBot, Dorian.winslow, AvicAWB, Scrizati, Isocliff, Plusorminuszero, Jhmmok, Bibcode Bot, Tehom2000, E.molgaard, Capslocking, ChrisGualtieri, Enyokoyama, StillFascinated, Anrnusna and Anonymous: 30

- **Extended theories of gravity** *Source:* https://en.wikipedia.org/wiki/Extended_theories_of_gravity?oldid=639623995 *Contributors:* I9Q79oL Trevj, NawlinWiki, Malcolma, JRSpriggs, Headbomb, AnomieBOT, Citation bot, Steve Quinn, Tom.Reding, Bibcode Bot, Barney the barney barney, Mogism, Mark viking and Mariana Espinosa Aldama

- **Gauge gravitation theory** *Source:* https://en.wikipedia.org/wiki/Gauge_gravitation_theory?oldid=676742087 *Contributors:* Michael Hardy, Gabbe, Drernie, MBisanz, Crasshopper, Teply, Stifle, Silly rabbit, CmdrObot, Adavidb, Moonriddengirl, TrulyBlue, Forbes72, Addbot, LaaknorBot, Xqbot, Gsard, Tom.Reding, Pmokeefe, Klbrain, ChrisGualtieri, Garuda0001 and Anonymous: 3

- **Gauge theory gravity** *Source:* https://en.wikipedia.org/wiki/Gauge_theory_gravity?oldid=639808633 *Contributors:* Michael Hardy, Giftlite, Chris Howard, JHCaufield, Teply, Tom.Reding, Quondum, Bibcode Bot, BG19bot, Anrnusna and Anonymous: 2

- **Gauge vector–tensor gravity** *Source:* https://en.wikipedia.org/wiki/Gauge_vector%E2%80%93tensor_gravity?oldid=638148270 *Contributors:* Tony1, Gaba p, Sourov0000, Qxfard and Monkbot

- **Gauss's law for gravity** *Source:* https://en.wikipedia.org/wiki/Gauss'{}s_law_for_gravity?oldid=676235011 *Contributors:* XJaM, Patrick, Michael Hardy, Giftlite, MFNickster, Edudobay, Versageek, Siddhant, Froth, Sbyrnes321, SmackBot, InverseHypercube, JorisvS, JRSpriggs, CBM, Cydebot, Reywas92, Kevinmon, Keith D, Yecril, Inflector, Paolo.dL, Fedosin, Excirial, SchreiberBike, BOTarate, Addbot, Ikara, Materialscientist, Tom.Reding, Miracle Pen, EmausBot, Quondum, SporkBot, Milad pourrahmani, Maschen, Zueignung, ChuispastonBot, ClueBot NG, Frietjes, PabloQC1979, Imgaril, F=q(E+v^B), SubratamindPal, Joker5974, PacWalker and Anonymous: 21

- **Gauss–Bonnet gravity** *Source:* https://en.wikipedia.org/wiki/Gauss%E2%80%93Bonnet_gravity?oldid=615268992 *Contributors:* Urhixidur, Rjwilmsi, Eeekster, Yobot, Paine Ellsworth, Throwmeaway, Raidr, Bibcode Bot, Enyokoyama and Anonymous: 3

- **Gravitational field** *Source:* https://en.wikipedia.org/wiki/Gravitational_field?oldid=673486898 *Contributors:* Voidvector, Axlrosen, SebastianHelm, Looxix~enwiki, Ahoerstemeier, Mxn, Smack, Bartosz, Giftlite, Wolfkeeper, Frencheigh, Klemen Kocjancic, JimJast, Discospinster, FT2, Vsmith, Bobo192, Hooperbloob, Oolong, Zachlipton, Stemonitis, Mpatel, Trlovejoy, The wub, DVdm, RussBot, Wimt, FyzixFighter, Sbyrnes321, SmackBot, Lestrade, KnowledgeOfSelf, Stepa, Hmains, Bluebot, Nakon, JorisvS, Ckatz, Digger3000, Makyen, Dicklyon, Alessandro57, Rracecarr, Biblbroks, Headbomb, Escarbot, Narssarssuaq, Bencherlite, FaerieInGrey, Zapp645, VoABot II, Avjoska, Someguy1221, Purgatory Fubar, Gerakibot, Paolo.dL, Godfinger, ClueBot, Xmilanz, Deviator13, Geoeg, LaosLos, Kbdankbot, Addbot, Fgnievinski, Lightbot, LGB, RHB100, HieronymousCrowley, DrTrigon, AnomieBOT, Galoubet, BWACKBEWWY, 78.26, Thehelpfulbot, FrescoBot, Lookang, DavidWTalmage, Baliballi, Drew R. Smith, Tom.Reding, Mean as custard, TjBot, WildBot, EmausBot, K6ka, JSquish, Stanford96, Zueignung, Aydin1884, SusikMkr, Dubadeca, MerllwBot, Helpful Pixie Bot, F=q(E+v^B), Justincheng12345-bot, YFdyh-bot, Suckdymick, Razibot, Manoj Tennyson, Isambard Kingdom and Anonymous: 64

- **Higher-dimensional Einstein gravity** *Source:* https://en.wikipedia.org/wiki/Higher-dimensional_Einstein_gravity?oldid=598890921 *Contributors:* Deb, Edward, Michael Hardy, Phys, Dmr2, Mpatel, MarSch, Ems57fcva, BradBeattie, Hillman, Salsb, Thiseye, FF2010, Caco de vidro, SmackBot, Nickst, Kdliss, Colonies Chris, BranStark, Headbomb, VolkovBot, XLinkBot, Addbot, Tassedethe, Jim1138, Omnipaedista, Erik9bot, Throwmeaway, ClueBot NG, BattyBot and Anonymous: 19

- **Higher-dimensional supergravity** *Source:* https://en.wikipedia.org/wiki/Higher-dimensional_supergravity?oldid=674711468 *Contributors:* Michael Hardy, Rich Farmbrough, Zazaban, Colonies Chris, R'n'B, Student7, Telecomtom, Davehi1, Yobot, AnomieBOT, LilHelpa, Omnipaedista, Hep thinker, Paine Ellsworth, Maschen, RockMagnetist, Snotbot, CaroleHenson and Anonymous: 7

- **Hoyle–Narlikar theory of gravity** *Source:* https://en.wikipedia.org/wiki/Hoyle%E2%80%93Narlikar_theory_of_gravity?oldid=640943750 *Contributors:* Bearcat, Bender235, Incnis Mrsi, Headbomb, Kudpung, Bearian, Yobot, Jim1138, Citation bot, Amaury, Tom.Reding, AvicAWB, Astrotech, ClueBot NG, Bibcode Bot, BG19bot, BattyBot and Anonymous: 5

- **Induced gravity** *Source:* https://en.wikipedia.org/wiki/Induced_gravity?oldid=681227802 *Contributors:* Kku, Phys, Lumidek, Jag123, HFarmer, Mpatel, GregorB, DVdm, YurikBot, Salsb, Ilmari Karonen, Colonies Chris, QFT, Radagast83, JorisvS, BBuchbinder, Aldis90, Alphachimpbot, SchreiberBike, Addbot, Mortense, Legobot, Charvest, FrescoBot, ZéroBot, Bibcode Bot, CitationCleanerBot, MaiyaH78 and Anonymous: 14

- **Loop quantum gravity** *Source:* https://en.wikipedia.org/wiki/Loop_quantum_gravity?oldid=683638404 *Contributors:* Bryan Derksen, The Anome, AstroNomer~enwiki, RK, Toby Bartels, Miguel~enwiki, Schewek, Ewen, Michael Hardy, TakuyaMurata, Islandboy99, GTBacchus, Mcarling, Looxix~enwiki, Ahoerstemeier, Cyp, Kimiko, Palfrey, Jordi Burguet Castell, Mxn, Charles Matthews, Sanxiyn, Maximus Rex, Phys, Omegatron, Finlay McWalter, Dmytro, Sdedeo, Astronautics~enwiki, Peak, Chris Roy, Mirv, Sverdrup, Kn1kda, Hadal, Jheise, Clementi, Connelly, Giftlite, Sj, Fastfission, Herbee, Anville, Dratman, Curps, JeffBobFrank, Jason Quinn, Gzornenplatz, C17GMaster, DÅ,ugosz, PhiloVivero, DefLog~enwiki, Gadfium, HorsePunchKid, Sam Hocevar, Lumidek, Tdent, Joyous!, M1ss1ontomars2k4, Eep², Poccil, Rich Farmbrough, Avriette, Pjacobi, Vsmith, MuDavid, Pavel Vozenilek, Bender235, ESkog, Clement Cherlin, Peter M Gerdes, Drhex, John Vandenberg, C S, Cmdrjameson, GTubio, Tweet Tweet, Slicky, Ral315, Lysdexia, Arthena, Xaphan9966, Wtmitchell, Greg Kuperberg, Count Iblis, Egg, Lee-Anne, Kazvorpal, Killing Vector, Linas, Merlinme, HFarmer, Sympleko, Hfarmer, Mpatel, GregorB, J M Rice, Ae7flux, Tjbk tjb, Alienus, Fleisher, Sjö, Rjwilmsi, Nightscream, Zbxgscqf, Bubba73, FlaBot, John Baez, Don Gosiewski, Smithbrenon, Chobot, Spasemunki, Bgwhite, Roboto de Ajvol, YurikBot, Wavelength, RobotE, Rt66lt, Hillman, DanMS, Chaos, Salsb, Welsh, Schmock, Crasshopper, Beanyk, Akashmitra, Bota47, JonathanD, Endomion, Modify, Petri Krohn, Ilmari Karonen, Caco de vidro, Benandorsqueaks, SmackBot, Bayardo, FlashSheridan, Unyoyega, Vald, JMiall, Chris the speller, IvanAndreevich, DHN-bot~enwiki, Colonies Chris, Chlewbot, Pepsidrinka, Chrylis, MegaHasher, TriTertButoxy, Lambiam, Vincenzo.romano, Loadmaster, Konklone, K, G-W, Kurtan~enwiki, Harold f, Will314159, Friendly Neighbour, Vyznev Xnebara, Ian Beynon, Myasuda, Gmusser, Rjm656s, Fournax, Headbomb, Nick Number, MichaelMaggs, Edokter, Byrgenwulf, Knotwork, Arch dude, Igodard, Yill577, WolfmanSF, Tonyfaull, Skylights76, Rickard Vogelberg, Gwern, AltiusBimm, Melamed katz, Vanished user 47736712, WJBscribe, Izno, KittyHawker, Sheliak, Maxzimet, AlnoktaBOT, Nxavar, Jackfork, Carlorovelli, Anotherak, SieBot, Keskival, AS, Robdunst, Hugh16, Senderista~enwiki, Bnsreenath, Caidh, Oxymoron83, Deattell, Swiebodzice, Sk8hack, Danthewhale, Martarius, Sfan00 IMG, Shaded0, Djr32, CohesionBot, JavierReynaldo, Arjayay, SchreiberBike, Pqnelson, Mjaniec, DumZiBoT, Ianbay, Neuralwarp, XLinkBot, Fastily, Tenner47, Arthur chos, Avoided, Tenderbuttons, Benplusnumber, Balungifrancis, Addbot, DOI bot, 15lsoucy, Tarosic, Debresser, SamatBot, Yobot, Ibayn, 4th-otaku, AnomieBOT, VanishedUser sdu9aya9fasdsopa, Archon 2488, Francois33, Citation bot, Xqbot, Imushfiq, MIRROR, Pra1998, Dumontierc, Omnipaedista, Franco3450, Rr2000, FrescoBot, Paine Ellsworth, Nunc aut numquam, Martlet1215, Citation bot 1, Jonesey95, Tom.Reding, Schiefesfragezeichen, ROMVLVS, Casimir9999, RobinK, Meier99, Dinamik-bot, Bj norge, ElPeste, Afteread, EmausBot, Detogain, John of Reading, Racerx11, GoingBatty, XinaNicole, Ensabah6, Uploadvirus, ZéroBot, Arbnos, Zueignung, WaterCrane, Crown Prince, LaurentRDC, Isocliff, Vodkacannon, Raidr, Helpful Pixie Bot, Titodutta, Bibcode Bot, BG19bot, Spaligo, KateWishing, PhnomPencil, Sylvain.maurin, Kecchina, Halfb1t, Brad7777, Fylbecatulous, Jimw338, MyTuppence, Mogism, LT-Woods, Andyhowlett, Jawa0, &reasNink, SomeFreakOnTheInternet, Tentinator, EvergreenFir, DimReg, Pedarkwa, Db9199 24, Anrnusna, Notspelly, Ntomlin1996, Monkbot, Isbromberg, Dspre, YeOldeGentleman, Tetra quark and Anonymous: 333

- **Lovelock theory of gravity** *Source:* https://en.wikipedia.org/wiki/Lovelock_theory_of_gravity?oldid=678619952 *Contributors:* Michael Hardy, Gabbe, E Wing, Sadads, Natsirtguy, Cuzkatzimhut, Wbrenna36, Omnipaedista, Mr legumoto, Maschen, Gastongiribet, Raidr, Mgvongoeden, Brad7777 and Anonymous: 5

- **Massive gravity** *Source:* https://en.wikipedia.org/wiki/Massive_gravity?oldid=682688693 *Contributors:* Denni, Phys, Lumidek, Rich Farmbrough, David Schaich, AdamSolomon, Mpatel, Fred Bradstadt, Conscious, SmackBot, Colonies Chris, Wesino, Anton Gutsunaev, Addbot, Yobot, Omnipaedista, Erik9bot, Trappist the monk, GoingBatty, Bibcode Bot, BG19bot, Anomalyfree, ChrisGualtieri, Fatio de Duillier, Monkbot and Anonymous: 4

- **Mechanical explanations of gravitation** *Source:* https://en.wikipedia.org/wiki/Mechanical_explanations_of_gravitation?oldid=678536386 *Contributors:* Michael Hardy, Charles Matthews, AnonMoos, Rgdboer, Polluks, Danski14, Keenan Pepper, GregorB, Rjwilmsi, Dlugosz, Schlafly, Grover cleveland, CmdrObot, TheTito, PamD, Mbell, Keraunos, D.H, KConWiki, Denis tarasov, BernardZ, Sheliak, LokiClock, Peregrinoerick, Noveltyghost, Duae Quartunciae, Addbot, Idnwiki, Systemizer, AnomieBOT, Materialscientist, Citation bot, Omnipaedista, Paine Ellsworth, Machine Elf 1735, Citation bot 1, Tom.Reding, Logical Cowboy, A930913, Malanoqa, Bibcode Bot, WithSelet, CitationCleanerBot, BattyBot, Cyril Thalpe Gamage, Fatio de Duillier, Dogov, SarahTehCat and Anonymous: 17

- **Metric-affine gravitation theory** *Source:* https://en.wikipedia.org/wiki/Metric-affine_gravitation_theory?oldid=674718535 *Contributors:* the speller, Myasuda, David Eppstein, Neo., Forbes72, Gsard, Maschen, Tritario and Anonymous: 1

- **Modified models of gravity** *Source:* https://en.wikipedia.org/wiki/Modified_models_of_gravity?oldid=674246845 *Contributors:* Rjwilmsi, SmackBot, Drunken Pirate, Levineps, Headbomb, Hqb, Forbes72, Ulric1313, RjwilmsiBot, Venustas 12, R150634l, Suslindisambiguator, Bibcode Bot, BattyBot, ChrisGualtieri, Monkbot, HFD90, Kangaete and Anonymous: 2

- **Nonsymmetric gravitational theory** *Source:* https://en.wikipedia.org/wiki/Nonsymmetric_gravitational_theory?oldid=650895966 *Contributors:* Torfason, Booyabazooka, Joy, Gene s, Pjacobi, Brim, Robweiller~enwiki, Siafu, Mpatel, Wisq, Rjwilmsi, Ems57fcva, YurikBot, Hillman, Eleassar, Bluebot, JorisvS, IronGargoyle, Xiaphias, Freederick, Vttoth, Headbomb, Cyktsui, TXiKiBoT, Hqb, Jan1nad, Vini 175, Addbot, Yobot, Dreamer08, Citation bot 1, Tom.Reding, RjwilmsiBot, Bibcode Bot, Anrnusna and Anonymous: 13

- **Nordström's theory of gravitation** *Source:* https://en.wikipedia.org/wiki/Nordstr%C3%B6m'{}s_theory_of_gravitation?oldid=674736202 *Contributors:* Charles Matthews, John Vandenberg, Jérôme, Jeff3000, Mpatel, MarSch, Ligulem, RE, Alfred Centauri, Wavelength, Hillman, ErkDemon, SmackBot, Colonies Chris, Titus III, Makyen, Novangelis, Fan-1967, WhoSaid?, CmdrObot, WillowW, Viscious81, Headbomb, Hyperlinker, Michael H 34, Rep07, Addbot, Download, AnomieBOT, Rubinbot, Tom.Reding, NAME XXX, WildBot, EmausBot, John of Reading, Maschen and Anonymous: 13

- **Rainbow Gravity theory** *Source:* https://en.wikipedia.org/wiki/Rainbow_Gravity_theory?oldid=674750034 *Contributors:* Bfpage, YohanN7, TJRC, Krzys Pe, BG19bot, JazonaArizer and Anonymous: 2

- **Scalar theories of gravitation** *Source:* https://en.wikipedia.org/wiki/Scalar_theories_of_gravitation?oldid=678522469 *Contributors:* The Anome, Jitse Niesen, BenRG, Intangir, Rich Farmbrough, Bookofjude, Cmdrjameson, Mpatel, Josh Parris, Ems57fcva, Physchim62, DVdm, Bgwhite, Hillman, Donald Albury, That Guy, From That Show!, Nbez, Colonies Chris, Ligulembot, Derek farn, WhiteHatLurker, GargoyleMT,

Headbomb, Jpod2, VolkovBot, Neparis, Addbot, Lightbot, AnomieBOT, Citation bot, Omnipaedista, Tom.Reding, Bibcode Bot, BG19bot, Stigmatella aurantiaca, GravityForce, Graviprop, Mfb and Anonymous: 20

- **Scalar–tensor–vector gravity** *Source:*https://en.wikipedia.org/wiki/Scalar%E2%80%93tensor%E2%80%93vector_gravity?oldid= 674751487*Contributors:*The Anome, Twang, Michael Snow, Gene s, Bbbl67, Pjacobi, Pearle, Mpatel, GregorB, Joke137, Rjwilmsi, Hillman, ChrisCapoccia, SmackBot, Poobarb, Drunken Pirate, JorisvS, Zzzzzzzzzz, Vttoth, Alaibot, Headbomb, Timstump, Icep, Goldenrowley, Paolo.dL, Addbot, DOI bot, Legobot, Yobot, Citation bot, J04n, Fortdj33, Tom.Reding, EmausBot, Ethaniel, Maschen, Bibcode Bot, Cyberderp, Tetraquark and Anonymous: 12

- **Supergravity** *Source:* https://en.wikipedia.org/wiki/Supergravity?oldid=668440740 *Contributors:* AxelBoldt, Michael Hardy, TakuyaMurata, Angela, Charles Matthews, Phys, Bevo, Robbot, Gandalf61, Giftlite, Herbee, LeYaYa, Fropuff, Moyogo, Jeremy Henty, Leonard G., Urvabara, Arivero, Masudr, Dmr2, Srbauer, Markryherd, Physicistjedi, Axl, Wtmitchell, Japanese Searobin, Linas, Kzollman, Mpatel, GregorB, Canderson7, Marasama, Gurch, LeCire~enwiki, Chobot, Roboto de Ajvol, Hillman, Conscious, E. Menay, Wimt, Smoggyrob, QmunkE, Ilmari Karonen, Caco de vidro, SmackBot, Melchoir, FlashSheridan, Vald, Chris the speller, Colonies Chris, QFT, BWDuncan, TheST, Kuru, Jim.belk, JarahE, Michaelbusch, Zero sharp, CapitalR, Jorbesch, Crichigno, CmdrObot, Myasuda, Equendil, Phatom87, Pyro95819, Mbell, WVhybrid, West Brom 4ever, Icep, Shlomi Hillel, Yill577, David Eppstein, N.Nahber, Andre.holzner, Mschel, EdBever, Freeboson, Wesino, WJBscribe, Fuenfundachtzig, Signalhead, Cuzkatzimhut, Jickle, Robdunst, WereSpielChequers, Caltas, Wing gundam, Paolo.dL, Oxymoron83, Lightmouse, JL-Bot, EmanWilm, RS1900, ClueBot, ArdClose, Mild Bill Hiccup, JavierReynaldo, Vivio Testarossa, Mastertek, Pqnelson, AnonyScientist, Truthnlove, Addbot, Some jerk on the Internet, Wentuq, Luckas-bot, Yobot, Bility, AnomieBOT, ArthurBot, Omnipaedista, Gsard, Hep thinker, FrescoBot, Paine Ellsworth, Pxpt, Tom.Reding, Casimir9999, Wornsear, EmausBot, Slightsmile, Wikipelli, HiW-Bot, ZéroBot, Cogiati, Arbnos, Quantumor, Terraflorin, Bbeehvh, ClueBot NG, Joefromrandb, Helpful Pixie Bot, Bibcode Bot, BG19bot, Altaïr, BattyBot, Jeremy112233, M0532062613, Jamesx12345, Mamzypig99, Bitprior, Monkbot, KasparBot and Anonymous: 75

- **Tensor–vector–scalar gravity** *Source:*https://en.wikipedia.org/wiki/Tensor%E2%80%93vector%E2%80%93scalar_gravity?oldid= 674736441*Contributors:*Ahoerstemeier, Anythingyouwant, Gene s, Bbbl67, Rich Farmbrough, Franjesus, Pjacobi, Mpatel, GregorB, Joke137, Aarghd-vaark, Rjwilmsi, Rangek, Alynna Kasmira, SCZenz, WAS 4.250, Alain r, SmackBot, Colonies Chris, Poobarb, Zsinj, JorisvS, Zzzzzzzzzz, CRGreathouse, Vttoth, MC10, Alaibot, MrXow, Headbomb, Icep, Goldenrowley, Yellowdesk, Ksara, Walknick, HEL, Alexbot, Good Ol-factory, Addbot, Yobot, Citation bot, J04n, Nagualdesign, Fortdj33, Tom.Reding, Night Jaguar, ThorX13, Suslindisambiguator, Maschen,Bibcode Bot, Fraulein451, Oxfard, HFD90 and Anonymous: 24

- **Whitehead's theory of gravitation** *Source:* https://en.wikipedia.org/wiki/Whitehead'[)s_theory_of_gravitation?oldid=680192927 *Contributors:* Goethean, Texture, Mboverload, Anythingyouwant, Ben Standeven, Mpatel, Rjwilmsi, Ligulem, DVdm, Hillman, Melchoir, Hmains, Colonies Chris, Bm gub, Katzmik, Djr32, Pfhorrest, Yobot, AnomieBOT, Ddsdls, Citation bot, Omnipaedista, Tom.Reding, Footnotes2plato, Bibcode Bot, Monkbot and Anonymous: 10

- **Yilmaz theory of gravitation** *Source:* https://en.wikipedia.org/wiki/Yilmaz_theory_of_gravitation?oldid=599457429 *Contributors:* Eiserlohpp, Pjacobi, Pavel Vozenilek, Adambro, Count Iblis, Mpatel, GregorB, Ligulem, Wragge, John Z, Hillman, That Guy, From That Show!, SmackBot, RaulMiller, Bluebot, Colonies Chris, Ligulembot, Cydebot, Headbomb, Wasell, Pawl Kennedy, Meilenweit, Addbot, DOI bot, Atethnekos, Legobot II, Citation bot, AzerRail, Sakimonk, Citation bot 1, Abductive, Bibcode Bot, BattyBot and Anonymous: 13

- **Theory of everything** *Source:* https://en.wikipedia.org/wiki/Theory_of_everything?oldid=683741317 *Contributors:* AxelBoldt, Paul Drye, CYD, The Anome, Eclecticology, Toby Bartels, Roadrunner, Zippy, Stevertigo, Lorenzarius, Michael Hardy, Rojelague, Nixdorf, TakuyaMurata, Karada, Skysmith, Kosebamse, CesarB, Anders Feder, Angela, Julesd, Salsa Shark, Ugen64, Poor Yorick, Evercat, Schneelocke, Feedmecereal, Timwi, Dcoetzee, Dysprosia, Jitse Niesen, Wik, Jakenelson, Omegatron, Raul654, Nnh, Kevin M C Harkess, UninvitedCompany, Fredrik, Altenmann, Nurg, Naddy, Gandalf61, Mirv, Academic Challenger, Rursus, Blainster, Caknuck, Wereon, Diberri, Pengo, Tobias Bergemann, Hookoovoo, Ancheta Wis, Dbenbenn, Mporter, Jabra, Ferkelparade, Bfinn, Xerxes314, Curps, Alison, FeloniousMonk, McGravin, Behnam, Gzornenplatz, JRR Trollkien, Steuard, Andycjp, Sonjaaa, Antandrus, Kim54, Tomruen, Lumidek, Gseshoyru, WpZurp, TJSwoboda, Zondor, JimJast, Discospinster, Rich Farmbrough, H0riz0n, Pjacobi, Vsmith, Pluke, Autiger, Mal~enwiki, Pavel Vozenlek, Floorsheim, El C, Lycurgus, Sourcecode, Oldsoul, PhilHibbs, Sietse Snel, Jpgordon, Atraxani, Smalljim, Slicky, LostLeviathan, Matpitka, Juesch, Danski14, Alansohn, Gary, DariuszT, ShardPhoenix, Kocio, Pion, Hdeasy, Bart133, Schaefer, BanyanTree, ClockworkSoul, Tycho, Suruena, Count Iblis, DV8 2XL, Gene Nygaard, Euphrosyne, Squidwina, Ott, Siafu, Roylee, Woohookitty, Mindmatrix, RHaworth, TigerShark, Savantnavas, MrDarcy, Mpatel, GregorB, Athletec64, Christopher Thomas, Aarghdvaark, Ashmoo, BD2412, Drbogdan, Rjwilmsi, Kinu, Strait, Lordsatri, Dennis Estenson II, HappyCamper, LjL, Bubba73, The wub, Yamamoto Ichiro, JohnDBuell, FayssalF, ColinJF, Wragge, Windchaser, Musical Linguist, Mindloss, RexNL, Gurch, Pete.Hurd, Lmatt, Diza, Zayani, Spencerk, Chobot, Sharkface217, DVdm, Hmonroe, Bgwhite, Ptah~enwiki, Ugha, Wavelength, Hillman, StuffOfInterest, Phantomsteve, Arado, John Smith's, Zigamorph, SpuriousQ, Jobe457, Stephenb, CambridgeBayWeather, Rsrikanth05, Vibritannia, Neilbeach, Salsb, Big Brother 1984, Anomalocaris, NawlinWiki, Joncolvin, ErkDemon, Trovatore, ETTan, Schrei, THB, Syrthiss, Wknight94, Richardcavell, FF2010, CWenger, Kevin, Caco de vidro, Katich5584, Banus, Sbyrnes321, Narkstraws, SmackBot, R.E. Freak, Kurochka, DuoDeathscyther 02, Bayardo, McGeddon, Delldot, Kintetsubuffalo, Portillo, Rmosler2100, Bluebot, Jjalexand, 7777777s, Silly rabbit, George Church, Colonies Chris, A. B., Cate rulz, Nicknitro71, Zsinj, TallyJoe, John Hyams, Jamse, Scott3, Jefffire, Serenity-Fr, Bilgrau, Avb, Rrburke, Addshore, DrL, Mr.LMNOP, Rassisi, Spanyard, Byelf2007, Nishkid64, Giovanni33, Soap, Cronholm144, Loadmaster, Stupid Corn, Benjaminlobato, FredrickS, SirFozzie, Waggers, Alexander Gieg, Geavep, Abel Cavaşi, Newone, Courcelles, Tubezone, Esn, Dave Runger, Valoem, JRSpriggs, Kurtan~enwiki, 0-8, Duduong, Friendly Neighbour, CRGreathouse, Geremia, Tkoeppe, Ken Gallager, DepartedUser2, Cydebot, Vanished user 2340rujowierfj08234irjwfw4, Ninguém, Steel, Peterdjones, Hebrides, David edwards, Michael C Price, Raoul NK, Wortzman, Ulnevets, Konradek, Mojo Hand, Raymond Feilner, Headbomb, Marek69, Inve40, Twejr, Duncan McB, KrakatoaKatie, Luna Santin, Gdo01, Byrgenwulf, Myanw, Knotwork, Len Raymond, JAnDbot, Barek, MER-C, Txomin, Inks.LWC, Matthew Fennell, Instinct, MoralMajority, Promking, Bongwarrior, VoABot II, JamesBWatson, JBKramer, DAGwyn, Theroadislong, Lenschulwitz, 28421u2232nfenfcenc, Peatbog, Allstarecho, Fang 23, Spellmaster, Philg88, Peter J Schoen, Denis tarasov, MartinBot, R'n'B, JCarlos, J.delanoy, Pharaoh of the Wizards, Maurice Carbonaro, LordAnubisBOT, Pyrospirit, AntiSpamBot, NewEnglandYankee, DadaNeem, Cometstyles, WJBscribe, Foofighter20x, Econofire, Squids and Chips, Germanium, Reelrt, ChaosCon343, Danwills, RingtailedFox, Jeff G., TXiKiBoT, Nxavar, Rei-bot, Vishal144, IllaZilla, Pouya sh, Corvus cornix, Michael H 34, Martin451, Cheffoxx, Betanon, BotKung, Everything counts, Popopp, MrMelonhead, Stephenmolesey, James McBride, Deanlsinclair, Pageman~enwiki, Monty845,

Logan, Kpa4941, PaddyLeahy, Dogah, SieBot, Tiddly Tom, Robdunst, Wing gundam, Gammanon, Bentogoa, Likebox, Tiptoety, SteakN-Shake, Momo san, Freeman501, BartekChom, Monkeyspangler, Lightmouse, Anakin101, Divinestuff, Carbogen, Ayleuss, Soporaeternus, ArepoEn, ClueBot, LAX, Cliff, Ian the Aussie, Monomath1, Boing! said Zebedee, Heldbacktheband, LonelyBeacon, Neverquick, Excirial, WikiZorro, Tamaratrouts, Wndl42, Brews ohare, PhySusie, Morel, Mastertek, Mikaey, 7, Crowsnest, Thinking Stone, TimothyRias, Pat-Dunphey, JKeck, XLinkBot, Bvssvni, Ougner, Truthnlove, YeAaMsLtA, Thatguyflint, Tayste, Balungifrancis, Addbot, Proofreader77, Some jerk on the Internet, Uruk2008, DOI bot, Couchie, Johnchang6868, Discrepancy, Mjamja, Bobtron5000, Fluffernutter, KaityJoe, MrOllie, Favonian, Barak Sh, F Notebook, Tide rolls, Scientryst, WikiDreamer Bot, Meisam, Blah28948, Yobot, Finiter, Ptbotgourou, Ezequiels.90, Jgmoxness, Amble, Mirandamir, RDemelo, AnomieBOT, ^musaz, Girl Scout cookie, 9258fahsflkh917fas, Theunify, Anxfisa, Kanat Abildinov, Materialscientist, Citation bot, Subhajit Ganguly, Fleaman5000, Amareto2, Addihockey10, Smk65536, Mlpearc, GrouchoBot, Rwmeo, Om-nipaedista, Shirik, RibotBOT, Fa.alt3r3g0, Fsdjfsdfk, Chaseroads, 🔲🔲, FrescoBot, Paine Ellsworth, Ribashka, Steven Avraham Rosten, Physics Explorer, Ottokar~enwiki, Tank hasmukh Khimjibhai, Tank theorist of everything, Hasmukh Khimjibhai Tank, DivineAlpha, Citation bot 1, Gil987, Three887, Tom.Reding, A8UDI, NarSakSasLee, Casimir9999, AndrewGrieder, Aknochel, IVAN3MAN, SchreyP, Noel Edward, Natwatchmaker, Weedwhacker128, Suffusion of Yellow, Koozedine, RjwilmsiBot, Specal ops, Afteread, DASHBot, Golumbo, EmausBot, Ikerus, Katherine, Dewritech, RA0808, K6ka, Zero939, Thecheesykid, Hhhippo, CanonLawJunkie, Traxs7, Arbnos, SporkBot, DanielBurn-stein, FinalRapture, Aatu Koskensilta, Staszek Lem, Sridattadev, M00se1989, Wiggles007, Andrushkkutza, Maschen, Vedoder, Donner60, GIAN PHIL, Davidaedwards, WHF Christie, Terra Novus, Matevz91, Isocliff, Sanno89, Cgt, Will Beback Auto, ClueBot NG, Stein Sivert-sen, ClaudeDes, Lord God Almighty, Hindustanilanguage, Helpful Pixie Bot, Nightingale.zj, B21O303V3941W42371, Bibcode Bot, Wiki13, Akashankitjain, Neutral current, Aranea Mortem, Stimulieconomy, Steven.w.kowalski, MathewTownsend, Flyerbri, GroupT, Megajakeroo, La marts boys, Zofo, LightandDark2000, Josepht404, Nickhwee, Davidyevgeny, Kingcircle, Vith Nix, Illuusio, Davidyevgenyroven, Quan-tumNico, Vladimir Leonov, Friek555, HesterShaw, Sol1, Phaedrx, Jwratner1, Jmassion, HeymynamesJon, Bigfootrobert, Elitousson, Md-sheraj, Kdmeaney, JaconaFrere, Somednguy4, Monkbot, LollyBear12, StacyPoyPie, Mujii loving, Mayojohns, Gronk Oz, Yoyosami, Hakan tomaşoğlu, Pfpguy, Cirksena, Svm sudhan, 39Debangshu, Quantalogos, KasparBot, Christos Theopoulos, Patrickmantonio, Quackriot and Anonymous: 519

47.8.2 Images

- **File:1919_eclipse_negative.jpg** *Source:* https://upload.wikimedia.org/wikipedia/commons/d/da/1919_eclipse_negative.jpg *License:* Public domain *Contributors:* F. W. Dyson, A. S. Eddington, and C. Davidson, "A Determination of the Deflection of Light by the Sun's Gravitational Field, from Observations Made at the Total Eclipse of May 29, 1919" *Philosophical Transactions of the Royal Society of London. Series A. Containing Papers of a Mathematical or Physical Character* (1920): 291-333, on 332. *Original artist:* F. W. Dyson, A. S. Eddington, and C. Davidson

- **File:1919_eclipse_positive.jpg** *Source:* https://upload.wikimedia.org/wikipedia/commons/3/37/1919_eclipse_positive.jpg *License:* Public domain *Contributors:* F. W. Dyson, A. S. Eddington, and C. Davidson, "A Determination of the Deflection of Light by the Sun's Gravitational Field, from Observations Made at the Total Eclipse of May 29, 1919" *Philosophical Transactions of the Royal Society of London. Series A. Containing Papers of a Mathematical or Physical Character* (1920): 291-333, on 332. *Original artist:* F. W. Dyson, A. S. Eddington, and C. Davidson

- **File:A_Swarm_of_Ancient_Stars_-_GPN-2000-000930.jpg** *Source:* https://upload.wikimedia.org/wikipedia/commons/6/6a/A_Swarm_of_Ancient_Stars_-_GPN-2000-000930.jpg *License:* Public domain *Contributors:* Great Images in NASA Description *Original artist:* NASA, The Hubble Heritage Team, STScI, AURA

- **File:Albert_Einstein_portrait.jpg** *Source:* https://upload.wikimedia.org/wikipedia/en/f/f7/Albert_Einstein_portrait.jpg *License:* PD-US *tributors:*

	http://images.google.com/hosted/life/628e99cf2e26233d.html *Original artist:*

	E. O. Hoppe. (1878-1972) Published on LIFE

- **File:Ambox_important.svg** *Source:* https://upload.wikimedia.org/wikipedia/commons/b/b4/Ambox_important.svg *License:* Public domain *Contributors:* Own work, based off of Image:Ambox scales.svg *Original artist:* Dsmurat (talk · contribs)

- **File:Apollo_15_feather_and_hammer_drop.ogg** *Source:* https://upload.wikimedia.org/wikipedia/commons/3/3c/Apollo_15_feather_and_hammer_drop.ogg *License:* Public domain *Contributors:* Taken from Spacecraftfilms.com DVD "Apollo 15: The Great Explorations Begin" *Original artist:* NASA

- **File:ArnowittDeserMisner2009_01.jpg** *Source:* https://upload.wikimedia.org/wikipedia/commons/1/11/ArnowittDeserMisner2009_01.jpg *License:* CC BY-SA 3.0 *Contributors:* Own work *Original artist:* Puzhok

- **File:Artist's_impression_of_the_pulsar_PSR_J0348+0432_and_its_white_dwarf_companion.jpg** *Source:* https://upload.wikimedia.org /wikipedia/commons/2/26/Artist%E2%80%99s_impression_of_the_pulsar_PSR_J0348%2B0432_and_its_white_dwarf_companion. jpg *License:* CC BY 4.0 *Contributors:* http://www.eso.org/public/images/eso1319e/ *Original artist:* ESO/L. Calçada

- **File:Black_Hole_Merger.jpg** *Source:* https://upload.wikimedia.org/wikipedia/commons/d/d1/Black_Hole_Merger.jpg *License:* Public domain *Contributors:* Taken from http://www.space.com/imageoftheday/image_of_day_060203.html credit is listed to NASA. *Original artist:* NASA

- **File:Black_Hole_Milkyway.jpg** *Source:* https://upload.wikimedia.org/wikipedia/commons/c/cd/Black_Hole_Milkyway.jpg *License:* CC BY-SA 2.5 *Contributors:* Gallery of Space Time Travel *Original artist:* Ute Kraus, Physics education group Kraus, Universität Hildesheim, Space Time Travel, (background image of the milky way: Axel Mellinger)

- **File:Calabi-Yau.png** *Source:* https://upload.wikimedia.org/wikipedia/commons/d/d4/Calabi-Yau.png *License:* CC BY-SA 2.5 *Contributors:* own work by Lunch
	http://en.wikipedia.org/wiki/Image:Calabi-Yau.png (english Wikipedia) *Original artist:* Lunch

47.8.3   Content license